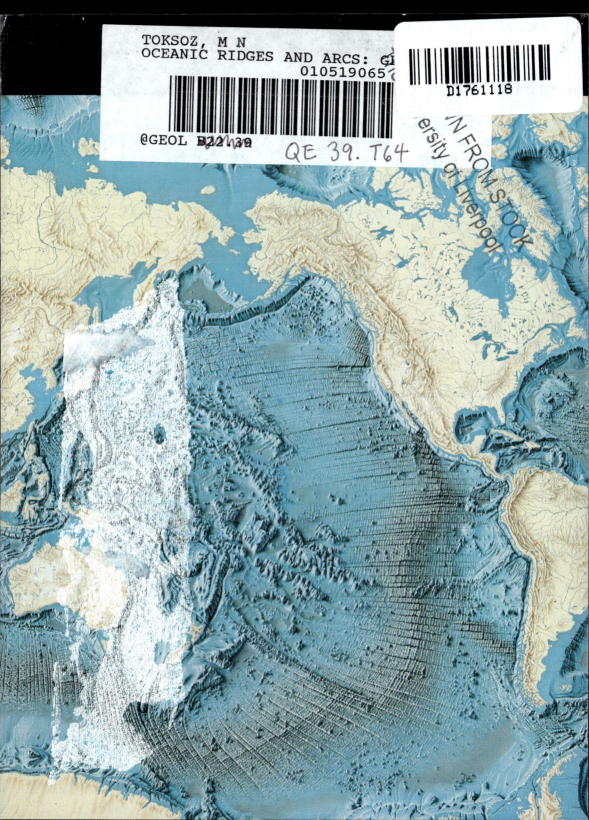

TOKSOZ, M N
OCEANIC RIDGES AND ARCS: GE
010519065
@GEOL B22139
QE 39. T64

D1761118

WN FROM STOCK
ersity of Liverpool

Geology

QE39.T64

Developments in Geotectonics 14

OCEANIC RIDGES AND ARCS

Developments in Geotectonics 14

UNIVERSITY LIVERPOOL LIBRARY

OCEANIC RIDGES AND ARCS
GEODYNAMIC PROCESSES

Edited by

M.N. TOKSÖZ
Department of Earth and Planetary Sciences, M.I.T., Cambridge, Mass., U.S.A.

S. UYEDA
Seismological Laboratory, California Institute of Technology, Pasadena, Calif., U.S.A.

and

J. FRANCHETEAU
Centre Océanologique de Bretagne, Brest, France

Selected papers from Tectonophysics

ELSEVIER SCIENTIFIC PUBLISHING COMPANY
Amsterdam — Oxford — New York   1980

ELSEVIER SCIENTIFIC PUBLISHING COMPANY
335 Jan van Galenstraat
P.O. Box 211, 1000 AE Amsterdam, The Netherlands

Distributions for the United States and Canada:

ELSEVIER/NORTH-HOLLAND INC.
52, Vanderbilt Avenue
New York, N.Y. 10017

Ocean Floor Map by M. Tanguy de Rémur
© Hachette — Guides Bleus

Library of Congress Cataloging in Publication Data

Main entry under title:

Oceanic ridges and arcs.

   (Developments in geotectonics ; 14)
   Bibliography: p.
   1. Submarine geology. 2. Geodynamics.
3. Plate tectonics. I. Toksöz, M. N., 1934-
II. Uyeda, Seiya, 1929-   III. Francheteau,
Jean. IV. Series.
QE39.026   551.4'608   79-25797
ISBN 0-444-41839-3

ISBN 0-444-41839-3 (Vol. 14)
ISBN 0-444-41714-1 (Series)

© Elsevier Scientific Publishing Company, 1980
All rights reserved. No part of this publication may be reproduced, stored in a retrieval
system or transmitted in any form or by any means, electronic, mechanical, photo-
copying, recording or otherwise, without the prior written permission of the publisher,
Elsevier Scientific Publishing Company, P.O. Box 330, 1000 AH Amsterdam,
The Netherlands

Printed in The Netherlands

PREFACE

Oceanic ridges and island arcs are the two most important expressions of the plate-tectonic processes. The new lithospheric plates from at the ridges, cool and thicken with time as they more away, and are generally subducted and consumed under island arcs. On the surface the formation, evolution and subduction of the lithospheric appears to be an orderly and natural phenomenon. Yet the processes involved are complex and of fundamental importance. On the one hand they are the surface expressions of the thermal and dynamic processes of the earth's mantle. On the other hand they are responsible for all the geological and tectonic phenomena.

To assess our knowledge of the oceanic ridges, subduction zones and associated geodynamic processes, three related Symposia were held in August, 1977 in Durham, England during the Joint General Assemblies of the International Associations of Seismology and Physics of the Earth's Interior (IASPEI) and Volcanology and Chemistry of the Earth's Interior (IAVCEI). The proceedings of these Symposia were published in three Special Issues of Tectonophysics: Numerical Modeling in Geodynamics (vol. 50(2/3), 1978, M.N. Toksöz, editor); Processes at Mid-Ocean Ridges (vol. 55(1/2), 1979, J. Francheteau, editor); and Processes at Subduction Zones (vol. 57(1), 1979, S. Uyeda, editor). "Oceanic Ridges and Arcs" has been prepared to present our current knowledge of the geodynamic processes of the ridges and subduction zones in a coherent form. The book contains nineteen papers selected from the three Special Issues and eight papers from other issues of Tectonophysics. As in any compilation, such an edited volume would have some gaps and redundancies. All efforts have been made to minimize these and to present a balanced and coherent view of the state of our knowledge.

The book begins with the most general chapters on the viscosity of the earth's mantle and global convection. Viscosity is a crucial parameter in geodynamic processes. Kinematic models of large scale convection illustrate a possible coupling between the oceanic ridges, plate motions, subduction zones and large scale flows in the earth's mantle. Differences between the thermal and viscous properties of the asthenosphere under oceans and continents are discussed in this section.

After the general articles, the sequence of chapters follows in order from oceanic lithosphere to oceanic ridges, subduction zones and back-arc basins. Chapters dealing with the ridges discuss the thermal, hydrothermal and magmatic processes. They demonstrate complexities and heterogeneities of different scales. Chapters on the oceanic crust and lithosphere cover both the geophysical and petrological models and effects of alterations and metamorphism.

The subduction-zone chapters start with the bending of the lithosphere and

752065

the thermal regime of subduction zones. Detailed studies of subduction zones under Japan and Peru are given. Petrological and geochemical data indicate the complexities of the magmatic processes associated with subduction.

A prominent feature behind many oceanic subduction zones is the marginal or back-arc basins. These are invariably related to the subduction process. In many cases they show active spreading. The geophysical and geological properties of these basins are described in several chapters along with possible models of their origin and formation.

The final three chapters are devoted to geodynamic processing involving continents. Spreading, subduction and lithospheric motions eventually lead to continental collision which is more complex than subduction under the arcs. These three chapters help to demonstrate the differences between these two types of place convergence.

The editors are grateful to the participants of the Symposia, the contributors to the Special Issues and to this volume.

M. Nafi Toksöz
Seiya Uyeda
Jean Francheteau

CONTENTS

Chapter 12. MAGMA MIXING AT MID-OCEAN RIDGES: EVIDENCE FROM
BASALTS DRILLED NEAR 22° N ON THE MID-ATLANTIC RIDGE
J.M. Rhodes, M.A. Dungan, D.P. Blanchard and P.E. Long

Chapter 13. LATE STAGE DEVELOPMENT OF MATURE ATLANTIC-TYPE
CONTINENTAL MARGINS
H.J. Neugebauer and T. Spohn

Chapter 14. AN ELASTIC-PERFECTLY PLASTIC ANALYSIS OF THE
BENDING OF THE LITHOSPHERE AT A TRENCH
D.L. Turcotte, D.C. McAdoo and J.G. Caldwell

Chapter 24. NUMERICAL STUDIES OF BACK-ARC CONVECTION AND
THE FORMATION OF MARGINAL BASINS
M.N. Toksöz and A.T. Hsui

Chapter 25. SUTURE ZONE COMPLEXITIES: A REVIEW
J.F. Dewey

Chapter 26. THE EASTERN MEDITERRANEAN AND THE LEVANT:
TECTONICS OF CONTINENTAL COLLISION
A. Nur and Z. Ben-Avraham

Chapter 27. FINITE ELEMENT MODELING OF LITHOSPHERE
DEFORMATION: THE ZAGROS COLLISION OROGENY
P. Bird

LIST OF AUTHORS

Chapter 1

GLACIAL ISOSTASY AND RELATIVE SEA LEVEL: A GLOBAL FINITE ELEMENT MODEL*

W.R. PELTIER, W.E. FARRELL and J.A. CLARK

Alfred P. Sloan Foundation Fellow, Department of Physics, University of Toronto, Toronto, Ont. M5S 1A7 (Canada)
Department of Materials Science and Mineral Engineering, University of California, Berkeley, Calif. 94720 (U.S.A.)
Department of Geological Sciences, Cornell University, Ithaca, N.Y. 14853 (U.S.A.)

SUMMARY

We review the subject of glacial isostatic adjustment and the structure of global models which have been developed to describe this phenomenon. Models of glacial isostasy have two basic ingredients: (1) an assumed deglaciation history, and (2) an assumed constitutive relation for the rheology of the planetary interior. Although both of these "functionals of the model" are imperfectly known, the parameters which are required to describe them are simultaneously constrained by observations of relative sea level. By comparing the model predictions of relative sea level for times subsequent to a major deglaciation event at a global distribution of sites, with the observed history of relative sea level at these sites, rather stringent bounds may be placed upon the model parameters.

INTRODUCTION

Observations of changes in the earth's shape produced by mass loads applied on its surface continue to provide a wealth of information concerning the physical properties of the interior. Such changes of shape are associated in general with both elastic and anelastic strains and with variations of the gravitational field. In order to deduce material properties from strain observations, assumptions must be adopted regarding the nature of the material of which the planet is composed. These assumptions are usually cast in the form of a constitutive relation connecting the stress τ and its time derivative $\dot{\tau}$ to the strain ϵ and its rate of change $\dot{\epsilon}$. The subject of the earth's rheology has always been one of the most controversial in geophysics.

Jeffreys has long maintained that an anelastic (lossy) rheology was important in seismology but only recently (Liu et al., 1976) have other seismologists become concerned with the effects that deviations from perfect

---

*Originally published as: Peltier, W.R., Farrell, W.E. and Clark, J.A., 1978. Glacial isostasy and relative sea level: a global finite element model. In: M.N. Toksöz (Editor), Numerical Modeling in Geodynamics. Tectonophysics, 50: 81—110.

Hookean elasticity have on elastic wave propagation. At first glance this may seem surprising since the fact that the earth's mantle *is* anelastic (at least for deviatoric stresses which persist for thousands of years) has been appreciated for the better part of this century. Qualitatively this knowledge was based upon the recognition that the earth's shape continued to change *after* the last major deglaciation of the surface had been completed (a process which began about $2 \cdot 10^4$ years ago and ended ca $0.5 \cdot 10^4$). In formerly glaciated regions (e.g., Fennoscandia, Canada) old strandlines were found to be located hundreds of metres above present-day sea level. Detailed investigation of the stratigraphic record (in Fennoscandia) indicated a more or less exponential uplift of the surface of the solid earth with respect to local sea level. The rate of uplift has been a decreasing function of time since deglaciation, but is not yet complete.

With the advent of radiocarbon dating (e.g., Libby, 1952) and its subsequent application to the quantification of relative sea level data the detailed global history of the relaxation of shape following deglaciation began to emerge. Walcott (1972) has discussed these data (with emphasis on the Laurentide uplift) in detail and we may summarize their dominant features by noting that where the ice was thickest the land is now elevated above sea level, but only a short distance away from the two main centres of glaciation old beaches are drowned.

The generally exponential character of the uplift data is strongly suggestive of a simple relaxation process, and it is not surprising that they were initially interpreted in terms of a viscous fluid model of the interior (Haskell, 1935, 1936, 1937; Vening Meinesz, 1937; Niskanen, 1948). This model is, of course, the antithesis rheologically of the Hookean elastic model which, at that time, had been found to accord well with seismic observation. In order to fit the observed relaxation times (ca 10^3 years) fluid models with molecular viscosities on the order of 10^{22} poise (cgs) were required. This magic number has existed in the literature for over 40 years. Our recent attempts to quantify the extent of our ignorance of this parameter have been motivated by several concerns. In the first instance we would like to know, as accurately as possible, how the viscosity varies with radius in the mantle. Secondly, and more fundamentally, we wish to understand the microphysics of the process(es) by which mantle material "creeps" in response to an applied stress. Finally, using "improved" viscosity models we wish to study the breakup and disintegration of the Pleistocene ice caps and the impact which this event had upon the rotation of the earth. These facets of the problem are of fundamental importance to climatology. In the remainder of this introduction we will discuss the first two questions in turn.

The transport coefficient for momentum, ν, (i.e., the viscosity) plays a decisive role in geodynamic models of long timescale processes in the planetary interior. For example, since the Rayleigh number for thermal convection, Ra, is such that $Ra \propto \nu^{-1}$, the higher the viscosity the less vigorous will be the convection forced by a given superadiabatic temperature gradient. As

ν increases the efficiency of convective heat transport decreases; thus the magnitude of this parameter has a direct impact upon the thermal history of the earth. If the viscosity were to increase sufficiently rapidly with depth then convection in the lower mantle would be inhibited. If one associates the downgoing slabs of plate tectonic theory with the descending limbs of cold thermal boundary layers of the global convective circulation, then seismic focal mechanism data (Isacks and Molnar, 1971) suggest that the slab encounters some resistance to continued vertical motion near a depth of 650 km. This resistance has been attributed by some to a rapid increase of viscosity, assumed to be a consequence of the spinel—post-spinel phase change. In order that convection be effectively inhibited from penetrating this hypothesized boundary an increase in viscosity of several orders of magnitude is required. Such a large viscosity increase at this depth ought to be detectable through careful analysis of the relaxation data.

Until rather recently (Cathles, 1975; Peltier, 1974; Peltier and Andrews, 1976) it has generally been believed that the isostatic adjustment data did in fact *require* a rapid increase of viscosity with depth. The strongest evidence was presented by McConnell (1968) through an analysis of the Fennoscandia data. In order to obtain resolution for lower mantle viscosity, however, his analysis made use of an inferred relaxation time for the $n = 2$ harmonic of the gravitational field of approximately 10^{13} sec. This inference was based upon two assumptions: (1) that the earth had a "genuine" non-hydrostatic equatorial bulge, and (2) that this bulge was itself caused by Pleistocene glaciation. The relaxation time was then deduced by interpreting historical changes in the length of the day as being a consequence of the collapse of the bulge. This interpretation led McConnell to suggest lower mantle viscosities on the order of 10^{25} poise.

McKenzie (1966) arrived at a similar conclusion which was again based upon an interpretation of the non-hydrostatic equatorial bulge. He assumed, as had been suggested in Munk and Macdonald (1960), that the excess bulge was a relic from a time when the earth was spinning faster than at present and its continued existence then suggested that the viscosity of the deep mantle must be extreme. This conclusion, and likewise McConnell's, was completely undermined by Goldreich and Toomre (1969) who showed that the non-hydrostatic bulge upon which both arguments were based was only an artifact of the spherical harmonic analysis in terms of which its existence had originally been suggested. The relaxation data from Fennoscandia did not, in themselves, suggest or even support the existence of a steep viscosity gradient in the upper mantle. McConnell (1968) elected not to fit his own long wavelength relaxation data, giving arguments as to why these should be inaccurate.

More recently Parsons (1972) has shown through a "resolving power" analysis of McConnell's original data that these are simply incapable of providing any information on the viscosity of the mantle at depths in excess of about 600 km. This is due to the relatively small horizontal scale of the Fennoscandian rebound. The usefulness of these data therefore fades at just those depths at which we become most interested. It is the existence of new data associated with the recovery of the larger scale Laurentide region (most of Canada) which has encouraged our re-examination of the isostatic adjustment process in an attempt to test the plate resistance hypothesis. Since the dominant wave number of the relaxation in this region corresponds to spherical harmonic degrees of the order of $n \cong 5$ whereas the Fennoscandia data provide information on $n \cong 16$ we expect that the Canadian data are potentially able ro resolve viscosity variations throughout the entire mantle. On the basis of analyses of this data Cathles (1975) and Peltier and Andrews (1976) have shown that the observed recovery is incompatible with a viscosity profile which is a rapidly increasing function of depth and in fact both studies found that a uniform mantle viscosity with $\nu \cong 10^{22}$ poise gives a reasonable fit to the data. These interpretations were based, however, upon theoretical models which were not consistent gravitationally in that the surface of the ocean was not constrained to remain an equipotential surface during and following deglaciation. One of the main purposes of the present paper is to summarize recent work which has remedied this defect and in the process has reinforced the original conclusions.

Although the inference of viscosity from relaxation data does not *require* a microphysical model of the creep mechanism, it is nevertheless true that the number which one assigns to this parameter may exhibit a dependence upon the model assumed. From a continuum mechanical point of view, microphysical models may be classified as either Newtonian or non-Newtonian. In the former case the relation between the stress and strain-rate tensors is linear and the coefficient of proportionality is the Newtonian viscosity, whereas in the latter the relation is non-linear and the effective viscosity is therefore stress dependent. When a Newtonian constitutive relation is employed then one is implicitly assuming that the creep mechanism is Herring—Nabarro or Coble or some other mechanism which leads to a linear relation between τ and $\dot{\epsilon}$. For the Herring—Nabarro mechanism (Herring, 1950) the equivalent Newtonian viscosity ν_N is:

$$\nu_N = (kTa^2/\alpha_0 D_0 \Omega) \exp[(E + pV_a)/kT] \qquad (1)$$

where k is Boltzmann's constant, T the absolute temperature, V_a the activation volume, a the grain radius (assumed constant), E the activation energy for self diffusion of the rate limiting species, p the pressure, Ω the atomic volume, and α_0 (≈ 5) is a dimensionless constant. The constitutive relation is then $\tau = \nu_N \dot{\epsilon}$. All microphysical mechanisms lead to expressions for the viscosity which are explicit functions of the thermodynamic coordinates. If we knew the way in which these coordinates varied with depth and correct values for the other parameters then we could deduce the viscosity depth profile directly (e.g., Sammis et al., 1977). Such, however, is unfortunately not the case. In interpreting the postglacial relaxation data what we are obliged to do is to suppress the explicit dependence of ν upon p, T, etc. and to include such dependence in the model through an assumed depth dependence. We then attempt to find the $\nu(r)$ which best fits the observations.

The existence of microphysical mechanisms which lead to linear macroscopic constitutive relations has provided some solace to those who would construct mathematical models of the stress relaxation associated with isostatic adjustment or of the mantle convective circulation. This is clearly due to the fact that a linear constitutive relation is easier to manage than a nonlinear one. However, recent experimental evidence (Ashby, 1972; Stocker and Ashby, 1973; Post and Griggs, 1973; Kohlstedt and Goetze, 1974) on the creep of mantle material (mostly olivine single crystals) suggests strongly that the rate limiting process is *not* linear. The experimental data rather suggests a power law creep equation with a stress exponent $m \approx 3$. Weertman and Weertman (1975) have reviewed the interpretation of these data and their implications in some detail.

A general form of the power law creep equation with $m = 3$ may be written as:

$$\dot{\epsilon} = \gamma(D/b^2)(\mu\Omega/kT)(\sigma/\mu)^3 \tag{2}$$

where the self diffusion coefficient D for the rate limiting species is

$$D = D_0 \exp\left[\frac{-E - pV_a}{kT}\right]$$

in which γ is a dimensionless constant, b is the length of the Burghers vector of the dislocation, and where μ is the elastic shear modulus. The remaining symbols are the same as in eq. 1. We may invert eq. 2 to write $\tau = \nu_{NN}\dot{\epsilon}$ where ν_{NN} is now a function of stress so that the constitutive relation is nonlinear. The laboratory data which suggest eq. 2 are, of necessity, taken for creep rates which are enormously in excess of those which are associated either with rebound or with convection (i.e., 10^{-6} s^{-1} as opposed to 10^{-16} s^{-1}). The interpretation therefore requires a huge extrapolation of the actual data.

In the models of isostatic adjustment described below we *assume* that the stress—strain relation is Newtonian. In fact we employ a model earth which is a simple linearly visco-elastic (Maxwell) solid. What we seek to accomplish is a direct test of this constitutive relation, and the test demands the resolution of two distinct questions. Firstly, is it possible with such a model to fit the observed relative sea level data? If the answer to this question is an unambiguous "yes" then we must furthermore attempt to understand the extent to which non-Newtonian models are capable of providing similar accord with the data. Here we shall only be concerned with an attempt to answer the first question. Brennan (1974) and Crough (1977) have made initial attempts to address the second but much further work remains to be done in this area.

The data which we must interpret concern the location with respect to present day sea level of raised or submerged beaches, the age of which may be determined by application of ^{14}C dating methods. These histories of relative sea level are a record of the manner in which the earth and its ice sheets and oceans responded to the cataclysmic redistribution of surface mass associated with the last major climatic change. They are the "seismograms" of ultra-low frequency geodynamics and — to continue the analogy — the deglaciation event was the equivalent "earthquake". In order to interpret these relative sea level data we must be able to construct the "synthetic seismograms" of the process. That is, given a certain model of deglaciation and a rheological model of the earth we need a theory which can predict relative sea level as a function of time everywhere where continent and ocean meet. The solution of this forward problem and of the inverse *problems* related to it (Peltier, 1976) are now practicable and the theoretical model and initial results of its application are reviewed below.

THE IMPULSE RESPONSE OF A MAXWELL EARTH

The simplest rheological model in terms of which *both* the instantaneous elastic and the long time scale viscous reactions to surface loading may be accommodated is that for a Maxwell solid. The stress—strain relation for a Maxwell medium is (Malvern, 1969):

$$\dot{\tau}_{kl} + \frac{\mu}{\nu}(\tau_{kl} - \tfrac{1}{3}\tau_{kk}\delta_{kl}) = 2\mu\dot{e}_{kl} + \lambda\dot{e}_{kk}\delta_{kl} \tag{3}$$

where τ_{kl} and e_{kl} are respectively the tensors for stress and strain and where μ and λ are the conventional Lamé parameters of elasticity. The dot denotes time differentiation. The first thing to note about eq. 3 is that it is linear. We can understand its physical content most simply by representing both tensors in terms of their Laplace transforms, i.e., by using the transform pair:

$$\tilde{\tau}_{kl} = \int\limits_{0}^{\infty} dt\, e^{-st}\, \tau_{kl} \tag{4}$$

$$\tau_{kl} = \frac{1}{2\pi i} \int\limits_{B} ds\, e^{st}\, \tilde{\tau}_{kl}$$

and similarly for e_{kl}. In eq. 4 B is the Bromwich path. In the Laplace transform domain of the imaginary frequency s eq. 3 becomes:

$$\tilde{\tau}_{kl} = \lambda(s)\, \tilde{e}_{kk}\delta_{kl} + 2\mu(s)\, \tilde{e}_{kl} \tag{5}$$

where $\lambda(s)$ and $\mu(s)$ are the following "compliances":

$$\lambda(s) = \frac{\lambda \cdot s + \mu K/\nu}{s + \mu/\nu} \; : \; \mu(s) = \frac{\mu \cdot s}{s + \mu/\nu} \tag{6}$$

where $K = \lambda + \tfrac{2}{3}\mu$ is independent of s. We note that in terms of these compliances the Laplace transformed constitutive relation has the same *form* as that for a Hookean elastic solid.

We can see clearly the way this material "works" by considering the two separate limits $s \to \infty$ and $s \to 0$ which correspond to short time scale and long time scale behaviour respectively (via the Tauberian theorems). Note that both s and μ/ν have the dimensions of inverse time. The "Maxwell time" is just $T_m = \nu/\mu$. For $t \ll T_m$ (i.e., $s \gg \mu/\nu$) we see from eq. 6 that $\lambda(s) \to \lambda$ and $\mu(s) \to \mu$ so that the material behaves as a Hookean elastic solid. In the opposite limit, $t \gg T_m$, $\lambda(s) \to K$, $\mu(s) \to 2\nu s$ and in the time domain the stress—strain relation becomes

$$\tau_{kl} = Ke_{kk}\delta_{kl} + 2\nu\dot{e}_{kl} \tag{7}$$

which is just that for an incompressible Newtonian fluid. The Maxwell solid is therefore "initially" elastic in its response to an applied stress and "finally" Newtonian viscous. The transition time between these asymptotic extremes is $T_m = 0$ (10^2 years). Thus for times in excess of a few hundred years after unloading the Maxwell earth behaves as a Newtonian viscous fluid. The constitutive relation for a standard linear solid (the most general visco-elastic model) differs from eq. 3 only in the addition of terms involving e_{kl} itself. The addition of such terms introduces a second viscosity coefficient associated with the Kelvin—Voigt element and there is thus dissipation associated with the short time scale response. Such rheologies are currently being investigated in the course of constructing earth models which have seismic velocity dispersion due to anelasticity (Liu et al., 1976; Kanamori and Anderson, 1977).

To solve the surface loading problem for Maxwell models we make use of the "Correspondence Principle" first described by Biot (1954) and Lee (1955). The principle is very simple to apply since it amounts to nothing more than a direct exploitation of the analogy between the visco-elastic constitutive relation (5) and that for a Hookean elastic solid. What we do is to solve an equivalent elastic problem many times for different values of the Laplace transform variable s to construct the "s-spectrum" of the solution. We then invert the spectrum to obtain its time domain form (Peltier, 1974).

The appropriate Laplace transformed and linearized field equations, respectively for the balance of momentum and for the perturbation of the gravitational potential are:

$$\underline{\nabla} \cdot \widetilde{\underline{\tau}} - \underline{\nabla}(\rho g \underline{u} \cdot \underline{e}_r) - \rho\underline{\nabla}\widetilde{\phi} + g\underline{\nabla} \cdot (\rho\widetilde{\underline{u}})\underline{e}_r = 0 \tag{8a}$$

$$\nabla^2\widetilde{\phi} = -4\pi G\underline{\nabla} \cdot (\rho\widetilde{\underline{u}}) \tag{8b}$$

where $\rho = \rho(r)$ is the density field in the basic hydrostatic equilibrium configuration, $g = g(r)$ is gravitational acceleration in the same state, $\underline{u}$ is the displacement field, ϕ the perturbation of the ambient gravitational potential, and G the gravitational constant. In eq. 8 $\underline{\nabla} \cdot (\rho\underline{u}) = \rho'$ has been substituted from the time integral of the continuity equation. The tilda, as before, represents implicit dependence upon the Laplace transform variable s. Thus $\underline{\tau}$ in eq. 8 is given by eq. 5. In eq. 8a the inertial force is suppressed because of the long time scale of the phenomenon in which we are interested. The second and fourth terms on the left hand side of eq. 8a require additional explanation. That the second term should appear in the elastic limit is well known (Love, 1911) and arises due to the existence of the hydrostatic pre-stress in the compressible medium. Since the ambient state is assumed hydro-static $\partial p/\partial r = -\rho g$ where $p(r)$ is the ambient pressure field. In an elastic dis-placement the material effectively transports its prestress with it. The fourth term in eq. 8a is the buoyancy force, since $\rho' = \underline{\nabla} \cdot (\rho \underline{u})$ from the continuity equation. If the background density field is adiabatic then this term should not appear in the viscous limit. It has been discussed by Cathles (1975) and by Peltier (1978).

We seek solutions to eq. 8 for $(\underline{u}, \phi)$ when the earth is deformed by gravi-tational interaction with a point mass load which is placed on its surface at $t = 0$ and instantaneously removed. If the physical properties of the interior $(\rho, \mu, \lambda, \nu)$ are functions of radius only then the response will be a function or (r, θ, s) only, where θ is the angular distance from the load. We may then expand $\underline{u}$ and ϕ as:

$$\tilde{\underline{u}} = \sum_{n=0}^{\infty} \left(U_n(r, s) P_n(\cos \theta) \underline{e}_r + V_n(r, s) \frac{\partial P_n(\cos \theta)}{\partial \theta} \underline{e}_\theta \right)$$

$$\tag{9}$$

$$\tilde{\phi} = \sum_{n=0}^{\infty} \phi_n(r, s) P_n(\cos \theta)$$

where $U_n(r, s)$, $V_n(r, s)$, $\phi_n(r, s)$ are spectral amplitudes for the harmonic dis-turbance of degree n and imaginary frequency s. Subject to appropriate boundary conditions on $r = a$ (the earth's surface) we may construct solu-tions for the spectral amplitudes in the form (Peltier, 1974):

$$\begin{vmatrix} U_n \\ V_n \\ \phi_{1,n} \end{vmatrix} = \phi_{2,n}(r) \cdot \begin{vmatrix} h_n(r, s)/g \\ l_n(r, s)/g \\ k_n(r, s) \end{vmatrix}$$

$$\tag{10}$$

where $\phi_n = \phi_{1,n} + \phi_{2,n}$ and $\phi_{2,n}$ is the perturbation of the gravitational po-tential due to the mass of the applied load. $\phi_{2,n}$ is independent of s since the applied load is assumed to have a $\delta(t)$ time dependence. The triplet of dimensionless scalars (h_n, l_n, k_n) thus constitute the non-dimensional spectral form of the impulse response of the system. They are the visco-elastic ana-logues of the so-called surface load "Love Numbers" of elasticity. Only the doublet (h_n, k_n) concerns us here. An example of the spectral surface $h_n(a, s)$ is shown in Fig. 1 for an earth model in which $\nu = \infty$ to a depth of 112.5 km (the lithosphere), $\nu = 10^{22}$ poise between the base of the lithosphere and the

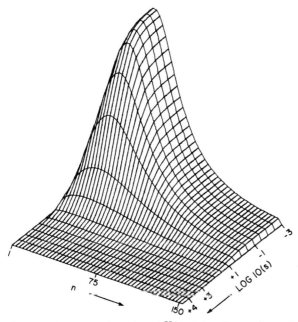

Fig. 1. The spectral surface $h_n^V(a, s)$ for the earth model described in the text which has a "lithosphere" of thickness $T = 112.5$ km at the surface $r = a$. Note that for large $n \gtrsim 150 h_n^V \equiv 0$ indicating that such short wavelength disturbances do *not* relax viscously. They are entirely supported by the lithosphere in which $\nu = \infty$.

core mantle boundary, and $\nu = 0$ throughout the core. The elastic structure of the model is "Gutenburg-Bullen A". In Fig. 1 we have in fact plotted $h_n^V(a, s) = h_n(a, s) - h_n^E$ where h_n^E is the large s elastic asymptote for each value of n (Peltier, 1974). In Fig. 1 it is clear that an additional asymptote exists for these spectra at small s.

It may be shown directly that these spectra have *exact* normal mode expansions of the form (Peltier, 1976):

$$h_n(s) = \sum_j \frac{r_j^n}{s + s_j^n} + h_n^E \tag{11}$$

where the s_j^n are a set of poles (a different set for each n) which lie on the negative real axis in the complex s-plane. The r_j^n are simply the residues at these poles and thus measure the extent to which a given normal mode is excited by the point forcing. A relaxation diagram showing the location of these poles for all n is shown in Fig. 2. The poles on this plot are marked with labels *MO*, *CO*, etc. which denote specific families which are distinguished from one another (Peltier, 1976) by the way in which the shear energy within them is distributed in radius. For example in *CO*, the fundamental core mode, the shear energy maximizes near the core—mantle boundary.

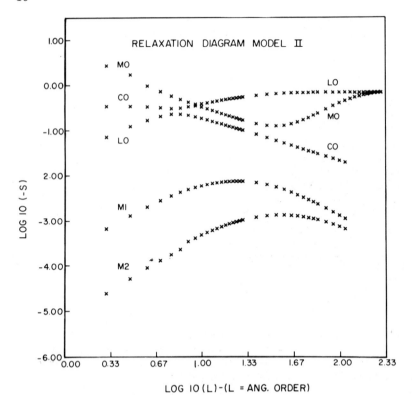

Fig. 2. Relaxation diagram for the earth model with lithosphere showing the location and multiplicity of the poles in the relaxation spectrum for each value of n. The ordinate is the logarithm of the inverse relaxation time and the relaxation times are all non-dimensionalized with a time scale of 10^3 years.

The above normal mode expansions of $h_n(s)$ and $k_n(s)$ have simple time domain representations:

$$h_n(t) = \sum_j r_j^n e^{-s_j^n t} + h_n^E \delta(t) \tag{12}$$

Remember that these are the non-dimensional spectral amplitudes for a temporally impulsive surface mass loading. If the load were allowed to remain on the surface for all $t > 0$ the time dependent harmonic amplitudes which result are obtained from the above simply by convolution with a Heaviside step function to give a form which has been previously labelled (Peltier and Andrews, 1976):

$$h_n^H(t) = \sum_j \frac{r_j^n}{s_j^n} (1 - e^{-s_j^n t}) + h_n^E = h_n^{H,V}(t) + h_n^E \tag{13}$$

Now notice that in the limit $t \to \infty$ these amplitudes are just:

$$\operatorname*{Lim}_{t \to \infty} h_n^H(t) = \sum_j \frac{r_j^n}{s_j^n} + h_n^E = \operatorname*{Lim}_{s \to 0} h_n(s) \tag{14}$$

The mechanism of isostatic compensation is thus clearly described in terms of these time dependent harmonic amplitudes. The number $\Sigma_j(r_j^n/s_j^n)$ for each wavenumber n describes the viscous contribution to the final isostatically adjusted amplitude of the disturbance. An example of the $h_n^{H,V}(t)$ is illustrated in Fig. 3 for the previous earth model.

We may obtain Green functions for various signatures of the response in terms of the Love Numbers h_n and k_n by summing infinite series as described by Farrell (1972) for the elastic problem and by Peltier (1974) for the viscoelastic equivalent. For example, the space and time dependent radial displacement of the original spherical surface is (for a unit applied point mass load):

$$u_r^H(\theta, t) = \frac{a}{M_e} \sum_{n=0}^{\infty} h_n^H(t) P_n(\cos \theta) \tag{15}$$

and the gravity anomaly on the deformed surface is:

$$\Delta g^H(\theta, t) = \frac{g}{M_e} \sum_{n=0}^{\infty} [n - 2h_n^H(t) - (n + 1)k_n^H(t)]P_n(\cos \theta) \tag{16}$$

whereas the perturbation of the ambient gravitational potential on the deformed surface is (M_e is the mass of the earth):

$$\phi^H(\theta, t) = \frac{ag}{M_e} \sum_{n=0}^{\infty} (1 + k_n^H - h_n^H)P_n(\cos \theta) \tag{17}$$

We will see in the next section that the latter is particularly important insofar as the prediction of relative sea level is concerned. Here we illustrate the Green functions in Fig. 4 where $u_r^{H,V}(\theta, t)$ for the previous earth model is shown. This function has been normalized by multiplication with "$a\theta$" to remove the geometric singularity and $u_r^{H,V}(\theta, t) = u_r^H(\theta, t) - u_r^E(\theta, t)$. If a lithosphere is not included in the model then $u_r^{H,V}$ is singular in the limit $t \to \infty$ (Peltier, 1978) so that without a lithosphere is not possible to calculate the set of isostatic Green functions which obtain in this limit. This singularity is simply a consequence of the physical fact that if the planet is everywhere viscous then a point mass placed on its surface will eventually sink to the planet's centre since this is the only position at which it will feel no net force. Since the gravity anomalies which we observe are associated with the extent of isostatic disequilibrium, in order to predict them with the model we require the infinite time Green function and thus need a lithosphere.

Given the Green functions determined as described above we may deduce the response of the earth model to the gravitational forcing associated with a surface load which has an arbitrary space—time dependence. Since the model is linear the solution under such circumstances may be obtained simply by invoking the principle of superposition. If we had a set of data which *exactly*

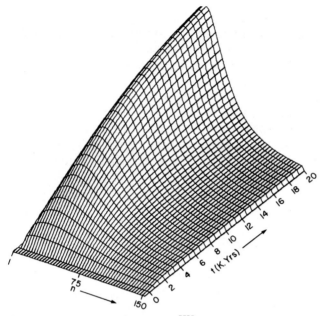

Fig. 3. The relaxation surface $h_n^{HV}(a, t)$ for the earth model with lithosphere. Note that for $h_n \overset{\sim}{>} 150 h_n^{HV}(a, t) \equiv 0$ so that sufficiently short wavelengths show no viscous relaxation. This was pointed out previously in connection with Fig. 1.

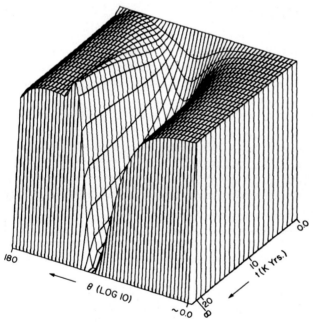

Fig. 4. Viscous part of the radial displacement Green function $u_r^{HV}(\theta, t)$ for the earth model with lithosphere. The effect of the geometric singularity at $\theta = 0$ has been removed for the purpose of illustration by multiplication with "$a\theta$".

recorded the time dependence of the variation of local radius $\Delta R(\theta, \phi, t)$ as a function of position on the surface (θ, ϕ) then the model prediction of this data would take the form:

$$\Delta R(\theta, \phi, t) = \iint d\Omega' \int_0^t dt' \, u_r(\theta/\theta', \phi/\phi', t/t')L(\theta', \phi', t') \tag{18}$$

where u_r is the kernel in the convolution integral, $d\Omega'$ the element of surface area, and L is the surface mass load functional which has the dimensions of mass per unit area (i.e., density $\times$ thickness). If the redistribution of mass on the surface takes place at a single instant of time then we can do the temporal part of the three variable convolution in eq. 18 analytically. The result is simply (Peltier and Andrews, 1976):

$$\Delta R^H(\theta, \phi, t) = \iint d\Omega' \, u_r^H(\theta/\theta', \phi/\phi', t)L(\theta', \phi') \tag{19}$$

For a more general history of load redistribution we may obtain the solution by the superposition of forms like eq. 19 which are appropriately weighted and phased in time.

Inspection of eq. 19 reveals two fundamental difficulties in applying the theory both of which are associated with L. Since the surface load must conserve mass and since it consists of two distinct parts associated respectively with the melting of the ice and the simultaneous filling of the ocean basins we may expand L as:

$$L = \rho_I L_I(\theta, \phi, t) + \rho_W L_O(\theta, \phi, t) \tag{20}$$

where ρ_I and ρ_W are the density of ice and water respectively, and L_I and L_O are the corresponding thickness. We assume that the hydrological cycle is closed so that mass conservation demands:

$$\int d\Omega \, \rho_W L_O = M_O = - \int d\Omega \, \rho_I L_I = - M_I \tag{21}$$

which is clearly an integral constraint on the model load history. Given $M_I(t)$, the time dependent mass flux to the ocean basins (assumed negative for melting), we can convert M_O to a uniform *equivalent* time dependent rise of sea level which we call:

$$S_{EUS} - \frac{M_O(t)}{\rho_W \cdot A_O} \tag{22}$$

where A_O is the surface area of the world ocean which we may assume to a good approximation is not affected by the sea level rise. Now S_{EUS} is an equivalent rise of sea level and we do not wish to suggest by eq. 22 that the sea level rise is uniform everywhere over the global ocean. Clearly it could not be uniform for then the "new" ocean would have a surface which was not everywhere an equipotential. *Given $M_I = -M_O$ we must in fact determine*

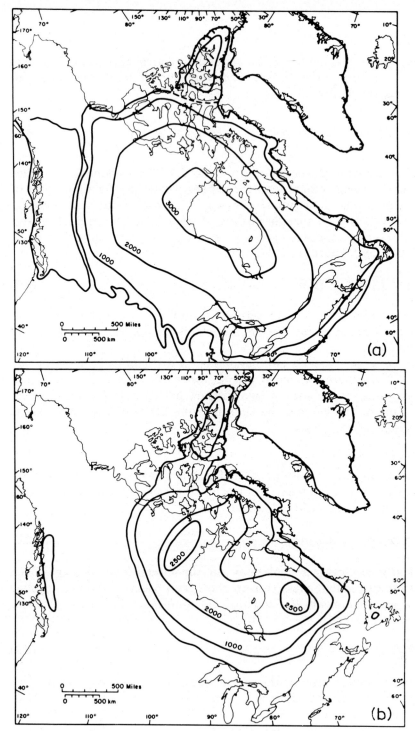

Fig. 5. The thickness of the Laurentide ice sheet in the model deglaciation history at two instants (a) Wisconsin maximum and (b) eight thousand years before present. Notice that the ice is assumed to retreat initially from over Hudson's Bay leaving high stands to the northwest and east. Such a history is required to fit the data on present day emergence *rates*.

the spatial distribution of M_O in the course of solving the problem. This will lead us to the Sea Level Equation which was described by Farrell and Clark (1976) and this will be discussed in the next section.

The second problem is associated with the determination of $L_I(\theta, \phi, t)$, the deglaciation chronology. In order to specify this function three distinct types of geological information are needed. We first assume that eq. 22 is correct to first order and as a measure of $S_{EUS}(t)$ we take relative sea level data from sites which are remote from the main centres of deglaciation. Such data indicate a net submergence which is on the order of 75—80 m since the last glacial maximum. Given $S_{EUS}(t)$ we compute $M_O(t)$ from eq. 22 then from mass conservation we get $M_I(t)$ directly. We next employ end moraine data which provide isochrons on the time dependent position of the edges of the major ice sheets. Knowing from this data the surface area of the major ice sheets as a function of time and the total mass which they must contain, the time dependent mass is partitioned among the major ice sheets in proportion to their instantaneous areas. Finally within each ice sheet and at each time we distribute the mass according to steady state ice mechanical arguments (Patterson, 1972) with allowance for other field data which make it possible to refine the distribution further. Such a "first guess" reconstruction of the major ice sheets (Fennoscandia and Laurentide) is tabulated in Peltier and Andrews (1976) and in Fig. 5a, b we show two time slices through the Laurentide ice history for 16 KBP and 8 KBP, respectively. In Fig. 6 we compare the corresponding $S_{EUS}(t)$ with the data described by

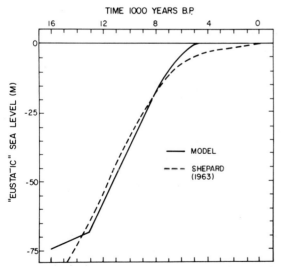

Fig. 6. The equivalent "eustatic" sea level curve for the deglaciation history tabulated in Peltier and Andrews (1976) compared to the observed eustatic curve of Shepard (1963). This constraint upon the deglaciation model is clearly satisfied.

Shepard (1963). Clearly our distintegration history fits this integral constraint quite accurately.

It must be emphasized that the ice sheet reconstruction described above is only approximate. The fact that we can make such a first guess, however, and one which we believe is reasonably accurate allows us to proceed with the solution of the forward problem. Given this deglaciation history we attempt to find, at first by guessing, the radial variation of viscosity which makes it possible to fit the observed relative sea level data. If we can find a viscosity profile which allows us to fit this data to some acceptable accuracy then we can proceed to refine *both* our knowledge of $M_I(\theta, \phi, t)$ *and* of $\nu(r)$ by employing the inverse theory outlined in Peltier (1976). The inverse problem is clearly non-linear because neither M_I nor $\nu(r)$ are known. We seek to deduce both directly from the relative sea level data by proceeding iteratively. We first fix M_I and refine ν; we then fix ν and refine M_I, continuing until (as we hope!) the process converges.

THE SEA LEVEL EQUATION

In the last section we left a major question unanswered and we turn to its resolution here. When the ice sheets melt and their meltwater is discharged to the ocean basins we must determine *where* in the oceans this water goes. This is clearly necessary if we wish to evaluate reponse integrals like eq. 18. Although we are able to obtain a reasonable a priori estimate for $L_I(\theta, \phi, t)$ as described in the last section we can as yet say little about $L_O(\theta, \phi, t)$ apart from the fact that it must conserve mass and thus satisfy the integral constraint eq. 22. By imposing the additional constraint that the surface of the ocean must be an equipotential surface at all times we may in fact deduce $L_O(\theta, \phi, t)$ *from* $L_I(\theta, \phi, t)$. Application of this constraint has led us (Farrell and Clark, 1976) to an explicit equation for the time variation of relative sea level locally. We review briefly the construction of this sea level equation in the remainder of this section.

Clearly the Green function for the forced variation of the gravitational potential must play a fundamental role in this discussion. We note that ϕ^H/g has the dimensions of length and begin by enquiring as to the meaning of this length scale. If we were to assume a single instant of melting then the net change of potential Φ at the earth's surface $r = a$ would be given by the convolution of ϕ^H with all mass loads as:

$$\Phi(\theta, \phi, t) = \rho_I \phi^H \underset{I}{*} L_I + \rho_W \phi^H \underset{O}{*} S \tag{23}$$

where the abbreviations $(\underset{I}{*})$ and $(\underset{O}{*})$ represent convolutions over the ice and oceans, respectively, and where we have replaced L_O by S, the local increase (decrease) of water thickness. The reason for this will become clear momentarily. The potential change Φ includes the effect due to the deformation of the surface of the planet since the function ϕ^H contains the Love number h_n.

This change in potential causes a change in the sea level with respect to the deformed surface of the solid earth in the amount (Farrell and Clark, 1976):

$$S = \frac{\Phi(\theta, \phi, t)}{g} + C \qquad (24)$$

where the constant C is determined to ensure conservation of mass. Equation 24 is valid for sufficiently small changes S of the local bathymetry. It is important to note that S, by construction, is the local variation of sea level with respect to the surface of the solid earth and thus is precisely the relative sea level which one observes. Substituting eq. 24 into eq. 23 then gives:

$$S = \rho_I \frac{\phi^H}{g_*} L_I + \rho_W \frac{\phi^H}{g_*} S + C \qquad (25)$$

To determine C we note that the integral over the surface area of the oceans of the product $\rho_W S$ must equal the instantaneous value of the total mass which has been lost by the ice sheets at time t. Thus:

$$\langle \rho_W S \rangle_O = \rho_W \left\langle \rho_I \frac{\phi^H}{g_*} L_I + \rho_W \frac{\phi^H}{g_*} S \right\rangle_O + \langle C \rangle_O \rho_W = -M_I(t) \qquad (26)$$

The minus sign on the r.h.s. of eq. 26 is required since $M_I(t) < 0$ for load removal as discussed in the last section. Since $C = $ constant thus $\langle C \rangle_O = C A_O$, thus:

$$C = -\frac{M_I(t)}{\rho_W A_O} - \frac{1}{A_O} \left\langle \rho_I \frac{\phi^H}{g_*} L_I + \rho_W \frac{\phi^H}{g_*} S \right\rangle_O \qquad (27)$$

and in eqs. 26 and 27 we have used $\langle \ \rangle_O$ to indicate an integral over the surface area of the oceans.

With C given by eq. 27, eq. 25 is an integral equation for S which we call the sea level equation. It is an integral equation since S appears not only on the l.h.s. but also in the convolution integrals on the right. Given the deglaciation history $L_I(\theta, \phi, t)$ and thus $M_I(t)$ we may deduce the history of relative sea level $S(\theta, \phi, t)$ at any point on the surface (θ, ϕ) by inverting this integral equation.

In order to make the solution of eq. 25 practicable within the context of current (and foreseeable) limitations upon computing resources we are forced to discretize our description of the phenomenon both in time and in space. What we do in practice is to divide the "active" surface area of the earth into a number of finite elements which cover all of the ocean basins and that portion of the surface area of the continents at which actual deglaciation occurs. In addition we discretize the system in time and allow in the process for the fact that the deglaciation history and simultaneous ocean loading are not instantaneous. We assume that the mass load upon the finite element with centroid at r' may be described by:

$$L(r', t) = \sum_{l=1}^{P} L_l(r')H(t - t_l) \tag{28}$$

where t_l $(l = 1, P)$ are a series of times which bracket the entire loading history at r' and the $L_l(r')$ are the loads applied or removed at the discrete times t_l. We allow the loads upon the "active" elements to change only at the times t_l and the times t_l are common to all active elements. This discrete approximation to the smooth functions $L(r', t)$ may be made arbitrarily accurate by choosing a set of t_l with sufficiently small $\Delta t = t_{l+1} - t_l$. In practice we sample the deglaciation history and the response at a uniform $\Delta t = 10^3$ years.

With the above discrete specification of the load history in time we may write eq. 25 in the form:

$$S(r, t) = \iint_{O} \rho_W S(r', t)G^E(r - r')dr' + \iint_{I} \rho_I I(r', t)G^E(r - r')dr'$$

$$+ \sum_{l=1}^{P} \iint_{O+I} L_l(r')G^{HV}(r - r', t - t_l)dr' - S_{EUS}(t) - K_C(t) \tag{29}$$

where we have expanded $\phi^H/g = G^E + G^{HV}$ into its elastic and viscous parts (Peltier, 1974; Peltier and Andrews, 1976) and the ice thickness has been represented simply as $L_I = I$. We note furthermore that in eqs. 28 and 29 L_l is a density weighted thickness. The function $S_{EUS}(t)$ corresponds to the first term in eq. 27, i.e.:

$$S_{EUS}(t) = \frac{\rho_I}{\rho_W A_O} \iint_{I} I(r', t)dr' \tag{30}$$

and $K_C(t)$ to the second term, i.e.:

$$K_C(t) = \frac{1}{A_O} [\iint_{O} dr'' \iint_{O} \rho_W G^E(r'' - r')S(r', t)dr'$$

$$+ \iint_{O} dr'' \iint_{I} \rho_I G^E(r'' - r')I(r', t)dr'$$

$$+ \iint_{O} dr'' \{ \sum_{l=1}^{P} \iint_{O+I} L_l(r')G^{HV}(r'' - r', t - t_l)dr' \}] \tag{31}$$

In order to solve eq. 29 we proceed as mentioned above by splitting the entire active area of the surface into a number of discrete finite elements. The area of the jth element we call E_j and on this area we assume that the load is piecewise constant in time. If r_i is the centroid of the ith element then the discrete version of eq. 29 is just:

$$a_{ip} = A_{ij}a_{jp} + B_{ij}b_{jp} + C_{iplj}c_{jl} - \frac{\rho_I}{\rho_W A_O} E_j b_{jp}$$

$$- \frac{1}{A_O} [E_i\{A_{ij}a_{jp} + B_{ij}b_{jp} + C_{iplj}c_{jl}\}] \tag{32}$$

in which a summation convention over repeated indices is implied such that $j = 1, N$ for terms involving a_{jp}; $j = N + 1, M$ for terms involving b_{jp}; $j = 1, M$ for terms involving c_{jl}; $l = 1, P_0$; $i = 1, N$. The upper limits in the summations are as follows:

N = the number of finite elements in the oceans
M = the total number of active elements (ocean plus ice)

$$(33)$$

P_0 = the number of time intervals ($\Delta t = 10^3$ years) since deglaciation commenced

The lower case matrices are load dependent and are defined as follows:

$a_{ip} = S(\underline{r}_i, t_p)$, total relative sea level change at $\underline{r}_i$ for time t_p
$b_{ip} = I(\underline{r}_i, t_p)$, *total* thickness of ice removed from r_i by time t_p $\quad(34)$
$c_{ip} = L_p(\underline{r}_i)$, incremental load change at $\underline{r}_i$ at time t_p

The upper case matrices depend only upon the geometry of the problem and the earth model. They are:

$$A_{ij} = \rho_W \iint_{E_j} G^E(\underline{r}_i - \underline{r}')d\underline{r}'$$

$$B_{ij} = \rho_I \iint_{E_j} G^E(\underline{r}_i - \underline{r}')d\underline{r}' = \frac{\rho_I}{\rho_W} A_{ij}$$

$$C_{iplj} = \iint_{E_j} G^{HV}(\underline{r}_i - \underline{r}', t_p - t_l)d\underline{r}' \tag{35}$$

In eq. 34 a_{ip} is an unknown matrix which is to be determined whereas b_{ip}, c_{ip} are input matrices which describe the deglaciation history.

The major computational expense in implementing the theory is connected with the calculation of the three "interaction matrices" A_{ij}, B_{ij}, C_{iplj}. These matrices depend only upon the geometric relations among the finite elements into which the "active" portion of the surface has been divided. In general, these interactions depend not only upon the separation between the field point $\underline{r}_i$ and the controid $\underline{r}_j$ of the element of area E_j and the magnitude of E_j, but also upon the *shape* of the element itself. In the interest of efficiency we have elected initially to suppress this shape dependence and to replace each finite element by a circular disk of equivalent area in calculating the matrix components. This assumption may of course be relaxed. We construct a table of disc factors as described by Farrell (1973) for an equivalent elastic problem and by Peltier and Andrews (1976) for the visco-elastic system. The normalization of the elements of this table is however different from that employed in these papers. For the visco-elastic problem the necessity of this was pointed out by Peltier and Andrews (1976). The new normalization factor is essentially the square of the angular separation between the field point and the centre of the disc (Clark, 1977). Linear interpolation in a regular table of disc factors is then employed to deduce the matrix components.

Given these geometry and earth model dependent matrices and the matrix specification of the deglaciation history we solve eq. 32 at each point in time by applying conventional relaxation methods. At each time t_p we make a first guess to the matrix a_{ip} and substitute this into eq. 32 to compute the residual. This residual is then fed back to update the first estimate and the process is continued until convergence is achieved. In practice three such iterations are normally sufficient (Farrell and Clark, 1976).

INITIAL RESULTS

Here we will discuss the results of applying the discrete form of the sea level eq. 32 in an initial series of calculations. These results have all been described elsewhere (Clark, 1978; Clark et al., 1978; Peltier, 1978) and the interested reader is referred to these articles for a more detailed discussion. The earth model employed is the first for which Green functions were derived in Peltier (1974) and later subjected to preliminary test in Peltier and Andrews (1976). It has an elastic structure which is Gutenburg-Bullen "A", an inviscid core and no lithosphere. Throughout the mantle the viscosity has a constant value of 10^{22} poise (cgs). It is the model which Cathles (1975) and Peltier and Andrews (1976) have claimed provides a good fit to a wide range of relative sea level data although this conclusion was in both cases based upon calculations which were not gravitationally self-consistent. The glacial history employed is that tabulated in Peltier and Andrews (1976) although this was refined somewhat by linear interpolation to give ice thickness at each active point at equispaced times separated by $\Delta t = 10^3$ years. This deglaciation history gives the equivalent "eustatic" sea level curve previously shown in Fig. 6.

The first output from the calculation is the global prediction of the relative sea level variation since the initiation of melting. It is *assumed* that the major ice sheets were in isostatic equilibrium before retreat commenced. Four time slices through the global solution are shown in Fig. 7. The sea level rise (fall) is contoured in metres. Clearly the rise of sea level in the "far field" of the ice sheets is not uniform as must be the case if the surface of the new ocean is to remain equipotential. The large fall of sea level relative to the surface of the solid earth in regions which were once ice covered (Fennoscandia, Laurentide) is also clearly evident. From this output we may

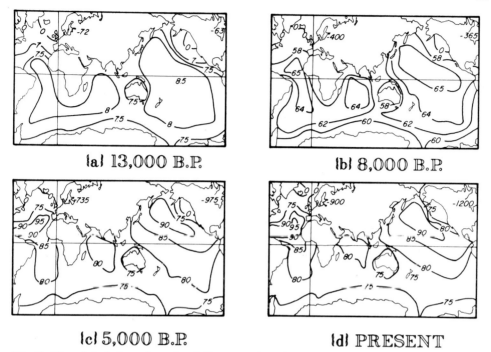

Fig. 7. The global rise of sea level (in metres) at four times subsequent to the onset of melting. Note the large negative values corresponding to a fall of local sea level in the vicinity of the Laurentide and Fennoscandia ice sheets. The rise of sea level is not uniform in the far field showing explicitly that the concept of eustatic sea level is of limited utility.

directly deduce relative sea level for any point on the surface for which we have observations available. Inspection of a complete set of output data (Clark et al., 1978) reveals that the surface may be divided roughly into six major regions within which there exists a fairly unique relative sea level signature. These regions are shown in Fig. 8 and their characteristics are as follows.

In I there is continuous emergence following deglaciation and these regions are those which were once under the ice sheets. In II there is continuous submergence due to the collapse of the proglacial forebulge (Peltier, 1974). In addition there is a rather narrow region separating I and II in which the relative sea level history is not monotonic, initial emergence is followed by submergence. This effect is due to the migration of the forebulge inwards (Peltier, 1974). In region III there is slight time dependent emergence and this occurs only over a very limited area of the globe (Clark et al., 1978). In IV there is continuous oceanic submergence and in V there is emergence which begins immediately upon the cessation of melting (Clark et al., 1978). Region VI consists of all continental shorelines which are sufficiently remote from the main deglaciation centres. Such locations are char-

22

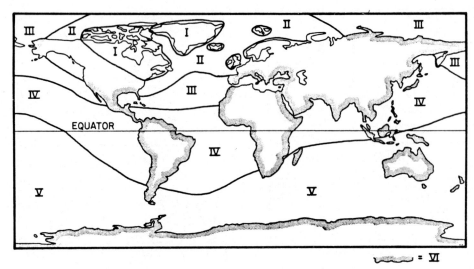

$= VI$

Fig. 8. The global extent of regions I—VI in each of which the sea level curve has a certain characteristic signature. The characteristics of this signature are described in the text.

acterized by emergence due to crustal tilting which is forced by the adjacent water load (Walcott, 1972). In the following paragraphs we compare briefly the predictions of the model with observed relative sea level data in each of these regions. The sources of the data shown on the following figures are listed in abbreviated form as an appendix to the main bibliography.

Figure 9 shows observations in region I respectively for the Laurentide (Ottawa Islands) and Fennoscandia (Oslo Fjord) regions and the superimposed predictions of the model for the two sites. At the former location the prediction is considerably in excess of the observed emergence indicating

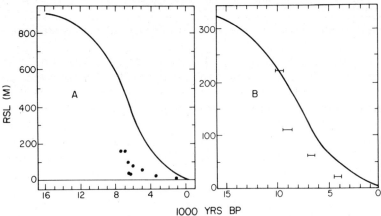

Fig. 9. Comparison of theory and observation for two sites in region I. A. Ottawa Islands, Hudson's Bay. B. Oslo Fjord, Norway.

either that the load removed over Hudson's Bay was somewhat large or that the rheology is incorrect or both. The fit to the data at the Oslo Fjord site is much better although the amount of emergence predicted is again somewhat in excess of the observation.

In Fig. 10 we compare theory with observation at a sequence of locations extending southward along the eastern seaboard of the continental U.S. in region II (Clark et al., 1978). Although continuous submergence is observed and predicted the model calculation gives excessive submergence, the error of fit being on the order of 100% for some sites. This again suggests that the load removed may have been too large or that the earth model is in error. In Fig. 11 the prediction is compared to observation at an equivalent location with respect to the Fennoscandia ice sheet, namely along the Atlantic coast of France. The dashed curve is for northern France (region II) and the solid curve for southern France (region III). All of the data along the French coast should lie between these two extremes and the gross error of fit is seen to be small (Clark, 1978).

In Fig. 12 we compare theory and observation at a sequence of locations near the edge of the ice sheet for the Cumberland Peninsula on Baffin Island (Canada). This comparison is from Clark (1978) and illustrates the characteristic non-monotonic nature of the relative sea level curves from such regions which was predicted by Peltier (1974) by direct inspection of the Green

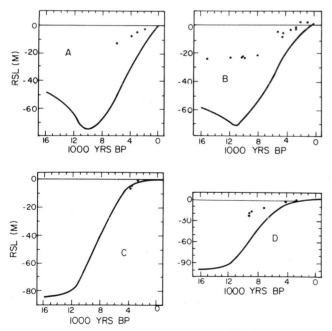

Fig. 10. Comparison of theory and observation for four sites along the eastern seaboard of the U.S. in region II. A. Brigantine, New Jersey. B. Virginia. C. Georgia. D. Bermuda.

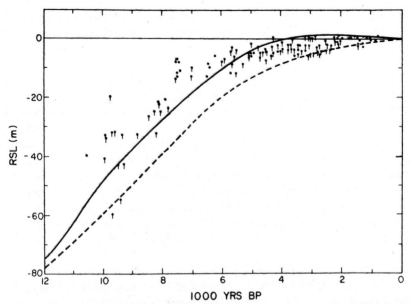

Fig. 11. Comparison of theory and observation for a segment of region II along the Atlantic coast of France. The solid curve is the prediction for sites in southern France (region III) while the dashed curve is for northern France (region II). The data from the French coast are seen to be bounded by these extremes.

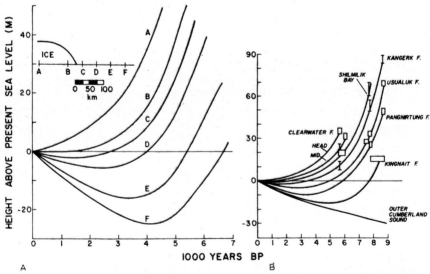

Fig. 12. Comparison of theory (A) with observations (B) at a sequence of sites near the edge of the ice sheet on the Cumberland peninsula of Baffin Island (boundary between region I and region II).

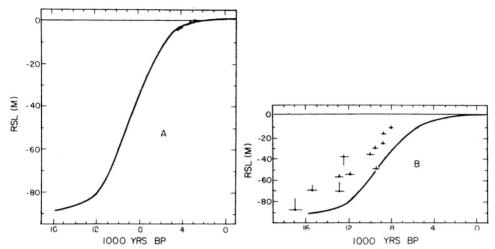

Fig. 13. Comparison of theory and observation at two sites in region III: (A) Florida, (B) Gulf of Mexico.

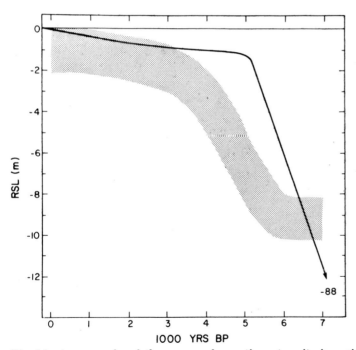

Fig. 14. An example of theory vs. observation at a site in region IV. This is for Oahu, Hawaii. The large divergence near $6 \cdot 10^3$ years before present is presumed to be associated with the volcanic eruption which took place at that time and is thus associated with local tectonics.

functions for the present earth model. Such sites are found on the boundary between regions I and II.

Examples of the comparison between theory and observation in region III are shown in Fig. 13 respectively for Florida and for the Gulf of Mexico. For the latter location the sea level record is extremely long and for both the fit of the theory to the observations is rather good. These comparisons were described previously in Clark et al. (1978) and some additional discussion is found in Peltier (1978).

As an example from region IV we show the comparison of theory and observation for Oahu, Hawaii (from Clark, 1978) in Fig. 14. Here again the magnitude of the misfit is acceptably small, the largest deviation occurring at roughly five thousand years before the present at the time of the last major volcanic eruption and therefore presumably due to local tectonics. In Fig. 15 (Clark, 1978) the relative sea level data from New Zealand (region V) is compared to the theoretical calculation. Here again the error of fit is observed to be small. Finally in Fig. 16 (Clark et al., 1978) we compare with theory the observations at a site in region VI. These data are for Recife, Brazil and illustrate the characteristic emergence at continental shorelines in the far field. The solid curve is the prediction for the coastline and the dashed curve for a site on the continental shelf 100 km east of Recife. The latter shows contin-

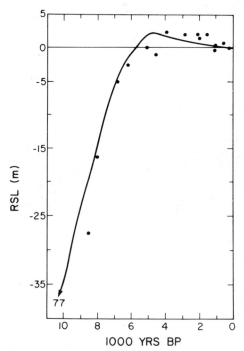

Fig. 15. Comparison of theory and observation for a site in region V. This is for New Zealand. The point is that even in the far field the fit to the observations is rather good.

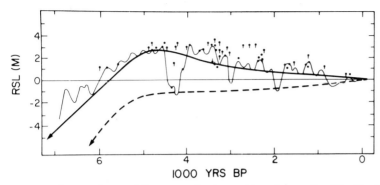

Fig. 16. Comparison of theory and observation for a site (Recife, Brazil) in region VI. This illustrates the characteristic emergence due to crustal tilting along continental margins in the far field.

uous submergence characteristic of region IV. The Fairbridge curve (light line) is bracketed by these two predictions.

DISCUSSION AND CONCLUSIONS

The comparison of theory and observation detailed in the last section is encouraging in that it has provided a positive test of both the rheology model and of the deglaciation history. As is clear from the way in which the theory has been constructed, the problem of glacial isostatic adjustment is highly non-linear in that the observed response (relative sea level data) is dependent both upon the rheology and upon the deglaciation history. If we are to discover *a* viscosity profile and *an* unloading history which are unique in the sense of being known within limits prescribed by our knowledge of the data then we must be able to perform an initial linearization of the problem with some confidence. We have done this by employing a "first guess" deglaciation history which is in accord with the best information currently available from Quaternary geology. That this first guess load history and a first guess viscosity profile (also conditioned by a priori knowledge) have led to substantial accord with the observations convinces us that we may proceed with the iterative refinement of both functionals in the manner outlined in Peltier (1976) and the refinement of $I(\theta, \phi, t)$ has already begun (Clark, 1977).

There remain certain characteristic misfits between theory and observation which will provide guidance as we proceed with this work. Most notably, the relative sea level data for the Laurentide region show rather large departures from theory *in the near field*. Although we expect that the incorporation of a lithosphere into the model (Peltier, 1978) may remedy some of the defects it is rather unlikely, indeed impossible, that the major errors are derivative of this source. Either the earth model is more fundamentally in error or the load history is incorrect. Because the large errors of fit are confined to the Laurentide region and do not exist in the vicinity of Fenno-scandia we suspect that the latter is more likely to be the case. These questions are under current consideration.

Because these characteristic misfits still remain in the near field of the Laurentide ice sheet we are as yet not able to confirm or to deny with complete confidence the conclusions previously drawn by Cathles (1975) and by Peltier and Andrews (1976). If "fine tuning" of these calculations does reconfirm these conclusions (as we suspect will be the case), namely that within the context of a Newtonian viscous rheology the viscosity of the mantle *must be* essentially constant in order to fit the rebound data, then the implications for mantle convection are important. Peltier (1972) has shown that the mixing length for convection in the mantle would then be on the order of the mantle thickness itself.

Aside from determining *a* mantle viscosity profile which allows us to fit the data we would also like to know the envelope of viscosity profiles which are capable of satisfying all of the data to within a standard error. This quantification of the extent to which the rebound information allows us to remain ignorant is perhaps the most important information which we could extract from the point of view of the solid state theory of microphysical creep mechanisms.

Here we have described only the constraints upon a Newtonian rheology which are implied by the relative sea level data themselves. There are at least two further observational data which are also capable of providing important information. The first of these, and perhaps the most important, are data concerning the present day gravity anomalies over the rebound centres. These data are particularly important because they provide information which is completely distinct from that afforded by the relative sea level data. Gravity data, in a real sense, "looks into the future" because such data essentially measure the degree of current gravitational disequilibrium and are therefore indicative of the amount of uplift (subsidence) remaining which is well known. The gravity data therefore "see" the long time tail of the relaxation spectrum whereas the sea level data do not. In order to predict the gravity anomaly with the present theory we *require* a model with a lithosphere (Peltier, 1978).

Besides the gravity data we may also expect to obtain interesting (although perhaps not so immediately useful) information by analysis of the non-tidal deceleration of rotation which is forced by the deglaciation event. The large shift in surface mass is clearly sufficient to produce a relatively large change in the components of the inertia tensor and thus a change in the length of the day. Although interesting in itself and in providing an additional constraint on deep mantle viscosity (O'Connell, 1971) this effect may be most interesting from a climatological point of view. It should be clear that there remains a significant amount of work to be done before we may properly claim a complete understanding of the phenomenon of glacial isostasy.

REFERENCES

Ashby, M.F., 1972. A first report on deformation mechanism maps. Acta Metall., 20: 887.

Biot, M.A., 1954. Theory of stress-strain relations in anisotropic visco-elasticity and relaxation phenomena. J. Appl. Phys., 25 (11): 1385.

Brennan, C., 1974. Isostatic recovery and the strain rate dependent viscosity of the earth's mantle. J. Geophys. Res., 79: 3993.

Cathles, L.M., 1975. The Viscosity of the Earth's Mantle. Princeton University Press, Princeton, New Jersey, 386 pp.

Clark, J.A., 1977. Global Sea Level Changes Since the Last Glacial Maximum. Thesis, Dept. of Geol. Sciences, University of Colorado, Boulder, Colo. (unpublished).

Clark, J.A., 1978. A numerical model of worldwide sea level changes on a visco-elastic earth. Proceedings of the Stockholm Conference on Earth Rheology and Late Cenozoic Vertical Movements (in press).

Clark, J.A., Farrell, W.E. and Peltier, W.R., 1978. Global changes in postglacial sea level: a numerical calculation. Quat. Res. (N.Y.), (in press).

Crough, S.T., 1977. Isostatic rebound and power law flow in the asthenosphere. Geophys. J. R. Astron. Soc., 50: 723.

Farrell, W.E., 1972. Deformation of the earth by surface loads. Rev. Geophys. Space Phys., 10: 761.

Farrell, W.E., 1973. Earth tides, ocean tides, and tidal loading. Phil. Trans. R. Soc. London, Ser. A, 274: 45.

Farrell, W.E. and Clark, J.A., 1976. On postglacial sea level. Geophys. J. R. Astron. Soc., 46: 647.

Goldreich, P. and Toomre, A., 1969. Some remarks on polar wandering. J. Geophys. Res., 74: 2555.

Haskell, N.A., 1935. The motion of a viscous fluid under a surface load. Physics, 6: 265.

Haskell, N.A., 1936. The motion of a viscous fluid under a surface load, 2. Physics, 7, 56:

Haskell, N.A., 1937. The viscosity of the asthenosphere. Am. J. Sci., 33: 22.

Herring, C., 1950. Diffusional viscosity of a polycrystalline solid. J. Appl. Phys., 21: 437.

Isacks, B. and Molnar, P., 1971. Distribution of stresses in the descending lithosphere from a global survey of focal mechanism solutions of mantle earthquakes. Rev. Geophys. Space Phys., 9: 103.

Kanamori, H. and Anderson, D.L., 1977. Importance of physical dispersion in surface wave and free oscillation problems: Review. Rev. Geophys. Space Phys., 15 (11): 105.

Kohlstedt, D. and Goetze, C., 1974. Low-stress high-temperature creep in olivine single crystals. J. Geophys. Res., 79: 2045.

Lee, E.H., 1955. Stress analysis in visco-elastic bodies. Q. Appl. Math., 13: 183.

Libby, W.F., 1952. Radiocarbon Dating. The University of Chicago Press, Chicago and London.

Liu, H.P., Anderson, D.L. and Kanamori, H., 1976. Velocity dispersion due to anelasticity: Implications for seismology and mantle composition. Geophys. J. R. Astron. Soc., 47: 41.

Love, A.E.H., 1967 (1911). Some Problems of Geodynamics. Dover Publications Inc., New York, 180 pp.

Malvern, L.E., 1969. Introduction to the Mechanics of a Continuous Medium. Prentice-Hall Inc., Englewood Cliffs, N.J.

McConnell, R.K., 1968. Viscosity of the mantle from relaxation time spectra of isostatic adjustment. J. Geophys. Res., 73: 7089.

McKenzie, D.P., 1966. The viscosity of the lower mantle. J. Geophys. Res., 71: 3995.

Munk, W.H. and MacDonald, G.J.F., 1960. The Rotation of the Earth. Cambridge University Press, Cambridge, Mass.

Niskanen, E., 1948. On the viscosity of the earth's interior and crust. Ann. Acad. Sci. Fenn., Ser. A3 (15): 22.

O'Connell, R.J., 1971. Pleistocene glaciation and the viscosity of the lower mantle. Geophys. J. R. Astron. Soc., 23: 299.

Parsons, B.E., 1972. Changes in the Earth's Shape. Thesis, Cambridge Univ., Cambridge (G.B.).

Patterson, W.S.B., 1972. Laurentide ice sheet: Estimated volumes during late Wisconsin. Rev. Geophys. Space Phys., 10: 885.

Peltier, W.R., 1972. Penetrative convection in the planetary mantle. Geophys. Fluid Dyn., 5: 47.

Peltier, W.R., 1974. The impulse response of a Maxwell earth. Rev. Geophys. Space Phys., 12 (4): 649.

Peltier, W.R., 1976. Glacial isostatic adjustment — II: The inverse problem. Geophys. J. R. Astron. Soc., 46: 669.

Peltier, W.R., 1978. Ice sheets, oceans, and the earth's shape. Proceedings of the Stockholm Conference on Earth Rheology and Late Cenozoic Movements (in press).

Peltier, W.R. and Andrews, J.T., 1976. Glacial isostatic adjustment — I: The forward problem. Geophys. J. R. Astron. Soc., 46: 605.

Post, R. and Griggs, D., 1973. The earth's mantle: evidence of non-Newtonian flow. Science, 181: 1242.

Sammis, C.G., Smith, J.C., Schubert, G. and Yuen, D.A., 1977. Viscosity-depth profile of the earth's mantle: effects of polymorphic phase transitions. J. Geophys. Res., 82: 3747.

Shepard, F.P., 1963. 35,000 years of sea level. In: Essays in Marine Geology. Univ. Southern California Press, 1.

Stocker, R.L. and Ashby, M.F., 1973. On the rheology of the upper mantle. Rev. Geophys. Space Phys., 11: 391.

Vening Meinesz, F.A., 1937. The determination of the earth's plasticity from post glacial uplift of Scandinavia: isostatic adjustment. Proc. K. Ned. Akad. Wet., 40: 654.

Walcott, R.I., 1972. Late Quaternary vertical movements in eastern North America: Quantitative evidence of glacio-isostatic rebound. Rev. Geophys. Space Phys., 10: 849.

Weertman, J. and Weertman, J.R., 1975. High temperature creep of rock and mantle viscosity. A. Rev. Earth Planet. Sci., 3: 293.

SEA LEVEL DATA SOURCES

Fig. 9A. Andrews, J.T. and Falconer, G., 1969. Can. J. Earth Sci., 6: 1263—1276.

Fig. 9B. In: T.C. Kenney, 1964. Geotechnique, 14: 203—230.

Fig. 10A. Stuiver, M. and Deddario, H.J., 1963. Science, 142: 951.

Fig. 10B. Harrison, W., Malby, R.J., Rusnak, G.A. and Terasmee, J., 1965. J. Glaciol., 73: 201—229.
Newman, W.S. and Rusnak, G.A., 1965. Science, 148: 1464—1466.

Fig. 10C. Wait, R.L., 1968. U.S. Geol. Surv., Prof. Pap., 600-D: 38—41.

Fig. 10D. Newman, A.C., 1971. Quaternaria, 14: 41—43.

Fig. 11. Ters, M., 1973. Assoc. Fr. Étude Quat., Suppl. au Bull., 36: 114—142.

Fig. 12B. Dyke, A., 1977. Unpublished Ph.D. Thesis, Univ. of Colorado, Dept. of Geography, Boulder, Colorado, 184 pp.

Fig. 13A. Scholl, D.W. and Stuiver, M., 1967. Geol. Soc. Am. Bull., 78: 437—454.

Fig. 13B. Curray, J.R., 1960. In: Recent Sediments N.W. Gulf of Mexico. Am. Assoc. Pet. Geol., pp. 221—266.

Fig. 14. Easton, W.H. and Olson, E.A., 1976. Geol. Soc. Am. Bull., 87: 711—719.

Fig. 15. Schofield, J.C., 1964. N.Z. J. Geol. Geophys., 7: 359—370.

Fig. 16. Fairbridge, R.W., 1976. Science, 191: 353—359.

Chapter 2

SUBDUCTION ZONE DIP ANGLES AND FLOW DRIVEN BY PLATE MOTION*

BRADFORD H. HAGER and RICHARD J. O'CONNELL

Department of Geological Sciences, Harvard University, Cambridge, MA 02138 (U.S.A.)

SUMMARY

Kinematic models of the large scale flow in the mantle accompanying the observed plate motions are calculated by neglecting thermal buoyancy forces. The large scale flow is therefore determined by the mass flux imposed by the moving plates. The energy and momentum equations decouple, and with the assumption of a radially symmetric Newtonian viscosity, the flow accompanying the plate motions can be obtained using harmonic analysis and propagator matrices. The resulting flow models predict remarkably well the observed dips of subducted slabs if the flow extends into the lower mantle. The plates drag along a thick boundary layer which should be included in models of the heating of subducted slabs.

INTRODUCTION

One very important consideration in modeling flow in the earth's mantle is that the extreme temperature dependence of effective rock viscosity, coupled with the large temperature gradient near the earth's surface, leads to the existence of a mechanical boundary layer, the lithosphere. The motion of the lithosphere implies a large-scale circulation due to the mass flux from the moving lithospheric plates themselves and viscous coupling between the plates and the underlying mantle. The motion of the plates should have a strong organizing influence upon the underlying flow, as has been shown by the experiments of Richter (1973; Richter and Parsons, 1975) and Parmentier and Turcotte (1976).

At present, numerical simulations of three-dimensional mantle convection are impractical. Complexities include the dependence of rheology on temperature, pressure, shear stress, mineralogy, and history; realistic distribution of heat sources; chemical differentiation; melting; phase changes; and the effects of the observed complex geometry. Even if realistic three-dimensional

---

*Originally published as: Hager, B.H. and O'Connell, R.J., 1978. Subduction zone dip angles and flow driven by plate motion. In: M.N. Toksöz (Editor), Numerical Modeling in Geodynamics. Tectonophysics, 50: 111—133.

calculations were feasible, the values of the model parameters are not well constrained for the earth. Thus, it is useful to make simpler models to investigate the effect and relative importance of complicating factors individually.

In the simple models presented here, we neglect the effect of temperature changes upon density and viscosity. This decouples the equations of motion from the energy equation and makes their solution much simpler. However, this decoupling removes the dynamics from the models. We use the observed plate motions as boundary conditions in our three-dimensional kinematic models without specifying the cause of the motion. This highly simplified model then permits us to isolate and understand the large scale flow accompanying the plates moving in their observed complex geometries.

The neglect of the buoyancy forces due to thermal and chemical heterogeneity is based on the observation that in dynamic models of convection with a moving boundary layer, the large-scale flow pattern is determined by the motion of the boundary layer even for boundary velocities as low as 3 cm/yr and Rayleigh number as great as 10^6 (Richter, 1973; Parmentier and Turcotte, 1976). Studies of the driving forces of plate motions suggest that the plates exert drag on the underlying mantle, due to concentration of buoyancy at the surface, rather than the mantle flow driving the plates as a result of distributed buoyancy (Forsyth and Uyeda, 1975; Solomon et al., 1975; Richardson et al., 1976). This is further justification for the applicability of our simple kinematic models. The neglect of the temperature dependence of the rheology is less easy to justify, but is done to make analytical modeling possible.

A major problem in evaluating numerical models of flow in the earth is that the plates themselves explain such a wide set of geophysical data that they effectively filter much of what is going on below them. However, there are some features of flow models which can be compared to observation to help assess the validity of the models. For example, the calculated normal stress at the base of the lithosphere can be compared to the observed bathymetry. Gravity anomalies can be calculated from perturbations in the radii of compositional boundaries. The shear stress at the base of the lithosphere can be calculated and compared to observations of lithospheric stress. Geochemical constraints in the form of turn-over times and segregation of source material may also be applied.

It is also possible to calculate the stress at any point in the mantle due to the flow entrained by the moving plates. That this flow pattern is dominated by the motion of the surface (Richter, 1973) indicates that the stresses in the mantle are probably not, on the average, much different in magnitude from the stresses calculated in our models. It is possible, then, to compare the computed deviatoric stress magnitudes to flow diagrams (e.g., Ashby and Verrall, 1977) to determine the probable flow mechanism. The consistency of the assumption of Newtonian viscosity can then be checked.

Mantle flow models generated using various viscosity models will be compared to these observations in another paper (Hager and O'Connell, in press).

Here we concentrate on another comparison with observation, the dip of subducted slabs.

The particle trajectories in mantle flow are observable under certain limited circumstances. In the plate tectonic model, deep earthquakes occur in slabs of subducted lithosphere. Thus the planar distributions of hypocenters in Benioff zones mark the trajectories of the subducted lithospheric slabs. The dips of the slabs, determined seismically, can be compared to the dips of the calculated velocity vectors of a flow model in order to evaluate the applicability of the model to the earth.

ANALYTIC ANALYSIS

To solve for the three-dimensional flow in the earth's mantle driven by the moving plates, we have ignored the effect of thermal buoyancy in order to calculate the large-scale flow pattern that is determined just by the mass flux imposed by the moving lithosphere. Since the Reynolds number is negligible for the earth, inertial forces can be ignored, and the viscous equations of motion have no time dependence. The models thus stimulate the flow accompanying the present-day plate motions.

The model rheology is Newtonian. Models by Parmentier et al. (1976) show that the flow pattern in some convection models is not much different for Newtonian and non-Newtonian rheologies. We also assume that viscosity is a function of radius only. This assumption may well neglect some important effects since the mantle temperature distribution is not radially symmetric, and viscosity is highly dependent upon temperature. The flow is assumed to be incompressible.

A brief summary of the mathematical method used to solve for the flow driven by the moving plates is given below. A more detailed explanation is given in the Appendix. The flow velocity and stress are expressed in terms of vector spherical harmonics. The Stokes equation can be transformed into two sets of coupled first order differential equations for each degree and order, one for poloidal and one for toroidal motions. These equations can then be solved analytically for each degree by the propagator matrix technique (Gantmacher, 1960; Gilbert and Backus, 1966; Cathles, 1975). The boundary conditions are free slip at the core—mantle boundary and the observed horizontal plate velocities at the surface. Once the Stokes equation is solved subject to these boundary conditions, the motion is determined everywhere.

HARMONIC EXPANSION OF PLATE VELOCITIES

The relative plate velocities used were those of Solomon et al. (1975). These are the relative motions of Minster et al. (1974) with slight modification. The North and South American plates are combined into one plate, and a Philippine plate is included, using the relative motion between Eurasia and the Philippine plates given by Fitch (1972). The relative motion between the Philippine and Pacific plates is poorly constrained due to the imprecision of the Eurasian—Philippine motion determination and due to the probability of spreading within the Philippine plate itself (Karig, 1971; Anderson, 1975). Because the Philippine plate is small, this uncertainty in its motion does not have a great effect on the global flow, but it does have important local effects.

The coefficients in the vector spherical harmonic expansion of surface velocities are given by appropriate integration over the sphere. For spheroidal coefficients, a_{slm}:

$$a_{slm} = \frac{1}{4\pi l(l+1)} \int_S \left(v_\theta \frac{\partial Y_l^m}{\partial \theta} + \frac{v\phi}{\sin \theta} \frac{\partial Y_l^m}{\partial \phi} \right) dS$$

and for toroidal coefficients, a_{tlm}:

$$a_{tlm} = \frac{1}{4\pi l(l+1)} \int_S \left(\frac{v_\theta}{\sin \theta} \frac{\partial Y_l^m}{\partial \phi} - v\phi \frac{\partial Y_l^m}{\partial \theta} \right) dS$$

Here Y_l^m is equal to $P_l^m(\cos \theta) \begin{bmatrix} \cos m\phi \\ \sin m\phi \end{bmatrix}$ and is normalized such that its mean square value is unity. The integrals were carried out numerically, using the trapezoidal rule on a $2° \times 2°$ grid. Calculations were done in double precision. Coefficients were calculated through degree and order twenty.

Included in Fig. 1 is the variance of each degree in the toroidal and poloidal expansions of plate velocities plotted against degree on a log-log scale. The first degree term in the toroidal expansion is omitted, since it corresponds to the net rotation of the lithosphere and is dependent upon the absolute reference frame used. It is surprising that the toroidal terms are almost as large as the spheroidal terms, since the toroidal components of velocity have no vertical motion associated with them and would not be expected to be excited, at least to first order, by thermal convection. Perhaps the large toroidal component in surface velocity is due to the extreme nonlinear effects of the mechanical boundary layer.

A departure from the general linear trend in the power spectrum is noticeable for the degree four and five terms. It is interesting to note that marginal stability calculations (Chandrasekhar, 1961, p. 244) show that modes 3—5 should be the most unstable for a body with the relative core size that the earth has.

We have compared the coefficients of the spheroidal velocity expansion with the gravity coefficients of Gaposhkin and Lambeck (1970); Gaposhkin (1974), and GEM-8 of Wagner et al. (1977) by computing correlation coefficients for each degree. The confidence levels for accepting a linear relation between flow and gravity are far below 90% except for the degree four terms. The relation for the degree four term is significant at the 90% confi-

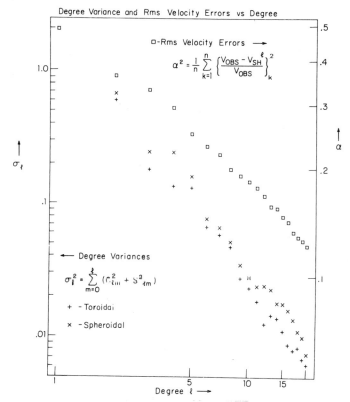

Fig. 1. Left scale: degree variance for spheroidal and toroidal terms in the spherical harmonic expansion of plate velocities plotted vs. degree. Right scale: root mean square velocity errors vs. degree.

dence level. Nevertheless, it is to be expected that, of twenty degrees tested, one degree should have this good a correlation, so the association of high power in the degree four terms with good correlation with the gravity field has no obvious significance.

Plotted on the right side of Fig. 1 is an index of the fit of the spherical harmonic expansion of plate velocities to the starting velocity model. The quantity:

$$\alpha_n = \left[\frac{1}{n} \sum_{k=1}^{n} \left(\frac{v_{\text{OBS}}^l - v_{\text{SH}}^l}{v_{\text{OBS}}} \right)^2 \right]^{1/2}$$

is plotted versus degree. Here v_{SH}^l is the velocity at a point calculated using the spherical harmonic expansion of plate velocities through degree l. The integral is computed numerically on a $20° \times 20°$ grid. The observed velocity, v_{OBS} is taken to be in a frame in which there is no net rotation of the lithosphere. By carrying out the expansion through degree 20, we have accounted for all but about 10% of the root mean square variance in the plate velocity.

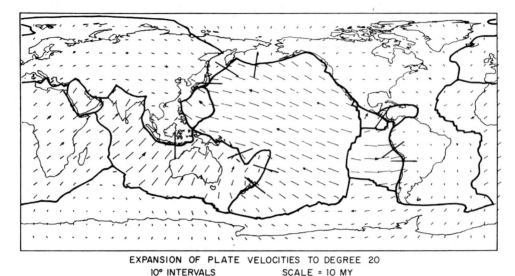

EXPANSION OF PLATE VELOCITIES TO DEGREE 20
10° INTERVALS   SCALE = 10 MY

Fig. 2. Representation of the spherical harmonic expansion of plate velocities through degree 20. Relative plate motions are from Solomon et al. (1975). The absolute reference frame is that in which the lithosphere has no net rotation. Displacement vectors are plotted on a 10° × 10° grid and show the instantaneous velocity extrapolated for 10 m.y. Also shown are the locations of sections through subduction zones used in the comparison of model flow direction with seismically determined slab dip.

Figure 2 shows the results of the spherical harmonic expansion of plate velocities through degree 20 plotted on a $10° \times 10°$ grid. The absolute reference frame is given by the condition that there be no net rotation of the lithosphere. Motions in the interior of the plates are well represented, although there are some edge effects. Even the motion of small plates, such as the Philippine and Cocos plates, is fairly well given by the spherical harmonic expansion of plate velocities through degree 20.

VISCOSITY MODELS

Although the determination of the viscosity structure of the earth is a problem in solid earth geophysics which has been studied for the past forty years, there is still disagreement as to the viscosity structure (see O'Connell, 1977, for discussion). Recent studies by Peltier and Andrews (1976) and Cathles (1975) conclude that the lower mantle has a fairly constant viscosity of 10^{22} poise. Cathles concludes that a low viscosity channel is present beneath the lithosphere, while Peltier finds the data to be better fit with no low viscosity channel.

Walcott (1973) concludes that the viscosity of the lower mantle is most likely in the range of $10^{23}-10^{26}$ poise, with a channel of lower viscosity above. If the viscosity of the lower mantle is on the order of 10^{25} P or higher, the flow driven by the moving plates is confined to the upper mantle (Hager and O'Connell, in press). McKenzie and Weiss (1975) have proposed the existence of a second lithosphere at a depth of about 700 km, which would also prevent flow between the upper and lower mantle.

In this paper we have used three radially symmetric viscosity models to investigate the effect of these proposed viscosity distributions on the flow driven by the moving plates. All three have a high viscosity layer of thickness 64 km (0.01a) of viscosity 10^{25} P to simulate the lithosphere. This is a conservative estimate of the thickness of the lithosphere and should give a lower bound to the amplitude of the flow driven by the moving plates. The "constant" viscosity model has a constant viscosity mantle of 10^{22} P. To isolate the effect of a low viscosity channel, a model similar to that proposed by Cathles (1975) is used, which includes a channel of $4 \cdot 10^{20}$ P and thickness

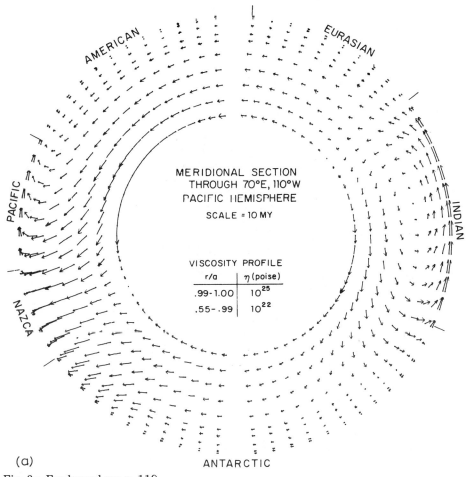

(a)

Fig. 3a. For legend see p. 119.

38

64 km at the base of the lithosphere in the "constant" model. A third model, the "rigid" model, constrains flow to the upper mantle by increasing the viscosity below 700 km ($r/a = 0.89$) in the "Cathles" model to 10^{25} P.

MANTLE FLOW MODELS

Figure 3 is a plot of the velocity vectors for the "constant" model for three orthogonal great-circle sections through the earth. Figure 3a shows a section through the 70°E and 110°W meridian. As can be seen from comparison with Fig. 2, the section passes through the Himalayas, near the Carlsberg Ridge, under Antarctica, across the Nazca plate close to the East Pacific rise, then on across the American and Eurasian plates. The section is viewed toward the Pacific hemisphere. The mantle flow dips beneath the Himalayas in the sense consistent with Asia overriding India.

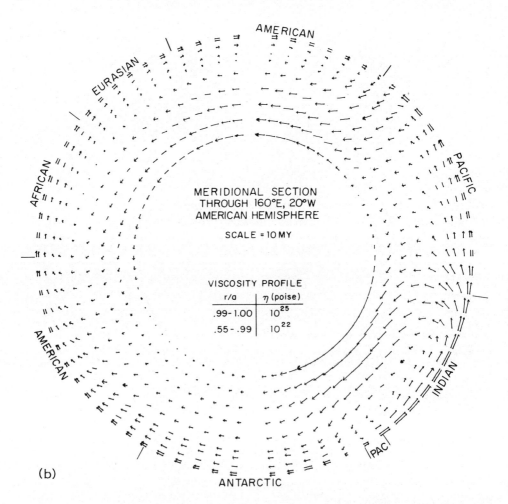

MERIDIONAL SECTION
THROUGH 160°E, 20°W
AMERICAN HEMISPHERE

SCALE = 10 MY

VISCOSITY PROFILE

r/a	η (poise)
.99 - 1.00	10^{25}
.55 - .99	10^{22}

(b)

There is a strong flow from near the core—mantle boundary under the East Pacific rise and strong vertical flow under the Carlsberg Ridge.

Figure 3b is a section through 160°E, 20°W, looking at the hemisphere including the Americas. This section passes through the Kamchatka area, through the Solomon Islands and New Zealand, under Antarctica, and along parts of the Mid-Atlantic Ridge, including Iceland. The flow sense is consistent with the direction of dip of the Benioff zones under Kamchatka and the Solomons. The vertical flow beneath the Mid-Atlantic Ridge is much slower than that shown above for the East Pacific rise.

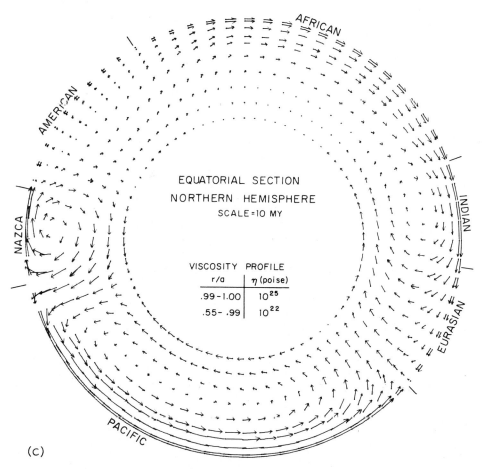

(c)

Fig. 3. Displacement vectors projected on great circle sections for the "constant" viscosity model. Displacement vectors are plotted at intervals of 5° and represent instantaneous velocities extrapolated for 10 m.y. a. Meridional section through 70°E, 110°W viewed toward the Pacific hemisphere. b. Meridional section through 160°E, 20°W viewed toward the American hemisphere. c. Equatorial section viewed toward the northern hemisphere.

Figure 3c is a section along the Equator, looking at the Northern Hemisphere. Interesting features include the consistency of the dip of the flows under Sumatra and South America with the observed direction of dip of the Benioff zones, the small cell under the Nazca plate, the strong vertical motion under the East Pacific rise, and the relative absence of shear under Africa.

SUBDUCTION ZONE DIP

Davies (1977) has suggested, on the basis of two-dimensional models of flow driven by a moving surface, that flow driven by the plates may be important in the determination of subduction zone dips. For the three great circle sections shown, the dip of the flow at convergent plate boundaries is at least qualitatively similar to the dip of the Benioff zones. One test of the usefulness of these models is to compare quantitatively the dips predicted by the flow models to the seismically determined dips.

The assumptions used in our models are weakest near the surface. The lithosphere is considered to be viscous, while an elastic-plastic rheology is probably more appropriate. Also, the truncation of the harmonic expansion at degree 20, although unimportant at depth due to the rapid decrease in amplitude of flow eigenfunctions with depth at high degree, is important near the surface. Thus we would not expect our predicted flow to be accurate in the top 100 km of the earth. In addition, it has been suggested that the dip of the upper several hundred kilometers of subducted slabs is determined by elastic effects (Isacks and Barazangi, 1977), although this conclusion might not apply in the case of very young slabs, such as those bordering Central and South America.

TABLE I

Subduction zone section parameters

Zone	Center Latitude	Center Longitude	Strike	v_0 (cm/yr)	v_s (cm/yr)	"Constant"	
						θ_{flow}	θ_c
Sunda	10°S	118°E	0°	0.78	5.88	83	76
New Zealand	40°S	178°E	135°	0.00	4.95	160	160
Tonga	26.5°S	176°W	109°	1.14	6.38	130	122
New Hebrides	16°S	166°E	70°	5.17	4.46	91	41
Japan	37°N	142°E	90°	1.96	7.53	121	107
Kurile	48°N	156°E	125°	0.61	6.54	126	121
Aleutian	51°N	179°E	3°	1.37	2.33	67	43
Middle America	11°N	88°W	45°	0.53	10.09	28	27
Peru	12°S	79°W	60°	0.40	9.31	22	21
Chile	21.5°S	71.3°W	89°	0.29	9.26	29	29

With these arguments in mind, we have restricted our comparisons to areas where old lithosphere reaches a depth greater than 200—300 km. With the exception of an ill-defined zone in the Calabrian area near Sicily, the seismic zones meeting these criteria are on the margins of the Indian and Pacific Oceans (Isacks and Molnar, 1971). We were able to obtain from the literature sections of the seismicity through the convergent boundaries of the Indian, Pacific, Cocos, and Nazca plates at approximately equal intervals. The locations of the sections used in comparison with flow dips are shown in Fig. 2. Information on the sources of hypocenter locations for these sections is given in Table I. Sections involving the Philippine plate were not used because the relative motion between the Philippine plate and the Eurasian and Pacific plates is very poorly constrained.

Figures 4a—j are plots of the hypocenter locations for the ten sections shown in Fig. 2 superimposed on the flow pattern predicted by the three viscosity models. In each figure, the top section is for the viscosity model with constant mantle viscosity of 10^{22} poise below the 10^{25} poise "lithosphere". The middle section is for the "Cathles" viscosity model. The bottom section is for the model which has an effectively rigid lower mantle, confining flow to the upper mantle. Flow vectors are plotted at intervals of $2.5°$. The upper 700 km of the earth is shown in each section.

The match between the direction of flow and the direction of the subducted slab given by the trend in hypocenters is fairly good for the "constant" model for most of the subduction zones. The match is usually improved by the inclusion of a low viscosity layer in the "Cathles" model. For the "rigid" viscosity model, the flow driven by the moving plates is often in a direction counter to that observed for the subducted slabs.

To test the hypothesis that the dip of the slab is determined by the flow

"Cathles"		"Rigid"			Source
θ_{flow}	θ_c	θ_{flow}	θ_c	$\theta_{seismic}$	
113	106	166	163	72	Fitch and Molnar (1970)
109	109	13	13	109	Ansell and Smith (1975)
130	121	11	10	131	Cohn (1975)
83	40	31	14	64	Isacks and Molnar (1971)
143	132	132	119	155	Cohn (1975)
141	138	148	145	132	Cohn (1975)
32	20	38	24	60	Engdahl (1973)
86	83	150	148	69	Dewey and Algermissen (1974)
19	18	82	79	4	Barazangi and Isacks (1976)
22	21	25	25	38	Barazangi and Isacks (1976)

SUNDA

NEW ZEALAND

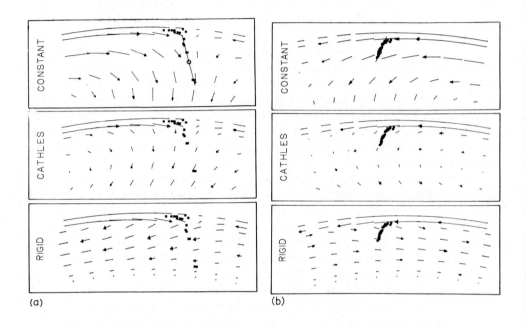

(a)

(b)

TONGA

NEW HEBRIDES

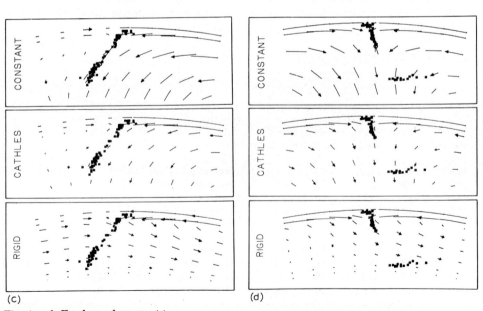

(c)

(d)

Fig. 4a—d. For legend see p. 44

JAPAN

KURILE

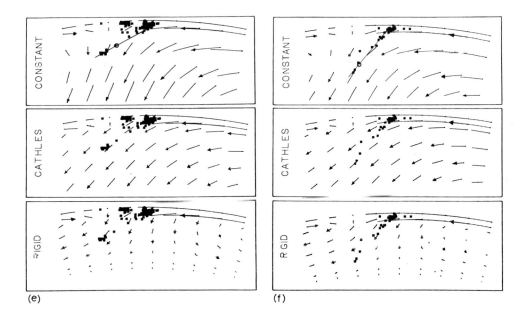

(e)

(f)

ALEUTIAN

MID AMER

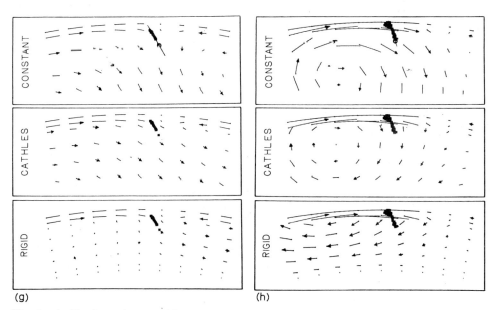

(g)

(h)

Fig. 4e—h. For legend see p. 44

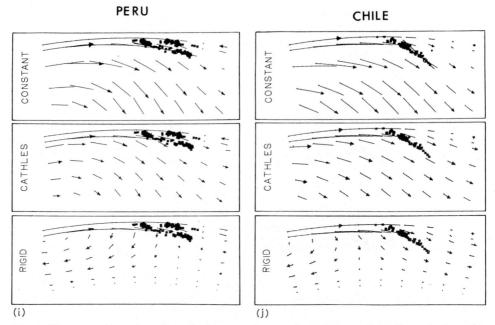

PERU CHILE

CONSTANT CATHLES RIGID

(i) (j)

Fig. 4. Flow vectors and earthquake hypocenters projected on sections through the subduction zones shown in Fig. 2. Flow vectors are instantaneous velocities extrapolated for 7 m.y. and are plotted at an interval of 2.5° in the upper 700 km of the earth.

pattern, we have measured a dip for each seismic zone and an average dip for the flow pattern, and calculated a linear regression. The dip assigned each seismic zone is shown by the solid lines in Figs. 4a—j. The average flow dip at the point circled was calculated by linear interpolation of the four adjacent points where the flow vector was plotted. For each viscosity model, we calculated the correlation coefficient:

$$r = \frac{\sum\limits_{i=1}^{10} (x_i - \overline{x})(y_i - \overline{y})}{(\sum (x_i - \overline{x})^2 \sum (y_i - \overline{y})^2)^{1/2}}$$

and the least squares slope m and intercept b in the relation:

$$y = mx + b$$

where y_i is the flow dip and x_i is the seismic zone dip of one of the ten subduction zone sections used. The significance of the correlation coefficient r is found by calculating "Student's" parameter $t = r[(n-2)/(1-r^2)]^{1/2}$ for $n = 10$, and comparing it to a one-sided Student's distribution with $n-2$ degrees of freedom (Cramer, 1946). The results of the regression are shown in Fig. 5.

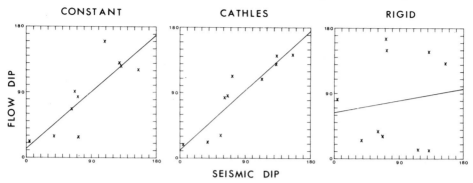

Fig. 5. Plots of the dip of the flow vs. the dip of the seismic zone for the subduction zones and viscosity models of Fig. 4.

For the "constant" viscosity model, the correlation coefficient $r = 0.83$ is significant to a confidence level of greater than 99%. For the "Cathles" model, the inclusion of the low viscosity zone improves the correlation coefficient to $r = 0.91$ which is significant at the 99.9% confidence level. The correlation breaks down when flow is confined to the upper mantle in the "rigid" model where $r = 0.14$, which can be considered to be significant at a confidence level of only 65%.

It has been suggested on the basis of shock wave data and velocity-density systematics that the lower mantle is enriched in iron with respect to the upper mantle (Anderson and Jordan, 1970). This chemical layering would create a barrier to vertical motion across the layer boundary and would inhibit mantle-wide convection. The existence of the chemical layering has been questioned in other studies using velocity-density systematics which suggest that the mantle is homogeneous with respect to iron content (Watt et al., 1975). Nonetheless, it is of interest to see how adding the constraint of no vertical flow across the boundary between the upper and lower mantle affects the flow.

To investigate this effect, we computed the flow for the "Cathles" model with vertical flow prohibited across the 700 km depth. The horizontal velocity and shear stress were constrained to be continuous across the boundary and the vertical velocity set equal to zero. The normal stress is discontinuous. The change in normal stress across the boundary can be interpreted physically as the stress induced by buoyancy forces caused by the displacement of the layer boundary by the flow.

The direction of the velocity vectors for this model is quite similar to those for the "rigid" model. The sections are not shown because of space limitations. The chief difference is that at depths of greater than about 350 km, the velocities have greater magnitude in the model which permits horizontal, but no vertical, flow at 700 km. For this model, the correlation coefficient r is 0.34, significant at a confidence level of only 82%.

DISCUSSION

It is remarkable that the correlation between the flow dips and the observed dips of the Benioff zones is so good for those models which allow whole mantle flow. Some caution is needed, however, in attaching significance to this correlation.

First, the flow models are time-independent. Since the energy equation has been ignored in these models and the Stokes equation is independent of time for low Reynolds number flow, the only time dependence in these models would enter through the time dependence of the boundary conditions. The models that are shown give the instantaneous velocity vectors for the present-day distribution of ridges and trenches. Since there is relative motion between the ridges and trenches, the velocity vector at a fixed point will change slowly with time. Thus the velocity vectors plotted are not particle trajectories.

This time dependence of the boundary conditions should have two effects on the dip angle. First, the change in the configuration of ridges and trenches will change the large scale flow driven by the moving plates. However, insofar as the rate of relative motion between ridges and trenches is small compared to the velocities of the subducted plates, the rearrangement of the large-scale flow pattern will take place on a time scale that is long in relation to the time it takes a subducted slab to sink through the upper mantle. Thus, neglecting the change in the large-scale flow pattern due to the relative motion of ridges and trenches should have only a minor effect on the subduction zone dips given by our models. The local flow in the region of a subduction zone should not change greatly due to the relative motion of a far away ridge during the time that a subducted slab can be identified by its seismic activity.

The second, more important, effect on the dip of a subducted slab which is not included in these time-independent models arises from the migration of the point of subduction. Even if each point of a subducted slab were sinking vertically, the dip of the slab would not be vertical if the location of the trench moved sufficiently rapidly.

Although it is not easy to determine the magnitude of this effect without making time-dependent models, we can set an upper bound on its magnitude. If a plate is being subducted with velocity v_s, a point on the slab travels a vertical distance $v_s \sin \theta$ and horizontal distance $v_s \cos \theta$ per unit time, where θ is the dip angle of the flow. If during this time the overriding slab is moving with velocity v_0, the trench will be displaced a distance v_0 over the subducted slab. If no change in flow pattern occurs during this time, the dip of the slab will be:

$$\theta_c = \tan^{-1} \frac{\sin \theta}{\dfrac{v_0}{v_s} + \cos \theta}$$

This is an upper bound to the effect of the moving trench on the dip of the slab, since the instantaneous flow driven by the plates will always be in a sense to reduce the change in dip induced by the trench motion.

To assess the importance of this effect, it is necessary to compare the relative magnitudes of the velocities of the overriding plate and the subducted plate. Examination of Fig. 2 shows that, in the absolute frame of reference used, the overriding plate is moving much slower than the subducted plate in a direction perpendicular to the trench for most subduction zones.

The values of v_0, v_s, θ, θ_c and θ_s, the seismic dip of the subduction zone, are given for each of the ten subduction zones in Table I. For most of the subduction zones, the upper bound to the change in dip caused by trench motion is not large. The correlations between flow dip and seismic dip are not changed significantly if the "corrected" dip angles are used. For the "constant" model, $r = 0.83$; for the "Cathles" model, $r = 0.89$, and for the "rigid" model, $r - 0.15$.

A second inadequacy of the models is that they do not include the density contrast of the downgoing slab. Since the slab is denser than the surrounding mantle, it would be expected to descend at a steeper angle than that predicted by the simple flow model. Including this effect should shift all of the flow dips closer to 90° (vertical), decreasing the slope of the regression line, but not affecting the correlation significantly. Work is in progress to determine the magnitude of this effect. However, it is difficult to see how it could substantially affect the correlation coefficient.

A third limitation of the models is that, since the viscosity distribution is radially symmetric, the effect of the higher viscosity of the slab is not included. As a result, the flow in the models resulting from the mass flux of the lithosphere itself does not form a concentrated plume moving at the speed of the subducted slab, but is accommodated by a slower, more diffuse downwelling. As a result, less vorticity is generated in the models at the "corner" between the surface and the subducted material. The effect of this vorticity would be to oppose the bending of the lithosphere around this corner. Thus, flow models omitting the high viscosity slab should predict steeper dips than those which include the slab. It is very difficult to assess the magnitude of this effect.

The three oversimplifications in the model just discussed work in opposite directions. Ignoring the trench motion and the high viscosity of the slab tend to make the dips predicted by the model too steep. These effects may have been partly cancelled by the neglect of the density contrast in the slab, which tends to make the model slab dip too shallow. The significance of the correlation between seismic dip and flow dip for the two models which allow flow in the lower mantle and the nearness of the slope of the regression line to unity suggest that the net effect of these simplifying assumptions is small.

CONCLUSIONS

Although the models developed here are kinematic, neglecting the buoyancy forces which must drive the plates, they incorporate the observed complex plate geometries and velocities as boundary conditions. They thus make it possible to isolate and understand the large scale flow due to the mass flux and viscous drag of the moving plates. Insofar as the large scale flow is dominated by the motions of the boundaries, these models should give a good representation of this flow. Small-scale thermal convection may be superimposed on this large scale flow.

For those models in which the viscosity structure and lack of chemical stratification permit flow to penetrate deeper than 700 km, the dips of the velocity vectors in the flow models match the dips of the Benioff zones remarkably well. The correlation between the flow dip and the seismic dip is statistically significant to better than the 99% confidence level.

We interpret this correlation to mean that the dips of subducted slabs are determined primarily by the large scale flow imposed by the plates moving in their observed geometry. The presence of the slabs does not change the flow direction significantly, and although the slabs may be important in the dynamics of mantle flow, they are oriented as if they were responding passively to the flow driven by the surface motion of the plates. This interpretation rules out convection confined to the upper mantle.

That the interactive global flow is important requires caution in local models of subduction zones. A thick boundary layer accompanies the moving plate in those models which successfully predict the dips of subducted slabs. This boundary layer may be important in the thermal evolution of slabs.

In areas of convergence in which both converging plates have similar crustal types it appears that the global flow determines which plate is underthrust. Examples include the Himalayas, the New Hebrides region, and the Tonga region. Thus geologic models in which the sense of subduction suddenly change (Dewey and Bird, 1970) seem to be unfeasible.

ACKNOWLEDGEMENTS

Sean Solomon kindly provided digitized plate boundary data and relative poles of rotation. Hager was supported by a National Science Foundation Graduate Fellowship for part of the time occupied by this research. This research was also supported by the National Science Foundation, grant EAR 75-22433.

APPENDIX

The equations of motion for a Newtonian fluid are transformed from second order differential equations into a set of coupled first order equations. This transformation is similar to the transformation of the second order equations describing the earth's elastic oscillations outlined by Alterman et al. (1959). The equations of motion of a Newtonian fluid are identical to those of an elastic body at zero frequency if strain rate and viscosities are substituted for strain and elastic moduli in the elastic equations. The derivation of the viscous equations is outlined on next pages.

The flow velocity and stress are expressed in terms of vector spherical harmonics as:

$$v_r = y_1^{lm} Y^{lm}$$

$$v_\theta = y_2^{lm} Y_\theta^{lm} + y_5^{lm} Y_\phi^{lm}$$

$$v_\phi = y_2^{lm} Y_\phi^{lm} - y_5^{lm} Y_\theta^{lm}$$

$$\tau_{rr} = y_3^{lm} Y^{lm}$$

$$\tau_{r\theta} = y_4^{lm} Y_\theta^{lm} + y_6^{lm} Y_\phi^{lm}$$

$$\tau_{r\phi} = y_4^{lm} Y_\phi^{lm} - y_6^{lm} Y_\theta^{lm}$$

In these equations, v_r, v_θ, and v_ϕ are the components of velocity in the radial, southerly, and easterly directions. The deviatoric normal stress in the radial direction is τ_{rr}. The components of radial shear in the southerly and easterly directions are $\tau_{r\theta}$ and $\tau_{r\phi}$. Y^{lm} are the surface spherical harmonics of degree l and order m normalized such that their root mean square is unity. θ and ϕ are colatitude and longitude.

$$Y^{lm} = P^{lm}(\cos \theta) \begin{bmatrix} \cos m\phi \\ \sin m\phi \end{bmatrix}$$

Also:

$$Y_\theta^{lm} \equiv \frac{\partial}{\partial \theta}(P^{lm})$$

and:

$$Y_\phi^{lm} \equiv \frac{1}{\sin \theta} \frac{\partial}{\partial \phi}(P^{lm})$$

The y_i^{lm} are functions of radius. Summation over the repeated superscripts l and m is implicit.

Both poloidal (spheroidal) and toroidal fields are necessary to describe an arbitrary displacement pattern. Coefficients y_1^{lm} through y_4^{lm} are associated with poloidal fields; y_5^{lm} and y_6^{lm} are associated with toroidal fields.

For incompressible flow with no vertical displacement of the boundaries, there is no perturbation of the gravity field. The Stokes equations can be transformed into the coupled first order differential equations:

$$\dot{y}_1^{lm} = -2y_1^{lm}/r + Ly_2^{lm}/r \tag{1}$$

$$\dot{y}_2^{lm} = -y_1^{lm}/r + y_2^{lm}/r + y_4^{lm}/\eta \tag{2}$$

$$\dot{y}_3^{lm} = 12\eta y_1^{lm}/r^2 - 6L\eta y_2^{lm}/r^2 + Ly_4^{lm}/r \tag{3}$$

$$\dot{y}_4^{lm} = -6\eta y_1^{lm}/r^2 + 2\eta(2L-1)y_2^{lm}/r^2 - y_3^{lm}/r - 3y_4^{lm}/r^2 \tag{4}$$

$$\dot{y}_5^{lm} = y_5^{lm}/r + y_6^{lm}/\eta \tag{5}$$

$$\dot{y}_6^{lm} = (L-2)\eta y_5^{lm}/r^2 - 3y_6^{lm}/r \tag{6}$$

Here $\dot{y}_i$ is dy_i/dr, $L = l(l+1)$ and η is the viscosity. The poloidal and toroidal equations are decoupled.

Similar reductions of the second order differential equations of motion to sets of coupled first order differential equations have been carried out for viscous flow by Takeuchi and Hasegawa (1965) and Kaula (1975), and for elastic oscillations by Alterman et al. (1959). Takeuchi and Hasegawa (1965) derived the poloidal equations, but some of the coefficients are misprinted in their paper. Kaula (1975) derived both the poloidal and the toroidal equations, but his torroidal equations contain a misprint. Equations 1—4 agree with the poloidal equations of Kaula (1975) and equations 1—6 agree with the equations of Alterman et al. (1959) for zero frequency and infinite Lame parameter λ, corresponding to incompressible flow.

These two coupled sets of equations could be solved numerically. However, with the change in variables:

$u_1 = y_1$

$u_2 = y_2$

$u_3 = ry_3/\eta$

$u_4 = ry_4/\eta$

$v_1 = y_5$

$v_2 = ry_6/\eta$

$\lambda = \ln(r/a)$

with a the radius of the earth, the sets of equations become

$$\frac{du^{lm}}{d\lambda} = A^l u^{lm} \tag{7}$$

$$\frac{dv^{lm}}{d\lambda} = B^l v^{lm} . \tag{8}$$

Here:

$$A^l = \begin{bmatrix} -2 & L & 0 & 0 \\ -1 & 1 & 0 & \eta^* \\ 2\eta^* & -6L\eta^* & 1 & L \\ -6\eta^* & 2(2L-1)\eta^* & -1 & -2 \end{bmatrix}$$

and:

$$B^l = \begin{bmatrix} 1 & 1/\eta^* \\ (L-2)\eta^* & -2 \end{bmatrix}$$

where $\eta^* = \eta/\eta_0$, and η_0 is a reference viscosity.

Equations 7 and 8 are homogeneous differential equations which can be solved analytically by the propagator matrix technique (Gantmacher, 1960; Gilbert and Backus, 1966; Cathles, 1975). The solution to eq. 7 for a layer in which A is constant is:

$$u^{lm}(\lambda) = \exp[(\lambda - \lambda_0)A^l u^{lm}(\lambda_0)] = P^l(\lambda, \lambda_0)u^{lm}(\lambda_0)$$

where $P^l(\lambda, \lambda_0)$ is the propagator matrix which propagates the vector u^{lm} from λ_0 to λ.

Boundary conditions

At the surface of the earth, the radial velocity is constrained to be zero and the observed horizontal plate motions are imposed. The core—mantle boundary is taken to be free slip. Then, at the core—mantle boundary:

$$u^{lm}_{r=c} = [0, u^{lm}_{2c}, u^{lm}_{3c}, 0]^T$$

and:

$$v^{lm}_{r=c} = [v^{lm}_{1c}, 0]^T$$

At the surface, $r = a$:

$$u^{lm}_{r=a} = [0, u^{lm}_{2a}, u^{lm}_{3a}, u^{lm}_{4a}]^T$$

and:

$$v^{lm}_{r=a} = [v^{lm}_{1a}, v^{lm}_{2a}]^T$$

Since u^{lm}_{2a} and v^{lm}_{1a} are the known coefficients in the spheroidal and toroidal expansions of the plate velocities of degree l and order m, the constraint that:

$$u^{lm}_a = P^l(\lambda a, \lambda c)u^{lm}_c$$

and similarly for v^{lm} leads to two sets of simultaneous equations which are solved for u^{lm}_c and v^{lm}_c. Then u^{lm} and v^{lm} can be determined at any radius by propagating these starting vectors upward.

REFERENCES

Alterman, Z.H., Jarosch, H. and Pekeris, C.L., 1959. Oscillations of the earth. Proc. R. Soc. London, Ser. A, 252: 80—95.
Anderson, D.L. and Jordan, T.M., 1970. The composition of the lower mantle. Phys. Earth Planet. Inter., 3: 23.
Anderson, R.N., 1975. Heat flow in the Mariana marginal basin. J. Geophys. Res., 80: 4043—4048.
Ansell, J.H. and Smith, E.G.C., 1975. Detailed structure of a mantle seismic zone using the homogeneous station method. Nature, 253: 518—520.
Ashby, M.F. and Verrall, R.A., 1977. Micromechanisms of flow and fracture, and their

relevance of the rheology of the mantle. Philos. Trans. R. Soc. London, Ser. A, 288: 59—95.

Barazangi, M. and Isacks, B.L., 1976. Spatial distributions of earthquakes and subduction of the Nazca plate beneath South America. Geology, 4: 686—692.

Cathles, L.M. III, 1975. The Viscosity of the Earth's Mantle. Princeton University Press, Princeton, N.J., 386 pp.

Chandrasekhar, S., 1961. Hydrodynamic and Hydromagnetic Stability. University Press, Oxford, 654 pp.

Cohn, S.N., 1975. Distribution of Earthquakes in the Tonga—Kermadec and Izu—Bonin—Japan—Kuril—Kamchatka Trench Systems. Thesis, Harvard College, Cambridge (unpublished).

Cramer, H., 1946. Mathematical Methods of Statistics. Princeton University Press, Princeton, N.J.

Davies, G.F., 1977. Viscous mantle flow under moving lithospheric plates and under subduction zones. Geophys. J. R. Astron. Soc., 49: 557—563.

Dewey, J.W. and Algermissen, S.T., 1974. Seismicity of the Middle America arc-trench system near Managua, Nicaragua. Bull. Seismol. Soc. Am., 64: 1033—1048.

Dewey, J.F. and Bird, F.M., 1970. Mountain belts and the new global tectonics. J. Geophys. Res., 75: 2625—2647.

Engdahl, E.R., 1973. Relocation of intermediate depth earthquakes in the central Aleutians by seismic ray tracing. Nature, Phys. Sci., 245: 23—25.

Fitch, T.J., 1972. Plate convergence, transcurrent faults and internal deformation adjacent to Southeast Asia and the Western Pacific. J. Geophys. Res., 77: 4432—4460.

Fitch, T.J. and Molnar, P., 1970. Focal mechanisms along inclined earthquake zones in the Indonesian—Philippine region. J. Geophys. Res., 75: 1431—1444.

Forsyth, D. and Uyeda, S., 1975. On the relative importance of the driving forces of plate motion. Geophys. J. R. Astron. Soc., 43: 163—200.

Gantmacher, F.R., 1960. The Theory of Matrices. Vols. 1 and 2. Chelsea Publishing Co., New York. (Translated from Russian by K.A. Hirsch.)

Gaposhkin, E.M., 1974. Earth's gravity field to the eighteenth degree and geocentric coordinates for 104 stations from satellite and terrestrial data. J. Geophys. Res., 79: 5377—5411.

Gaposhkin, E.M. and Lambeck, K., 1970. 1969 Smithsonian standard Earth (2). Spec. Rep., 315, Smithsonian Astrophysical Observ., Cambridge, Mass., 93 pp.

Gilbert, F. and Backus, G.E., 1966. Propagator matrices in elastic wave and vibration problems. Geophysics, 31: 326—332.

Hager, B.H. and O'Connell, R.J., 1978. Kinematic models of large scale flow in the earth's mantle. J. Geophys. Res., in press.

Isacks, B.L. and Barazangi, M., 1977. Geometry of Benioff zones: lateral segmentation and downwards bending of the subducted lithosphere. In: M. Talwani and W.C. Pitman, III (Editors), Island Arcs, Deep Sea Trenches, and Back-Arc Basins. American Geophysical Union, Washington, D.C., 480 pp.

Isacks, B. and Molnar, P., 1971. Distribution of stresses in the descending lithosphere from a global survey of focal-mechanism solutions of mantle earthquakes. Rev. Geophys. Space Phys., 9: 103—174.

Karig, D.E., 1971. Structural history of the Mariana Island arc system. Geol. Soc. Am. Bull., 82: 323—344.

Kaula, W.M., 1975. Product-sum conversion of spherical harmonics with application to thermal convection. J. Geophys. Res., 80: 225—231.

McKenzie, D. and Weiss, N., 1975. Speculations on the thermal and tectonic history of the earth. Geophys. J.R. Astron. Soc., 42: 131—174.

Minster, J.B., Jordan, T.H., Molnar, P. and Haines, E., 1974. Numerical modeling of instantaneous plate tectonics. Geophys. J. R. Astron. Soc., 36: 541—576.

O'Connell, R.J., 1977. On the scale of mantle convection. Tectonophysics, 38: 119—136.

Parmentier, E.M. and Turcotte, D.L., 1976. Studies of thermal convection beneath a rigid lithosphere. EOS, Trans. Am. Geophys. Union, 57: 329.

Parmentier, E.M., Turcotte, D.L. and Torrance, K.E., 1976. Studies of finite amplitude non-Newtonian thermal convection with application to convection in the earth's mantle. J. Geophys. Res., 81: 1839—1846.

Peltier, W.R. and Andrews, J.T., 1976. Glacial-isostatic adjustment — I. The forward problem. Geophys. J. R. Astron. Soc., 46: 605—646.

Richardson, R.M., Solomon, S.C. and Sleep, N.H., 1976. Intraplate stress as an indicator of plate tectonic driving forces. J. Geophys. Res., 81: 1847—1856.

Richter, F.M., 1973. Dynamical models for sea floor spreading. Rev. Geophys. Space Phys., 11: 223—287.

Richter, F.M. and Parsons, B., 1975. On the interactions of two scales of convection in the mantle. J. Geophys. Res., 80: 2529—2541.

Solomon, S.C., Sleep, N.H. and Richardson, R.M., 1975. On the forces driving plate tectonics: inferences from absolute plate velocities and intraplate stress. Geophys. J. R. Astron. Soc., 42: 769—801.

Takeuchi, H. and Hasegawa, Y., 1965. Viscosity distribution in the earth. Geophys. J. R. Astron. Soc., 9: 503—508.

Wagner, C.A., Lerch, F.J., Brownd, J.E. and Richardson, J.A., 1977. Improvement in the geopotential derived from satellite and surface data (GEM 7 and 8). J. Geophys. Res., 82: 901—913.

Walcott, R.J., 1973. Structure of the earth from glacio-isostatic rebound. In: F.A. Donath (Editor), Annual Reviews of Earth and Planetary Sciences. Annual Reviews, Inc., Palo Alto, Calif., pp. 15—37.

Watt, J.P., Shankland, T.J. and Mao, N., 1975. Uniformity of mantle composition. Geology, 3: 91—94.

Chapter 3

THERMAL AND MECHANICAL STRUCTURE OF THE UPPER MANTLE: A COMPARISON BETWEEN CONTINENTAL AND OCEANIC MODELS*

C. FROIDEVAUX [1], G. SCHUBERT [2] and D.A. YUEN [2]

[1] *Laboratoire de Physique des Solides, Université Paris-Sud, Orsay (France)*
[2] *Department of Geophysics and Space Physics, University of California, Los Angeles, Calif. (U.S.A.)*

SUMMARY

Temperature, velocity and viscosity profiles for coupled thermal and mechanical models of the upper mantle beneath continental shields and old ocean basins show that under the continents, both the tectonic plate and the asthenosphere, are thicker than they are beneath the oceans. The minimum value of viscosity in the continental asthenosphere is about an order of magnitude larger than in the shear zone beneath oceans. The shear stress or drag underneath continental plates is also approximately an order of magnitude larger than the drag on oceanic plates. Effects of shear heating may account for flattening of ocean-floor topography and heat flux in old ocean basins.

INTRODUCTION

Plate tectonics has stressed the importance of the concepts of lithosphere, the rigid surface layer of the earth which is divided into a mosaic of moving plates, and of asthenosphere, a softer layer under the plates facilitating the mechanical decoupling with the lower mantle. The concept of an asthenosphere was derived long ago from the study of isostasy. On the other hand, the physical structure of the upper mantle has been discussed either on the basis of seismic velocity distributions derived from observations (Press, 1972; Knopoff, 1972; Jordan, 1975), or on the basis of purely thermal models (McKenzie, 1967; Sclater and Francheteau, 1970; Turcotte and Oxburgh, 1972). Both of these latter approaches have led to the conclusion that the physical structure is different under oceans and under continents. It is, however, difficult to make quantitative comparisons between seismological data and theoretical thermal profiles, since this involves a precise description of the petrological state of the mantle. For oceanic plates alone, other geophysical data such as topography have been used to test theoretical thermal mod-

---

*Originally published as: Froidevaux, C., Schubert, G. and Yuen, D.A., 1977. Thermal and mechanical structure of the upper mantle: a comparison between continental and oceanic models. In: S. Uyeda (Editor), Subduction Zones, Mid-Ocean Ridges, Oceanic Trenches and Geodynamics. Tectonophysics, 37: 233—246.

els (Sorokhtin, 1974; Davis and Lister, 1974; Parsons and Sclater, in preparation; Oldenburg, 1976). Convective models have not focused on the distinction between oceanic and continental regions.

Here we present models of the upper mantle wherein both the thermal and flow-velocity profiles are calculated on the basis of coupled thermal and mechanical equations satisfying simple geophysical boundary conditions (Froidevaux and Schubert, 1975; Schubert et al., 1976). This is illustrated in Fig. 1 which shows the oceanic model on the left, and the continental one on the right. The boundary conditions indicated in the figure, are that the temperature and horizontal velocity at the surface be T_0 and u_0, and that the horizontal velocity at great depth be zero. In the continental case the heat flows at the surface q_0 and at great depth q_∞ are also imposed, whereas in the oceanic case, the temperature at great depth T_∞ and the vanishing of the vertical velocity at the surface are specified. The physical quantities to be calculated are also listed in the figure; they are temperature T, horizontal velocity u, shear stress τ and viscosity, and, for the oceanic case only, vertical velocity v. Except for τ, these quantities vary with depth y. In the oceanic case they all vary with horizontal distance from the ridge x, i.e. with age x/u_0. Given a petrologic model of the mantle, seismic velocity profiles can also be computed.

All models are based on the law of deformation of olivine (Post, 1973; Kohlstedt et al., 1976):

$$\dot{\epsilon} = \frac{B_3 \tau^3}{T} \exp\left(-\frac{E^* + pV^*}{RT}\right) \tag{1}$$

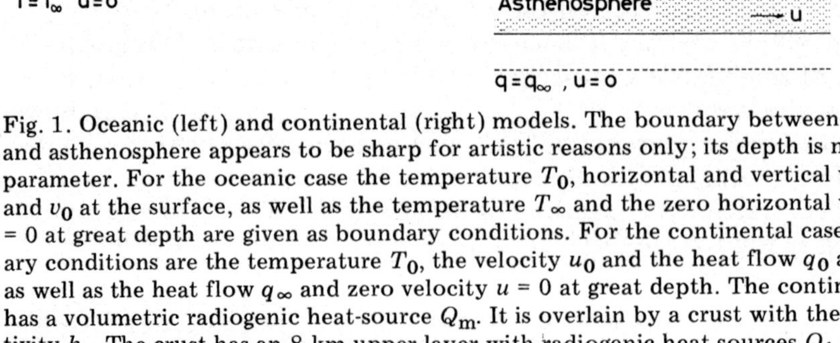

Fig. 1. Oceanic (left) and continental (right) models. The boundary between lithosphere and asthenosphere appears to be sharp for artistic reasons only; its depth is not a specified parameter. For the oceanic case the temperature T_0, horizontal and vertical velocities u_0 and v_0 at the surface, as well as the temperature T_∞ and the zero horizontal velocity $u = 0$ at great depth are given as boundary conditions. For the continental case the boundary conditions are the temperature T_0, the velocity u_0 and the heat flow q_0 at the surface, as well as the heat flow q_∞ and zero velocity $u = 0$ at great depth. The continental mantle has a volumetric radiogenic heat-source Q_m. It is overlain by a crust with thermal conductivity k_c. The crust has an 8 km upper layer with radiogenic heat sources Q_1 and a 32 km lower layer with unspecified radiogenic heat sources Q_2. The quantities listed on the right-hand side of both the oceanic and continental diagrams are those to be computed.

where T is the absolute temperature, p is the pressure, $\dot{\epsilon} \equiv \frac{1}{2}\, du/dy$ is the strain rate, and R is the gas constant. The proportionality constant B_3 and the activation energy E^* have been measured in the laboratory, whereas the activation volume V^* has to be estimated. The justification for a model based on olivine is that this mineral is the most abundant in the upper mantle and it deforms more readily than other minerals in most mantle-derived peridotite samples (Boullier and Nicolas, 1973). The coupling between velocity and temperature comes about through the T dependence of eq. 1 and through the viscous-heating term $2\tau\dot{\epsilon}$ in the temperature equation. In the continental case the temperature equation also contains a radioactive-heating term Q_m and a vertical-conduction term. The continental mantle is overlain by a crust divided in two layers. Radioactive heat generation Q_1 in the upper 8 km is a given quantity, whereas the volumetric heat production Q_2 for the 32 km thick lower crust has to be calculated to satisfy the boundary conditions. In the oceanic case the temperature equation contains both horizontal and vertical advection of heat, vertical conduction of heat and viscous dissipation. Since horizontal advection of heat is a dominant effect, radioactive heating has been neglected in this presentation. The vertical velocity in the oceanic model requires consideration of another mechanical equation, the continuity equation. These models imply a mechanical decoupling between the surface, moving at velocity u_0, and deeper regions. They do not fix, a priori, the depth of the shear zone nor the magnitude of τ. Since no return flow is assumed at shallow depth, the momentum equation simply states that τ is independent of y; in the oceanic case it will however vary with x, the horizontal coordinate or age.

The system of equations just described contains no time dependence. The models should thus apply in regions where steady state might exist. We assume this to be the case for oceans and for old shields (Polyak and Smirnov, 1968). The quasi-equality of heat flow for old continental shields and old ocean basins has been considered a paradox. Our models should tell us something about this problem. They should also lead to an interesting comparison of the lithosphere—asthenosphere structures of both regions. Such comparisons will be presented in Section 3, after the most striking characteristics of the models have been reviewed for continents in Section 1 and for oceans in Section 2. Further details and mathematical formulations are given in Froidevaux and Schubert (1975) and Schubert et al. (1976).

1. TEMPERATURE AND VELOCITY UNDER OLD CONTINENTAL SHIELDS

The values of the rheological parameters in eq. 1 depend on the amount of volatiles, particularly water, present in the mantle (Carter and Ave 'Lallemant, 1970; Stocker and Ashby, 1973). It is therefore useful to test the behavior of our thermo-mechanical model when B_3, E^* and V^* have values appropriate to wet or dry olivine. This is illustrated in Fig. 2, where solutions are drawn for extreme values of V^* (11 and 30 cm^3/mole) for a continental system with mantle radioactivity but without heat flow from great depth. The hatched areas in Fig. 2 indicate the range of solutions for intermediate values of V^*. Other parameter values are listed in the figure caption. The temperature rises monotonically to a final value depending mainly upon V^*. This dependence reflects the contribution of shear heating to the thermal state. The velocity drops from its surface value, 5 cm/year, to zero in a shear zone or asthenosphere beneath the rigid plate or lithosphere. The depth of the shear zone is shallow or the lithosphere is thin when the pressure dependence in eq. 1 is strong, i.e. when V^* is large. A larger activation volume also yields a thinner asthenosphere. These characteristics are shown by the dotted bands representing the shear zones for the various solutions. The temperature and velocity profiles are relatively insensitive to whether the mantle is wet or dry. One remarkable property of all these solutions, is that the viscosity reaches about the same minimum value of 10^{21} poise in the shear zone, or asthenosphere. Thermo-mechanical coupling leads to self-adjustment of the temperature and of the depth of the asthenosphere, so that the values of the rheological parameters do not influence the value of the viscosity minimum.

What variables in our continental models have a major influence on the viscosity value in the asthenosphere? There are two: the amount of radioactive heat sources in the mantle and the plate velocity. The effects of radiogenic heating in the mantle Q_m are seen in Fig. 3 which presents continental solutions for $u_0 = 2$ cm/year with $Q_m = 0$ HGU and $Q_m = 0.06$ HGU (1 HGU $= 10^{-13}$ cal. cm^{-3} · sec^{-1}). The addition of radioactivity raises the temperature

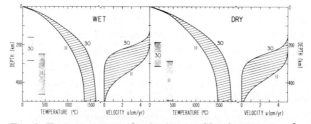

Fig. 2. Temperature and velocity profiles for a wet and a dry olivine continental mantle containing radioactive heat sources. The parameter values for the wet case are $B_3 = 5.41$ · 10^{-15} cm^{3}sec^{5}K/gm^3, $E^* = 95$ kcal./mole (Post, 1973), and for the dry case $B_3 = 6.45$ · 10^{-13} cm^{3}sec^{5}K/gm^3, $E^* = 125$ kcal./mole (Kohlstedt and Goetze, 1974). The activation volume V^* was 11 or 30 cm^3/mole and the corresponding solutions are labelled accordingly. The hatched area indicates the range of solutions for intermediate values of V^*. In all of these figures, the mantle thermal conductivity model $k_m(T,z)$ of Schatz and Simmons (1972) is used. The continental crust has a top layer of 8 km with $Q_1 = 5$ HGU and a bottom layer of 32 km with the computed Q_2 value quoted in the text. The mantle radioactivity content $Q_m = 0.06$ HGU is specified down to 500 km depth. Boundary conditions are $T_0 = 0°C$, $q_0 = 1.05$ HFU, $u_0 = 5$ cm/year, $q_\infty = 0$ HFU. The computed shear stress τ is 9.3 bars ($V^* = 11$ cm^3/mole) or 50.9 bars ($V^* = 30$ cm^3/mole) for the wet case, and 17.1 bars ($V^* = 11$ cm^3/mole) or 63.2 bars ($V^* = 30$ cm^3/mole) for the dry case. The dotted bands indicate the depth range of the asthenosphere for a particular value of V^* in cm^3/mole. The asthenosphere is defined as the depth range between $u = 0.95\ u_0$ and $u = 0.05\ u_0$.

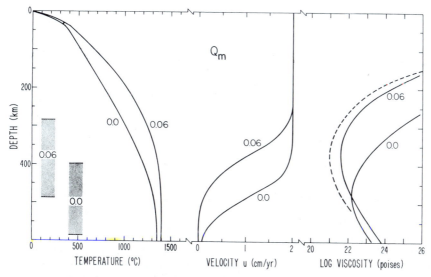

Fig. 3. Temperature, velocity, and viscosity profiles for a dry olivine, continental mantle. Q_m = 0.0 HGU or 0.06 HGU, V^* = 11 cm^3/mole, u_0 = 2 cm/year, and q_∞ = 0.0 HFU. For the models with mantle radioactivity, the heat sources extend down to 500 km depth. The computed shear stresses are 27.2 bars (Q_m = 0.06 HGU) and 114 bars (Q_m = 0.0 HGU). The dashed curve is the viscosity profile for V^* = 11 cm^3/mole, Q_m = 0.06 HGU and u_0 = 5 cm/year (τ = 17.1 bars).

at all depths, moves the asthenosphere upward and reduces the value of the minimum viscosity by an order of magnitude. By comparing solutions in Figs. 2 and 3 for V^* = 11 cm^3/mole and Q_m = 0.06 HGU one finds that an increase in plate velocity from 2 to 5 cm/year lowers the viscosity by another factor of 4. The viscosity profile for V^* = 11 cm^3/mole, Q_m = 0.06 HGU and u_0 = 5 cm/year is shown as the dashed curve in Fig. 3. The computed shear stress τ, i.e. the drag under the continental plate, also decreases with increases in mantle radioactivity and plate velocity. For the cases in Fig. 3 with u_0 = 2 cm/year, the τ values are 114 bars for Q_m = 0 HGU and 27.2 bars for Q_m = 0.06 HGU; in the latter case τ drops to 17.1 bars when u_0 is increased to 5 cm/year. This decrease of τ with u_0 is not as strong as that found by Froidevaux and Schubert (1975) in situations where viscous heating was the dominant effect in the mantle and τu_0 was almost constant in the velocity range investigated.

Comparisons with the earth have been attempted and the results are shown in Fig. 4. Since the activation volume is not known we imposed an additional constraint on the model, namely that the geotherm is required to pass through the data points (shaded areas in Fig. 4) derived from phase equilibria in kimberlite nodules from the South African shield (Boyd, 1973; MacGregor and Basu, 1974). Only the unperturbed branch (Green and Guéguen, 1974) of this experimentally determined geotherm is used. Solutions for wet (Post,

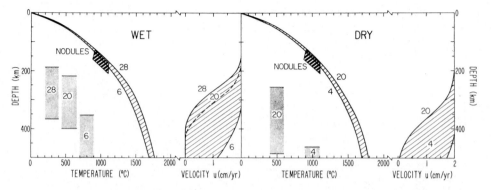

Fig. 4. 'Realistic' models of the upper mantle beneath continental shields. The shaded region delineates the approximate range of the pyroxene geotherms (MacGregor and Basu, 1974). Computed models of the thermal and mechanical structures are shown for a range of activation volumes for wet and dry olivine. Geotherms satisfying the paleotemperatures are relatively insensitive to the value of activation volume. An additional mechanical constraint based on the interpretation of sheared nodules (Boullier and Nicolas, 1973) is imposed on the depth of the top of the shear zone. For wet olivine and $q_\infty = 0.1$ HFU, the activation volumes for the solutions which best satisfy *both* the thermal and mechanical constraints lie between 20 and 28 cm^3. mole^{-1}. For dry olivine acceptable solutions could not be found. The computed shear stresses are 88.8 bars ($V^* = 28$ cm^3/mole), 71.9 bars ($V^* = 20$ cm^3/mole), 62.5 bars ($V^* = 6$ cm^3/mole) for wet olivine, and 88.1 bars ($V^* = 20$ cm^3/mole) and 62.4 bars ($V^* = 4$ cm^3/mole) for dry olivine.

1973) and dry (Kohlstedt and Goetze, 1974) olivine rheological parameters are obtained with $q_\infty = 0.1$ HFU, the imposed heat flow from the deep mantle. This value of q_∞ corresponds to an estimate of the adiabatic temperature gradient. The temperature solutions in Fig. 4, which approximately satisfy the nodule data, are relatively insensitive to the value of the activation volume. Thus the model geotherms alone do not provide much of a constraint on the value of V^*. However, since the sheared nodules in kimberlites are thought to originate at depths of about 180 km or somewhat deeper (Boullier and Nicolas, 1973), one can use the depth of the shear zone in the model solutions as an additional constraint on acceptable models and appropriate values of V^*.

For wet olivine, a V^* between 20 and 28 cm^3/mole would give models with acceptable temperatures and shear zones starting at depths between about 225 km and 180 km, respectively. For dry olivine, no solutions could be found which simultaneously satisfied the constraints in both temperature and depth to the top of the shear zone (depth at which $u/u_0 = 0.95$). In the case of dry olivine V^*-values of 16 and 20 cm^3/mole yield models with shear zones starting at depths of about 305 km and 260 km, respectively.

The viscosity minima for both the wet and dry solutions have values of about 10^{21} poise. The shear stresses for these solutions are about 80 bars, in agreement with values proposed by Goetze and Kohlstedt (1973) on the basis of dislocation studies in mantle-derived peridotite nodules. The heat generation values Q_2 in the lower crust range from 0.75 to 0.8 HGU, about 40—35% lower than averaged values for gabbro. In these solutions, the granitic upper crustal layer had Q_1 = 5 HGU and Q_m = 0.06 HGU was specified for the mantle down to 500 km depth.

In conclusion, if a steady-state regime is assumed to exist under old continental shields, our coupled thermomechanical model is capable of satisfactorily predicting the physical structure of the upper mantle. The finite-thickness asthenosphere underlying the rigid lithosphere is a consequence of the strong T and p dependences of the mechanical constitutive equation (1). Under old continents the depth of the asthenosphere is determined mainly by V^*, whereas its 'softness' is essentially governed by Q_m and u_0. Shear-wave velocity profiles computed on the basis of solid-state data for pure olivine and our model geotherms are flat. This agrees with seismological data and emphasizes that the seismological low-velocity zone, which does not exist under old continental shields, is not equivalent to the shear zone or asthenosphere.

2. TEMPERATURE AND VELOCITY UNDER OLD OCEANIC BASINS

For oceanic plate models, the two-dimensional flow field u and v and temperature profile depend not only upon depth but also upon distance from the ridge as pictured in Fig. 1. The boundary conditions do not include the surface heat flow, since this quantity is not well known. Instead we have imposed the temperature T_∞ at great depth. This quantity is not known either, so that several values between 1200°C and 1600°C have been used. This boundary condition has the advantage of making a comparison with purely thermal models simple. All models in this and the remaining parts of the paper are based upon a dry olivine rheology.

The temperature profiles are dominated by heat advection, i.e. by lithospheric and asthenospheric cooling. This is particularly true for high T_∞-values and slow plates. For a rather cool asymptotic temperature T_∞ = 1200°C and a high plate velocity u_0 = 10 cm/year, Fig. 5 shows solutions for two ages, 10 and 150 m.y. For 10 m.y. the T-profile is almost identical to the purely thermal solution calculated for the same variable thermal conductivity (dotted curve). The velocity profiles $u(y)$ and $v(y)$ are seen to depend on the value of V^*; the shear zone is shallower when V^* is large, as was found in the continental models. In Fig. 5 the hatched areas indicate again the range of solutions for intermediate values of V^*. The viscosity minimum is less than 10^{21} poise; for T_∞ = 1600°C it drops below 10^{20} poise. The vertical velocity v increases from zero to an asymptotic value in the region where the horizontal velocity drops to zero, i.e. in the asthenosphere. The finite value of v at great depth provides a mass flow necessary for the accretion of the thickening lithosphere. For an age of 150 m.y. Fig. 5 shows that the temperature profile departs strongly from the purely thermal cooling solution (dotted curve). This is due to shear heating, an effect which is stronger for

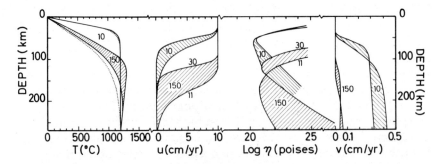

Fig. 5. Temperature, viscosity, horizontal and vertical velocity fields of an oceanic lithosphere and asthenosphere model for a dry olivine rheology. $T_\infty = 1200°C$ and $u_0 = 10$
cm/year. Solutions are shown for two extreme values of V^*, 11 and 30 cm³/mole, and
for two ages, 10 and 150 m.y. They are indicated in the figure by the numbers adjacent
to the curves. For an age of 10 m.y., temperature profiles are insensitive, at the accuracy
of our graph, to the value of V^*. This is not so for the velocity curves. The shaded area
between two such curves represents the range of solutions for intermediate values of V^*.
For an age of 150 m.y. the temperature profiles depend on V^*. The upper curve is for
$V^* = 30$ cm³/mole and the lower one is for $V^* = 11$ cm³/mole. The dotted curves denote
the temperature profiles for simple boundary layer cooling with variable thermal conductivity. The conduction solution for 10 m.y. is similar to the $V^* = 11$ cm³/mole and V^*
$= 30$ cm³/mole curves. The stresses developed in these oceanic models are 11.8 bars (V^*
$= 11$ cm³/mole), 34.0 bars ($V^* = 30$ cm³/mole) for 10 m.y. and 25.8 bars ($V^* = 11$ cm³/
mole), 75.5 bars ($V^* = 30$ cm³/mole) for 150 m.y.

larger values of V^*. The viscosity in the asthenosphere is somewhat higher
beneath old oceans. At 150 m.y. the upward velocities are an order of magnitude smaller than at 10 m.y.

Shear heating generates marked departures from a simple $\sqrt{age}$ dependence
for derived quantities like lithospheric thickness, topography and heat flow.
This is illustrated in Fig. 6 for the first two of these geophysical quantities.
The lithospheric thickness, defined as the depth at which u falls to 95% of
the surface value u_0, increases with age, but not as fast as a $\sqrt{age}$ law
which would yield a straight line on this plot. As mentioned earlier the
thickness is smaller for larger V^*-values. For each value of V^* the upper and
lower curves correspond to u_0-values of 10 cm/year and 2 cm/year, respectively. Shear heating is stronger for larger u_0-values; consequently the lithospheric plate is somewhat thinner. The topography of the ocean floor represented in Fig. 6 by the water-depth also departs from a $\sqrt{age}$ law indicated
by the dashed straight line. The departure is very pronounced for $V^* = 30$
cm³/mole. The experimental points redrawn from Parsons and Sclater (in
preparation) could be fitted for a V^*-value just above 11 cm³/mole. This
topography model reflects the gradual cooling of the lithosphere and asthenosphere; it was computed for a thermal expansion coefficient $\alpha = 4 \cdot 10^{-5}$ K^{-1}.
No data are available for the lithospheric thickness as defined here. The
depth of origin of sheared peridotite nodules might in the future give some

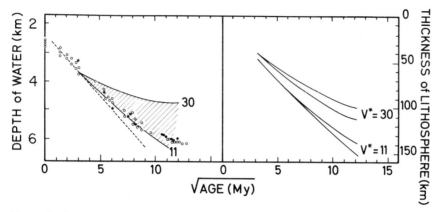

Fig. 6. Bathymetry of the ocean floor and lithospheric thickness (depth at which u/u_0 = 0.95) as a function of $\sqrt{\text{age}}$ for a dry olivine rheology with $T_\infty = 1200°C$, and $V^* = 11$ and 30 cm^3/mole. For the depth of water plot, $u_0 = 10$ cm/year and $\alpha = 4 \cdot 10^{-5}$ K^{-1}. The shaded region represents the bathymetry solutions for intermediate values of V^*. The dashed lines are the ocean floor depths for simple boundary layer cooling. The data points are taken from Parsons and Sclater (in preparation). For the thickness of the lithosphere curves, the value of u_0 distinguishes the solid curves with the same value of V^*; the thicker lithosphere is for $u_0 = 2$ cm/year, the thinner one is for $u_0 = 10$ cm/year.

constraints. For now we can only say that the depth to the seismic low-velocity zone inferred from surface and body wave data increases monotonically up to ages as old as 150 m.y. (Leeds, 1975; Sipkin and Jordan, in preparation). The ocean floor heat-flow data, on the other hand are too controversial to warrant a detailed comparison. Our models of q_0 for $T_\infty = 1200°C$ also show a certain amount of flattening, i.e., a departure from an (age)$^{-1/2}$ proportionality for heat flow. Viscous dissipation provides additional heat to the old ocean floor. Stress values increase by factors of 2–5 for the cases shown above as age increases. They are less than 80 bars for solutions based on the rheology of dry olivine. For a wet olivine rheology, stresses and shear heating effects are reduced by factors of 2–3. They are also reduced for larger T_∞-values, $T_\infty = 1200°C$ being the value applicable to Figs. 5 and 6.

In summary, for young ages thermo-mechanical coupling beneath oceans is negligible and the T-profiles are close to the solutions for simple boundary-layer cooling. Note however that although the isotherms for such purely thermal solutions are parabolic, they would not yield a parabolic lithosphere because of the pressure dependence of the viscosity implied by eq. 1. A second point to emphasize is that departures from $\sqrt{\text{age}}$ laws at old ages are a result of distortion of the geotherms by shear heating. This explanation of known geophysical behavior (see topography) is more satisfactory than that based on the ad-hoc representation of the lithosphere by a slab of constant thickness (Parsons and Sclater, in preparation). The stage is now set for a comparison of continents and oceans.

3. A DIRECT COMPARISON OF CONTINENTAL AND OCEANIC MODELS BASED ON A DRY OLIVINE RHEOLOGY

So far we have shown that the one-dimensional flow field under old continental shields, as well as the two-dimensional flow field under oceanic plates, and their associated temperature fields can be modelled on the basis of simple geophysical boundary conditions and physical properties corresponding to the thermal conductivity and rheology of olivine as measured in the laboratory.

A comparison of the situations under old continents and under old oceanic basins can be undertaken. This requires an appropriate matching of boundary conditions. We shall proceed in the following way. First we take T_0 = 0°C, q_0 = 1.05 HFU (1 HFU = 10^{-6} cal. cm^{-2}. sec^{-1}) for continental shields and compute solutions corresponding to q_∞ = 0 HFU. This yields a computed temperature T_∞ at great depth under old continents. We then compute oceanic solutions for the same T_0- and T_∞-values, and of course for the same surface velocity u_0. All this is carried out under the assumption that mantle rheological properties are identical to those of dry olivine and are the same

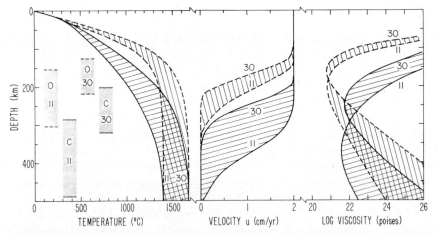

Fig. 7. A comparison of the thermal and mechanical structures between old oceanic basins (150 m.y.) and continental shields. Dry olivine rheology, u_0 = 2 cm/year, q_∞ = 0.0 HFU and equal asymptotic temperatures, T_∞, are required for both tectonic models. Q_m = 0.06 HGU is specified for the continental mantles down to 500 km depth. The effect of an increase in V^* is greater on continental models, which decouple at a greater depth than the oceanic ones. The viscosity minima for the oceanic cases are an order of magnitude lower than the continental values. The stresses in the continental cases (124.0 bars, V^* = 30 cm^3/mole; 27.2 bars, V^* = 11 cm^3/mole) are an order of magnitude greater than those found for the oceanic solutions (10.7 bars, V^* = 30 cm^3/mole; 4.9 bars, V^* = 11 cm^3/mole). Shaded areas indicate the range of solutions for intermediate values of V^*: shading with subvertical lines is for oceanic solutions; shading with subhorizontal lines is for continental solutions.

under ocean basins and under continental shields. Since the activation volume V^* has not been measured in the laboratory, we present solutions for $11 \leqslant V^* \leqslant 30$ cm^3/mole.

In the oceanic models, where horizontal and vertical advection of heat dominates, we shall neglect the thin basaltic crust as well as mantle radioactivity. However the continental models not only include a radioactive crust with $Q_1 = 5$ HGU and a lower crust with a Q_2 value which has to be computed, but also a radioactive upper mantle with $Q_m = 0.06$ HGU, a reasonable average value. We realize that the true value of Q_m is not well determined; for example Sorokhtin (1974) has argued that its value could be zero. In any case, although the specific comparison discussed here is based on a continental model with non-zero Q_m we can use the results of Fig. 3 to infer the effect of a smaller or zero value of Q_m beneath continents. Shear heating is, of course, included in both oceanic and continental models.

Figure 7 gives continental and oceanic solutions for $u_0 = 2$ cm/year. For $V^* = 11$ cm^3/mole and 30 cm^3/mole, the continental solutions have $T_\infty = 1392°$C, $Q_2 = 1.04$ HGU and $T_\infty = 1665°$C, $Q_2 = 0.59$ HGU, respectively. The corresponding oceanic geotherms for an age of 150 m.y. have surface heat-flow values of 0.82 and 0.98 HGU, respectively. Both the surface heat flows in the oceanic cases and the lower continental crustal radioactivity have reasonable values when compared with geophysical and geochemical data.

What are the most striking points emerging from a comparison of the solutions in Fig. 7? Firstly, we see that the oceanic geotherm is everywhere higher than the corresponding continental one and thus has a much steeper gradient at shallow depths. This is basically due to the existence of a thick radioactive crust for the continents. As a consequence the lithospheric thickness is larger under continental shields than under old oceanic basins. This is pictured by the position of the top boundary of the dotted bands representing the depth range of the asthenosphere (defined by $0.05 \leqslant u/u_0 \leqslant 0.95$). The asthenosphere beneath continents is thicker than that under oceans. The value of the viscosity minimum in the continental asthenosphere is essentially an order of magnitude larger than that in the oceanic shear zone. Similarly the drag under the continental plate is about an order of magnitude larger than that beneath the oceanic plate. The drag is given by the computed value of shear stress τ which is 4.9 and 27.2 bars for the oceanic and continental cases, respectively, when $V^* = 11$ cm^3/mole; corresponding τ-values for $V^* - 30$ cm^3/mole are 10.7 and 124 bars. These results compare favorably with those of Forsyth and Uyeda (1975) and Solomon et al. (1975) who suggest a larger drag under continental plates on the basis of analyses of plate motions.

To discuss the relative importance of the various heat sources beneath the oceans and continents we have constructed Table I from the solutions of Fig. 7. In order to make vertical heat advection into the oceanic lithosphere negligible we now define the base of the oceanic lithosphere as the depth for

TABLE I

Comparison of thermal states under old oceans and continental shields

	$V* = 11$ cm^3/mole		$V* = 30$ cm^3/mole	
	ocean (150 m.y.)	continent	ocean (150 m.y.)	continent
Lithospheric thickness L (km)	140	245	95	185
Sublithospheric temperature $T_L(°C)$	1225	1185	1190	1260
Temperature below the asthenosphere T_∞ (°C)	1392	1392	1665	1665
Heat flux into the lithosphere q_L(HFU)	0.28	0.19	0.63	0.37
Contribution of lithosphere radioactivity to q_0(HFU)	—	0.86	—	0.68
Contribution of advection inside the lithosphere to q_0(HFU)	0.54	—	0.35	—
Surface heat flow q_0(HFU)	0.82	1.05	0.98	1.05
Contribution of shear heating τu_0 to q_L(HFU)	0.0074	0.041	0.016	0.19

which $u = 0.995\ u_0$. For consistency we apply the same definition of lithospheric base to the continental case, even though there is no advection in the continental models. Given this definition we can read the entries in Table I for lithospheric thickness and sublithospheric temperature directly from Fig. 7. The heat flux into the base of the lithosphere, q_L, is computed as the product of thermal conductivity and temperature gradient at the appropriate depth from the detailed solutions. The difference between surface heat flow q_0 and q_L is either the contribution of lithosphere radioactivity to q_0 in the continental case or the contribution of horizontal advection inside the lithosphere to q_0 in the oceanic case.

Table I shows that a substantial fraction (1/3 to 2/3) of the surface heat flow in the old ocean floor comes from the cooling of the lithosphere as it thickens and moves away from the ridge. The remainder of the surface heat flow derives from asthenospheric cooling and viscous dissipation τu_0. The source of the continental surface heat flux is radiogenic heating in the crust (55—70%) and in the mantle (26%), and viscous dissipation (19—4%). These results apply to the case of simple decoupling of the plates from the mantle below. The effects of a return flow inside the asthenosphere are under study and will be described elsewhere.

Sclater and Francheteau's (1970) purely conductive model for old oceanic lithosphere does not include advection of heat. Our results show that this is still important even for an age of 150 m.y. and must be included when the "equality" of heat flow between continental shields and oceanic basins is discussed. One should notice finally that the sub-asthenospheric temperature is considerably higher (200—400°C) than the temperature just under the plates. In the literature these two quantities are often indiscriminately equated, just as the constant-thickness slab introduced by McKenzie (1967) for mathematical convenience is often presented as the lithosphere.

CONCLUDING REMARKS

This paper has shown that it is possible to construct satisfactory thermo-mechanical models of the upper mantle beneath both continents and oceans. Under the oceans, the plates are found to be thinner, the asthenosphere softer and hence the drag smaller than under the continents. The comparison of continental and oceanic models with each other and with the real earth would be improved by including radioactivity under the oceans, by accounting for a non-zero heat flux from great depth and by testing the effects of a return flow. Further improvement would require additional data on possible lateral heterogeneities in the distribution of radioactive heat sources and in the rheological properties of the mantle.

ACKNOWLEDGMENT

This work was partly supported by the Earth Sciences Section, National Science Foundation, NSF Grant GA 40749 and by the National Aeronautics and Space Administration, NSG 7002.

REFERENCES

Boullier, A.M. and Nicolas, A., 1973. Texture and fabric of peridotite nodules from kimberlites. In: P.H. Nixon (editor), Lesotho Kimberlites. Lesotho National Dev. Corp., Maseru, pp. 57—66.
Boyd, F.R., 1973. A pyroxene geotherm. Geochim. Cosmochim. Acta, 37: 2533—2546.
Carter, N.L. and Ave 'Lallemant, H.G., 1970. High temperature flow of dunite and peridotite. Geol. Soc. Am. Bull., 81: 2181—2202.
Davis, E.E. and Lister, C.R.B., 1974. Fundamentals of ridge crest topography. Earth Planet. Sci. Lett., 21: 405—413.
Forsyth, D. and Uyeda, S., 1975. On the relative importance of the driving forces of plate motion. Geophys. J. R. Astron. Soc., 43: 163—200.
Froidevaux, C. and Schubert, G., 1975. Plate motion and structure of the continental asthenosphere: A realistic model of the upper mantle. J. Geophys. Res., 80: 2553—2564.
Goetze, C. and Kohlstedt, D.L., 1973. Laboratory study of dislocation climb and diffusion in olivine. J. Geophys. Res., 78: 5961—5971.
Green, H.W. II. and Guéguen, Y., 1974. Origin of kimberlite pipes by diapiric upwelling in the upper mantle. Nature, 249: 617—620.
Jordan, T.H., 1975. The continental tectosphere. Rev. Geophys. Space Phys., 13: 1—12.
Knopoff, L., 1972. Observation and inversion of surface wave dispersion. Tectonophysics, 13: 497—519.
Kohlstedt, D.L. and Goetze, C., 1974. Low-stress, high-temperature creep in olivine single crystals. J. Geophys. Res., 79: 2045—2051.

Kohlstedt, D.L., Goetze, C. and Durham, W.B., 1976. Experimental deformation of single crystal olivine with application to flow in the mantle. In: R.G.J. Strens (editor), The Physics and Chemistry of Minerals and Rocks. Wiley, New York, pp. 35—50.

Leeds, A.R., 1975. Lithospheric thickness in the western Pacific. Phys. Earth Planet. Inter., 11: 61—64.

MacGregor, I.D. and Basu, A.R., 1974. Thermal structure of the lithosphere; a petrologic contribution. Science, 185: 1007—1011.

McKenzie, D.P., 1967. Some remarks on heat flow and gravity anomalies. J. Geophys. Res., 72: 6261—6273.

Oldenburg, D.W., 1975. A physical model for the creation of the lithosphere. Geophys. J. R. Astron. Soc., 43: 425—451.

Parsons, B. and Sclater, J.G., in preparation. An analysis of the variation of ocean floor heat flow and bathymetry with age.

Polyak, B.G. and Smirnov, Y.B., 1968. Relationship between heat flow and the tectonics of continents. Geotectonics, 4: 205—213.

Post, R.L., 1973. The Flow Laws of Mount Burnett Dunite. Ph. D. thesis, Univ. of Calif., Los Angeles.

Press, F., 1972. The earth's interior as inferred from a family of models. In: E.C. Robertson (editor), The Nature of the Solid Earth. McGraw-Hill, New York, pp. 147—171.

Schatz, J.F. and Simmons, G., 1972. Thermal conductivity of earth materials at high temperatures. J. Geophys. Res., 77: 6966—6983.

Schubert, G., Froidevaux, C. and Yuen, D.A., 1976. Oceanic lithosphere and asthenosphere: thermal and mechanical structure. J. Geophys. Res., 81: 3523—3540.

Sclater, J.G., 1972. New perspectives in terrestrial heat flow. Tectonophysics, 13: 257—291.

Sclater, J.G. and Francheteau, J., 1970. The implication of terrestrial heat-flow observations on current tectonic and geochemical models of the crust and upper mantle of the earth. Geophys. J. R. Astron. Sco., 20: 509—537.

Sipkin, S.A. and Jordan, T.H., in preparation. Lateral heterogeneity of the upper mantle determined from travel times of multiple ScS.

Solomon, S.C., Sleep, N.H. and Richardson, R.M., 1975. On the forces driving plate tectonics: Inferences from absolute plate velocities and intraplate stress. Geophys. J. R. Astron. Soc., 42: 769—803.

Sorokhtin, O.G., 1974. Global Evolution of the Earth. "Science" edition, Moscow, in Russian.

Stocker, R.L. and Ashby, M.F., 1973. On the rheology of the upper mantle. Rev. Geophys. Space Phys., 11: 391—426.

Turcotte, D.L. and Oxburgh, E.R., 1972. Mantle convection and the new global tectonics. Ann. Rev. Fluid Mech., 4: 33—68.

Chapter 4

A GEOPHYSICAL, GEOCHEMICAL, PETROLOGICAL MODEL OF THE SUB-MARINE LITHOSPHERE[1]

Y. BOTTINGA * and L. STEINMETZ

*Laboratoire de Géochimie et Cosmochimie ***, Institut de Physique du Globe, Université Paris VI, 75230 Paris (France)*
*Laboratoire de Géophysique Interne ****, Institut de Physique du Globe, Université Paris VI, 75230 Paris (France)*

SUMMARY

A schematic but quantitative geochemical, petrological, model of the sub-marine lithosphere and its genesis is given. With this model we calculate numerically, a priori, the geophysical characteristics of the lithosphere, its acoustic properties, density, oceanic heat flow and ocean bottom topography. Comparison with observational data for these characteristics shows good agreement. Particular attention is given to anomalous upper mantle in the vicinity of spreading centres. Compressional and shear wave velocity distributions are given in tabular form for the submarine lithosphere as a function of age. Comparison between observations for V_p, V_s and the calculated acoustic properties suggests that the lower marine lithosphere is anisotropic. Possible thickening of layer 3 with age is discussed. Melt distribution in the ridge axial region has been evaluated. All calculations were done for a plate velocity of 1 cm y^{-1}.

1. INTRODUCTION

Within the framework of plate tectonics one would like to calculate the properties of the marine lithosphere, produced by a spreading centre, via a simultaneous solution of the conservation equations for mass, momentum and energy transport in the upper mantle. This requires of course a quantitative knowledge of the petrology for this region. At present the realization of such an ambitious programme is not possible, because a detailed simultaneous solution of the three conservation equations for a real earth poses many problems and our quantitative knowledge of upper mantle petrology and rheology is incomplete. This does not mean that one has unlimited free-

---

[1] Originally published as: Bottinga, Y. and Steinmetz, L., 1979. A geophysical, geochemical, petrological model of the sub-marine lithosphere. Tectonophysics, 55: 311—347.

* Present address: Université de Nice, Parc Valrose, 06034 Nice (France).
** Laboratoire associé au CNRS No. 196.
*** Laboratoire associé au CNRS No. 195.

dom in speculating about the nature of the marine lithosphere. Observations of seabottom topography, oceanic heat flow, the composition and thickness of layer 2, the thickness of layer 3 and its inferred composition, the known seismic characteristics of the marine lithosphere, the known rheology of olivine and our qualitative understanding of the petrology of peridotite, impose stringent constraints on plate models. Likewise the thermodynamic and thermal properties of potential upper mantle materials restrict considerably one's freedom in the construction of models of the marine lithosphere.

Unfortunately, our information on the properties of the lithosphere is not only incomplete but there are also conflicting pieces of evidence. For example the high P-wave velocities of about 9 km s^{-1} underneath spreading centres, reported by Woollard (1975) and Neprochnov and Rykunov (1970), do not agree with the generally, observed or inferred, reduced acoustic properties for these regions. The results obtained with pyroxene geothermometry on ultramafic nodules are controversial. Certain observations of the trace element and the radioactive isotope composition of mid-oceanic ridge and volcanic island basalts cannot be integrated in the classical picture of plate tectonics in a straightforward fashion. These ambiguous, conflicting, or poorly understood observations are neglected in this paper. Instead the attention is focussed on those aspects which are reasonably non-controversial.

The purpose of this paper is to present a model of the marine lithosphere, in which is integrated all more or less well established information, available for this part of the earth. In the construction of this model extensive use was made of the experience obtained with previous models, i.e., Langseth et al. (1966), Sclater and Francheteau (1970), Bottinga (1974), Bottinga and Allègre (1976) and Parsons and Sclater (1977). Various features of this model have already been discussed by Bottinga and Allègre (1978) in connexion with partial melting under spreading ridges. Steinmetz et al. (1977) also used this model in their discussion of the seismic structure of young marine lithosphere.

2. DESCRIPTION OF THE MODEL

In the model material wells up below a spreading centre to form the lithosphere. During the upward motion of this material partial fusion takes place, giving rise to a melt in a porous medium. The porosity of the medium is at least in part due to the partial melting process. The melt phase will tend to escape as a consequence of its low density, as compared with the residual solid, its low viscosity (Kushiro et al., 1977), and the easily deformable solid matrix (Kohlstedt et al., 1976). The melt is chemically different from the solid, hence differential movement of melt and solid will result in chemical differentiation. The shape of the upwelling column and the velocity distribution in it are not known, but are assumed to be as the primary flow field given in Fig. 1. The liquid in the upwelling mass has a velocity in excess of the primary flow velocity. Locally the liquid is supposed to be in equilibrium

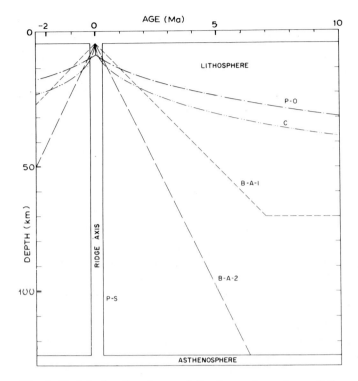

Fig. 1. Models for the shape of the lower boundary of the upper mantle region where flow is essentially horizontal. *P—O*: Parker and Oldenburg (1973) and Oldenburg (1975); *C*: Crough (1975); *B—A—1*: Bottinga and Allègre (1973, 1976) and Bottinga (1974); *B—A—2*: Bottinga and Allègre (1978) and this work; *P—S*: Parsons and Sclater (1977) and Sclater and Francheteau (1970); the width of the dike crossing the lithosphere at the ridge axis has not been specified by these authors.

with the enclosing solid. Liquid in the central 24 km wide (12 km half width) column of the uprising material escapes via a system of fractures, which reaches to about 25 km depth below the ocean bottom, to erupt as mid-oceanic ridge basalt (MORB). Melt in the uprising mass, but outside the central 24 km wide column, moves upwards until it gets trapped somewhere below the Moho in the upper mantle, because of a lack of permeability or because it solidifies as a result of cooling due to a local ambient temperature lower than its solidus temperature. In summary, at a certain depth partial melting starts in the uprising mantle; part of the melt escapes and erupts at the seabottom in the axial zone of the ridge and is used in the construction of the oceanic crust. Another part is trapped in the upper mantle, giving rise to a sub-Moho zone enriched in volatiles and low melting point material. Below this zone and above the level at which partial melting starts there is a region depleted in the low melting point fraction.

In the model the primary flow field has been selected somewhat arbitrarily, this point will be discussed in Section 5. Because of buoyancy partial melt will move upward with respect to the already upward moving solid. For reasons of continuity, we have that when the liquid moves up with a velocity V_L, with respect to the primary velocity V, the residual solid should move downward with a velocity V_S with respect to V. Mass balance requires that:

$$\rho_L F \overline{V}_L + \rho_S (1 - F) \overline{V}_S = 0 \tag{1}$$

The energy conservation equation is:

$$\frac{\partial}{\partial t}(T + FL/c_p) = \text{div} \cdot [\lambda \text{ grad } T - \{\overline{V} + F\overline{V}_L + (1 - F)\overline{V}_S\} T$$

$$- F(\overline{V} + \overline{V}_L)L/c_p] + q/c_p \tag{2}$$

where: ρ_L, ρ_S = liquid and solid density, respectively; F = fraction of melted material; L = latent heat of fusion; c_p = specific heat at constant pressure; λ = thermal diffusivity; T = temperature; q = radioactive heat generation per gram of rock per second (see Bottinga and Allègre, 1976).

In eq. 1 a distinction is made between the densities of the solid and liquid phases, in order to calculate V_S once V_L is known. In eq. 2 the difference between liquid and solid densities is only of second order importance and is neglected. It is also assumed that the liquid and solid have the same specific heat. As is usual in eq. 2, the temperature, pressure and composition dependence of the density was ignored. Not incorporated in eq. 2, for reasons of simplicity, is the adiabatic heat effect associated with the vertically moving convection currents; this is a correction to the term between the curly brackets. In the actual calculations this correction was applied. One of the difficulties in solving eq. 2 is the uncertainty in evaluating F. The relationship between F, temperature and chemical composition should be given by a petrologic model for the upper mantle.

As did Ringwood (1966) in the pyrolite model, we have assumed that one can express upper mantle chemical compositions in terms of two components: peridotite and tholeiite. These two components are not considered as end members. The chemical composition of the solid phase in the upper mantle is given by X_S peridotite and $1 - X_S$ tholeiite. Compositions for which $X_S > 1$, contain algebraically a negative quantity of tholeiite and are interpreted as being harzburgitic or dunitic. Compositions with $X_S < 0$ contain a negative quantity peridotite. Compositions with $X_S < -0.05$ are interpreted as being basaltic, enriched in alkalimetals and incompatible elements, while when $X_S \simeq 0$ we deal per definition with an oceanic tholeiite (MORB). With this model and the known phase relations for peridotite and tholeiite (Ito and Kennedy, 1967; Cohen et al., 1967), we can calculate F as was described previously by Bottinga (1974) and Bottinga and Allègre (1976, 1978).

Once F is evaluated one may calculate V_L in the manner indicated by Frank (1968). Of course V_L is larger than zero. In the computations we did not use the Frank (1968) equation to obtain a precise estimate of V_L; in the course of the numerical solution of eq. 2 it was noticed that a precise value of V_L was not needed as long as $|V_L| \gg |V|$. In our calculations we assumed that $V_L = 3$ cm y^{-1} in the frame moving with velocity V. Knowing V_L and F, V_S can be obtained from eq. 1.

A discussion of the numerical solution of eq. 2 is given in Bottinga and Allègre (1976, 1978) and will not be repeated here. The present model is as described by Bottinga and Allègre (1978) and is in the following details different from that in Bottinga and Allègre (1976):

(1) A different flow field was adopted.

(2) A somewhat more ultrabasic mantle composition was used; $X_S = 0.85$ instead of $X_S = 0.75$ previously.

(3) The basalt—eclogite transition occurs over a 300°C temperature interval.

(4) An error in the printout of F was corrected.

Points 1—3 will be discussed in Section 5. The physical and petrological constants used in this paper are reproduced in Table I.

3. RESULTS AND COMPARISONS WITH OBSERVATIONS

In this section our calculated results are compared with available observations. Subsequently are discussed the calculated results for temperature, composition (chemical and mineralogical), density and partial melt distribution in the top 100 km of the mantle, from the ridge axis to 1500 km away from there. The plate velocity in all these calculations was 1 cm y^{-1}. These results are used in Section 4 to calculate these acoustic properties for the same upper mantle region.

3.1. Temperature field

The calculated temperature field close to the ridge axis is given in Fig. 2, while the temperature distribution out to a distance of 1500 km from the ridge axis is plotted in Fig. 3. The temperature in the upper mantle at the ridge axis is independent of the vertical upwelling speed as long as it is larger than 1 cm y^{-1} (Allègre and Bottinga, 1974; Bottinga, 1974). In the upper mantle region stretching from close to the ridge axis to about 100 km away and at depth less than 100 km, the temperature in our model depends weakly on the plate velocity, i.e., the temperature distribution in a plate strip with a certain age will be somewhat different for different plate velocities. This is due to the fact that our primary flow field (Fig. 1) is independent of the plate velocity as far as flow directions are concerned, and that vertical convection currents occur over a finite width in our model. In a more realistic model the temperature distribution in this region close to the ridge axis will also be influenced by frictional heating and will thus in this

TABLE I

Characteristics of the oceanic lithosphere

	Layer 1	Layer 2	Layer 3a	Layer 3b	Layer 4	Dimensions
Thickness [a]	0.5	1.5	1.5	3.0	120 [b]	km
Density (STP)	2.0 [c]	2.5 [c]	2.96 [d]	3.15 [e]	3.37 [f]	g cm^{-3}
Thermal conductivity ×10^3	2 [g]	7 [h]	6 [h]	6 [h]	7.4 [j]	cal s^{-1} cm^{-1} deg^{-1}
Thermal expansion [j] ×10^5	3.2	3.2	3.2	3.44	3.5	deg^{-1}
Compressibility ×10^4	—	12.5 [r]	12.5 [r]	9.0 [s,t]	7.5 [v]	kb^{-1}
Radiogenic heat [k] ×10^{14}	—	1.96	1.96	0.39	0.39	cal g^{-1} s^{-1}
Composition	sediment	basaltic rubble [l]	basalt + gabbro [m]	0.8 harzburgite + 0.2 serpentinite [m]	0.85 peridotite + 0.15 basalt [n]	

	Basalt		Eclogite	Peridotite			Dunite	Dimensions
	liquid	solid		plag.	spinel	garnet		
Density	2.60 [p]	2.97 [r]	3.55 [s]	3.30 [s]	3.33 [u]	3.36 [u]	3.31 [s]	g cm^{-3}
Thermal expansion ×10^5	4.0 [p]	3.2 [v]	3.0 [v]	3.5 [v]	3.5 [v]	3.5 [v]	3.5 [w]	deg^{-1}
Compressibility ×10^4	65.0 [q]	12.0 [r]	5.3 [s]	5.8 [u]	5.8 [u]	5.8 [u]	7.9 [w]	kb^{-1}
V_p	2.4 [q]	6.67 [r]	8.43 [s]	8.26 [u]	8.35 [u]	8.48 [u]	8.39 [w]	km s^{-1}
V_s	—	3.62 [r]	4.91 [s]	4.60 [u]	4.65 [u]	4.68 [u]	4.86 [w]	km s^{-1}
$(\partial V_p/\partial T)_P$ ×10^4	−8.0 [q]	−7.5 [x]	−7.0 [x]	−4.8 [x]	−4.8 [x]	−4.8 [x]	−4.95 [w]	km s^{-1} deg^{-1}
$(\partial V_p/\partial P)_T$ ×10^2	1.2 [q]	1.2 [x]	1.08 [s]	1.11 [u]	1.11 [u]	1.08 [u]	1.03 [w]	km s^{-1} kb^{-1}
$(\partial V_s/\partial T)_P$ ×10^4	—	−4.0 [x]	−3.8 [x]	−3.7 [x]	−3.7 [x]	−3.7 [x]	−3.53 [w]	km s^{-1} deg^{-1}
$(\partial V_s/\partial P)_T$ ×10^2	—	4.5 [r]	3.9 [s]	4.2 [u]	4.2 [u]	4.2 [u]	3.72 [w]	km s^{-1} kb^{-1}

Specific heat of mantle material $C_p = 0.244 + 3.62 \cdot 10^{-5} \cdot T - 7.27 \cdot 10^3 \cdot T^{-2}$ cal g⁻¹ deg⁻¹ [y]

Latent heat of fusion of mantle material $L = 100$ cal g⁻¹ [z]

Notes

[a] Schematic values, c.f. Christensen and Salisbury (1975).

[b] Parsons and Sclater (1977).

[c] Estimated.

[d] Gabbro density, see Birch (1966).

[e] Weighted average of the densities of serpentine (20%) and Harzburgite (80%).

[f] Density calculated for a mantle composition as given in the Table I.

[g] From the compilation by Lee and Uyeda (1965).

[h] Kawada (1966).

[i] Temperature weighted average of Schatz and Simmons (1972).

[j] Estimation of Bottinga and Allègre (1976), based on data compiled by Skinner (1966).

[k] See Bottinga and Allègre (1976), based on data by Tatsumoto et al. (1964).

[l] Hyndman et al. (1976).

[m] Allègre et al. (1973).

[n] See text, juvenile layer 4 is estimated to consist of 60% olivine, 20% orthopyroxene, 10% clinopyroxene, and 10% spinel, in the stability field of spinel peridotite; c.f., Green and Lieberman (1976).

[p] Calculated by Bottinga and Weill (1970), observed by Murase and McBirney (1973).

[q] Murase and McBirney (1973).

[r] Christensen (1972a).

[s] Christensen (1974), Birch (1966).

[t] Christensen (1972b).

[u] Estimated from mineral data compiled by Birch (1966), Ahrens (1973), and Green and Lieberman (1976).

[v] Estimated from data compiled by Skinner (1966).

[w] Kumazawa and Anderson (1969).

[x] Estimated from data by Kumazawa and Anderson (1969), Frisilio and Barsch (1972), Soga (1967), Ahrens (1973) and Christensen (1975).

[y] This is the specific heat of diopside (Kelley, 1960).

[z] C.f., Bottinga and Allègre (1976, 1978).

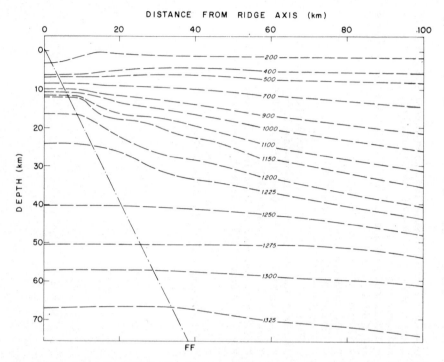

DISTANCE FROM RIDGE AXIS (km)

Fig. 2. Temperature field in the upper mantle close to the ridge axis. Depth is measured with respect to the top of layer 2. Isotherms are marked in degrees centigrade. The line marked *FF* corresponds to the boundary *B—A—2* of Fig. 1.

way depend on the plate velocity. But from a practical point of view, at distances greater than 100 km away from the ridge axis, plate strips with the same age but moving with different velocities have the same temperature distribution.

The penetration of seawater into the newly formed oceanic crust at the ridge axis has been discussed at great length by Lister (1974). Observational evidence for this phenomenon has been published by Williams et al. (1974), Detrick et al. (1974) and Jehl (1975). To take this effect into account, we have assumed that in the interior valley of the rift seawater penetrates the lithosphere to 6 km below seabottom, causing the top 6 km of the lithosphere to be cooled instantaneously to 400°C, from the ridge axis to 12 km away from the axis. Temperatures at 6 km below seabottom are affected by this assumption to about 50 km away from the ridge axis in our model.

Temperatures at depth less than 6 km below seabottom are not plotted in Fig. 3, because our results were calculated at the nodes of a square grid with 6 km distance between them. Hence vertical inhomogeneity, such as occurs in the oceanic crust, could not be adequately modeled. However, measured temperature gradients on Iceland and Sao Miguel (Azores) by Pal-

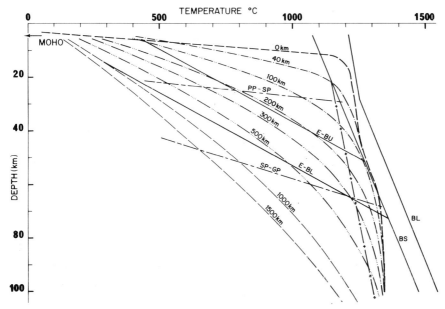

Fig. 3. Phase relations and temperature in the top of the upper mantle. Depth is with respect to the top of layer 2. Temperature distributions are plotted for 0, 40, 100, 200, 300, 500, 1000 and 1500 km from the ridge axial plane. *BL* = basalt liquidus; *BS* = basalt solidus; *PP—SP* = phase transition plagioclase—spinel peridotite; *SP—GP* = phase transition spinel—garnet peridotite; *E—BU* = upper boundary of the transition interval for the phase change eclogite—basalt; *E—BL* = lower boundary for this phase change; — + — + — = beginning of hydrous melting in a peridotite mantle containing 0.1 wt.% water, see text for further discussion.

mason (1967) and Ade-Hall et al. (1974), respectively, are in agreement with the calculated temperature distributions. The magnetotellurically inferred temperature and temperature gradient in the crust and upper mantle at Iceland (Hermance and Grillot, 1974) are also in agreement with our results as given in Figs. 2 and 3.

In Fig. 4, we have plotted our calculated model heat flow on a diagram given to us by Parsons and Sclater (1977) with their compilation of observations. The agreement between observations and calculations is satisfactory. Close to the ridge axis the calculated heat flow is seriously influenced by the way one incorporates the thermal aspects of seawater interaction with the newly formed oceanic crust. Another factor influencing model heat flow values at ages less than 40 Ma is the model flow field for the upper mantle close to the ridge centre.

3.2. Compositional and mineralogical changes in the lithosphere

In Fig. 3, we have also plotted the basalt—eclogite phase relationships and the stability fields of garnet, spinel and plagioclase peridotite. The precise

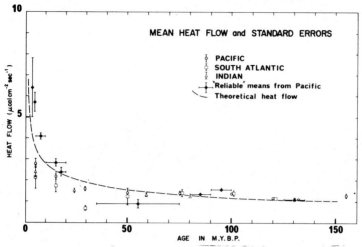

Fig. 4. Heat flow calculated (- - - - - -) and observations as reported by Parsons and Sclater (1977): the filled in symbols, and by Sclater and Francheteau (1970): open symbols.

location of these boundaries is controversial, in particular at temperatures below 1000°C. Recent discussions of certain of these boundaries have been published by O'Hara et al. (1971), Herzberg and Chapman (1976) and Obata (1976). In the absence of definitive data we have retained the peridotite phase boundaries as they were published by Green and Ringwood (1970). Figure 3 shows how in a cooling lithosphere the phase boundaries drift upward when the age of the plate increases.

We have assumed that the basaltic component of our binary petrologic model will undergo an eclogitic phase transition. Of course we are well aware of the schematic nature of this approximation. It does not mean that we literally interpret the upper mantle to be a mechanical mixture of a peridotitic and a basaltic component. Such an interpretation would be naive in view of the relatively high temperature in the upper mantle facilitating reactions between these two components. However, to keep track of the phase relations and the energetic aspects of phase transitions in the upper mantle material, with its varying chemical composition, we have considered the two components separately. There is considerable uncertainty on the temperature, pressure and chemical composition dependence of the basaltic—eclogite phase transition. But it is sure that this transition occurs over a temperature interval. The boundaries we have chosen for this temperature interval are compatible with and inspired by the results summarized by Ringwood (1975). In Section 5 we return to this point.

In Fig. 5b five composition profiles are plotted for the top of the upper mantle at 0, 15, 30, 45 and 60 km from the axial plane of the spreading centre. The compositional variations occurring at depth less than 64 km are due to partial melting and the upward migration of the melt. At distances

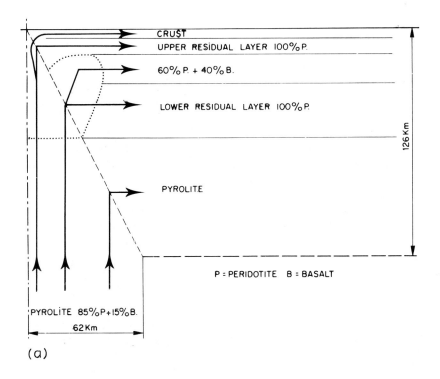

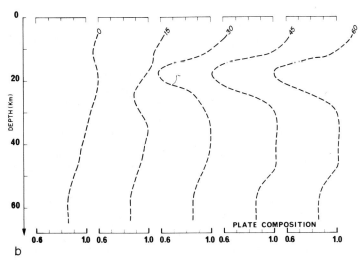

Fig. 5. a. Schematic representation of the submarine lithosphere. *P* = peridotitic material; *B* = basaltic (eclogitic) material. b. Calculated compositional variation with depth in the submarine plate at 0, 15, 30, 45 and 60 km from the ridge axial plane. See text for explanation of the compositional scale.

less than 30 km from the spreading centre axial plane, the chemical variations are also influenced by the flow field of the bulk material.

Partial melting occurs in our model as a consequence of pressure decrease.
The temperature variation with depth in the upwelling material, at depth
below which partial melting starts, is given by an adiabatic distribution (see
Fig. 3, temperature profile 0). The adiabatic gradient is much steeper than
the melting point gradient, this results in partial melting at a depth where
our calculated adiabatic temperature becomes equal to the local solidus temperature. The trace element geochemistry of mid-ocean ridge basalts indicates that these rocks were produced by 8—20% partial melting (Joron et al.,
1976). We refer the reader to papers by Allègre et al. (1977) and Minster et
al. (1977) for a rigorous discussion on the application of trace element geochemistry in the determination of the evolution of volcanic rocks. In order
to have such degrees of partial melting, 64 km below ocean bottom is about
the maximum depth at which partial melting can start in an upwelling
mantle model. If one assumes that no liquid escape occurs until the crystal
mush arrives close to the surface, or if one allows also frictional heating in
the ascending column, then the level at which partial melting starts will be
even shallower. If one insists on having partial melting starting at a greater
depth then the degree of partial melting will become greater than 20%, using
our petrologic model. As a matter of fact REE concentrations in MORB are
consistent with basalt being a partial fusion product of a spinel peridotite
(Allègre et al., 1973; Frey and Green, 1974). This indicates that the partial
melting to produce MORB starts at depth less than about 68 km (see Fig. 3).

The primary flow field (Fig. 1) has as a consequence that the upwelling
material closest to the ridge axial plane experiences the greatest degree of
partial fusion, resulting in a very residual (harzburgitic) upper mantle just
below Moho (see Figs. 5a, b).

This residual zone is intruded by liquids and volatiles produced by partial
melting, also in the upwelling material but outside the central column with
12 km half width. When the primary flow changes its direction from vertical
to horizontal (see Fig. 1) these liquids and volatiles continue to move upwards because of their buoyancy and become eventually trapped just below
Moho. Hence in the initial residual zone, a layer interstitially enriched in
volatiles and the low melting point component is formed, as indicated by
profiles at 20 and 45 km from the ridge in Fig. 5b. The brown hornblende
mylonite, outcropping at St. Paul's Rocks (Melson et al., 1972) may be
a sample of this layer. Below this low melting point fraction enriched layer,
there is still more residual material, because partial melting occurred to a
depth of about 64 km. But below 64 km depth there is a region not affected
by the production of MORB.

The ophiolite model for the top of the oceanic lithosphere indicates a
very residual sub-Moho zone of tectonized harzburgite. In certain cases
(see for example Juteau, 1974) it has been observed that this residual harzburgite grades downward into a peridotite with a less residual character.

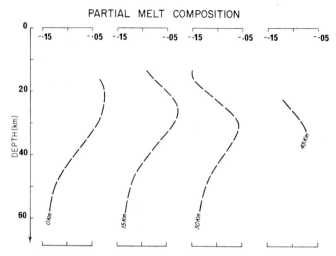

PARTIAL MELT COMPOSITION

Fig. 6. Composition of anhydrous partial melt, close to the ridge axis, at 0, 15, 30 and 45 km from the ridge axial plane; same compositional scale as in Fig. 5b.

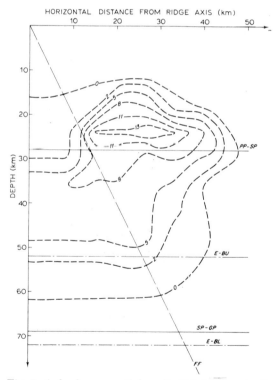

Fig. 7. Anhydrous partial melt distribution in the upper mantle close to the ridge axis. Contours are marked in wt.% partial melt. Other symbols are as in Figs. 2 and 3.

In Fig. 6 we have plotted the volatile free composition of the liquid present in the upper mantle underneath the crest of an active ridge, at distances of 0, 15, 30 and 45 km from the axial plane. The melt distribution is plotted in Fig. 7. This distribution is influenced by the primary flow field, by the escape of liquid from the central column underneath the ridge crest to erupt as MORB at the ridge axis, by the permeability of the upper mantle and by the temperature.

3.3. *Density distribution and seabottom depth*

In Fig. 8 we have plotted iso-density contours for the top 100 km of the upper mantle. These densities are computed from the calculated composition and temperature, the depth, and the constants given in Table I. The low density region beneath the ridge is conspicuous. Melt accumulation taking place somewhat off ridge axis causes also a density low. The calculated densities for the lithosphere are in general lower than the commonly cited values ranging from 3.33 to 3.50 g cm^{-3}. Only in lithosphere older than 100 Ma, we have locally densities in excess of 3.33 g cm^{-3}. In Fig. 9 is plotted the average density for an upper mantle strip reaching to 120 km depth and at the

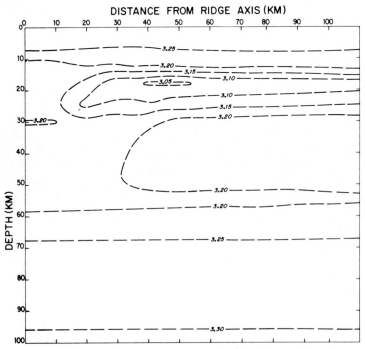

Fig. 8. Density contours for the upper mantle close to the ridge axis. Density values are in g cm^{-3}.

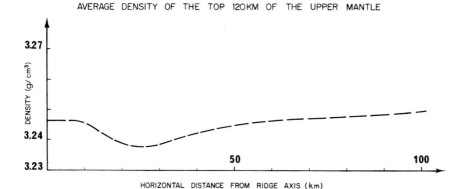

Fig. 9. Average density of the top 120 km of the upper mantle close to the ridge axis.

indicated distance from the ridge axis. At distances between 60 and 200 km from the ridge axial plane the steady increase in the average density is mainly due to thermal contraction. Beyond 200 km the basalt—eclogite phase transition for the basaltic component of our model plate contributes significantly to the density variation. But beyond 1000 km from the ridge axis the density variations are again chiefly due to thermal contraction.

Figure 10a is from Parsons and Sclater (1977), with their observed North Atlantic bottom depth; also plotted are depths calculated with our model. We used the Lambeck (1977) contraction model for these depth calculations; the numerical difference between the results calculated with the contraction models of Lambeck (1977) and Sclater and Francheteau (1970) is minor. The differences between the observations, the Parsons and Sclater (1977) calculated depth, and our model depth are small; the maximum depth difference between the two models is 250 m, which is comparable to the noise in the observations.

Figure 10b shows our calculated ocean bottom depth against the square root of lithospheric age. For ages less than 2.5 Ma the topography does not obey the linear square root age relationship of Davis and Lister (1974). This was observed by Tréhu (1975) for the Mid-Atlantic Ridge. Of course our model does not account for the formation of the central valley of the Mid-Atlantic Ridge; factos influencing the formation of such a central valley have been analysed by Tapponier and Francheteau (1977). In any case, an isostatic model with a uniform compensation depth cannot be valid close to the ridge axis. For the age interval 2.5—100 Ma the square root age—ocean-bottom depth relation is well obeyed with an average slope of 344 m [age (Ma)]$^{-1/2}$. This is similar to the slopes reported by Parsons and Sclater (1977) and Tréhu (1975). For ages less than 16 Ma, Tréhu (1975) observed for the Mid-Atlantic Ridge variations in the ocean bottom age relation which resemble variations shown by our model.

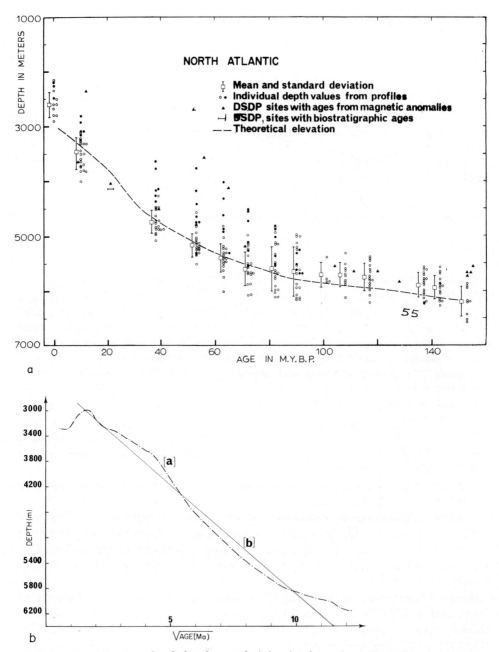

Fig. 10. a. Ocean bottom depth for the north Atlantic, the various observations are as compiled by Parsons and Sclater (1977) and the curve was calculated with the present model. b. Ocean bottom depth, calculated with the present model, plotted against the square root of age. The average slope is given by the curve [b], it amounts to $344 \text{ m (Ma)}^{-1/2}$. [a] marks a local steep part in the model oceanic bottom profile with a slope of 552 m $(\text{Ma})^{-1/2}$. Plate velocity is 1 cm y^{-1}.

4. ACOUSTIC PROPERTIES

4.1. *P and S wave velocity model calculations*

Calculated P and S wave velocities are given in Figs. 11, 12a, b. To compute these velocities we used our model values of the compositional, mineralogical, partial melt, temperature and pressure variables. First the medium was supposed to be totally solid and the velocities were calculated with the constants given in Table I, as simple arithmetic averages of the velocities of the components. The pure components are acoustically not too different, hence arithmetic averaging does not introduce important errors as compared with the accuracy of the determined lithospheric velocities.

The presence of a melted fraction was taken into account with the method of O'Connell and Budiansky (1977). These authors have calculated in a self consistent manner the acoustic properties of a solid medium containing cracks filled with liquid, as a function of the crack density and the aspect ratio of the cracks. To estimate these parameters for our model, as a function of the degree of partial fusion and the amount of liquid present, we have assumed that for a small degree of melting (abt. 1%), mainly clinopyroxene is involved; at moderate degrees of partial melting (abt. 10%) clinopyroxene, spinel, and orthopyroxene are affected; at large degrees of partial melting (>15%) all the minerals (olivine, both pyroxenes, and spinel) are affected. We have taken the average grain-size to be 3 mm. Such a dimension is suggested by ultramafic nodules found in oceanic island basalts. To

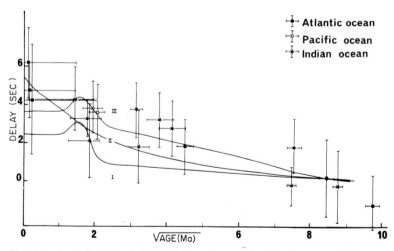

Fig. 11. A plot of the S-delay times observed by Duschenes and Solomon (1977). Results obtained with the model proposed by these authors are plotted as curve II. Curve I gives delays calculated with the anhydrous version of the present model; curve III gives the S-delays calculated for a hydrous (0.1 wt.% water) version of the present model. See text for discussion.

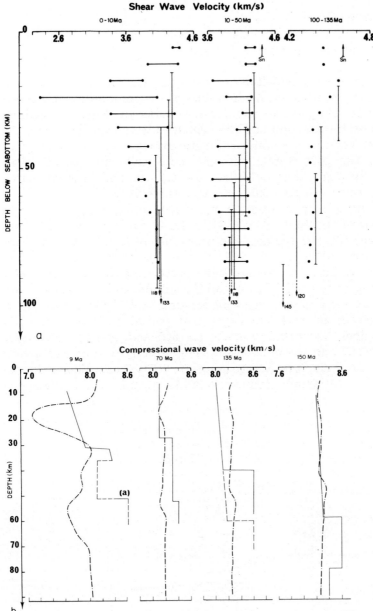

b

Fig. 12. a. Shear wave velocities deduced by Forsyth (1975, 1977), from Rayleigh wave observations in lithosphere of 0—10, 10—50 and 100—135 Ma; these velocities are indicated by vertical lines spanning the appropriate depth range. These velocities are to be compared with our calculated ranges of S-wave velocities (horizontal lines) for the indicated age zones. Calculations were done for a model plate containing 0.1 wt.% water. Upward pointing arrows near the top of the diagram give Sn velocities reported by Hart and Press (1973). b. Observed (——— and ------) profiles for P velocities in the oceanic lithosphere for various ages. These profiles are to be compared with our calculated profiles (— · — · — · —) for a 0.1 wt.% water containing lithosphere. Note the increase in the observed P-wave velocities at about 50 km below the top of layer 2, a corresponding increase is absent in the data of Forsyth (1975, 1977), see Fig. 12a.

estimate the crack density and the aspect ratio we suppose the upper mantle to consist of equidimensional cubic grains with 3 mm edges, hence there are about 40 cubic grains per cubic centimeter. Approximately 10% of these grains are clinopyroxene, hence initial partial melting will create about 24 fluid filled cracks with an aspect ratio of $1.4 \cdot 10^{-2}$ for 1% partial fusion. O'Connell and Budiansky (1977) defined the crack density as:

$$\epsilon = \frac{2N}{\pi} \left\langle \frac{A^2}{P} \right\rangle$$

where N equals the number of cracks per cubic centimeter, A is the surface of a crack and P its perimeter. The initial mineralogical composition of the unfused upper mantle is taken to be 10% clinopyroxene, 20% orthopyroxene, 10% spinel, and 60% olivine. With this information we have estimated the aspect ratio and crack density for various degrees of partial melting (Table II).

O'Connell and Budiansky (1977) considered various endmember models, two of which are relevant to the present discussion, namely saturated isolated and saturated isobaric cracks. These latter cracks are filled with fluid, which can flow between cracks as a result of local pressure differences, generated by the passing of a sound wave. Cracks will behave like this when they are interconnected and when the frequency $\omega < (\kappa/\eta)(c/a)^3$, where ω is the frequency of the soundwave, κ is the bulk modulus of the uncracked medium, c/a is the aspect ratio and η is the viscosity of the fluid in the crack. Experimental data on silicate melt viscosity under upper mantle conditions (Kushiro et al., 1977), theoretical analysis of the pressure dependence of silicate melt viscosity (Waff, 1977) and the compositional dependence of silicate melt viscosities (Bottinga and Weill, 1972) indicate that the viscosity of liquid produced by small degrees of partial melting in the upper mantle

TABLE II

Acoustic characteristics of partially melted upper mantle material

Degree of partial melt	Aspect ratio	Crack [a] density	$\overline{V}_s/V_s$ [b]	Q_s [c]	$\overline{V}_p/V_p$ [b]	Q_p [d]
1%	0.014	0.1	0.92	500	0.96	1000
10%	0.07	0.2	0.82	100	0.92	500
15%	0.05	0.5	0.43	10	0.80	50

[a] For definition see text.
[b] Barred quantities are for the fluid filled cracked medium, unbarred quantities for the uncracked solid medium.
[c] Q_s is the quality factor for shear wave propagation.
[d] Q_p is the quality factor for compressional wave propagation.

will be smaller than 100 poise. This gives $\omega < 10^4$ Hertz, while the frequency of a seismic signal is about 3 Hertz. Hence we are dealing with saturated isobaric cracks, except when the degree of partial fusion becomes small and the cracks are not any longer interconnected. In that case the O'Connell and Budiansky (1977) model of isolated saturated cracks becomes appropriate. Using figs. 2 and 6 of O'Connell and Budiansky, we have obtained estimates of the acoustic properties of partially melted mantle material listed in Table II.

Of course we are well aware of the approximate nature of these estimates. However, certain of our assumptions will lead to an underestimation of the effect of partial melting on the acoustic properties rather than an overestimation. For instance, the shearing occurring beneath the ridge crest will result in grain-sizes smaller than our estimated 3 mm. High temperature peridotites are mineralogical not isotropic. If this is also true for the upper mantle then even smaller degrees of partial melting will cause cracks to be interconnected.

4.2. S-wave travel times for trajectories across the lithosphere

Duschenes and Solomon (1977) have determined S-wave travel times across the oceanic lithosphere. They looked at earthquakes occurring in oceanic lithosphere of various ages, the epicentres were located from the P arrivals, using the Jeffreys—Bullen tables and an S-delay time was calculated with respect to Jeffreys—Bullen travel times for each observation of the earthquake. Seismic station corrections were applied and, if possible, depths to hypocentres were reduced to a common value. Figure 11 is a plot of the observed S delays of Duschenes and Solomon (1977) as a function of the age of the lithospheric epicentral region. Duschenes and Solomon (1977) argue that the S-delay differences for earthquakes in different age zones of the lithosphere are generated along the lithospheric ray path between the earthquake source region and the asthenosphere. In Fig. 11, the absolute value of the S delays is not important, the significant aspect is the difference between the S delays as a function of age of the lithosphere containing the earthquake source.

To check the compatibility of these observations with our model, we have calculated P- and S-travel times for trajectories across lithosphere of different ages in the same way as was done by Duschenes and Solomon (1977). Curve I, in Fig. 11 represents our calculated delays. Curve II was calculated by Duschenes and Solomon for a wet peridotite model of the lithosphere, proposed by these authors. The relative delays shown by curve I for ages between 0 and 20 Ma are smaller than the averages of the observations and also smaller than the delays for the Duschenes and Solomon model (curve II).

The model we have developed is essentially a dry peridotite model, because independent evidence discussed by Bottinga and Allègre (1976,

1978) indicates that there is virtually no water in the MORB source region in the upper mantle. Moreover the activity of the little bit of water there is in the upper mantle is severely diminished by the presence of other volatiles like CO_2. From a petrological point of view the water activity in the MORB source region is too small to be of interest. However, it is known that even minute degrees of partial melting have an important influence on the acoustic properties of the medium (Spetzler and Anderson, 1968). Rheologically, small degrees of partial melting seem to be without significant consequences (Auten et al., 1974).

The presence of a small quantity of water in the upper mantle, will affect the upper mantle solidus temperature. Kushiro et al. (1968) have determined solidus temperatures for peridotite as a function of pressure, under conditions that $P_{H_2O} = 0$ and $P_{H_2O} = P_{total}$. Their results are not detailed enough to indicate precisely the solidus temperature as a function of water fugacity and total pressure. From this work we have taken a solidus temperature—depth relationship of $T_S = 1100 + 2z$, where T_S is in degrees centigrade and z is in kilometers depth, and the partial water pressure is smaller than the total pressure. The admisssible values of z are restricted to the interval 20—150 km depth. The resulting solidus temperature has been plotted on Fig. 3, which was used to determine the extent of the region affected by incipient melting. Estimating that the S velocities are influenced by this as by 1% partial melting, one finds an S velocity for this region of 4.1 km s^{-1} as observed by Forsyth (1975). The S delays calculated for this modified model are plotted in Fig. 11, curve *III*, they are in closer agreement with the observations than before. The data of Duschenes and Solomon (1977) are not precise enough to differentiate their model (curve *II*) from our model (curve *III*), close to the ridge axis. Error limits for the location of earthquakes occurring close to the ridge axis are ±30 km in horizontal distance and ±2 s for origin times, i.e., close to the origin of the x axis of Fig. 11, the uncertainty in the earthquake location is ±1.7 units.

4.3. *Lithospheric shear and compressional wave velocities*

The velocity of the Sn phase corresponds to the maximum S velocity in the lithosphere (Stephens and Isacks, 1977; Mantovani et al., 1977). In Fig. 12a we have plotted mean Sn velocities for age zones of the Atlantic lithosphere larger and smaller than 50 Ma, reported by Hart and Press (1973). Also are shown the S velocities occurring in our model as a function of plate age. For plate ages smaller than 100 Ma our calculated maximum velocity is somewhat smaller than observed by Hart and Press (1973), but for ages greater than 100 Ma the difference between calculated maximum velocities and observed values becomes unimportant.

Rayleigh wave studies are another source of shear wave velocity distributions for the oceanic lithosphere. The most detailed data and analyses have been published by Forsyth (1975, 1977) for the Pacific ocean; in this work

TABLE III

Shear wave velocity distribution (km s⁻¹) in the submarine upper mantle

The heavily printed numbers indicate velocities affected by a small degree of hydrous partial melting as explained in the text. The boundary marked by "m" indicates the Moho. Between the Moho and the line marked "h" the mantle composition is harzburgitic. Below the boundary "h" the upper mantle is enriched in basaltic material and volatiles, this zone stretches to the superior "r" boundary. The vertical partition marked ++++, signifies the beginning of the transformation of basaltic material into eclogitic material. The line marked "s" separates the plagioclase peridotite stability field from the spinel peridotite stability field; plagioclase peridotite is stable above "s". The region between the two lines marked "r" is residual in petrologic character, i.e. harzburgitic and dunitic. The boundary marked "g" indicates the upper stability limit for garnet peridotite. In the region lying between the ridge axial plane and the dotted (·····) boundary the concentration of partial melt exceeds 1%.

Distance to ridge axis in kilometers

Depth km	3	15	27	39	57	93	129	183	231	309	405	507	627	759	897	1053	1491
0	3.75	3.75	3.75	3.75	3.75	3.75	3.75	3.75	3.75	3.75	3.75	3.75	3.75	3.75	3.75	3.75	3.75
6	4.39	4.33	4.28	4.28	4.31	4.33	4.38	4.42	4.44	4.47	4.49	4.50	4.52	4.53	4.53	4.54	4.56
12	3.91	4.12	4.28	4.29	4.32	4.36	4.39	4.42	4.44	4.46	4.48	4.50	4.52	4.53	4.54	4.55	4.56
18	3.73	3.74	3.35	3.75	3.81	3.84	3.88	3.90	4.04	4.20	4.33	4.43	4.53	4.60	4.65	4.66	4.67
24	3.59	2.60	2.30	3.36	3.99	4.05	4.06	4.11	4.16	4.30	4.38	4.45	4.52	4.57	4.58	4.60	4.61
30	3.70	3.36	3.58	3.66	3.96	4.29	4.32	4.35	4.37	4.40	4.43	4.45	4.47	4.49	4.50	4.51	4.53

36	3.46	3.46	3.60	3.85	3.88	4.20	4.23	4.26	4.29	4.33	4.35	4.38	4.41	4.43	4.44	4.46	4.49
42	3.62	3.62	3.62	3.88	3.91	3.92	4.24	4.26	4.28	4.32	4.35	4.37	4.39	4.42	4.43	4.45	4.48
48	3.67	3.67	3.63	3.92	3.94	3.94	3.96	4.27	4.29	4.32	4.34	4.37	4.39	4.41	4.43	4.44	4.50
54	3.80	3.80	3.77	3.84	3.84	3.85	3.87	3.90	4.23	4.29	4.34	4.39	4.41	4.43	4.48	4.49	4.52
60	3.88	3.88	3.88	3.88	3.88	3.89	3.90	3.92	3.95	4.30	4.36	4.38	4.43	4.45	4.47	4.48	4.51
66	3.94	3.94	3.94	3.94	3.94	3.95	3.96	3.98	4.04	4.36	4.38	4.40	4.42	4.43	4.45	4.47	4.50
72	4.03	4.03	4.03	4.03	4.03	4.03	4.03	4.04	4.04	4.05	4.37	4.39	4.41	4.42	4.44	4.45	4.48
78	4.03	4.03	4.03	4.03	4.03	4.03	4.04	4.04	4.04	4.05	4.37	4.38	4.40	4.41	4.43	4.44	4.47
84	4.04	4.04	4.04	4.04	4.04	4.04	4.04	4.04	4.05	4.05	4.06	4.38	4.39	4.41	4.42	4.43	4.46
90	4.05	4.05	4.05	4.05	4.05	4.05	4.05	4.05	4.05	4.06	4.06	4.37	4.39	4.40	4.41	4.42	4.45
96	4.06	4.06	4.06	4.06	4.06	4.05	4.05	4.06	4.06	4.06	4.07	4.37	4.38	4.40	4.41	4.42	4.44
102	4.07	4.06	4.06	4.06	4.06	4.06	4.06	4.06	4.06	4.07	4.07	4.08	4.38	4.39	4.40	4.41	4.42
108	4.08	4.07	4.07	4.07	4.07	4.07	4.07	4.07	4.07	4.07	4.07	4.08	4.38	4.39	4.40	4.40	4.42
114	4.08	4.08	4.08	4.08	4.08	4.08	4.08	4.08	4.08	4.08	4.08	4.08	4.09	4.39	4.40	4.40	4.41
120	4.08	4.08	4.08	4.08	4.08	4.08	4.08	4.08	4.08	4.08	4.09	4.09	4.09	4.39	4.39	4.40	4.40
126	4.09	4.09	4.09	4.09	4.09	4.09	4.09	4.09	4.09	4.09	4.09	4.09	4.09	4.09	4.09	4.09	4.09

velocity distributions with resolution lengths are given for different plate ages. These data are here reproduced in Fig. 12a and compared with our model velocities. The agreement between the observations and the model values is satisfactory, except for the depth interval 10—65 km of the age zone 0—10 Ma. In this region the model predicts considerable smaller values than observed by Forsyth (1975). However, it should be appreciated that the resolution of the Forsyth (1975) observations does not permit detection of small scale phenomena as shown by the model. Further the S delays discussed in Section 4.2. demand according to Duschenes and Solomon (1977) even smaller S velocities in this region than calculated with our model.

The Rayleigh wave data of Weidner (1974) and Girardin and Jacoby (1977) are in general agreement with the Forsyth (1975, 1977) data and also with our model. Jacoby and Girardin (1977) being less conservative in their error estimates than Girardin and Jacoby (1979), have published S-wave velocity distributions compatible with their Rayleigh wave data, for the age zone 0—10 Ma, which are in agreement with our model velocities over the total depth range. In particular they report the existence of a zone of reduced S-wave velocities for a lithospheric age zone comprised between 3 and 8 Ma in the northern Atlantic, which could correspond to the partial melt enriched region drawn in Fig. 7. This reduced S-wave velocity zone extends, while gradually diminishing in importance, into the 10—20 Ma age old plate.

The lower part of the diagram in Fig. 12a shows that Forsyth's (1975, 1977) observations are smaller than our model values, in particular for the 100—135 Ma age zone. This is at least in part due to the fact that these velocities are averages over depth ranges reaching into the low velocity zone while our model values reflect only a local property.

In Table III are tabulated the shear wave velocities calculated for our model upper mantle. Figures 11 and 12a were derived from this table. The heavily printed numbers in Table III give velocities which were affected by the small degree hydrous partial melting.

In Fig. 12b the compressional wave velocities observed in various age zones of the oceanic lithosphere are compared with calculated model velocities. The 9 Ma profile (Steinmetz et al., 1977) shows a qualitative agreement with the model values. The relative velocity high at about 30 km depth and the decrease in velocity below this zone are noticeable in the observations and the model. At about 50 km depth the model predicts a velocity decrease in contradiction with the increase shown by the observations. Steinmetz et al. (1977) have pointed out that instead of an increase, the change in velocity could seismologically also be interpreted as a very abrupt decrease, because the change was only observed with reflected waves. However, the model cannot reproduce the velocity change demanded by the observations. Quantitatively the model values at depth less than 50 km tend to be too small by about 0.2 km s^{-1} when compared with the observations. The 70 Ma observed profile (Orcutt and Dorman, 1977) is qualitatively in agreement with the model, but quantitatively the model is 0.1—0.2 km s^{-1} too slow. The set of data of Steinmetz and Hirn (1973) for the 135 Ma profile (Fig. 12b) run between Madeira and the Canary Islands, is not yet sufficiently complete to allow a full interpretation (see also Steinmetz, 1977). But apparent velocities of 8.6 km s^{-1} are evident, the probable depth for this high velocity extends from 40 to 65 km below the seabed. Down to a depth of 60 km, the 150 Ma profile, observed by Asada and Shimamura (1976) is conform to the model. But the velocity high in the observations between 60 and 80 km depth is not seen in the model.

All the observed data show a velocity increase at depth between 40 and 60 km. Bottinga et al. (1976) have already drawn attention to this feature and have concluded that it is highly unlikely that this velocity increase is due to a solid—solid phase transition in a peridotitic lithosphere. Our model supports this conclusion, only the relatively small increase observed by Orcutt and Dorman (1977) in the 70 Ma profile could possibly be due to the transition spinel peridotite—garnet peridotite.

Fig. 12a shows that the shear wave velocities observed by Forsyth (1975, 1977) tend to decrease in the depth range where the P increase occurs. But the absolute value of the S-wave velocity decrease is less than the absolute value of the corresponding P increase. We think that all these observations may be best explained by the occurrence of acoustical anisotropy caused by the orientational anisotropy of olivine crystals.

The observations plotted in Fig. 12b were obtained from seismic refraction profiles, which run in one particular direction, while the shear wave data are averages over all possible directions and over depth intervals which may be greater than the thickness of the high P-wave velocity layer. The elastic properties of single olivine crystals (Fo 92.7) measured by Kumazawa and Anderson (1969) indicate that the relative P-wave anisotropy is much more important than the relative S-wave anisotropy. Preliminary data by Steinmetz (1977) for the Atlantic between Madeira and the Canary Islands suggest also that the velocity increase shown in the 135 Ma profile (Fig. 12b) is due to anisotropy. This conclusion of an anisotropic lower lithosphere, is also supported by the data of Forsyth (1975, 1977). Bottinga et al. (1976) reached the same conclusion and we refer the reader to this paper for a more detailed discussion of this point.

Solomon and Julian (1974) have shown that lateral heterogeneity in the P-wave velocity field for the upper mantle region below a spreading centre, could be the cause for the apparent non-orthogonality of nodal planes associated with mechanisms derived for earthquakes occurring in this region. Earthquakes located in the neighbourhood of the ridge axis in our model will display this effect too. In Fig. 13, we have traced rays for earthquake sources at 0, 15 and 24 km horizontal distance from the ridge axial plane, and at Moho depth; emergence angles were increased in 5° steps. For epicentres on the ridge axis focalization of the rays takes place, at 45° from the vertical, but this does not affect the orientation of the nodal plane directions. For off-

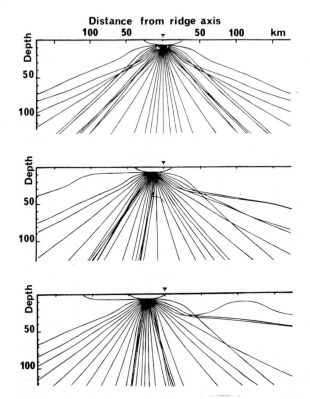

Fig. 13. P-wave ray tracing for earthquakes occurring 6 km below the top of layer 2 at the ridge axis (top diagram), at 15 km from the ridge axis (centre diagram) and at 24 km from the ridge axis (bottom diagram). The angle of incidence of the traced rays increases in steps of 5°.

ridge axis earthquakes one may notice a general downward refraction of the rays; rays leaving the source region under angles of 40—50° show on the average a 15° downward inflection. Taking into account the refraction of rays leaving towards the right and left hand sides of Fig. 13, one obtains easily a decrease of the angle at which the nodal planes intersect of the order of 30°. Ray tracing has shown that nodal plane rotations, similar to those observed in nature, can be obtained for epicentres located maximal 30 km away from the ridge axis. The model calculated rotations of the nodal planes depend strongly on the depth of the earthquake source. Because the error in the localization of oceanic ridge earthquakes is about 30 km, it is not feasible at present to differentiate our model, with liquid accumulation just off the axial plane, from the Solomon and Julian (1974) model, with liquid accumulation in the axial plane of the ridge.

Our calculated P-wave velocity field for the upper mantle region is given in Table IV. Figures 12b and 13 are derived from the data listed in this table. The general petrologic nature of the submarine lithosphere is also indicated

TABLE IV

Compressional wave velocities (km s^{-1}) in the submarine upper mantle

Depth km	Distance to ridge axis in kilometers																
	3	15	27	39	57	93	129	183	231	309	405	507	627	759	897	1053	1491
0	6.00	6.00	6.00	6.00	6.00	6.00	6.00	6.00	6.00	6.00	6.00	6.00	6.00	6.00	6.00	6.00	6.00
6	8.11	8.11	8.07	8.07	8.09	8.10	8.14	8.16	8.18	8.19	8.20	8.21	8.22	8.23	8.23	8.24	8.25
12	7.76	7.84	7.86	7.88	7.90	7.96	8.00	8.05	8.06	8.09	8.12	8.15	8.17	8.18	8.20	8.21	8.22
18	7.68	7.52	6.73	6.93	7.02	7.06	7.13	7.16	7.36	7.59	7.77	7.92	8.04	8.15	8.22	8.23	8.25
24	7.44	6.32	6.00	6.83	7.38	7.47	7.49	7.55	7.63	7.84	7.95	8.05	8.14	8.21	8.24	8.25	8.28
30	7.45	7.09	7.31	7.33	7.81	7.95	7.98	8.03	8.05	8.10	8.13	8.16	8.18	8.20	8.22	8.23	8.26
36	7.19	7.20	7.32	7.74	7.81	7.93	7.96	8.00	8.03	8.07	8.10	8.13	8.16	8.18	8.20	8.22	8.25
42	7.34	7.34	7.33	7.75	7.82	7.83	7.95	7.98	8.01	8.05	8.08	8.11	8.14	8.16	8.18	8.20	8.24
48	7.27	7.27	7.26	7.84	7.83	7.84	7.85	7.98	8.00	8.03	8.06	8.09	8.12	8.15	8.17	8.19	8.35
54	7.39	7.39	7.39	7.62	7.62	7.63	7.65	7.70	7.84	7.92	8.00	8.06	8.09	8.12	8.25	8.28	8.31
60	7.61	7.61	7.60	7.66	7.67	7.68	7.69	7.73	7.77	7.94	8.02	8.05	8.18	8.21	8.23	8.26	8.30
66	7.76	7.76	7.76	7.76	7.77	7.78	7.79	7.81	7.84	8.10	8.12	8.15	8.18	8.20	8.22	8.25	8.29
72	7.96	7.96	7.96	7.96	7.96	7.97	7.97	7.98	7.98	8.00	8.12	8.15	8.17	8.19	8.20	8.22	8.28
78	7.98	7.98	7.98	7.98	7.98	7.98	7.99	7.99	8.00	8.01	8.13	8.15	8.17	8.19	8.21	8.23	8.27
84	8.00	8.00	8.00	8.00	8.00	8.00	8.00	8.01	8.01	8.02	8.03	8.15	8.17	8.19	8.21	8.23	8.26
90	8.02	8.02	8.02	8.02	8.02	8.02	8.02	8.02	8.03	8.03	8.04	8.16	8.18	8.19	8.21	8.23	8.26
96	8.04	8.04	8.04	8.04	8.04	8.04	8.04	8.04	8.04	8.05	8.06	8.17	8.18	8.20	8.21	8.23	8.25
102	8.06	8.06	8.06	8.06	8.06	8.06	8.06	8.06	8.06	8.07	8.07	8.08	8.19	8.20	8.22	8.23	8.25
108	8.08	8.08	8.08	8.08	8.08	8.08	8.08	8.08	8.08	8.08	8.09	8.09	8.20	8.21	8.22	8.23	8.25
114	8.10	8.10	8.10	8.10	8.10	8.10	8.10	8.10	8.10	8.10	8.11	8.11	8.12	8.22	8.23	8.23	8.25
120	8.12	8.12	8.12	8.12	8.12	8.12	8.12	8.12	8.12	8.12	8.12	8.13	8.13	8.23	8.24	8.24	8.25
126	**8.14**	**8.14**	**8.14**	**8.14**	**8.14**	**8.14**	**8.14**	**8.14**	**8.14**	**8.14**	**8.14**	**8.14**	**8.14**	**8.14**	**8.14**	**8.14**	**8.14**

See Table III for explanation of the markings in this table.

in Table IV. It is obvious that these petrologic boundaries are not marked by great acoustic contrasts. The heavily printed numbers in Table IV, indicate velocities affected by a small degree of hydrous partial melting below the anhydrous solidus.

5. DISCUSSION AND CONCLUSIONS

5.1. *Moho velocity in young lithosphere*

Christensen and Salisbury (1975) have pointed out that more than 50% of the measured Pn velocities in oceanic lithosphere, younger than 15 Ma, are smaller than 8.0 km s^{-1}, with 7.2 km s^{-1} not being uncommon. Anomalous upper mantle velocities, close to the ridge axis could be due to a simple temperature effect. However, this explanation is unlikely to be valid for lithosphere older than 5 Ma, and in particular for the case that cooling of the total oceanic crust takes place by means of seawater penetration at the ridge axis.

Another possible cause for Pn velocities smaller than 8.0 km s^{-1} is the occurrence of anisotropy, initially proposed by Hess (1964). In this model that feature has not been taken into consideration quantitatively. Still another explanation for the anomalous low Pn velocities close to the ridge axis is the possible presence of partially melted rock. Such an explanation is only reasonable for the region really close to the spreading centre; like the temperature explanation it fails for lithosphere older than 5 Ma. In the following paragraphs we discuss possible chemical causes for the occurrence and eventual dissappearance of anomalously low Pn velocities.

Low Pn velocities may be caused by partial serpentinization of the upper mantle immediately below Moho and occurring at small distances from the ridge axis. In Section 3.2. we have shown that the lithosphere just below Moho may be compositionally layered. It was explained that the residual harzburgite beneath the Moho is interstitially enriched in low melting point material, which changes gradually from volatile rich to basaltic when going downward. At about 30 km below the seafloor the basalt enriched zone gives way to a residual upper mantle (see Figs. 5a and b). The temperature just below Moho is well in excess of 450°C, which is about the upper stability limit of serpentine, when this compositional gradation is formed. But as the lithosphere moves away from the ridge axial zone and becomes cooler, the harzburgite containing aqueous vapour transforms partially into serpentine when its temperature becomes less than 450°C. The extent of the serpentinization depends of course on the availability and water activity of the aqueous vapour. For degrees of serpentinization varying from 10 to 20% the amount of water involved ranges from 1 to 2%; this means a local enrichment in water by a factor of 10—20 with respect to the water concentration in the MORB source region. Before serpentinization takes place this water enrichment has no significant acoustical consequences because the temperature is in any case well below the hydrous harzburgite solidus (Kushiro et al., 1968). The P-wave velocity of the harzburgite below the volatile enriched zone will depend on how much interstitial basaltic material is locally present; this may be considerable (see Fig. 5b).

It is difficult to determine a priori to what height in the upper mantle the water vapour coming from below will penetrate the sub-Moho harzburgite, or to what height the basaltic material will rise. These phenomena depend among other things upon the local porosity distribution, for which we do not have an adequate model. The degree of basaltic enrichment as a function of depth as given in Fig. 5b is an illustration of a concept rather than a quantitative representation. The petrologic model we have used in this paper is not detailed enough to treat the volatile phase separately from the basaltic material.

If the aqueous vapour manages to rise to Moho height then partial serpentinization of the cooling lithosphere results in a gradual thickening of the seismically perceived layer 3. Observational evidence for the thickening with age of layer 3 has been compiled by Goslin et al. (1972). The observations by Lewis and Snydsman (1977) and Steinmetz et al. (1977) for the lower limit of the thickening layer 3 and our calculated 450°C isotherm (Fig. 2) for the upper mantle coincide virtually. Another way to obtain thickening of layer 3 is by means of seawater penetration to the appropriate depth, see Lister (1974). It is also possible that the water vapour rising from below does not reach the level of the Moho, in that case a velocity inversion will develop below normal Pn velocities of about 8.0 km s^{-1}. The thickening of layer 3 as can occur in our model is similar to the serpentinization of layer 3 envisioned by Hess (1962). It is like the process described by Le Pichon et al. (1973). In less detail this process has also been discussed by Steinmetz et al. (1977).

Arguments pro and contra the occurrence of serpentine in layer 3 have been given by Le Pichon (1969), Bottinga and Allègre (1973) and Christensen and Salisbury (1975). The latter authors assert that the Poisson ratios for partially serpentinized peridotites are "much higher than those observed from oceanic crustal seismic measurements". According to them (Christensen and Salisbury, 1975, fig. 15) a 20% serpentinized peridotite at 1 kb and room temperature has $V_p \simeq 7.45$ km s^{-1}, $V_s \simeq 4.15$ km s^{-1}, and thus $\sigma \simeq 0.275$. When these values are compared with the tabulation of observations given by Christensen and Salisbury (1975, Table I) one notices that the Christensen and Salisbury measurements support rather than refute the hypothesis of a partially serpentinized layer 3. These quoted values for V_p, V_s and σ are quite close to those observed at shallow depth in young lithosphere (see also Figs. 12a, b and Tables III and IV).

A frequently reported objection to the presence of partially serpentinized peridotite in layer 3 is that the serpentinite dredged from the ocean floor has magnetic and acoustic properties different from those observed for layer 3. This is true, but the serpentinite dredged from the ocean floor is virtually always nearly completely serpentinized harzburgite, as such it should not have much in common with the 10—20% serpentinized harzburgite believed to be present in the lower part of layer 3. In the process which leads to the outcropping of the lower part of layer 3 further serpentinization takes place most likely. There is nothing in the detailed isotopic studies of Wenner and Taylor (1971) on serpentinite which contradicts such a sequence of events.

The upper mantle zone enriched in basaltic material has P-wave velocities less than that of pure residual harzburgite. This can be the cause of anomalous upper mantle velocities in young lithosphere. In older lithosphere this anomalous characteristic will disappear when the temperature is low enough for the transformation of the basaltic material into eclogitic material. The extent and depth to which this takes place progresses with the age of the plate. Our calculations indicate that anomalous upper mantle is restricted to lithosphere younger than 20 Ma. This is in agreement with the evidence given by Christensen and Salisbury (1975).

5.2. Basalt—eclogite phase transition and mantle composition

The data of Table I show that of all the potential solid—solid phase transitions in the upper mantle, only the basalt—eclogite transition can affect significantly the acoustic properties and the local density of the lithosphere. In previous models (Bottinga, 1974; Bottinga and Allègre, 1976) the temperature interval over which this transition occurs at a given pressure was ignored, and a sharp transition temperature was used. This has as consequence a seabottom topographic effect, when the lithosphere has cooled sufficiently for this transition to take place, which has not been observed. A 300°C temperature interval for this transition has as result that the topographic effect is spread out and not observable in the noise of seabottom topography data (see Figs. 10a, b).

Bottinga and Allègre (1976) have remarked that the acoustic properties of peridotite and the variation of temperature with depth at the top of the oceanic mantle, result in a sub-Moho negative acoustic wave velocity gradient. This is contrary to the observed facts. It must mean that the sub-Moho lithosphere has a chemical or mineralogical composition which varies with depth. The model presented here has only negative sub-Moho acoustic wave velocity gradients for young lithosphere. When our model lithosphere cools with age and attains the upper basalt—eclogite transition temperature, the basaltic component occurring interstitially in the harzburgite just below Moho transforms progressively to eclogitic material. This causes an increase in seismic velocities large enough to offset the negative temperature effect.

Our choice of the initial chemical composition of the upwelling mantle material of 85% peridotite and 15% tholeiite was influenced by various considerations. In the process of upwelling, partial melting due to pressure decrease, should produce enough basaltic liquid with a MORB composition to construct oceanic layers 2 and 3a. V_s and V_p of our model lithosphere should be similar to the actual observations. The initial composition of 75% peridotite and 25% basalt, used by Bottinga (1974) and Bottinga and Allègre (1976) resulted in somewhat too low seismic velocities. Previously, too much mantle trapped liquid was produced, causing topographic effects which were not observed in reality. The present initial composition has as consequence that a smaller quantity of liquid is produced and thus also a smaller quantity of liquid is trapped in the upper mantle.

5.3. *Primary flow field and plate thickness*

Our present model has a primary flow field different from previous models (see Fig. 1). The flow field has an influence on the heat flow and the seabottom topography of the young lithosphere and on the distribution in space of liquid produced during upwelling. Our previous flow field (Bottinga, 1974; Bottinga and Allègre, 1976) caused a too high heat flow for plate ages less than 30 Ma, as is the case for the models of Parker and Oldenburg (1973), Oldenburg (1975) and Crough (1975). Analogous effects are seen in the computed seabottom depth. By restricting the width over which upwelling occurs, we restricted the region containing liquid produced by partial fusion and we obtained a model surface heat flow and oceanic bottom depth, which are conform to the observations compiled by Parsons and Sclater (1977). Of course one cannot narrow down the width over which upwelling takes place too much, because then the upwelling material has to flow at speeds which are rheologically unreasonable.

Plate thickness in the present model was taken to be as derived by Parsons and Sclater (1977) from thermal considerations. Rheological, seismological and thermal plate thickness should be distinguished. Rheologically, a plate thickness of 126 km is unreasonable (Tozer, 1972; Kohlstedt et al., 1976; Bottinga and Allègre, 1976). Laboratory (Auten et al., 1976) and theoretical evidence (O'Connell and Budiansky, 1977) indicates that rheological and seismological plate thicknesses can be quite different, because the cause for one does not have to affect the existence of the other; see also Goetze (1977) and Section 4.2.

We think that the thermally determined plate thickness of Parsons and Sclater (1977) is a measure of the depth over which vertical heat transport is predominantly conductive. In other words, flow in the upper mantle to a depth of about 130 km (the Parsons and Sclater plate thickness) is predominantly horizontal, but not necessarily uniform. Such a medium can be modeled as being made up of horizontal layers, each layer moving with its own horizontal velocity and heat transfer taking place vertically across the layers. Steady state vertical heat transport in this medium is equivalent to conductive heat transport across a composite slab of horizontal layers, each layer having its own thermal properties, provided that horizontal conductive heat transport is insignificant in comparison with horizontal convective heat transfer. As was shown by Carslaw and Jaeger (1959, section 12.8) the heat

flow out the top of a composite halfspace has a reciprocal square root time dependence. Therefore we think that while the Parsons and Sclater (1977) model reproduces almost perfectly the observations, it does not prove the physical reality of a 130 km thick rigid plate.

5.4. *Conclusions*

When the model is compared with the observations one notices that:

(1) Observed temperatures and temperature gradients are like the corresponding model values.

(2) Model heat flow agrees with the observed heat flow.

(3) The mineralogical composition and its variation with depth in the upper mantle, inferred from the ophiolite model and from outcrops like St. Paul's Rocks, correspond well to predictions of the model.

(4) The degree of partial fusion, deduced geochemically for MORB, agrees with the model calculated value.

(5) Petrologically and geochemically inferred depth of the MORB source region is in agreement with the model calculation.

(6) The quantity of basalt produced at the ridge axis is like in the model.

(7) The average density in the lithosphere tends to be lower than suggested by certain seismic observations.

(8) Calculated sea bottom topography agrees with observations compiled by Parsons and Sclater (1977).

(9) S-delay times reported by Duschenes and Solomon (1977) for the marine lithosphere are reasonably well reproduced by the model.

(10) Calculated and observed shear wave velocity distributions for the oceanic lithosphere are similar.

(11) Calculated and observed P-wave velocity distributions agree also reasonably well, except at depth below 50 km where most observations show an increase in V_p not predicted by the model and not observed for V_s.

(12) The model partial melt distribution in the upper mantle close to the ridge axis is in agreement with observed seismic wave attenuation, rotation of nodal planes for earthquake source mechanisms and inferred shallow level of isostatic compensation (Bowin and McKenzie, 1976).

(13) The model gives explanations for acoustic wave velocity diminution close to the ridge axis, generally known as anomalous mantle.

(14) The observed thickening of layer 3 is a possible but not necessary, consequence of processes taking place in the ridge axial region.

(15) It is suggested that the lower lithosphere is seismically anisotropic (see also Bottinga et al., 1976; Hirn, 1977). The orientation of this anisotropy is different from the well known shallow sub-Moho anisotropy.

The described model is to a large extent free of adjustable parameters, the properties listed in Table I were taken from the literature. Arbitrary aspects of the model are the flow field, the mantle porosity distribution, seawater interaction with the oceanic lithosphere at the ridge axis, and the width of the upper mantle zone below the ridge crest from which basaltic liquid can escape to erupt at the ridge axis. Features which are not yet quantitatively understood are the thermally determined plate thickness and its relation to the rheological plate thickness. These two thicknesses are not completely independent of each other but are probably related via the compensation depth, used in the isostatic calculations for the seabottom topography.

ACKNOWLEDGEMENTS

We would like to acknowledge the discussions with our collegues A. Hirn, K. Lambeck, and N. Shimizu of I.P.G., Paris, and R.J. O'Connell and A. Dziewonski at Harvard. In particular we thank B. Parsons and J.G. Sclater of M.I.T., who permitted us kindly the usage of two of their diagrams. Y.B. wishes also to acknowledge his stay at the department of Geological Sciences, Harvard University, during which most of this contribution was written.

REFERENCES

Ade-Hall, J.M., Aumento, F., Muecke, G.K., McDonald, A., Hyndman, R.D., Quinto, J. and Lowrie, W., 1974. Deep drilling on Sao Miguel, Azores: preliminary results. EOS, Trans. Am. Geophys. Union, 55: 454 (abstr.).

Ahrens, T.J., 1973. Petrological properties of the upper 670 km of the earth's mantle; geophysical implications. Phys. Earth Planet. Inter., 7: 167—186.

Allègre, C.J. and Bottinga, Y., 1974. Tholeiite, alkali basalt and ascent velocity. Nature, 252: 31—32.

Allègre, C.J., Montigny, R. and Bottinga, Y., 1973. Cortège ophiolitique et cortège océanique, géochimie comparée et mode de genèse. Bull. Soc. Géol. Fr., XV: 461—478.

Allègre, C.J., Treuil, M., Minster, J.F., Minster, J.B. and Albarède, F., 1977. Systematic use of trace elements in igneous processes, Part I. Contrib. Mineral. Petrol., 60: 57—75.

Asada, T. and Shimamura, H., 1976. Observation of earthquakes and explosions at the bottom of the western Pacific: structure of oceanic lithosphere revealed by longshot experiment. Geophys. Monogr., Am. Geophys. Union, 19: 135—154.

Auten, T.A., Gordon, R.B. and Stoker, R.L., 1974. Q and creep in the mantle. Nature, 250: 317—318.

Birch, F., 1966. Compressibility; elastic constants. In: S.P. Clark, Jr. (Editor), Handbook of Physical Constants. Geol. Soc. Am., Mem., 97: 107—173.

Bottinga, Y., 1974. Thermal aspects of seafloor spreading, and the nature of the suboceanic lithosphere. Tectonophysics, 21: 15—38.

Bottinga, Y. and Allègre, C.J., 1973. Thermal aspects of seafloor spreading and the nature of the oceanic crust. Tectonophysics, 18: 1—17.

Bottinga, Y. and Allègre, C.J., 1976. Geophysical, petrological and geochemical models of the oceanic lithosphere. Tectonophysics, 32: 9—59.

Bottinga, Y. and Allègre, C.J., 1978. Partial melting under spreading ridges. Philos. Trans. R. Soc. London, 288: 501—525.

Bottinga, Y. and Weill, D.F., 1970. Densities of liquid silicate systems calculated from partial molar volumes of oxide components. Am. J. Sci., 269: 169—182.

Bottinga, Y. and Weill, D.F., 1972. The viscosity of magnetic silicate liquids: a model for calculation. Am. J. Sci., 272: 438—475.

Bottinga, Y., Steinmetz, L. and Allègre, C.J., 1976. Acoustic wave velocity anisotropy in the oceanic lithosphere. Bull. Soc. Géol. Fr., XVIII: 941—947.

Bowin, C. and McKenzie, D.P., 1976. The relationship between bathymetry and gravity in the Atlantic ocean. J. Geophys. Res., 81: 1903—1915.

Carslaw, H.S. and Jaeger, J.C., 1959. Conduction of Heat in Solids. Clarendon Press, Oxford.

Christensen, N.I., 1972a. Compressional and shear wave velocities at pressures to 10 kb for basalt from the east Pacific rise. Geophys. J., 28: 425—429.

Christensen, N.I., 1972b. The abundance of serpentinites in the oceanic crust. J. Geol., 80: 709—719.

Christensen, N.I., 1974. Compressional wave velocities in possible mantle rocks to pressures of 30 kb. J. Geophys. Res., 79: 407—412.

Christensen, N.I., 1975. Ultrasonic velocities in minerals and rocks at high pressures and temperatures. Geol. Soc. Am., Annu. Meet., p. 1026 (abstr.).

Christensen, N.I. and Salisbury, M.H., 1975. Structure and constitution of the lower oceanic crust. Rev. Geophys. Space Phys., 13: 57—86.

Cohen, L.H., Ito, K. and Kennedy, G.C., 1967. Melting and phase relations in anhydrous basalt up to 40 kb. Am. J. Sci., 265: 475—518.

Crough, S.T., 1975. Thermal model of the oceanic lithosphere. Nature, 256: 388—390.

Davis, E.E. and Lister, C.R.B., 1974. Fundamentals of ridge topography. Earth Planet. Sci. Lett., 22: 60—66.

Detrick, R.S., Williams, D.L., Mudie, J.D. and Sclater, J.G., 1974. The Galapagos spreading centre: bottom water temperatures and the significance of geothermal heating. Geophys. J., 38: 627—637.

Duschenes, J.D. and Solomon, S.C., 1977. Shear wave travel time residuals from oceanic earthquakes and the evolution of the oceanic lithosphere. J. Geophys. Res., 82: 1985—2000.

Forsyth, D.W., 1975. The early structural evolution and anisotropy of the oceanic upper mantle. Geophys. J., 43: 103—162.

Forsyth, D.W., 1977. The evolution of the upper mantle beneath mid-oceanic ridges. Tectonophysics, 38: 89—118.

Frank, F.C., 1968. Two component flow model for convection in the earth's upper mantle. Nature, 220: 350—352.

Frey, F.A. and Green, D.H., 1974. The mineralogy, geochemistry and origin of lherzolite inclusions in Victorian basanites. Geochim. Cosmochim. Acta, 38: 1023—1060.

Frisilio, A.L. and Barsch, G.R., 1972. Measurement of single crystal elastic constants of bronzite as a function of pressure and temperature. J. Geophys. Res., 77: 6360—6384.

Girardin, N. and Jacoby, W.R., 1979. Rayleigh wave dispersion along the Reykjanes Ridge. In: J. Francheteau (Editor), Processes at Mid-Ocean Ridges. Tectonophysics, 55: 155—171.

Goetze, C., 1977. A brief summary of our present day understanding of the effect of volatiles and partial melt on the mechanical properties of the upper mantle. In: M.H. Manghnani (Editor), High pressure Research: Applications to Geophysics. Academic Press, New York, N.Y., pp. 3—23.

Goslin, J., Beuzart, P., Francheteau, J. and Le Pichon, X., 1972. Thickening of the oceanic layer in the Pacific ocean. Mar. Geophys. Res., 1: 418—427.

Green, D.H. and Lieberman, R.C., 1976. Phase equilibria and elastic properties of a pyrolite model for the oceanic upper mantle. Tectonophysics, 32: 61—92.

Green, D.H. and Ringwood, A.E., 1970. Mineralogy of peridotite compositions under upper mantle conditions. Phys. Earth Planet. Inter., 3: 359—371.

Lister, C.R.B., 1974. On the penetration of water into hot rock. Geophys. J., 39: 465–509.

Mantovani, E., Schwab, F., Liao, H. and Knopoff, L., 1977. Interpretation of oceanic Sn. (Preprint.)

Melson, W.G., Hart, S.R. and Thompson, G., 1972. St. Paul's Rocks, Equatorial Atlantic: petrogenesis, radiometric ages, and implications on seafloor spreading. Geol. Soc. Am., Mem., 132: 241–271.

Minster, J.F., Minster, J.B., Treuil, M. and Allègre, C.J., 1977. Systematic use of trace elements in igneous processes. Part 2. Contrib. Mineral. Petrol., 61: 49–77.

Murase, T. and McBirney, A.R., 1973. Properties of some common rocks and their melts at high temperatures. Geol. Soc. Am. Bull., 84: 3563–3592.

Neprochnov, Yu. and Rykunov, L.N., 1970. Experimental data on the high velocity layer in the upper mantle. Dokl. Akad. Nauk SSSR, 194: 80–82.

Obata, M., 1976. The solubility of Al_2O_3 in orthopyroxenes and plagioclase peridotites and spinel pyroxenite. Am. Mineral., 61: 804–816.

O'Connell, R.J. and Budiansky, B., 1977. Viscoelastic properties of fluid saturated cracked solids. (Preprint.)

O'Hara, M.J., Richardson, S.W. and Wilson, G., 1971. Garnet peridotite stability and occurrence in crust and mantle. Contrib. Mineral. Petrol., 32: 48 –68.

Oldenburg, D.W., 1975. A physical model for the creation of the lithosphere. Geophys. J., 43: 425–451.

Orcutt, J.A. and Dorman, L.M., 1977. An oceanic long range explosion experiment. J. Geophys., 43: 257–263.

Palmason, G., 1967. On heat flow in Iceland in relation to the Mid-Atlantic Ridge. In: S. Björnsson (Editor), Iceland and Mid-Oceanic Ridges. Soc. Sci. Islandica, Publ., 38: 111–127.

Parker, R. and Oldenburg, D.W., 1973. Thermal model of ocean ridges. Nature, 242: 137–139.

Parsons, B. and Sclater, J.G., 1977. An analysis of the variation of ocean floor heat flow and bathymetry with age. J. Geophys. Res., 82: 803–827.

Ringwood, A.E., 1966. Mineralogy of the mantle. In: P.M. Hurley (Editor), Advances in Earth Science. Mass. Inst. Technol. Press, Cambridge, Mass., pp. 357–399.

Ringwood, A.E., 1975. Composition and Petrology of the Earth's Mantle. McGraw-Hill, New York, N.Y., 672 pp.

Schatz, J.F. and Simmons, G., 1972. Thermal conductivity of earth materials at high temperatures. J. Geophys. Res., 77: 6966–6983.

Sclater, J.G. and Francheteau, J., 1970. The implications of terrestrial heat flow observations on current tectonic and geochemical models of the crust and upper mantle of the earth. Geophys. J., 20: 509–542.

Skinner, B.F., 1966. Thermal expansion. In: S.P. Clark (Editor), Handbook of Physical Constants. Geol. Soc. Am. Mem., 97: 78–96.

Soga, N., 1967. Elastic constants of garnet under pressure and temperature. J. Geophys. Res., 77: 4938–4944.

Solomon, S.C. and Julian, B.R., 1974. Seismic constraints on ocean ridge mantle structure: anomalous fault plane solutions from first motions. Geophys. J., 38: 265–285.

Spetzler, H. and Anderson, D.L., 1968. The effect of temperature and partial melting on velocity and attenuation in a simple binary system. J. Geophys. Res., 73: 6051–6060.

Steinmetz, L., 1977. Sondages sismiques dans le Manteau supérieur. Disccusion d'un Modèle de Dorsale océanique. Thesis, University of Paris.

Steinmetz, L. and Hirn, A., 1973. A long range seismic experiment in the oceanic domain: Madeira to Canary Islands. 1st Annu. Meet. European Geophys. Soc., Zürich, p. 99 (abstr.).

Steinmetz, L., Whitmarsh, R.B. and Moreira, V.S., 1977. Upper mantle structure beneath the Mid-Atlantic Ridge, north of the Azores based on observations of compressional waves. Geophys. J., 50: 353–380.

Hart, R.S. and Press, F., 1973. Sn velocities and the composition of the lithosphere in the regionalized Atlantic. J. Geophys. Res., 78: 407—411.

Hermance, J.F. and Grillot, L.R., 1974. Constraints on temperatures beneath Iceland from magnetotelluric data. Phys. Earth Planet. Inter., 8: 1—12.

Herzberg, C.T. and Chapman, N.A., 1976. Clinopyroxene geothermometry of spinel lherzolite. Am. Mineral., 61: 626—637.

Hess, H.H., 1962. History of the ocean basins. In: A.E.J. Engel et al. (Editors), Petrological Studies. Buddington Memorial Volume. Geol. Soc. Am., Boulder, Colo., pp. 599—620.

Hess, H.H., 1964. Seismic anisotropy of the upper mantle under oceans. Nature, 203: 629—631.

Hirn, A., 1977. Anisotropy in the continental upper mantle: possible evidence from explosion seismology. Geophys. J., 49: 49—58.

Hyndman, R.D., Aumento, F., Melson, W.G., Hall, D.C.J.M., Bougault, H., Dimitriev, L., Fischer, J.F., Flower, M., Howe, R.C., Miles, G.A., Robinson, P.T. and Wright, T.L., 1976. Seismic structure of the oceanic crust from deep drilling on the Mid-Atlantic Ridge. Geophys. Res. Lett., 3: 201—204.

Ito, K. and Kennedy, G.C., 1967. Melting and phase relations in a natural peridotite to 40 kb. Am. J. Sci., 265: 211—217.

Jacoby, W. and Girardin, N., 1977. The development of the lithosphere at Reykjanes Ridge, based on Rayleigh wave dispersion. (Preprint).

Jehl, V., 1975. Le Métamorphisme et les Fluides associés des Roches océaniques de l'Atlantique. Thesis, University of Nancy.

Joron, J.L., Bougault, H., Treuil, M. and Allègre, C.J., 1976. Etude géochimique des roches magmatiques de la zone FAMOUS et de l'archipel des Açores. Bull. Soc. Géol. Fr., XVIII: 811—818.

Juteau, T., 1974. Les Ophiolites des Nappes d'Antalya, Pétrologie d'un Fragment de l'ancienne Croûte océanique Téthysienne. Thesis, University of Nancy.

Kawada, K., 1966. Studies of thermal states of the earth. Part 17. Bull. Earthquake Res. Inst., Tokyo Univ., 44: 1071—1091.

Kelley, K.K., 1960. Contributions to the data of theoretical metallurgy. U.S. Bur. Mines, Bull., 584.

Kohlstedt, D.L., Goetze, C. and Durham, W.B., 1976. Experimental deformation of single crystal olivine with application to flow in the mantle. In: R.G.J. Strens (Editor), The Physics and Chemistry of Minerals and Rocks. Wiley, New York, N.Y., pp. 35—49.

Kumazawa, M. and Anderson, O.L., 1969. Elastic moduli, pressure derivatives and temperature derivatives of single crystal olivine and single crystal forsterite. J. Geophys. Res., 74: 5961—5972.

Kushiro, I., Syono, Y. and Akimoto, S., 1968. Melting of a peridotite at high pressure and water pressures. J. Geophys. Res., 73: 6023—6029.

Kushiro, I., Yoder, H.S. and Mysen, B.O., 1976. Viscosity of basalt and andesite melts at high pressure. J. Geophys. Res., 81: 6351—6356.

Lambeck, K., 1977. Thermal contraction and ocean ridge topography. (Preprint.)

Langseth Jr., M.G., Le Pichon, X. and Ewing, M., 1966. Crustal structure of mid-ocean ridges. J. Geophys. Res., 71: 5321—5355.

Lee, W.H.K. and Uyeda, S., 1965. Review of heat flow data. In: W.H.K. Lee (Editor), Terrestrial Heat Flow. Geophys. Monogr., Am. Geophys. Union, 8: 87—190.

Le Pichon, X., 1969. Models and structure of the oceanic crust. Tectonophysics, 7: 385—401.

Le Pichon, X., Francheteau, J. and Bonnin, J., 1973. Plate Tectonics. Elsevier, Amsterdam, 300 pp.

Lewis, B.T.R. and Snydsman, W.E., 1977: Structure of the crust and upper mantle of the northern Cocos plate, 0—12 m.y. (Preprint.)

Stephens, Ch. and Isacks, B.L., 1977. Toward understanding Sn: normal modes of Love waves in an oceanic structure. Bull. Seismol. Soc. Am., 67: 69—78.

Tapponnier, P. and Francheteau, J., 1977. Necking of the lithosphere and mechanics of slowly accreting plate boundaries. (Preprint.)

Tatsumoto, M., Hedge, C.E. and Engel, A.E.J., 1965. K, Rb, Sr, Th, U and the ratio of $^{87}Sr/^{86}Sr$ in oceanic tholeiitic basalts. Science, 150: 886—888.

Tozer, D.C., 1972. The present thermal state of the terrestrial planets. Phys. Earth Planet. Inter., 6: 182—197.

Tréhu, A.M., 1975. Depth versus $(age)^{1/2}$: A perspective on mid-ocean ridges. Earth Planet. Sci. Lett., 27: 287—304.

Waff, H.S., 1975. Pressure induced coordination changes in magmatic liquids. Geophys. Res. Lett., 2: 193—196.

Weidner, D.J., 1974. Rayleigh wave phase velocities in the Atlantic Ocean. Geophys. J., 36: 105—139.

Wenner, D.B. and Taylor Jr., H.P., 1971. Temperatures of serpentinization of ultra mafic rocks based on $^{18}O/^{16}O$ fractionation between coexisting serpentine and magnetite. Contrib. Mineral. Petrol., 32: 165—185.

Williams, D.L., Von Herzen, R.P., Sclater, J.G. and Anderson, R.N., 1974. The Galapagos spreading centre: lithospheric cooling and hydrothermal circulation. Geophys. J., 38: 587—608.

Woollard, G.P., 1975. The interrelationships of crustal and upper-mantle parameter values in the Pacific. Rev. Geophys. Space Phys., 13: 87—137.

Chapter 5

FINE STRUCTURE OF THE LOWER OCEANIC CRUST ON THE COCOS PLATE*

BRIAN T.R. LEWIS and W.E. SNYDSMAN

Department of Oceanography and Geophysics Program, University of Washington, Seattle, Wash. 98195 (U.S.A.)

SUMMARY

From seismic refraction data on the Northern Cocos Plate, it is found that the crust thickens with age. Assuming that processes of crustal formation have remained constant over at least the past 10 m.y., the seismic data indicate that the thickening is caused by a gradual transformation of the top 2 km of the upper mantle into material having crustal-like velocities. Serpentinization is a possible mechanism. The data also suggest that the crust—mantle interface may be laterally variable, in some places being relatively sharp and in others gradational. Upper mantle anisotropy appears to decrease from about 0.6 km/sec near the rise axis to 0.3 km/sec at 10 m.y.

INTRODUCTION

Syntheses of seismic refraction data in the Pacific suggest that the lower oceanic crust thickens by about 2 km with increasing age (Goslin et al., 1972; Christensen and Salisbury, 1975). This result was obtained from an analysis of published values of "layer 3" thicknesses versus age. This effect can be seen clearly in a few individual experiments crossing the East Pacific Rise, possibly the best example being that of Shor et al. (1968) across the Juan de Fuca Ridge. This is a rather strange result since it implies that some process is operating in the lower crust or upper mantle that is systematically creating the effect of making the acoustically defined crust thicken with age.

This process is not the result of a lowering of upper crustal velocities since these velocities have been shown to increase with age (Houtz and Ewing, 1976). This increase in upper crustal velocity with age explains the apparent thinning of layer 2 (Christensen and Salisbury, 1975) and if no other processes were operating this might cause an apparent thinning of the crust. Changes of "layer 3" velocities with age also do not explain the apparent

---

*Originally published as: Lewis, B.T.R. and Snydsman, W.E., 1979. Fine structure of the lower oceanic crust on the Cocos Plate. In: J. Francheteau (Editor), Processes at Mid-Ocean Ridges. Tectonophysics, 55: 87—105.

thickening since Bibee and Shor (1976) have shown that "layer 3" velocities (about 6.9 km/sec) remain relatively unchanged with age. These results imply that the thickening must involve a process in the lowermost section of the oceanic crust or the uppermost mantle.

Lewis and Snydsman (1977) suggested that the thickening could be due to a transformation of upper mantle rocks into rocks having velocities close to or less than "layer 3" velocities, thereby acoustically thickening the lower crust. In the 1977 paper, we presented data from three seismic refraction lines on the Cocos Plate and suggested that the data indicated the development of a low velocity zone at the base of the crust. In this paper, we will present additional data from the Cocos Plate supporting this hypothesis together with an analysis of the changes in upper mantle velocity with age and a look at the evidence for layering in the upper crust.

THE COCOS PLATE SEISMIC EXPERIMENT

This experiment was conducted in 1973 and 1974 on the Cocos Plate between the Orozco and Clipperton fracture zones and extended from the East Pacific Rise to the Middle America Trench. In addition to bathymetric and magnetic data, 40 refraction profiles were shot (see Fig. 1 for locations). An analysis of magnetic data by Lynn and Lewis (1976) shows that the magnetic anomalies fan away from the rise axis. The average half-spreading rate was found to be about 4.4 cm/yr. Isochrons determined from the magnetic anomalies are shown in Fig. 2.

Most of the seismic data were obtained on five long-range telemetering buoys built at the University of Washington. These buoys were generally

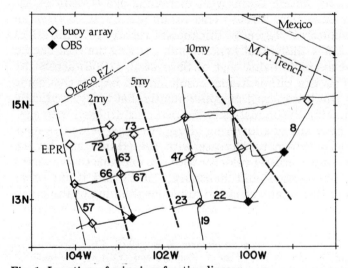

Fig. 1. Location of seismic refraction lines.

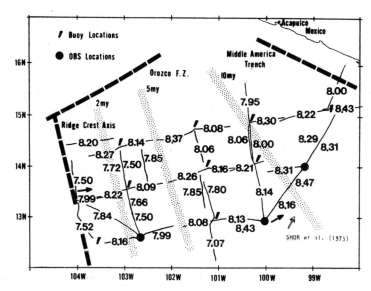

Fig. 2. Upper mantle velocities determined in this experiment together with isochrons from magnetic anomalies.

deployed as a free-floating array about 2 km long to provide apparent velocity data. Two University of Washington ocean bottom seismometers (Lister and Lewis, 1976) were deployed twice, first near the trench and later near the rise axis. Repeated attempts to deploy one on the axis failed because the concrete anchor broke on impacting the rocky bottom. Deficiencies in the tape modulation system (since rectified) prevented digitization of these records. However, first arrivals were picked from analog records.

The quality of the buoy records varies considerably, depending in part on surface currents and possibly on background seismic activity. In this paper only those data are presented which have good signal-to-noise ratios and low drift of the buoy array (less than 2 km over the period of the shooting). The numbered lines in Fig. 1 represent these data.

RESULTS

In order to determine the velocity structure of the lower oceanic crust from refraction data, it is necessary to determine the upper crustal velocities. These will be discussed briefly. Since changes in the bottom of the oceanic crust may be related to changes at the top of the mantle, the age dependence of the upper mantle velocities will also be discussed. Following these sections results relevant to lower crustal velocities and crustal thickening are presented.

Upper crustal velocities

Traditionally, marine refraction profiles have been interpreted by fitting the first arrivals with a small number of straight lines (usually two or three) and computing layer velocities and depths by using the slope-intercept method. This has resulted in the concept of a layered oceanic crust consisting of layers 1, 2 and 3. The validity of this approach may be tested by examining the behavior of amplitudes and array apparent velocities. Applications of amplitudes are given by Helmberger (1977), Orcutt et al. (1975) and Kennett (1977). They generally find that velocity gradients give an improved fit to the upper crustal amplitude data when compared to models having constant velocity layers. Apparent velocities across an array (or $dT/d\Delta$) can be used to detect rapid changes in velocity with depth since these manifest themselves as discontinuities in the slope of the travel time curve.

Measurements of apparent velocity across the 2 km long buoy array for lines 57, 67 and 73 (see Fig. 1) are shown in Fig. 3. These data strongly suggest that the velocity in the upper crust increases gradationally rather than discontinuously, and therefore that the concept of a "layered" crust may be an artifact of the method of interpretation. The data do not suggest the presence of any well-defined first order discontinuities in the upper crust. The data for lines 67 and 73 contain jumps in apparent velocity correspond-

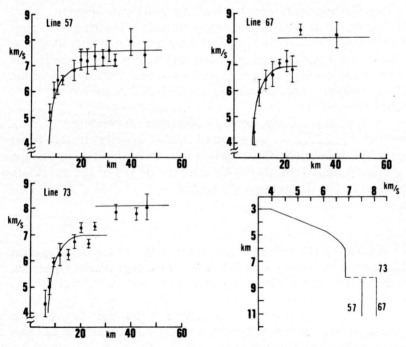

Fig. 3. Apparent velocities across 2 km array of buoys for lines 57, 67 and 73.

ing to the change from crust to mantle refractions indicating that this velocity change is sufficiently rapid to cause a triplication in the travel time curve. However, the data from line 57 (on the rise axis) indicate a more gradual crust mantle transition.

Change in upper mantle velocities with age

Most of the upper mantle velocities (P_n) determined from this experiment have been reported by Snydsman et al. (1975). However, no attempt was made in this paper to quantitatively describe the changes due to both anisotropy and age.

The most general form of small anisotropy can be described approximately by a function of the form (Backus, 1965):

$$\Delta V(\phi) = A \sin(2\phi) + B \cos(2\phi) + C \sin(4\phi) + D \cos(4\phi)$$

where ΔV is the departure from the mean velocity, ϕ is an angle with respect to some reference, usually north, and A, B, C and D are coefficients to be determined. Morris et al. (1969) and Raitt et al. (1969) have found the 4ϕ terms to be small and Shor et al. (1970) have shown that the 2ϕ terms are generally oriented with respect to magnetic anomalies, with the maximum velocity being perpendicular to the strike of the anomalies. In this experiment we do not have sufficient azimuthal coverage to directly determine the principal velocity directions. However, the velocities in Fig. 2 generally show high velocities perpendicular to the magnetic anomalies and low velocities parallel to the strike. A Scripps anisotropy station (Shor et al., 1973) is located near the east end of line 22 and shows the maximum velocity to be roughly perpendicular to isochrons (see Fig. 2).

If we now assume that the high velocity is normal to the strike of the magnetic anomalies for the whole plate, then we can determine the amount of anisotropy by solving the equation $(V - \overline{V}) = A \cos 2(\lambda - \theta + \pi/2)$, where V is the observed velocity: $\overline{V}$ is the mean velocity; A is the amplitude of anisotropy; λ is the azimuth of the line; θ is the azimuth of the isochron, and we have velocities in at least two directions.

This method was applied to the data in Fig. 2 by using all the data in a few age bands (0—1, 1—4, 4—8, 8—12 m.y.) to estimate $\overline{V}$ and A and the variances of these quantities. The results are shown in Fig. 4 by plotting the anisotropy *(A)* and the maximum and minimum velocities as a function of age. From these data, it appears that the anisotropy near the rise axis is almost twice as large as it is at 10 m.y. (0.6 km/sec compared to 0.3 km/sec). The change in anisotropy appears to occur mostly by an increase in the velocity parallel to the isochrons.

Thickening of the crust and lower crustal velocities

Possibly the most compelling evidence that the crust on the Cocos Plate is thickening with age comes from the P_n intercept times. This is because the

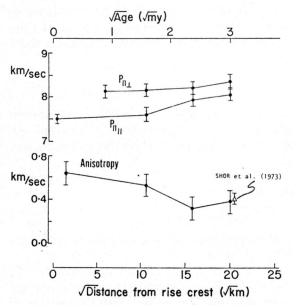

Fig. 4. Variation of upper mantle anisotropy with age.

intercept times are obtained with a minimum of interpretation. The P_n intercept times from this experiment are shown in Fig. 5 (from Lewis and Snydsman, 1977). The travel time through the water has been removed from the intercept times in order to show more clearly the increase with age. On nearly all lines, the sediment thickness is zero or very small and therefore no sediment corrections were made. The intercept times increase from about 1.0 sec at age zero to 1.5 sec at 10 m.y. About 0.1 sec of this amount is probably due to the increase of P_n velocity with age leaving 0.4 sec unexplained.

To demonstrate that the increase in intercept time is not the result of a decrease in crustal velocity, an average crustal velocity $\overline{V}_c$ was obtained by fitting the observed travel times with a simple layered model and computing $(\sum_i V_i\, h_i)/(\sum_i h_i)$ (after Woollard, 1975). Fig. 5 shows that the increasing intercept times represent real changes in crustal thickness.

The $\overline{V}_c$ data coupled with more detailed crustal velocity models (see Fig. 20) strongly suggest that the apparent crustal thickening is occurring at the crust—mantle boundary. The determination of the detailed velocity structure at this interface is complicated because, in general, the information about this interface is contained in secondary arrivals of the refraction seismogram.

We have used the best data from this experiment to investigate the velocity behavior in the lower crust. To interpret the refraction data, we first constructed synthetic seismograms for several types of models using the Fuchs and Müller (1971) reflectivity method. This allowed us to deter-

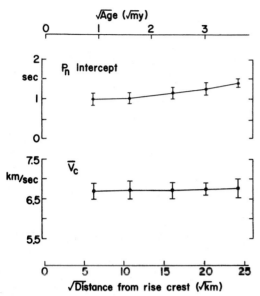

Fig. 5. Variation of P_n intercept times and average crustal velocity with age. Intercept times have the travel time in the water subtracted.

mine the effects on the seismogram of layering, low velocity zones, gradients in the upper crust and a transition zone between crust and mantle. Fig. 6 shows the comparison of synthetic seismograms from four types of models. Model A corresponds to a layered crust with a few thick layers corresponding to layer 2, 3a and 3b. In model B, we have approximated a linear velocity gradient in the upper crust with many thin layers. The selection of the gradient in this model was based on the model derived from the apparent velocity data in Fig. 3. Model C has the same upper crust as model B and a low velocity zone at the base of the crust. Model D is similar to B in the upper crust, but has a transition zone between the lower crust and upper mantle and a thinner crust than B. The amplitudes in the synthetic seismograms have all been normalized by using a distance-dependent multiplier $(R/R_0)^{1.5}$ (R = distance, R_0 = distance at 10 km). The synthetics were computed for phase velocities between 4 and 10 km/sec and a source function having one cycle of a 5 Hz sine wave.

From an inspection of these synthetic seismograms, we note the following. The differences between the first arrival travel times for models A, B and C are less than a few hundredths of a second and, therefore, it is impossible to distinguish between these models on the basis of first arrival travel times. The difference between these models is most apparent in the second arrivals, particularly in the arrival times of the wide angle reflections off the crust—mantle boundary (P_mP). At large ranges, the apparent velocity of the P_mP branch is asymptotic to the highest apparent velocity of P_g and, there-

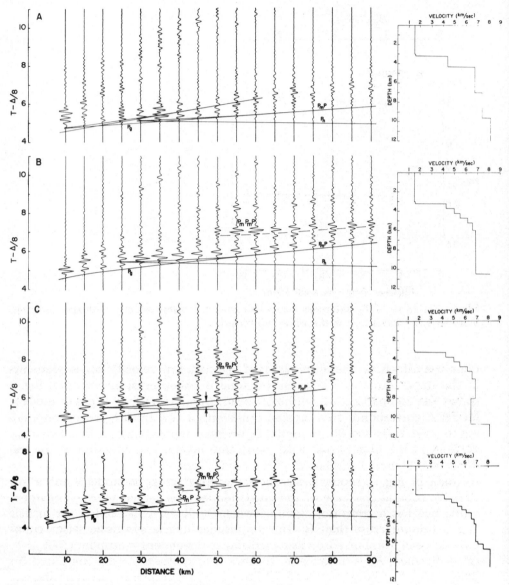

Fig. 6. Synthetic seismograms for four types of models.

fore, to the highest velocity zone in the crust having a thickness greater than a few wavelengths. Thus, the slope of the P_mP branch at large ranges is diagnostic of the difference between the lower crust of models A and B. The difference between models without and with a low velocity zone at the base of the crust (models B and C) is the offset between P_mP and P_g. Helmberger and Engen (1978) have shown that downdip of the Moho can produce a similar effect whereas updip produces the opposite effect, that is, the P_mP branch will arrive early, showing that one has to use this result with care.

Models B and C in Fig. 6 show another large amplitude coherent late branch that arrives after P_mP. We have labeled this branch P_mP_mP because the travel times of these arrivals correspond to internal crustal reflections of P_mP. The ray paths for this branch are shown schematically in Fig. 7. This branch starts at twice the distance of the P_mP, P_n critical point and mimics P_mP. The parameters that control the amplitude of this phase are not yet clear to us.

Model D in Fig. 6 has a 1 km thick transition zone between the lower crust and the mantle. The crust in this model is about 2 km thinner than Models A, B and C and produces synthetic seismograms similar to data observed near the rise axis (see, for example, line 73, Fig. 13). This model does not produce a clearly identifiable P_mP branch at large ranges, but does produce a clear P_mP_mP branch similar to the data in Fig. 13.

There are several other model parameters which influence the character of the observed seismograms. In the crustal low velocity zone model, the amplitude of P_g can be considerably reduced by making the thickness of the layer above the low velocity zone small. This can cause an apparent disappearance of P_g before the P_n crossover. Another parameter which affects the character of the seismograms is the P_n velocity. As the velocity contrast between the crust and mantle is reduced, the P_mP, P_n critical point and the P_g, P_n crossover distance move to greater ranges. Data illustrating both these effects and the effect of distance from the rise axis are shown in Fig. 8. This figure shows an expanded view of the first few cycles of data from lines 66, 63 and 22. The midpoints of these lines are about 60, 110 and 340 km from the rise axis and the data have been corrected for shot size and distance. Line 63 is parallel and lines 66 and 22 are perpendicular to the rise axis. Line 66 shows an orderly change in velocity and amplitude between P_g and P_n which is similar to model D in Fig. 6. Line 22 shows a rapid decrease in amplitude and apparent disappearance of P_g beyond about 22 km and a clear P_mP, P_n

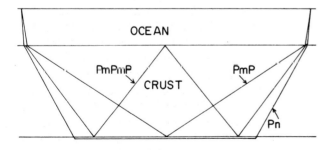

Fig. 7. Schematic diagram of the P_n, P_mP and P_mP_mP ray paths.

116

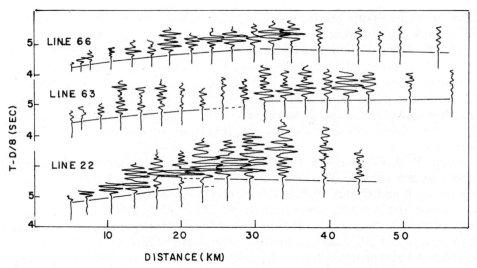

Fig. 8. Comparison of first arrivals from three lines at increasing distance from the rise axis. Line 22 is most distant. Note the apparent disappearance of P_g on line 22 compared to strong P_g on line 66.

critical point near 20 km. This rapid decrease in the amplitude of P_g can be accounted for by a thin layer overlying a low velocity zone. Line 63 also appears to show an apparent disappearance of P_g, but not as marked as line 22. The high amplitudes associated with the P_n critical point are beyond about 30 km for line 63 as compared with 20 km for line 22. This effect is mostly due to the low P_n velocity of line 63 compared to the higher velocity of line 22.

As can be seen from the synthetic seismograms in Fig. 6 and the data in Fig. 8, a relatively large number of parameters need to be adjusted to fit any one line of data. By comparing models obtained by ray tracing with the synthetic seismograms, we found that a fairly accurate prediction of the synthetic seismograms could be made by using amplitude information contained in the ray tracing technique. In our ray tracing program, we increment the ray parameter p in uniform steps and use $dp/d\Delta$ (where Δ is the range and $dp/d\Delta$ is an estimate of the number of rays emerging at the surface per unit range) to determine relative amplitudes. This is calculated for all branches of the travel time curve. This technique can be used iteratively and with great speed to arrive at a model which fits the data. In those cases where the data suggest the presence of a low velocity zone, either by an offset of P_mP with respect to P_g or the disappearance of P_g before the crossover point, an assumption has to be made about either the velocity in or the thickness of the low velocity zone. In the models to be presented, we constrained the velocity in the low velocity zone to be between 6.5 and 6.7 km/sec and adjusted the thickness to give the best fit to the travel times and amplitudes of P_g and P_mP. In a few cases, synthetic seismograms have been computed

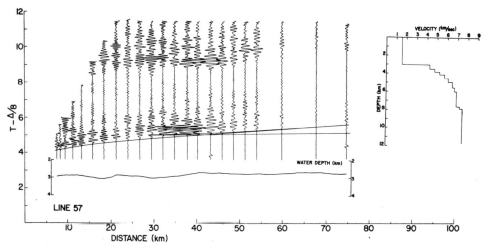

Fig. 9. Record section of line 57. Amplitudes corrected for charge weight and range (see text),

for these models for comparison with the data. In these computations, the phase velocity range and source function were the same as used in Fig. 6.

In Figs. 9—19, record sections of data are shown going from the rise axis to about 10 m.y. (see Fig. 1 for locations). Two amplitude normalization schemes have been used in plotting the data. Most of the data have been normalized by correcting for the shot size and applying a range-dependent

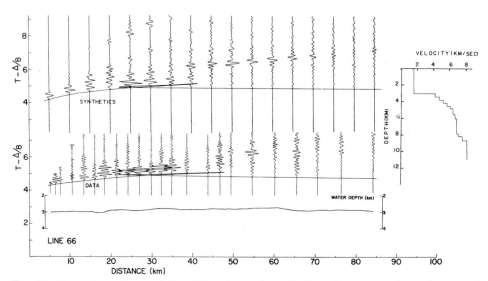

Fig. 10. Record sections of line 66 data, and synthetic seismograms from the model. Amplitudes corrected for charge weight and range (see text).

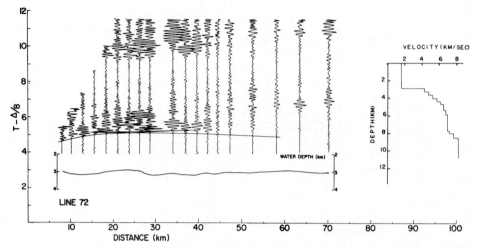

Fig. 11. Record section for line 72. Amplitudes corrected for charge weight and range (see text).

factor. This correction is $(R/R_0)^1(W_0/W)^{-1/3}$, where $R_0 = 10$ km, W_0 is the charge weight at 10 km and W is the charge weight at distance R. The rest of the data have been normalized by scaling each seismogram to the maximum amplitude event in the seismogram. The figure captions indicate which method was used. Next to each data set is also shown the model we think fits the data best and the travel time curve for this model is shown superimposed on the data.

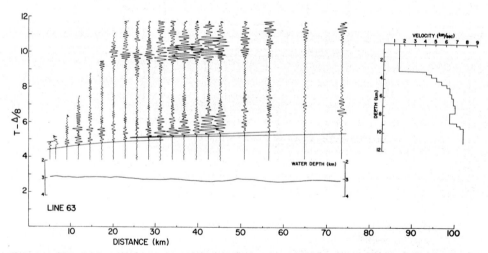

Fig. 12. Record section for line 63. Amplitudes corrected for charge weight and range (see text).

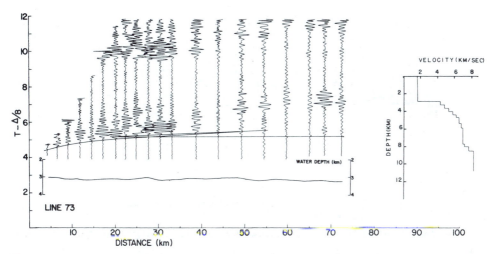

Fig. 13. Record section for line 73. Amplitudes corrected for charge weight and range (see text).

The data in Figs. 9—19 are ordered in terms of increasing distance from the rise axis. We note that lines 57, 66, 72 and 73 (within about 140 km of the rise axis) can be fit with a crustal thickness of about 5 km and a transition zone about 1 km thick between crust and mantle. Lines 66, 72 and 73 show evidence of the $P_m P_m P$ branch, but line 57 does not. As we proceed to older crust, lines 67, 47, 19 and 8 have clear $P_m P$ branches that are offset from the P_g branch. There does not seem to be any systematic relationship between line direction and the offset, indicating that at least some of this offset may

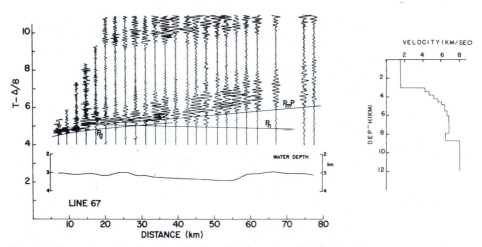

Fig. 14. Record section for line 67. Amplitudes normalized to largest event on each trace.

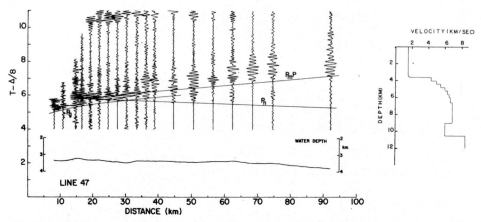

Fig. 15. Record section for line 47. Amplitudes normalized to largest event on each trace.

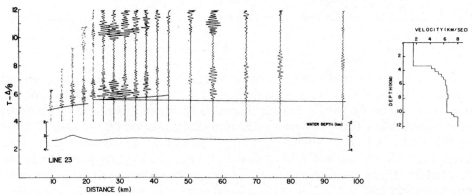

Fig. 16. Record section for line 23. Amplitudes corrected for charge weight and range (see text).

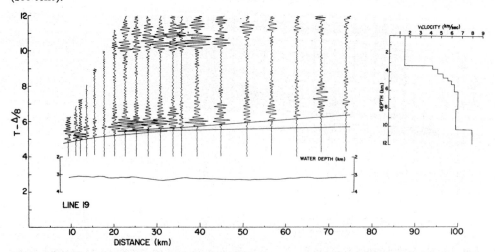

Fig. 17. Record section for line 19. Amplitudes corrected for charge size and range (see text).

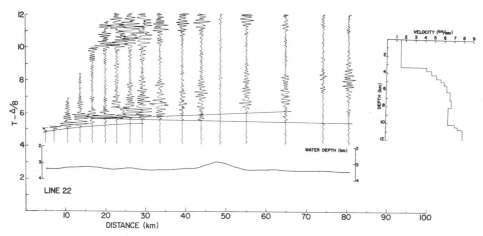

Fig. 18. Record section for line 22. Amplitudes corrected for charge size and range (see text).

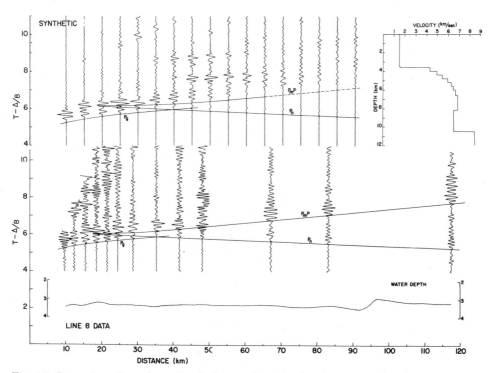

Fig. 19. Record sections from line 8 data, and synthetic seismograms for the model. Data amplitudes normalized to largest event on each trace (see text).

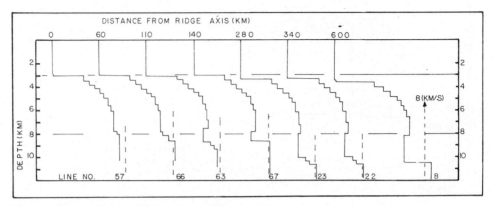

Fig. 20. Changes in crustal structure with age.

be due to a low velocity zone. The low velocity zone offset may be enhanced or decreased by dip as suggested by Helmberger and Engen (1978). Lines 22, 23 and 63 do not have a clear P_mP branch, but the rapid decrease in amplitude and apparent disappearance of P_g (see also Fig. 8) are suggestive of a low velocity zone overlain by a thin lid.

Figure 20 shows a selection of these models plotted against increasing distance from the rise axis. From a comparison of these models, we conclude that the crust in this area is thickening by the development of a low velocity region at the base of the crust. Further, since the crustal section down to about 8 km (the depth to Moho at the rise axis) remains relatively

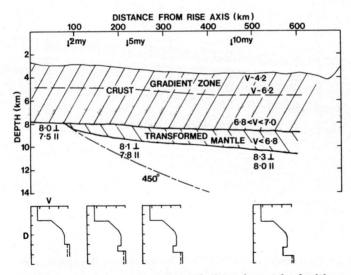

Fig. 21. Cartoon illustrating the evolution of crustal velocities.

unchanged with age, we suggest that the thickening is occurring by the alteration of upper mantle material into material having crustal-like velocities. It also appears that the lower crust may be laterally variable, in some places having a sharp crust—mantle boundary and in others a transition zone between crust and mantle.

Figure 21 shows a schematic summary of these results. In this figure, we have included an estimate of a 450°C isotherm from the cooling model of Parker and Oldenburg (1973).

CONCLUSIONS AND DISCUSSION

From this study of the Northern Cocos Plate, we conclude that:

(1) The crust thickens systematically with increasing age. This conclusion is independent of the interpretation of a low velocity zone at the base of the crust.

(2) From a comparison of synthetic seismograms with the data, we interpret the thickening to be due to the development of a zone in the lower crust which has a velocity somewhat lower than that in the middle crust. If we assume that processes of crustal formation have been constant over the last 10 m.y., then we infer that this zone is caused by the transformation of mantle rocks into rocks having crustal-like velocities.

(3) The data suggest that the lower oceanic crust may be laterally heterogeneous, in some places having a sharp crust—mantle boundary and in some places being gradational.

(4) The degree of upper mantle anisotropy appears to decrease by a factor of two away from the rise axis, having a value of about 0.6 km/sec at zero age and about 0.3 km/sec beyond about 2 m.y.

(5) The data strongly suggest that the velocity-depth function in the upper crust is better represented by a linear gradient than constant velocity layers. No evidence is found for any large first order discontinuities in the upper crust.

Assuming that processes of crustal formation have remained constant over at least the past 10 m.y., we then infer that the crustal thickening occurs by a transformation of mantle rocks into rocks having crustal-like velocities. If this is correct, then there is a large body of circumstantial evidence that suggests that the process causing the transformation is the hydration of the ultramafic rocks in the top 2 km or so of the upper mantle. Some of this evidence is as follows:

(1) Analysis of heat flow data (for example, Anderson et al., 1977) indicates that in youthful crust, convective processes are effective for cooling the crust. The depth to which the convective system transports seawater is not yet clear, but the wavelength of the surface expression of these convective systems (about 7—10 km) suggests that the depth of these convective systems is at least 3.5 km (Ribando et al., 1976), assuming constant permeability with depth, and may be deeper if the permeability decreases with depth.

/(2) The increase of the upper crustal velocity with age has been interpreted as due to the sealing of cracks with age (Houtz and Ewing, 1976). It is these cracks, probably caused by initial cooling of the magma and the development of a propagating crack front (Lister, 1974), that provide the paths for water penetration into the crust.

(3) Measurements of velocities on dry versus wet basalts and gabbros by Christensen and Salisbury (1975) indicate that cracks are not completely closed until a pressure of about 2 kbars. This pressure corresponds to a depth of about 10 km below sea level (Woollard, 1975). These experiments maintained the pore water pressure at atmospheric pressure. If the pore water pressure is increased (for example, by hydrostatic pressure), the cracks would remain open to greater pressure.

(4) In most ophiolite suites, the lower most section of ultramafics is serpentinized (Clague and Straley, 1977). There is some debate, however, about whether the serpentinization occurred before or after emplacement.

The explanation of the crustal thickening in terms of serpentinization of the topmost mantle does raise some problems, principally regarding the amount of water available for the reaction. In order to lower the mantle velocity from about 8 to 6.7 km/sec requires at least 20—30% serpentinization. The water for this reaction would have to pass through the middle of the crust. It can be shown that 1% of water in cracks will lower the P velocity by about 0.3 km/sec (see for example, Anderson et al., 1974, for calculations regarding fluid-filled oriented cracks). Since "layer 3" velocities do not change significantly with age (Bibee and Shor, 1976) (within a total scatter of about 0.4 km/sec), and "layer 3" velocities correspond approximately to an expected gabbroic composition (Christensen and Salisbury, 1975), it is unlikely that more than 1% of water is available in the lower crust for serpentinization. The only way one can achieve 20—30% serpentinization is if the convective water circulation systems continually flush out some of the products of serpentinization (for example MgO and SiO_2) leaving room for the reaction to proceed. This is because of the volume increase involved in the serpentinization reaction (see for example, Wyllie, 1971).

Finally, one point upon which we speculate is that the increased mantle anisotropy observed near the rise axis (about 0.3 km/sec) may be caused by cracks oriented perpendicular to the axis that have resulted from thermal contraction. The cracks could be oriented if other stresses, such as plate-driving forces, are acting perpendicular to the ridge and close cracks parallel to the axis. If the increased anisotropy is due to oriented cracks and these cracks do contain water, then as the temperature falls below about 450°C one would get serpentinization of the uppermost mantle.

ACKNOWLEDGMENTS

This research was supported by National Science Foundation grant DES72-01554AO and Office of Naval Research contract N00014-75-C-0502.

REFERENCES

Anderson, D.L., Minster, B. and Cole, D., 1974. The effect of oriented cracks on seismic velocities. J. Geophys. Res., 79: 4011—4015.

Anderson, R.N., Langseth, M.G. and Sclater, J.G., 1977. Mechanisms for heat transfer through the floor of the Indian Ocean. J. Geophys. Res., 82: 3391—3410.

Backus, G.E., 1965. Possible forms of seismic anisotropy of the uppermost mantle under oceans. J. Geophys. Res., 70: 3429—3439.

Bibee, L.D. and Shor, Jr. G.G., 1976. Compressional wave anisotropy in the crust and upper mantle. Geophys. Res. Lett., 3: 639—642.

Christensen, N.I. and Salisbury, M.H., 1975. Structure and constitution of the lower oceanic crust. Rev. Geophys. Space Phys., 13: 57—86.

Clague, D.A. and Straley, P.F., 1977. Petrologic nature of the oceanic Moho. Geology, 5: 133—136.

Fuchs, K. and Müller, G., 1971. Computation of synthetic seismograms with the reflectivity method and comparison with observations. Geophys. J. R. Astron. Soc., 23: 417—433.

Goslin, J., Benzart, B., Francheteau, J. and LePichon, X., 1972. Thickening of the oceanic layer in the Pacific Ocean. Mar. Geophys. Res., 1: 418—427.

Helmberger, D.V., 1977. Fine structure of an Aleutian crustal section. Geophys. J. R. Astron. Soc., 48: 81—90.

Helmberger, D.V. and Engen, G., 1978. Fine structure of an oceanic crustal section near the East Pacific Rise. Bull. Seismol. Soc. Am. 68: 369—382.

Houtz, R. and Ewing, J., 1976. Upper crustal structure as a function of plate age. J. Geophys. Res., 81: 2490—2498.

Kennett, B.L.N., 1977. Towards a more detailed seismic picture of the oceanic crust and mantle. Mar. Geophys. Res., 3: 7—42.

Lewis, B.T.R. and Snydsman, W.E., 1977. Evidence for a low velocity zone at the base of the oceanic crust. Nature, 226: 340—344.

Lister, C.R.B., 1974. On the penetration of water into hot rock. Geophys. J. R. Astron. Soc., 39: 465—509.

Lister, C.R.B. and Lewis, B.T.R., 1976. An ocean bottom seismometer suitable for arrays. Deep-Sea Res., 23: 113—124.

Lynn, W.S. and Lewis, B.T.R., 1976. Tectonic evolution of the Northern Cocos Plate. Geology, 4: 718—722.

Morris, G.B., Raitt, R.W. and Shor, Jr. G.G., 1969. Velocity anisotropy and delay-time maps of the mantle near Hawaii. J. Geophys. Res., 74: 4300—4316.

Orcutt, J., Kennett, B., Dorman, L. and Prothero, W., 1975. A low velocity zone underlying a fast spreading rise crest. Nature, 256: 475—476.

Parker, R.L. and Oldenburg, D.W., 1973. Thermal model of ocean ridges. Nature, 242: 137—139.

Raitt, R.W., Shor, Jr., G.G., Francis, T.J.G. and Morris, G.B., 1969. Anisotropy of the Pacific upper mantle. J. Geophys. Res., 74: 3095—3109.

Ribando, R.J., Torrance, K.E. and Turcotte, D.L., 1976. Numerical models for hydrothermal circulation in the oceanic crust. J. Geophys. Res., 81: 3007—3012.

Shor, Jr., G.G., Dehlinger, P., Kirk, H.K. and French, W.S., 1968. Seismic refraction studies off Oregon and Northern California. J. Geophys. Res., 73: 2175—2194.

Shor, Jr., G.G., Menard, H.W. and Raitt, R.W., 1970. Structure of the Pacific Basin. In: A.E. Maxwell (Editor), The Sea, Vol. 4, Part 2. Wiley-Interscience, New York, N.Y., pp. 3—27.

Shor, Jr., G.G., Raitt, R.W., Henry, M., Bentley, L.R. and Sutton, G.H., 1973. Anisotropy and crustal structure of the Cocos Plate. Publicada con la ayuda del Consejo National de Ciencia y Tecnologia, 13: 337—362.

Snydsman, W.E., Lewis, B.T.R. and McClain, J., 1975. Upper mantle velocities on the Northern Cocos Plate. Earth Planet. Sci. Lett., 28: 46—50.

Woollard, G.P., 1975. The interrelationships of crustal and upper mantle parameter values in the Pacific. Rev. Geophys., 13: 87—138.

Wyllie, P.J., 1971. The Dynamic Earth: Textbook in Geosciences. Wiley, New York, N.Y., 416 pp.

Chapter 6

ESTIMATORS FOR HEAT FLOW AND DEEP ROCK PROPERTIES BASED ON BOUNDARY LAYER THEORY*

C.R.B. LISTER

Departments of Geophysics and Oceanography, University of Washington, Seattle, Washington 98195 (U.S.A.)

SUMMARY

Boundary-layer theory has predicted that the topographic profile of mid-ocean ridges should become a straight line when plotted against square root of age. Many ocean ridges follow the 'square-root law' of topography from 0.5 to 80 m.y., and the heat-flow prediction of the theory should hold in this age range also. Heat flow (q) should vary inversely as root age even when the diffusivity of the rock is a strong function of temperature. Topographic decay is related to q by $q = dh/dt \times \alpha_i \kappa/k_0$, where the isostatically modified expansion coefficient α_i and the diffusivity κ are sampled at about 0.7 of initial temperature, but the thermal conductivity is the k_0 at surface temperature. The best heat-flow data older than 2 m.y. agree with the estimator $12(t)^{-1/2}$ HFU with t in m.y., and the numerical constant is well within the plausible range for the rock parameters of ultra-basic rocks. Measurements on young crust do not agree with the estimator even in sedimented areas and the total heat output of ridge crests must still be calculated from theory. The best models are two-dimensional and embody 'thermal balance' boundary conditions at the origin, but the simple one-dimensional asymptotic estimate of $24(t)^{-1/2}$ HFU for the mean flux to age t in m.y. is accurate to better than 1% beyond 25 m.y.

INTRODUCTION

Many models of the temperature distribution in spreading oceanic lithosphere have been discussed in the literature. They began with a solution for the temperature field in a uniform horizontal flow, in the form of a series of separated functions (McKenzie, 1967). This was followed by a numerical investigation that justified the one-dimensional flow approach through the boundary layers that develop in high Prandtl number convective flow (Oxburgh and Turcotte, 1968). Most of the numerical models do not resolve the temperature distribution near the ridge crest because of the coarseness of the computer grids (Le Pichon and Langseth, 1969), while the simple analytic models cannot take account of the two-dimensional flow associated with upwelling. Since the heat-flow measurements made on young oceanic

---
*Originally published as: Lister, C.R.B., 1977. Estimators for heat flow and deep rock properties based on boundary layer theory. In: A.M. Jessop (Editor), Heat Flow and Geodynamics. Tectonophysics, 41: 157—171.

crust obviously do not agree with the theoretical models if reasonable rock parameters and magma-like source temperatures are used (Sleep, 1969), it was not clear that any understanding had been gained. However, analysis of the topography of the ridges has proved considerably more fruitful.

It has been known for some time that the mid-ocean ridges are not associated with significant free-air gravity anomalies, and therefore that the excess topographic height must be compensated by lowered densities somewhere in the upper 100 km or so (Talwani et al., 1965). It was Sleep (1969) who showed that the excess height agreed with that expected for rock cooling from a reasonable magma-like source temperature with a reasonable expansion coefficient, and that the contraction curve had approximately the right form to fit the topography. There are, however, many other plausible ways of satisfying the gravity data, and there has even been dispute over whether it is 'reasonable' to use the cubical-expansion coefficient of rock in the calculation. Since the rocks are finding their isostatic level, the rock *density* is the controlling factor, and it is immaterial to the height calculation how the horizontal contraction has been taken up. Even when this is accepted, there are difficulties in comparing a computed curve, based on several estimated parameters, with noisy topographic data (Sclater and Francheteau, 1970). The parameters cannot be optimized readily, nor can the validity of various models be tested.

Two more steps had to be made before the theoretical predictions and the topographic data could be compared critically. First it was necessary to find a simple asymptotic solution for the thermal field that might apply over a range of lithospheric ages. Parker and Oldenburg (1973) showed that one family of two-dimensional solutions satisfied a simpler one-dimensional (vertical) heat-conduction equation within a fraction of a m.y. from the intrusion origin. Davis and Lister (1974) then applied the one-dimensional asymptotic equation to the problem of topography, and showed not only that the cooling models predicted a drop in topographic height proportional to the square root of time, but that this was, indeed, satisfied by real topographic data to a remarkable accuracy over a large range of age. Since all the rock parameters boil down to a single kernel, and this is simply the slope of a straight-line plot, deviations from the asymptotic solution can be assessed readily. Full two-dimensional series solutions converge rapidly near the ridge crest, and then diverge again at great ages when the effect of their fixed bottom boundary temperature becomes apparent. Over a range of nearly 80 m.y. both theory and data are in agreement, and this range includes the region where the heat-flow measurements disagree with theory.

The problem of too-low heat-flow measurements could be resolved satisfactorily only when evidence from the measurements themselves led to a plausible mechanism for their bias and scatter. Iceland is a segment of spreading center and is justly famed for its geothermal areas. Speculations about the possible role of water circulation in the transport of heat in young oceanic crust were discussed from the earliest days of ridge crest heat-flow measurement. Appropriately, they appeared in print with an Icelandic author (Palmason, 1967). However, consideration of the distribution of heat-flow values relative to topography and sediment cover did not come until Lister (1972) showed that some fortunately placed measurements were consistent only with convective heat transport in the crust. He also adduced evidence that substantial sediment cover acted as an impermeable barrier to hydrothermal flow, so that a large enough area devoid of outcrops should permit the true heat flow to be measured by the probe technique. Thus, the thermal output predicted by theoretical cooling models is no longer in conflict with the measurements, and good estimates of the heat loss from lithosphere of moderate age may soon be available. It is therefore time to consider in some detail what information on upper-mantle materials is contained in relation between heat flow and topographic slope, and what boundary-layer theory predicts about the distribution of real heat output with oceanic age.

HEAT FLOW AND TOPOGRAPHIC SLOPE

For ocean lithosphere older than about 0.5 m.y., the thermal solutions satisfy a one-dimensional vertical heat-flow equation to a good approximation (Parker and Oldenburg, 1973). Then the full heat-flow equation for a uniform horizontal flow:

$$\rho c_p u \, \partial T/\partial x = k \, \partial^2 T/\partial x^2 + k \, \partial^2 T/\partial z^2 + H \tag{1}$$

can be simplified to:

$$(u/\kappa) \, \partial T/\partial x = \partial^2 T/\partial z^2$$

where ρ is density, c_p is specific heat at constant pressure, u is lateral flow velocity, T is temprature, k is thermal conductivity, H is internal heat production, κ is diffusivity, and the coordinate system is shown in Fig. 1. For solutions to belong to the asymptotic family, H must be negligible, and lateral gradients must be small enough for $\partial^2 T/\partial x^2$ to be small compared to $\partial^2 T/\partial z^2$. Since $\partial x/\partial t = u$ (t is time), the most useful form of the simplified equation is:

$$(1/\kappa) \, \partial T/\partial t = \partial^2 T/\partial^2 z^2 \tag{2}$$

This equation, with appropriate boundary conditions, describes the cooling of an infinite half space. If the initial temperature is uniform with depth (T_1), the surface temperature is zero and κ is constant, the solution is:

$$T = T_1 \, \mathrm{erf}[z/2(\kappa t)^{1/2}]$$

(Carslaw and Jaeger, 1959, p. 59). The surface heat flow of this solution is:

$$q = k(\partial T/\partial z)_{z=0} = k T_1 (\pi \kappa t)^{-1/2}$$

while the topographic height h is given by:

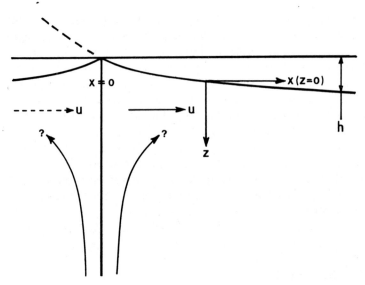

Fig. 1. Diagram of the coordinate system and boundary-condition location for thermal models of sea-floor spreading. The height drop from the crestal level (h) is not included in the thermal analysis, and the dashed lines indicate the mathematical assumptions inherent in eq. 1. The upwelling lines marked by ? indicate the extent of the mathematical idealization (see text: section on crestal mechanism effects).

$$h = \alpha_i \int_0^\infty (T_1 - T)\, dz = 2\alpha_i T_1 (\kappa t/\pi)^{1/2}$$

where α_i is the effective cubical expansion coefficent, modified for isostatic adjustment (Davis and Lister, 1974). The slope of topography against time is then:

$$dh/dt = \alpha_i T_1 (\kappa/\pi t)^{1/2}$$

so, finally:

$$(1/q)\, dh/dt = \alpha_i \kappa/k = \alpha_i/\rho c_P \tag{3}$$

or:

$$q = (\rho c_P/\alpha_i)\, dh/dt$$

Thus, the heat flow is *proportional to the slope of topographic height against time*, and the initial temperature does not enter into the relationship.

This simple result can be obtained also by direct physical argument. The approximation used in deriving the asymptotic equation (2) is the same as saying that each vertical column of lithosphere can be considered to have insulated sides. The heat that escapes through the top must come from cooling the material below, and the cooled material must shrink. Thus, the ratio between the drop in height and the integrated heat flux to produce that drop is simply the ratio between the expansion coefficient and the heat capacity per unit volume, whatever the temperature distribution in the cooling rock. The ease of this argument suggests that it may be possible to derive a relationship between heat flow and topographic slope for the more general case where the rock parameters are not constant, but are functions of temperature.

VARYING PARAMETERS

The variation of rock thermal conductivity with temperature has been discussed fairly extensively in the literature, as it is important in estimating static or average thermal profiles in planetary interiors. For the transient cooling problem the parameter of greatest interest is the diffusivity (eq. 2), and direct measurements have shown that this varies even more than the conductivity because of the increase of specific heat with temperature. For common rock-forming minerals, the diffusivity drops by about a factor of 3 between room temperature and $1200°K$ (Fujisawa et al., 1968), and then levels off. The pressure effect is also considerable, about a 50% linear increase between zero pressure and 30 kbar, corresponding to a depth of 100 km. The analysis given here cannot handle pressure effects because the mathematical simplification depends on the square root of time similarity relation for a thickening cooled layer. However, significant cooling only reaches 100 km after about 80 m.y., and the diffusivity values at high temperatures are all relatively small, so that neglect of the pressure effect should not generate serious error for younger ages. Let us therefore examine what *can* be deduced from an analysis of temperature effects alone, since it is not untuitively obvious what kind of average parameter should be used, either for diffusivity or expansion coefficient.

The more general form of eq. 2 is:

$$(\rho c_P) \, \partial T/\partial t = \partial/\partial x \, (k \, \partial T/\partial x) \qquad (4)$$

where ρ, c_P and k are all now function of the temperature T. The transformation $\theta = 1/k_0 \int_0^T k dT$ can be used to convert this equation back into a form similar to eq. 2:

$$[1/\kappa(\Theta)] \, \partial\Theta/\partial t = \partial^2\Theta/\partial z^2 \qquad (5)$$

where Θ is just a modified temperature scale (Carslaw and Jaeger, 1959, pp. 10—11). The further transformation $y = zt^{-1/2}$ reduces this to the simple differential equation:

$$-[y/2\kappa(\Theta)] \, \Theta'(y) = \Theta''(y) \qquad (6)$$

where the primes indicate full derivatives. Appropriate solutions of this equation satisfy the boundary conditions $\Theta(0) = 0$, $\Theta(\infty) = \Theta_1$, $\Theta' \to 0$ and $\Theta'' \to 0$ as $y \to \infty$, but an analytic form does not appear to be available for arbitrary functions κ. The equations for q and dh/dt can be derived as before:

$$q = (k\ \partial T/\partial z)_{z=0} = k_0 t^{-1/2}\Theta'(0)$$

$$h = (\rho_1 - \rho_w)^{-1} \int_0^\infty (\rho - \rho_1)\,dz \qquad \text{(Davis and Lister, 1974)}$$

where $\rho(\Theta)$ is $\rho_1 + \rho_1 \int_{\Theta_1}^\Theta - \alpha(\Theta)\,d\Theta$, the subscript 1 indicates properties at the initial temperature Θ_1, the expansion coefficient α is also a function of temperature, and ρ_w is sea-water density.

$$dh/dt = d/dt[t^{1/2}(\rho_1 - \rho_w)^{-1} \int_0^\infty (\rho - \rho_1)\,dy]$$

$$= \tfrac{1}{2}\rho_1 t^{-1/2}(\rho_1 - \rho_w)^{-1} \int_0^\infty \int_\Theta^{\Theta_1} \alpha(\Theta)\,d\Theta dy$$

$$(1/q)\,dh/dt = \tfrac{1}{2}\rho_1[k_0(\rho_1 - \rho_w)\,\Theta'(0)]^{-1} \int_0^\infty \int_\Theta^{\Theta_1} \alpha(\Theta)\,d\Theta dy$$

After three integrations by parts and one use of the different equation, this can be transformed into the more useful equation:

$$(1/q)\,dh/dt = \rho_1 k_0^{-1}(\rho_1 - \rho_w)^{-1}\{\alpha_0\kappa_0 + \int_0^{\Theta_1} [\Theta'/\Theta'(0)]\,\partial(\alpha\kappa)/\partial\Theta \cdot d\Theta\}$$

where $\Theta'(0)$ is the value of the derivative at $\Theta = y = 0$. The definite integral on the RHS can be converted back into a form dependent on real temperature T by using the relations:

$$\Theta' = d\Theta/dy = dT/dy \cdot d\Theta/dT = k/k_0 \cdot dT/dy = kT'/k_0$$

$$\Theta'(0) = (d\Theta/dy)_0 = (dT/dy)_0 = T'(0)$$

whence:

$$(1/q)\,dh/dt = \rho_1 k_0^{-1}(\rho_1 - \rho_w)^{-1}\{\alpha_0\kappa_0 + \int_0^{T'} (k/k_0)[T'/T'(0)]\,\partial(\alpha\kappa)/\partial T \cdot dT\} \quad (7)$$

When $\alpha\kappa$ is constant, $\partial(\alpha\kappa)/\partial T$ is zero and the relation agrees with the result derived above: $(1/q)\,dh/dt = \alpha_i/\rho c_P$

The form of the expression for variable parameters shows that the value of $\alpha\kappa$ observed in the field data are somewhere between the surface-temperature value and the initial temperature value, as expected, but not simply that of the mean temperature. The surface-temperature value $\alpha_0\kappa_0$ is modified by the integral over the temperature range of $\partial(\alpha\kappa)/\partial T$ multiplied by the

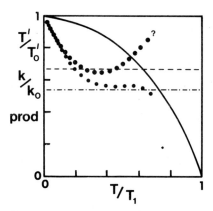

Fig. 2. Plots of three non-dimensionalized ratios versus non-dimensionalized temperature. Solid curve: vertical temperature derivative for the constant parameter one-dimensional thermal solution $T = T_1 \operatorname{erf}[\frac{1}{2}z\,(\kappa t)^{-1/2}]$, mean value: dashed line; large dots: conductivity of polycrystalline dunite, $Fo_{91}Fa_9$ (Holt, 1975); small dots: product of the preceding ratios, mean value: dash-dot line.

temperature-scale correction factor k/k_0, and by the solution dependent parameter $T'/T'(0)$. This last starts at unity for $T = 0$, and then decreases to zero when $T = T_1$ or $y = \infty$. An idea of its variation can be obtained from the thermal solution for constant parameters $T = T_1 \operatorname{erf}(\frac{1}{2}y\pi^{-1/2})$, although the result applies strictly only to cases where the $\alpha\kappa$ variation is small. Here $T'/T'(0) = \exp(-y^2/4\kappa)$ while $T/T_1 = \operatorname{erf}(\frac{1}{2}y\kappa^{-1/2})$ so that a plot of $\exp(-r^2)$ against $\operatorname{erf}(r)$ gives the variation of the dimensionless parameter against dimensionless temperature. It is given in Fig. 2, and an idea of its use can be seen in the following argument. If $\alpha\kappa$ varied linearly with temperature, $\partial(\alpha\kappa)/\partial T$ would be a constant, and therefore the amount of the total change applicable to the integral would be simply the mean value under the curve of Fig. 2, or about 0.7. Again, since the variation is linear, the final value of $\alpha\kappa$ to be taken is that at about 0.7 of the initial temperature, or around 900°C for the lithosphere.

The temperature-scale correction factor k/k_0 modifies this result somewhat by additional weighting of the curve. Reliable data on the variation of thermal conductivity of mantle material with temperature is not available, as we do not know what its composition or mineralization is. However, values presented for a polycrystalline dunite ($Fo_{91}Fa_9$) by Holt (1975) may be considered as reasonably representative: k decreases from about 0.010 cal $cm^{-1}s^{-1}\,°C^{-1}$ at 0°C to a minimum of 0.0065 at 500°C and then rises gradually toward 0.008 at 900°C. This k/k_0 curve is also presented in Fig. 2, together with the approximate product of it and the ratio $T'/T'(0)$. The mean value of the combined curve is clearly about 0.55, and represents a minimum estimate for the integral in eq. 6. This is because we have allowed k to vary without considering the effect of the variation of κ on the solution temperature gradient T'. Where the diffusivity is low, the local temperature gradient in the cooling solution should be high. In fact, this qualitative physical argument suggests that the effect of varying κ on the $T'/T'(0)$ curve should be about equal and opposite to the factor k/k_0 if the specific heat remains approximately constant. In real materials, κ drops more rapidly with temperature than k because of the increase of specific heat with temperature. Hence, the best estimate of the mean factor modifying $\partial(\alpha\kappa)/\partial T$ in the integral is closer to 0.7 than 0.55, and it is quite insensitive to variations in the conductivity k.

Therefore, as a *general rule of thumb*, provided parameter variations are not so large that the temperature distribution departs radically from the error function, or so uneven that means are inappropriate, the rock properties determined by topographic and heat-flow data are the $\alpha\kappa$ at about *0.7 of the initial temperature*, but the k_0 is the conductivity at the surface temperature.

$$(1/q)\,\mathrm{d}h/\mathrm{d}t \simeq \rho_1 k_0^{-1}(\rho_1 - \rho_w)^{-1}\alpha(0.7T_1)\,\kappa(0.7T_1) \tag{8}$$

It might seem surprising that the topographic slope should sample materials at such a high temperature. What is happening physically can be demonstrated by considering two error-function curves drawn at a small time interval, as in Fig. 3. Although the temperature *gradient* is greatest near the surface, the largest *temperature change* occurs at some depth, in fact the depth of the maximum of $p\exp(-p^2)$. Not unexpectedly, the maximum occurs at $p = 0.7$ (Carslaw and Jaeger, 1959, p. 485). The importance of the more rigorous analysis is in showing that the ratio of topographic slope to heat flow is a solution insensitive parameter, and therefore the best one to use in investigation rock properties. Both the surface heat flow and the

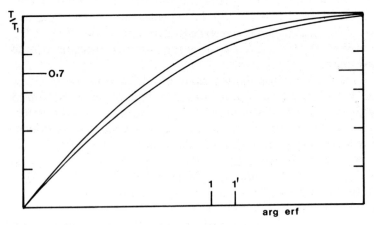

Fig. 3. Two time-successive temperature plots for the one-dimensional constant-parameter thermal solution described in Fig. 2. The *1* and *1′* indicate the scales of the error function argument for the two curves; the argument axis also serves as an arbitrary depth axis.

topographic slope predictors contain $T'(0)$ to first order; only their ratio eliminates first-order dependence on the exact form of the solution for varying parameters.

ASYMPTOTIC PREDICTION FOR HEAT FLOW

Having analyzed the relationship between heat flow, topographic slope and rock parameters, the heat-flow prediction can be calculated and compared with real data. The age range for which the theory is applicable is the range in which the topography obeys the square root of time law. As much slopes are usually expressed as $s = $ km $(\text{m.y.})^{-1/2}$, an expression for dh/dt should be derived in terms of s and time in root m.y.:

$$h \quad = s(t)^{1/2} \qquad\qquad \text{km}$$

$$dh/dt \quad = \tfrac{1}{2}s(t)^{-1/2} \qquad\qquad \text{km/m.y.}$$

$$\qquad = 1.58 \cdot 10^{-9}\, s(t)^{-1/2} \qquad\qquad \text{cm/s } (t \text{ in m.y.})$$

The expression for q can now be written out (eq. 8):

$$q = k_0(\rho_1 - \rho_w)\, dh/dt\, [\rho_1 \alpha(0.7T_1)\, \kappa(0.7T_1)]^{-1} \qquad\qquad \text{cal cm}^{-2}\,\text{s}^{-1}$$

$$\quad = 1.58 \cdot 10^{-3} \cdot s(t)^{-1/2}(1 - \rho_w/\rho_1) \cdot k_0/\alpha(0.7T_1)\,\kappa(0.7T_1) \quad \text{HFU}$$

so that q is proportional to $t^{-1/2}$, to a series of constants that are determinable, in principle, for any rock assemblage, and to the slope constant of the ocean floor in the region.

A mean slope constant of 0.39 km $(\text{m.y.})^{-1/2}$ applies to most ocean floor younger than 80 m.y. (Davis and Lister, 1974), but the problems of determining k_0, $\alpha(900°\text{C})$ and $\kappa(900°\text{C})$ for reasonable rock assemblages are beyond the scope of this paper. Minor changes in chemical composition have a large effect on the parameters, even for the same mineral: for example, $\alpha(900°\text{C})$ for pure forsterite olivine is $4.7 \cdot 10^{-5}$, but for olivine containing 10% fayalite it is only $3.9 \cdot 10^{-5}$ (Skinner, 1966). In addition, olivine is an anisotropic mineral that can cause the rock to have significantly different k_0-values if the surface material posesses a grain orientation, and different κ-values if the grains are also oriented at depth. The behavior of κ in the 001-direction with temperature reported by Kanamori et al. (1968) for 18% Fa is significantly different from that of a forsterite ceramic (Fujisawa et al., 1968). The value at 900°C for the 001-direction in a single crystal is 0.014 cm^{2}s^{-1}, while the ceramic reads about 0.008 at moderate pressure (20 kbar). This is probably due to the higher radiative heat-transfer component in the more transparent single crystal, and such hard-to-estimate factors may be important in the mantle. Recent data from Holt (1975) on a natural dunite of $Fo_{91}Fa_9$ composition gives a value of about 0.070 cm^{2}s for κ at 900°C, and about 0.010 cal cm^{-1}s^{-1} °C^{-1} for k_0 at 0°C. Because no α was determined for this material, an approximate estimate for a dunite mantle can be made as well by taking $\alpha\kappa$-values for pure forsterite at $900°C$ (Skinner, 1966; Fujisawa et al., 1968) and the k_0 for 'North Carolina Dunite' at low temperature (Clark, 1966). These values are $\alpha = 4.7 \cdot 10^{-5} \, °C^{-1}$, $\kappa = 0.008 \, cm^2 s^{-1}$, $k_0 = 0.012 \, cal \, cm^{-1} s^{-1} \, °C^{-1}$, $\rho_1 = 3.3 \, g \, cm^{-3}$, $\rho_w = 1.0 \, g \, cm^{-3}$, so that:

$$q = 12(t)^{-1/2} \qquad \text{HFU (}t\text{ in m.y.)} \qquad (9)$$

The substitution of the k_0 and $\kappa(900°C)$ for Holt's (1975) dunite would not change this result appreciably. The expansion coefficients and thermal diffusivities of the various pyroxenes are similar to those for olivine, with somewhat lower α's and higher κ's leading to a similar $\alpha\kappa$ product. However, a numerical parameter anywhere between 8 and 20 HFU $(m.y.)^{1/2}$ would be plausible for mantle material.

The heat-flow data have given a good deal of difficulty historically, but, now that the role of hydrothermal circulation is partially understood, the best measurements can be selected. Some recent results are given in Fig. 4, plotted on a $(t)^{-1/2}$ graph so that the asymptotic relation shows up as a straight line through the (right hand) origin. Values derived from measurements on well-sedimented ocean floor lie very close to a $12t^{-1/2}$ line, down to a mean of measurements on the west flank of the Juan de Fuca ridge at about 3.3 m.y. The congruence of the numerical factor (12) is largely accidental. Nevertheless, even well-sedimented ocean floor of age less than 1

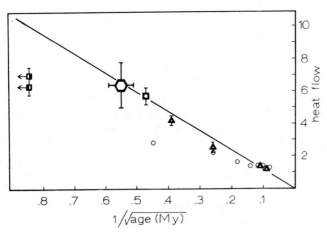

Fig. 4. Average values of heat flow plotted with respect to $(\text{age})^{-1/2}$. Values are as follows: Light circles: East Pacific (Sclater and Francheteau, 1970) *data unselected for sediment cover*; heavy triangles: East Pacific (Parsons and Sclater, personal communication) *selected for sediment cover*; heavy square: Explorer Ridge, sedimented east flank (Davis and Lister, unpublished); large hexagon: Juan de Fuca Ridge, sedimented flank 47°N (Lister, 1972); half-filled squares: Juan de Fuca Ridge, sedimented (Davis and Lister, unpublished).

m.y. does not yield mean values close to the curve, as the points that belong off to the left of Fig. 4 at an abcissa of 2 demonstrate (Davis and Lister, unpublished data). This is probably due in part to the presence of rock outcrops and hot springs near enough to the measurements to cause error, and in part to the time taken for a quasi-steady state to be established in the hydrothermal zone after the violent initial activity. In any case, the data show that we are still dependent on theoretical models for estimates of the heat output near the ridge crest itself.

CRESTAL MECHANISM EFFECTS

As mentioned in the introduction, there has been considerable dispute in the literature about the crestal contribution to the overall heat flow emitted from a spreading center. Numerical models tend to have a grid size larger than the whole region of interest, and the boundary conditions of the analytical steady-flow solutions have been arbitrary, sometimes for simplicity and sometimes in an attempt to fit mean heat-flow measurements. Doubt about the fit to real processes in this important region has led to a general reluctance to accept the heat-flow estimates of the one-dimensional solution. The problem is simply that of stating a boundary condition for the uniform flow model that corresponds in a reasonable way to the basic physics of real magma or mush injection.

The difficulty that arises is best understood by referring to Fig. 1. At a real spreading center, material rises as magma, mush, or low-viscosity plastic solid somewhere near the axis, and then spreads laterally. In the mathematical model, embodied by definition in eq. 1, material flows uniformly from $x = -\infty$ to $x = +\infty$, passing an entirely arbitrary line $x = 0$ that we call the origin. The boundary condition at $x = 0$ has to reflect, as well as possible, a real process of intrusion without being able to model real two-dimensional flow, magma segregation or any of the other geological complexities. Simply defining an arbitrary temperature will not do, since McKenzie (1967) demonstrated that it can lead to a solution with an infinite integrated heat flux through the top surface. The important parameter at the boundary $x = 0$ is the total heat supply into positive x-space. This is in two components, the advected heat $u\rho c_P T$ and the conducted heat $k \cdot \partial T/\partial x$, so that the sum of these should be set equal to the *heat supply*. A *minimum* value for the heat supplied by new material is $u\rho c_P T_M$, since it can hardly be emplaced, even intermittently, at temperatures below the melting point T_M. A *maximum* value is $u\rho(c_P T_1 + L)$, the sum of supply at some observed or computed initial temperature T_1 and the entrained latent heat of melting L. The problem is treated in detail by Sleep (1975), and for the purpose of estimating heat flow the intermediate boundary condition $u\rho c_P T_1 = u\rho c_P T + k \cdot \partial T/\partial x$ is sufficiently precise. This is especially so because the crustal basaltic and gabbroic zone that begins by being fully molten is probably cooled hydrothermally (Lister, 1972), so that the conductive solution

138

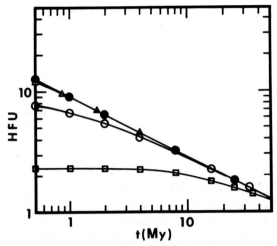

Fig. 5. Heat flow against age compared for several thermal models of spreading lithosphere. Spreading rate 3 cm yr^{-1}, $\kappa = 0.008$ cm^{2}s^{-1}, $k = 0.0066$ cal cm^{-1}s^{-1} °C^{-1}. Filled circles: McKenzie (1967); straight line: asymptotic one-dimensional solution; triangles: Davis and Lister (1974) 'thermal balance' model; open circles: Lister, 1972; open squares: Lubimova and Nikitina (1975). The low k assumed for these calculations generates a different asymptote from the $12(t)^{-1/2}$ suggested by the data (Fig. 4).

applies only to the material below. The whole heat contribution of a crust cooled rapidly from the molten state is *in addition* to that provided by the conductive solution, from the point where the hydrothermal processes stabilize.

The simple 'thermal balance' boundary condition given above was used by Davis and Lister (1974) to test for crestal topographic effects using a series solution with constant thermal parameters. The ridge crest point was found to be depressed a mere 50 m below the square root of time straight line, and the lack of a significant deviation agreed well with the data from smooth segments of the East Pacific Rise. Therefore, it is the heat flow computed from this model that should be compared with the prediction of the asymptotic one-dimensional solution. Figure 5 shows the results. Points are plotted for two other solutions of minor interest, neither of which can match observed topography, and also for McKenzie's (1967) solution that diverges at the origin. The 'thermal balance' and one-dimensional solutions match within 5% at 0.5 m.y. and are indistinguishable beyond 1 m.y. The close congruence of the asymptotic solution with the more accurate two-dimensional calculation can also be seen in Table I where mean heat flows between 0 and 25 m.y. are compared. Thus the effect of two-dimensional heat flow near the spreading center does not affect the viability of the asymptotic solution as an estimator of the mean heat flow of a large region. The large-age limit for the estimator is the age at which topography deviates substantially from the square-root of time law; as Fig. 4 shows, the measured values do not deviate at 100 m.y.

TABLE I

Mean heat flow from 0 to 25 m.y.

	HFU
McKenzie (1967) boundary	∞
One-dimensional approximation (analytic)	3.620 *
Thermal-balance boundary (computed analytic)	3.605
Lister (1972) cooled model (analytic) **	3.026
Lubimova and Nikitina (1975) model (computed analytic) **	1.919

Spreading rate 3 cm yr^{-1}, κ = 0.008 cm^2 s^{-1}, $k*$ = 0.0066 cal cm^{-1} s^{-1} °C^{-1}

* These calculations do not agree with the $12(t)^{-1/2}$ estimator because of the k value assumed, but there is no point in running them again.
** These models include arbitrary cooling at the crest and do not obey the square root of time law of topography in the crestal region.

CONCLUSIONS

The relationship between heat flow and topographic slope has been analyzed carefully for the partially generalized case of arbitrarily temperature variable parameters. The results show that the diffusivity and thermal-expansion coefficients of the upper mantle are sampled at about 0.7 of the initial temperature (T_1), but the surface conductivity also enters into the relation. The low sensitivity of the ratio to rock properties above 0.7 T_1 may help to explain why the theory works well even when it is known that the pressure coefficients of the rock parameters are not negligible. The rock at great depths is also the rock at the highest temperature, and changes in the parameters at a given temperature due to the thickening of the lithosphere have only a small effect on the results. The reasoning applies also to the square-root law of topography itself, since the formula for h derived from the analysis given here is:

$$h = 2\rho_1 t^{1/2}(\rho_1 - \rho_w)^{-1} T'(0)[\alpha_0\kappa_0 + \int_0^{T_1} (k/k_0)[T'/T'(0)] \, \partial(\alpha\kappa)/dT \cdot dT]$$

The additional factor $T'(0)$ is also insensitive to changes in the form of the solution at the deep end, due either to pressure induced changes or increasing initial temperature. However, $T'(0)$ is very sensitive to changes in diffusivity at near-surface temperatures, and κ varies rapidly with temperature between 0 and 400°C because of the change in specific heat with temperature of the closely-bound rock lattices. Hence the top boundary temperature may be a critical parameter in establishing the $(t)^{1/2}$ topographic slope. Variations due to differing hydrothermal regimes in different crustal structures may explain the remarkable variations in topographic slope found for the East Pacific Rise by Davis and Lister (1974).

The boundary layer theory's prediction for heat flow is that it should vary as $(t)^{-1/2}$, and the numerical factor could be computed for different upper-mantle mineral assemblages if the relevant properties of the minerals were known. The thermal conductivity between room temperature and 200°C, together with measurements of the expansion coefficient and diffusivity at about 900°C would be most useful for speculations about the composition. The best heat-flow data, taken from stations on well-sedimented ocean crust many kilometers from rock outcrops and possible hot springs, agrees with the estimator $12(t)^{-1/2}$ HFU where t is in m.y. Comparison of the asymptotic estimator with the mean heat flow computed from a two-dimensional series solution shows that the simple estimator is accurate to better than 1% for ages greater than 25 m.y. The failure of the asymptotic solution to account for lateral heat flow near the injection zone is not as serious as the failure of many models to account for either the full original heat supply or the conducted component through their $x = 0$ origin. The use of a thermally balanced computational model confirms that the simple estimator is quite accurate enough for calculating the mean heat flux through oceanic crust younger than 80 m.y. The processes that occur under older ocean floor are still obscure, but those areas are well-sedimented and accurate field data could be obtained.

ACKNOWLEDGEMENTS

This work was supported by National Science Foundation Grant DES 73-06593 A 01. I wish to thank Earl Davis and William Pearson for their assistance with running the computations of heat-flow.

REFERENCES

Carslaw, H.S. and Jaeger, J.C., 1959. Conduction of heat in solids. Oxford University Press., 2nd edit.

Clark, S.P., Jr., 1966. Thermal conductivity. In S.P. Clark, Jr. (editor), Handbook of Physical Constants. Geol. Soc. Am., Mem., 97: 461—482.

Davis, E.E. and Lister, C.R.B., 1974. Fundamentals of ridge crest topography. Earth Planet. Sci. Lett., 21: 405—413.

Fujisawa, H., Fujii, N., Mizutani, H., Kanamori, H. and Akimoto, S., 1968. Thermal diffusivity of Mg_2SiO_4, Fe_2SiO_4, and NaCl at high pressures and temperatures. J. Geophys. Res., 73: 4727—4733.

Holt, J.B., 1975. Thermal diffusivity of olivine. Earth Planet. Sci. Lett., 27: 404—408.

Kanamori, H., Fujii, N. and Mizutani, H., 1968. Thermal diffusivity of rock forming minerals from 300° to 1100°K. J. Geophys. Res., 73: 595—605.

Le Pichon, X. and Langseth, M.G., 1969. Heat flow from the mid-ocean ridges and seafloor spreading. Tectonophysics, 8: 319—344.

Lister, C.R.B., 1972. On the thermal balance of a mid-ocean ridge. Geophys. J.R. Astron. Soc., 26: 515—535.

Lubimova, E.A. and Nikitina, V.N., 1975. On heat-flow singularities over mid-ocean ridges. J. Geophys. Res., 80: 232—243.

McKenzie, D.P., 1967. Some remarks on heat flow and gravity anomalies. J. Geophys. Res., 72: 6261—6273.

Oxburgh, E.R. and Turcotte, D.C., 1968. Mid-ocean ridges and geotherm distribution during mantle convection. J. Geophys. Res., 73: 2643—2661.

Palmason, G., 1967. On heat flow in Iceland in relation to the mid-Atlantic ridge. In: S. Bjornsson (editor), Iceland and Mid Ocean Ridges. Soc. Sci. Islandica, Publ., 38: 111—127.

Parker, R.L. and Oldenburg, D.W., 1973. Thermal model of ocean ridges. Nature, 242: 137—139.

Sclater, J.G. and Francheteau, J., 1970. The implications of terrestrial heat-flow observations on current tectonic and geochemical models of the crust and upper mantle of the earth. Geophys. J.R. Astron. Soc., 20: 509—542.

Skinner, B.J., 1966. Thermal expansion. In: S.P. Clark Jr. (editor), Handbook of Physical Constants. Geol. Soc. Am., Mem., 97: 78—96.

Sleep, N.H., 1969. Sensitivity of heat flow and gravity to the mechanism of sea-floor spreading. J. Geophys. Res., 74: 542—549.

Sleep, N.H., 1975. Formation of oceanic crust: some thermal constraints. J. Geophys. Res., 81: 4037—4042.

Talwani, M.B., Le Pichon, X. and Ewing, M., 1965. Crustal structure of the mid-ocean ridges, 2. Computed model from gravity and seismic refraction data. J. Geophys. Res., 70: 341—352.

Chapter 7

QUALITATIVE MODELS OF SPREADING-CENTER PROCESSES, INCLUDING HYDROTHERMAL PENETRATION*

C.R.B. LISTER

Departments of Geophysics and Oceanography, University of Washington, Seattle, Washington 98195 (U.S.A.)

SUMMARY

Observations on active geothermal areas suggest that water can penetrate rapidly into hot rock, and this is confirmed by semi-quantitative theory. Upwelling of magma material at an oceanic spreading center places hot rock close to an ample supply of cold water. Clues to the nature of the interaction can be found in the structure of ophiolite suites, and in simple physical induction on the nature of the upwelling. The subsurface dike injection zone seems to be dominated by conductive cooling and to provide a barrier against water penetration as long as it is heated by liquid magma in a chamber below. As soon as the magma has crystallized into a cumulate, water penetration proceeds rapidly to the base of the crust. It is probably stopped by meeting an ultramafic layer of olivine phenocrysts separated from the upwelling melt. The olivine crystals may be oriented by a fluidized-bed process near the spreading axis, forming a thin acoustically anisotropic layer. Below, mantle material depleted by partial melting grades back to primitive composition at depth. The mantle material should remain viscous enough to upwell in a broad zone, and the coupling of distributed upwelling to a thick rigid crust formed close to the spreading axis is responsible for the central valley and block tectonics characteristic of slow-spreading ridges. At fast spreading rates, the larger magma chamber decouples a thinner rigid surface layer and smooth topography results.

INTRODUCTION

Evidence has been accumulating that water penetration into young oceanic crust is a major factor in determining the thermal and physical state of the rocks. Heat-flow measurements have shown conclusively that the thermal flux through sediment ponds near the ridge crest is far too low to be compatible with new lithosphere formed at reasonable temperature (Lister, 1972). Nevertheless, careful analysis of the decay of topographic height with age has shown that this *is* compatible with a lithosphere of reasonable

*Originally published as: Lister, C.R.B., 1977. Qualitative models of spreading-center processes, including hydrothermal penetration. In: S. Uyeda (Editor), Subduction Zones, Mid-Ocean Ridges, Oceanic Trenches and Geodynamics. Tectonophysics, 37: 203—218.

rock parameters formed at the melting point (Davis and Lister, 1974).
The *distribution* of heat-flow values over medium-scale topography was the
vital evidence that confirmed the existence of hydrothermal circulation in
the basement rocks (Lister, 1972). Detailed investigations since then have
suggested that the lateral scale of the convection is about 8 km between
peaks (Williams et al., 1974) and that the fluctuations are too smooth for
the water flow to be confined to widely spaced fractures as suggested by
Bodvarsson and Lowell (1972). If permeability is distributed through the
rocks, then the convection cells should have a similar aspect ratio to the 1.6
found for λ/d in laboratory experiments (Elder, 1965). This would imply
active circulation to a depth of 5 km, so that the early thermal history of
the entire oceanic crust may be dominated by geothermal systems.

Highly metalliferous sediments are found near active ocean ridges
(Bostrom and Petersen, 1969; Dymond et al., 1973; Piper, 1973) and the
geochemical flux needed to generate them should be associated with the
circulation of hydrothermal waters. Diffusion of even the more mobile
chemical species is slow in the silicate rock matrix, and therefore complex
processes of chemical change, such as those discussed by Wolery and Sleep
(1976), require that the water and rock be in intimate contact.This means
that the permeability should be distributed through the rock as many cracks
a short distance apart. A mechanism for the formation of this kind of permeability has been given by Lister (1974). The boundary between the
cracked, permeable rock and relatively undisturbed hot rock can be treated
as a propagating thermal front. Rapid cooling and thermal contraction of the
rock permit the development of horizontal tension even under high overburden pressure, and the cracks propagate. The theory predicts water penetration rates of the order of 60 m year^{-1}, based on very approximate rock
parameters and a cracking temperature of 800°K. An independent estimate
based on the thermal output, spacing and apparent lifetimes of New Zealand
geothermal areas suggests that a rate of 2 m year^{-1} may be more appropriate
(Lister, 1976).

It seems as if a plausible general mechanism for water penetration has
been established, and that natural rate of the process has been estimated to
within one or two orders of magnitude. The most salient feature of the
results is this rapid rate of penetration: meters per year. Geothermal areas
should have a highly active phase lasting only a few thousand years, during
which time the thermal output should be large. After active penetration
has ceased, circulation may continue, but the thermal flux is then merely
the conductive flux from below. It is as if the conduction-cooling boundary of the new lithosphere were moved down to the base of the crust by the
penetration of water. This paper is an attempt to divine the consequences
of such a sudden cooling process for the mechanism of generation of new
sea floor.

GENERALIZED OCEAN CRUST STRUCTURE

Discussion of the mechanisms operating at a spreading center requires at least a working model for the rock types and textures characteristic of each depth range in the crust and lithosphere. It is clear even from a cursory examination of any ridge crest bathymetric profile that reality is not represented well by any scheme of simple layering. Secondary volcanism and faulting are largely responsible for the rugged topography commonly observed, but the existence of a few smooth sections of the East Pacific Rise suggests that these processes are not primary or essential to the generation of a basic crust and mantle structure. Therefore a layered model can be used as the starting point for a generalized discussion so long as its layers are clearly understood to be quite variable at best, and possibly dissected by other rock types.

The relationship of ophiolite suites to oceanic crust has been discussed by Coleman (1971), and the actual structures by Moores and Vine (1971). There is general agreement between the seismic velocity structure observed in the deep ocean and the pressure-corrected velocities measured on ophiolite samples (Christensen and Salisbury, 1975). Though considerable reconstruction is needed to generate a simple suite of layers, and some idealization of the original crust may be inevitable, four distinct units can be discerned. At the top there are the expected sediments and pillow basalts, often in a melange. Under this is usually a "sheeted complex" consisting of essentially 100% dikes, and where the individual dikes are discernible throughout the 1—2 km thick layer. Below are gabbro masses showing the cumulate structures usually associated with the orderly cooling of large magma bodies. The basal layer is ultramafic rock showing little sign of more than partial melting, and varying from highly silica-depleted dunite to less depleted harzburgite. These layers have been depicted in Fig. 1 with two additions. The first is the sequence of geothermal convective systems that penetrate at least the crustal layers in young ocean floor. The second is the grading associated with separation and settling of the components in magma generated by the partial melting of "primitive" mantle material as it upwells under a spreading center.

The crustal layer of basaltic composition must be offset by depleted ultramafic rocks that grade back toward the original mantle composition in the layer labeled as peridotite. However, it is unlikely that primitive material could experience enough partial melting from pressure release alone to leave a residue of pure olivine dunite. Such materials are more likely to be made by the settling of olivine crystals from the fluid in a magma chamber. A possible mechanism considers the escape of a small fraction of partial melt from its host matrix. Whatever the initial method of separation on a microscopic scale, the liquid must eventually upwell through relatively narrow channels or fissures. The channeling is assisted by the energy release associated with the upwelling of a fluid of lower density than the host rock. If the density contrast were 0.6 g cm^{-3}, for example, upwelling from a depth of 30 km would result in the liberation of 40 cal. cm^{-3} of frictional

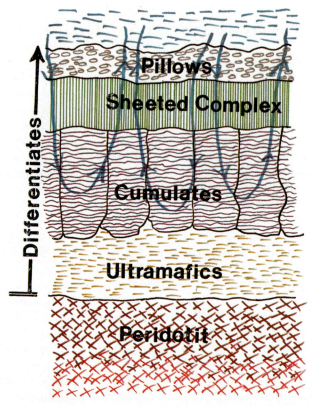

Fig. 1. Idealized ocean crust structure. The differentiates are materials that separate from upwelling magma containing phenocrysts. Mantle material depleted by partial melting is arbitrarily labeled 'peridotite'. Hydrothermal circulation is indicated by the blue arrows.

heating, enough to melt out the channels. The relatively high velocity flow in these channels would be capable of carrying dense phenocrysts up with the liquid. As soon as the fluid is dispersed in a magma chamber these crystals would settle to form a layer of dunite, and this would happen before crystals from the cooling of the melt could be deposited. Thus, the "differentiation boundary" refers to the redistribution of components from the fluid mush that separates from mantle material after pressure release, and not to the graded depletion due to partial melting.

THERMAL CONSTRAINTS AND THE SHEETED COMPLEX

Having generated a magma of basaltic composition by the pressure-release melting of asthenosphere material, one must begin to consider its interaction with the cold ocean boundary above. The mechanisms associated with the extrusion of pillow lavas are well enough understood not to need discussion here. The sheeted complex represents the real quasi-steady-state boundary between magma and cold water, and is a thick enough layer of uniformly solid material to dominate the mechanics of new oceanic plate. It is obvious how continuous but not steady extension, and continuous upwelling of magma, can form a structure composed of nothing but dikes. The thermal regime associated with this process is, however, highly dependent on whether or not water percolation is important.

Consider first the case of pure conductive cooling. Material is supplied intermittently at the melting point, and deposits its latent heat and specific heat of cooling into the walls. Mathematical modeling of this process is tricky but has been discussed extensively by Sleep (1975). The model assumes constant horizontal velocity for the material as soon as emplaced, and takes an average of the temperature at the emplacement axis through the cycle of dike intrusion and cooling. It cannot model the existence of a large magma chamber, because vertical heat transport by convection maintains a nearly constant heat flux at the top of the chamber while the crystals settle at the bottom. The zone of dike intrusion is represented fairly well, and the point at which the magma just remains molten between dike injections is given accurately by the intersection of the solidus isotherm with the axis of spreading. This is a lower limit for the extent of the sheeted complex because no dike "chilled margins" can be generated when accretion of solid rock by freezing is continuous. The smooth transition between solidified dike and permanent magma is an important feature of the mechanism because it can account for the re-splitting of the most recent dike rather than quasi-random dike injection in tensioned crust above a magma chamber. The maximum thickness of the sheeted complex is a function only of rock/magma properties and spreading rate: it is of the order of 700 m at 5 cm year^{-1}, 1100 m at 3 cm year^{-1} and over 3 km at 1 cm year^{-1} (Sleep, 1975, fig. 1). The point at which chilled margins become visible on the dikes depends greatly on the rate-of-freezing differences needed to produce noticeable grain size changes, and on the actual thickness of the dikes. It is unlikely to require that the geologically indentifiable sheeted complex be less than 70% of the above quoted thicknesses.

The conductive model is capable of producing a sheeted complex, but one of thickness limited by spreading rate. If water percolation is an important contributor to heat transfer, the sheeted complex could be considerably thicker, in fact, almost indefinitely thick. Comparison between the thicknesses observed in ophiolite suites and those expected is not practical, since it is unlikely that reliable spreading rates will ever be estimated for the slivers of oceanic crust now found on land. Although there is good reason to suppose that the sheeted complex is the magnetic recording layer that generates the marine anomalies observed at the ocean surface, determination of the thickness of the main magnetized layer is difficult. The surface pillow lavas should be sufficiently randomized by the mechanics of the extrusion process not to have a significant magnetic expression at a distance: attempts at modeling a magnetization to fit the field observed near the ocean floor tend to confirm this (Larson et al., 1974). The gabbros below the sheeted

complex are cooled by the downward propagation of nearly horizontal isotherms, and reversals in their magnetization should also have little surface
expression. Thus the sheeted complex should comprise the magnetic layer,
but the thickness of the layer affects only the amount of "overshoot" in the
anomaly amplitude that would be observed above a sharp reversal transition.
A thicker magnetized layer should generate smoother anomalies, and this
tendency should be reinforced by the narrower magnetized blocks associated
with the slower spreading rate. Unfortunately, there are variations in the
sharpness of the magnetization boundary generated at a real ridge crest, since
the dike-injection center may move about. There are enough variations in the
"sharpness" of Pacific anomalies generated at roughly similar spreading rates
(Pitman et al., 1968) to make quantitative estimates of magnetic layer thickness impractical. All one can conclude is that water percolation, though
necessary to account for low axial valley heat-flow measurements, does not
seem to play a major rôle in the formation of the sheeted complex if the
thicknesses observed in ophiolite suites are representative of crust generated
at moderate spreading rates.

The lack of a dramatic cooling of the sheeted complex by penetrating
hydrothermal convection may appear strange when one considers the high
penetration speeds calculated even for very low cracking temperatures
(Lister, 1974). However, the stress situation in the sheeted complex departs
radically from the simple one-dimensional model, and, in fact, represents an
extreme case of the "pressure-case" dominated penetration discussed qualitatively by Lister (1976). The one-dimensional model assumes that each
cooled layer is of infinite extent and thus cannot contract at all without

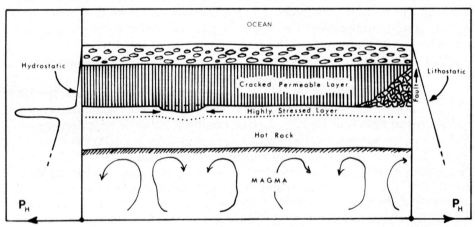

Fig. 2. Diagram of the stress system in a surface layer partially penetrated by water, when
faulted regions are under normal lithostatic pressure (graph of horizontal pressure at
right). The lower *hydrostatic* pressure in the permeable region must be compensated by a
highly stressed layer below so that horizontal forces balance. Incipient instability in the
cracking front boundary (shown at left center) is inhibited by stress concentration in the
thinner, warmer, and distorted stressed layer.

cracking. The ends of a ridge segment are unlikely to offer any resistance to large-scale shrinkage of the crustal layer (e.g. Turcotte, 1974). Any tension developed in surface layers must be balanced by compression below, and, if magma can upwell in major shrinkage cracks, the compressive layer must support the full difference between the lithostatic and hydrostatic pressures, integrated down to the depth of penetration. A schematic diagram of the stress situation is given in Fig. 2. The thin layer cool enough to sustain high compressive stress tends to stabilize the system against geometric instability as long as a magma heat source is maintained below. Any convective cell that penetrates deeper than average causes increased local stress, and therefore local compressive creep, and also higher local thermal flux (Fig. 2). The presence of the magma chamber prevents any large-scale thickening of the compressed layer and so permits a kind of equilibrium to be established with only partial water penetration.

MAGMA CHAMBER COOLING

The presence of an equilibrium impermeable layer between the circulating seawater and the magma chamber permits a crude estimate to be made of the crystallizing time of the cumulates. The hydrothermal systems maintain the top of the impermeable layer at near seawater temperature, while the base is at the solidus of basaltic magma (say $1200°C$). The conductivity of basalt is about $4.5 \cdot 10^{-3}$ cal. cm^{-1} s^{-1} °C^{-1} (Clark, 1966), so that a 1 km thick layer would pass a heat flow of 54 HFU or 1700 cal. year^{-1}. Assuming a latent heat of about 300 cal. cm^{-3}, 3 km of magma would be frozen in 50,000 years by heat losses through the top alone. On fast spreading ridges, say 5 cm year^{-1}, the magma chamber would extend about 2.5 km on each side of the crest, since lateral heat losses are small. On slow-spreading ridges, lateral losses to nearby penetrative hydrothermal systems should dominate, and, in spite of a thicker impermeable layer, only a small volume of magma may be present. If the rate is slow enough, conductive cooling alone would prevent the existence of any magma chamber at all, and spreading would have to become highly episodic (Sleep, 1975).

The overall effect of partial water penetration into the sheeted complex should be to somewhat thicken the layer at high spreading rates, and reduce the width of the magma chamber underneath it. The effects at slow spreading ridges should be expressed mainly by higher lateral heat losses at depth, since the narrow magma chamber top would prevent even quasi-equilibrium conditions from developing.

PENETRATIVE CONVECTION

As soon as the cumulate has crystallized, there is no longer a magma-convection heat source at the base of the sheeted complex. Cooling begins to propagate downwards, and the stabilizing influence of the heat boundary is lost. Under these conditions, three-dimensional "cold-finger" convective structures should develop, and propagate downward slightly more slowly than the one-dimensional model prediction for their cracking temperature (Lister, 1976). Rates similar to those estimated for the New Zealand geothermal areas should prevail: $0.2-20$ m year^{-1}. Even the slowest of these rates is substantially faster than sea-floor spreading, and cold water should penetrate to the depth of the Mohorovicic discontinuity within a negligible increment of distance from the ridge crest. The various mechanisms that may terminate water penetration have been discussed at length (Lister, 1974). The only point that need be added is the probable influence of the change in rock compositions at the base of the differentiated crustal layer. As mentioned above, a transition from gabbro to ultramafic material should occur at the base of the magma chamber. Olivine has the property of being readily attacked by water and metamorphosed into serpentine and brucite, with a substantial volume increase (Coleman and Keith, 1971). It is most likely that this process, with or without a contribution from static fatigue of the rock, causes closing of the cracks and cessation of penetration.

PLATE THICKNESS AND UPWELLING

The thermal constraints of the sheeted complex, and a great deal of circumstantial evidence, suggest that the divergence in the velocity field of the top surface of the lithosphere is confined to a very narrow region. This requires that all the material flux needed to supply new crust upwells in the same narrow zone. Since the bulk of the oceanic crust is made of cooled magma products (Fig. 1) there is no difficulty in the requirement: the viscosity of a basaltic magma is many orders of magnitude lower than that of lithosphere or asthenosphere. A different situation applies below the crust: upwelling asthenosphere material will have a lowered viscosity because of the partial-melting process, but not sufficiently so to upwell in a very narrow column or plume. Moreover, the supply of asthenosphere material must be derived at a depth below the partial-melting zone, so that the upwelling must begin by being relatively broad.

How much mantle material flows horizontally away from the ridge crest with the lithosphere is difficult to estimate, since it depends on the distribution of viscosity in the upwelling. All realistic thermal models of the lithosphere itself imply that it thickens with age (e.g. Oxburgh and Turcotte, 1968; Lister, 1972; Parker and Oldenburg, 1973), and begins by being quite thin. Viscous drag on the underside of it entrains a considerable additional volumetric flow with the moving surface, and the existence of a large overall volumetric upwelling is confirmed by the considerable volume of pressure-release partial melt. The details of the flow pattern are well beyond this discussion, but the surficial mechanism is clearly affected by the fact that the mantle upwelling is broader than the accretion zone of a rigid crust.

If a rigid skin develops the full spreading velocity at the axis, but the full material supply is not upwelled there, then conservation of mass can be satisfied only by the presence of an axial valley (Deffeyes, 1970). Segments of the skin are uplifted where they reside on the flank of the upwelling, and, since there is a lateral gradient of upwelling, they must also be tilted outward. The existence of such tilting has been verified by acoustic reflection profiling of turbidite sediments on the Juan de Fuca ridge (Davis and Lister, in press). It is well known that median valleys are normal on slow-spreading ridges and absent on fast-spreading ridges, with the Juan de Fuca ridge at 3 cm year^{-1} showing transitional behavior. Two aspects of the mechanism of spreading can account for this: the decoupling effects of a magma chamber, and the reduction of entrained mantle flow due to lower viscosity beneath fast-spreading ridges.

The Juan de Fuca ridge is particularly interesting because the rate of sedimentation seems to influence the crestal structure. Over most of the ridge, a central valley exists only as a small depression of 200 m or so. At the north end, where sediments flood onto the spreading center from the nearby continental margin, there is an enormous central valley reaching a depth of almost 5 km beneath the sediment pile (Davis and Lister, in press). An intriguing possibility, suggested by the thermal and water-percolation analysis above, is that the sediment blanket prevents the formation of the usual quasi-stable impermeable sheeted complex as there is too little surficial conductive cooling. The paradoxical result is that cooling extends deeper than usual near the injection zone because of penetrative hydrothermal convection. This removes the buffering magma chamber, and increases the thickness of the mantle layer flowing with the plate, so that the transitional ridge shows the tectonic features of a fully developed slow-spreading ridge. There is some evidence that cooling has penetrated to such a depth that the spreading of that central valley has slowed almost to a standstill, and the center of activity has jumped to the west (Davis and Lister, in press).

COMPLETED QUALITATIVE MODELS

Now that the constituent mechanisms of spreading have been developed to a sufficient qualitative level, it is possible to assemble diagrams showing the whole system in operation. The fast-spreading ridge is the most straightforward and is shown in Fig. 3. A weak "central valley" exists in the mantle layer, but is decoupled from the surface by the magma chamber. Water penetration is limited to the upper part of the sheeted complex by thermal stabilization until crystallization of the cumulates is complete. At this point, rapid penetrative convection of water cools the whole crustal layer, and may extend a short distance into the ultramafic material. Once the depth of water circulation in the crust has stabilized, conductive cooling of the upper mantle begins at that reference depth, and the lithosphere thickens according to the square-root law of thermal cooling (Parker and Oldenburg, 1973; Davis and Lister, 1974).

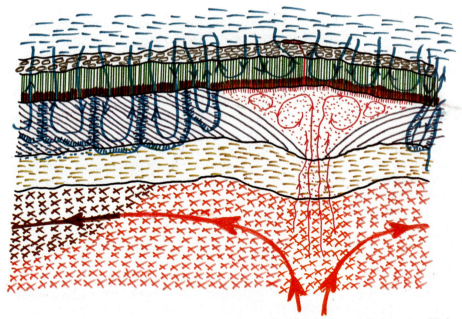

Fig. 3. Completed picture of the main processes active at a fast spreading center. High temperatures are indicated by orange, red and pink, except in the ultramafic layer. Generalized flow directions are indicated by arrows in the mantle and in the water-saturated region. Because the magma chamber isolates the thin rigid crust from the upwelling trough in mantle material, the system can spread smoothly without significant surface tectonics. Orientation of the olivine crystals in the ultramafic layer occurs due to stretching of the central region while it is fluidized by upwelling magma.

The case of the slow-spreading ridge is more speculative because of its complexity, and the probable importance of some of the non-uniform processes that are hard to incorporate into a simple model. An attempt to picture the mechanism has been made in Fig. 4. Here a small magma chamber is cooled both laterally and at the top by nearby water circulation. The lateral closeness of the water penetration, and the width of the magma chamber, are both exaggerated in the figure relative to what they should be at a really slow spreading ridge (1 cm year^{-1}). The basic difference between Fig. 4 and Fig. 3 is just the spatially more rapid cooling of the new material because of the slower spreading rate. The dramatic difference in tectonic structure is due to the coupling of differential upwelling in the mantle material to the

Fig. 4. Simplified and somewhat speculative picture of spreading at a slow rate. Colors as in Fig. 3. A narrow magma chamber fails to isolate the rapidly developed thick rigid crust from the axial trough inherent in broad mantle upwelling, and block tectonics result. A broader zone of partial melting than of crustal magma retention will cause secondary intrusion and volcanism near the spreading center. This process is not shown because of the pictorial complexity it would introduce.

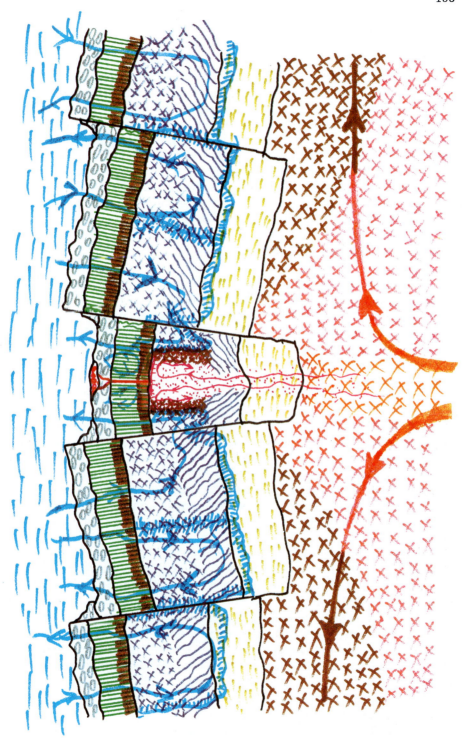

thick rigid crustal blocks. Instead of a smooth surface trending almost uniformly down from the ridge crest, there is large-scale block faulting that develops the ridge-and-valley topography so characteristic of the Mid-Atlantic ridge. The fact that upwelling and partial melting should occur in a mantle region much wider than the accretion zone of the crust suggests that secondary volcanism and/or intrusion may be an important process. It was not practical to include this in the diagram due to the graphic difficulties even in a multicolor presentation. However, one minor point that was included is the difference in the cumulate structure between material laterally accreted by cooling of the walls, and material deposited by crystal settling. The former is essentially a dikeless sheeted complex that should exist as a narrow transition region even at a fast spreading center. At a slow spreading ridge it would be a major feature of the gabbroic layer.

TEST OF THE THERMAL PREDICTION

The principal prediction of any model of spreading that incorporates water percolation is the concentration of heat output near the ridge crest. Not only is the high thermal flux from the incipient cooling of the mantle transferred to the ocean, but the entire heat content of the crust is removed by the penetrative convection of water. This prediction cannot be tested readily in the open ocean because of the rapid dispersal of any output of hot water. Only the largest thermal plumes that are generated during the highly active phase of hydrothermal penetration can be detected (Lister, 1972; Scott et al., 1974). However, the water temperatures around one active ridge in a closed basin have been investigated by Detrick et al. (1974), and data from their fig. 4 can be reinterpreted to demonstrate the ridge crest output. Four near-bottom potential temperature profiles from a region just south of the ridge crest are replotted in Fig. 5. The variations between profiles are largely due to lack of synopticity: current meter data in the area (Detrick et al., 1974) show that water can move up to 50 km in 23 days, while the profiles are up to 25 km apart and taken over a comparable period of time.

The average history of water in the basin can be modeled by supply at a given source potential temperature over a topographic sill, rapid spreading out over the floor of the basin, and then a gradual rise toward the outflow level at an approximately constant flow rate. As the deep basin is stratified by upward increasing potential temperature, the horizontal eddy diffusion rate should be much larger than the vertical eddy diffusion rate. Then each horizontal layer in a restricted basin, such as that south of the ridge crest, can be considered laterally mixed, while the vertical mixing should be quite restricted. This is a crude approximation, but can be improved upon only if synoptic temperature and current meter data are available for the whole basin. Where the horizontal cross-section of the basin is relatively constant, the upward flow velocity should be fairly constant also. In the absence of

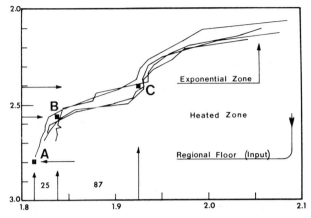

Fig. 5. Potential temperature profiles (°C) against depth (km) from near the Galapagos Rift Zone in the Panama basin, replotted from Detrick et al. (1974). The mean ridge crest height is at the level of point C, while point A is approximately the basin-floor level south of the spreading center. Point B is placed arbitrarily at the inflexion in the profiles. Temperature differences AB and BC are shown in millidegrees C.

significant bottom heating, the temperature profile would be exponential downwards if the effective diffusivity were constant (Carslaw and Jaeger, 1959). There is such a region in Fig. 5 for depths above the mean ridge crest at 2400 m (point C). High gradients in the mid-water that can flow through the basin above several sills decay downward into a nearly isothermal zone.

Below the depth of the mean ridge crest, the temperature profiles show two quasi-linear regions: steeply changing temperature just below the ridge crest (BC), and then a more gradual decrease toward the floor of the basin (AB). The striking linearity of two of the profiles could be a random artifact, but the crude flow model would actually show this if the heat output of thermally shrinking topography were added to the water. Each incremental drop in the height of cooling lithosphere requires the release of an approximately constant amount of heat, since the shrinkage at any place is $\alpha \Delta T$ while the heat loss is $\rho c \Delta T$: they are linearly proportional to the local ΔT. Here α is expansion coefficient (corrected for isostatic compenstion), ρ is density, c is specific heat and ΔT is a temperature change. If the heat release from the change in topographic height is added to the water layer at that height its temperature will rise a constant amount per unit time (spreading rate and upward advection approximately constant).

If there were no hydrothermal heat release from the crustal layer, the linear profile would join the base of the exponential temperature profile due to diffusion from above. Instead there is a strong thermocline in the depth range of the ridge crest, indicating a large localized heat output. In semi-quantitative terms, a topographic height drop of 230 m produces a rise of 25 millidegrees on the flank, but a crestal height drop of only 150 m produces a heating of 87 millidegrees (Fig. 5). The height drop due to rapid cooling of the crustal layer near the spreading axis is not observed in the topography. The heat-loss equivalent in topographic height can be estimated by matching the total heat content of the basaltic layer with simple cooling of mantle material. If 6 km of mantle with an α of $5.6 \cdot 10^{-5}$ (Lister, in press) cool through $1500°C$ (the approximate equivalent heat content of molten basalt), the topography would drop by 500 m. Thus the heating of the crestal region is really equivalent to a total topographic drop of 650 m, bringing the ratio of the topographic drops, 2.8, much closer to the ratio of temperature rises, 3.5.

The above analysis is very crudely approximate, but it does predict the generation of a thermocline at the ridge crest depth when a circulating closed basin contains a spreading center. The only alternative explanations for such an observed thermocline are that it is the result of highly discontinuous feed of cold water over the basin sill, or that it is produced by a large contemporary lava flow at the ridge crest. Both these alternatives could be ruled out if a reinvestigation of the area demostrated a similar pattern after several years.

CONCLUSIONS

The first-order model of a spreading center incorporating the effects of rapid water penetration into hot rock appears to be consistent with the observed data. Moderate-sized magma chambers are permitted near the axis at fast spreading rates because water does not penetrate close to the hot isothermal top boundary of magma. Nevertheless, the entire crustal section is fully cooled within a few km of the ridge crest even on the East Pacific Rise, and so is consistent with the normal crustal layering found by seismic refraction measurements near the crest (Snydsman et al., 1975).

Another hitherto puzzling aspect of the seismic data may become more comprehensible. Strong velocity anisotropy is observed in the suboceanic mantle near the ridge crest, but diminishes on older litosphere (Snydsman et al., 1975). An important feature of the models in Figs. 3 and 4 is the layer of olivine crystals that develops at the base of the magma chamber. This layer is fluidized near the axis by the upwelling of magma through it, and it is also being stretched in the direction of spreading. These conditions are ideal for the alignment of anisotropic crystals, and the long axis of the crystals becomes parallel to the stretching. Since the long axis of olivine crystals is also the direction of the highest seismic velocity (N.H. Christensen, personal communication, 1974), the highest velocities should be observed transverse to the ridge. This is the pattern that is observed, and the anisotropy decreases with age by an increase in the velocity observed parallel to the spreading axis, while the transverse velocity remains approximately constant (Snydsman et al., 1975). The model is consistent with this also, since the anisotropic layer is a thin one, underlain by relatively isotropic material. Near the ridge crest, the temperature of the underlying layer is still high and the velocity is reduced substantially by increased temperature (N.H. Christensen, personal communication, 1976). The fastest velocity observed is the anisotropically low velocity of the surface mantle layer. As cooling penetrates into the mantle below the anisotropic layer, the velocity there rises and is observed as soon as it exceeds that in the layer above. In the fast direction, the anisotropic layer remains faster than even fully cooled mantle material below, so that the observed velocity does not change.

Well-known laws of heat conduction (Sleep, 1975) and the qualitative theory of water penetration into hot rock have been applied to the surficial mechanisms at an oceanic spreading center. Together with simple ideas about the derivation of magma from upwelling asthenosphere material. they have produced a physically workable ridge crest mechanism. The first-order predictions of the model are consistent with marine geophysical data, at least for the case of fast spreading ridges. Further testing can be accomplished by unravelling some of the complexities at slow spreading centers, such as the Mid-Atlantic Ridge, and careful comparison of model predictions with the structures observed in ophiolite suites.

ACKNOWLEDGEMENTS

This work was supported by National Science Foundation grant DES 73-06593-A01.
I wish to thank S. Uyeda for the moral encouragement that caused this work to be undertaken.

REFERENCES

Bodvarsson, G. and Lowell, R.P., 1972. Ocean-floor heat flow and the circulation of interstitial waters. J. Geophys. Res., 77: 4472—4475.
Bostrom, K. and Petersen, M.N.A., 1969. The origin of aluminum ferro-manganoan sediments in areas of high heat flow on the East Pacific Rise. Mar. Geol., 7: 427—447.
Carslaw, H.S. and Jaeger, J.C., 1959. Conduction of Heat in Solids. Oxford Univ. Press, 2nd ed., 388 pp.
Christensen, N.I. and Salisbury, M.H., 1975. Structure and constitution of the lower oceanic crust. Rev. Geophys. Space Phys., 13: 57—86.
Clark, S.P. Jr., 1966. Thermal conductivity. In: S.P. Clark Jr. (editor), Handbook of Physical Constants. Geol. Soc. Am. Mem., 97: 461—482.
Coleman, R.G., 1971. Plate-tectonic emplacement of upper-mantle peridotites along continental edges. J. Geophys. Res., 76: 1212—1222.
Coleman, R.G. and Keith, T.E., 1971. A chemical study of serpentinization — Burro Mountain, California. J. Petrol., 12: 311—328.
Davis, E.E. and Lister, C.R.B., 1974. Fundamentals of ridge crest topography. Earth. Planet. Sci. Lett., 21: 405—413.
Davis, E.E. and Lister, C.R.B., in press. Tectonic structures on the Juan de Fuca ridge. Geol. Soc. Am. Bull.
Deffeyes, K.S., 1970. The axial valley: a steady-state feature of the terrain. In: H. Johnson and B.L. Smith (editors), Megatectonics of Continents and Oceans. Rutgers Univ. Press., Camden, N.J., pp. 194—222.

Detrick, R.S., Williams, D.L., Mudie, J.D. and Sclater, J.G., 1974. The Galapagos spreading center: bottom water temperatures and the significance of geothermal heating. Geophys. J.R. Astron. Soc., 38: 627—637.

Dymond, J., Corliss, J.B., Heath, G.R., Field, G.W., Dasch, E.J. and Veeh, H.H., 1973. Origin of metalliferous sediments from the Pacific Ocean. Geol. Soc. Am. Bull., 84: 3355—3372.

Elder, J.W., 1965. Physical processes in geothermal areas. In: Terrestrial Heat Flow. Am. Geophys. Union Monogr., 8: 211—239.

Larson, R.L., Larson, P.A., Mudie, J.D. and Spiess, F.N., 1974. Models of near-bottom magnetic anomalies on the East Pacific Rise crest at 21°N. J. Geophys. Res., 79: 2686—2689.

Lister, C.R.B., 1972. On the thermal balance of a mid-ocean ridge. Geophys. J.R. Astron. Soc., 26: 515—535.

Lister, C.R.B., 1974. On the penetration of water into hot rock. Geophys. J.R. Astron. Soc., 39: 465—509.

Lister, C.R.B., 1976. Qualitative theory on the deep end of geothermal systems. 2nd U.N. Geothermal Symp. Vol.: in press.

Lister, C.R.B., in press. Estimators for heat flow and deep rock properties based on boundary layer theory. Tectonophysics.

Moores, E.M. and Vine, F.J., 1971. The Troodos massif, Cyprus, and other ophiolites as oceanic crust, evaluation and implications. Philos. Trans. R. Soc. London, Ser. A, 268: 443—466.

Oxburgh, E.R. and Turcotte, D.C., 1968. Mid-ocean ridges and geotherm distribution during mantle convection. J. Geophys. Res., 73: 2643—2661.

Parker, R.L. and Oldenburg, D.W., 1973. Thermal model of ocean ridges. Nature, 242: 137—139.

Piper, D.Z., 1973. Origin of metalliferous sediments from the East Pacific Rise. Earth Planet. Sci. Lett., 19: 75—82.

Pitman, W.C., III, Herron, E.M. and Heirtzler, J.R., 1968. Magnetic anomalies in the Pacific and sea-floor spreading. J. Geophys. Res., 73: 2069—2085.

Scott, R.B., Rona, P.A., McGregor, B.A. and Scott, M.R., 1974. The TAG hydrothermal field. Nature, 251: 301—302.

Sleep, N.H., 1975. Formation of oceanic crust: some thermal constraints. J. Geophys. Res., 80: 4037—4042.

Snydsman, W.E., Lewis, B.T.R. and McClain, J., 1975. Upper-mantle velocities on the northern Cocos plate. Earth Planet. Sci. Lett., 28: 46—50.

Turcotte, D.L., 1974. Are transform faults thermal contraction cracks? J. Geophys. Res., 79: 2573—2577.

Williams, D.L., Von Herzen, R.P., Sclater, J.G. and Anderson, R.H., 1974. The Galapagos spreading center: lithospheric cooling and hydrothermal circulation. Geophys. J. R. Astron. Soc., 38: 587—608.

Wolery, T.J. and Sleep, N.H., 1976. Hydrothermal circulation and geochemical flux at mid-ocean ridges. J. Geol: in press.

Chapter 8

HYDROTHERMAL CONVECTION AT SLOW-SPREADING MID-OCEAN RIDGES[1]

UDO FEHN and LAWRENCE M. CATHLES *

Department of Geological Sciences, Harvard University, Cambridge, Mass. 02138 (U.S.A.)
Ledgemont Laboratory, Kennecott Copper Corporation, Lexington, Mass. 02173 (U.S.A.)

SUMMARY

The discrepancy observed between measured heat flow data and the heat flow predicted by thermal models of mid-ocean ridges is commonly explained by the presence of hydrothermal convection in young oceanic crust. Numerical modelling of fluid flow through porous media has been used to investigate what permeability, depth of penetration and mass fluxes are necessary to produce conductive heat flow distributions compatible with observed heat flow data at spreading centers. The results presented here were calculated for oceanic crust between 0 and 2 m.y. old at a mid-ocean ridge with a half-spreading rate of 1 cm/yr.

The calculations show that theoretical and observed heat flow near mid-ocean ridges can be brought into better agreement if non-uniform rather than uniform permeability is assumed in the oceanic crust. If the crustal permeability is uniform, the percentage of heat flow values ($\geqslant 25\%$) which are increased by upwelling flow above the predicted values is significantly higher than that observed at mid-ocean ridges ($<10\%$). If — due to faulting, for example — zones of high permeability exist in crust of low permeability, upwelling flow can be concentrated and the area of increased heat flow can be greatly reduced. In the latter case, the percentage of sea floor near mid-ocean ridges ($\geqslant 90\%$) where heat flow is depressed below the predicted values corresponds well with observed heat flow distributions near active spreading centers.

Average temperatures in crust with hydrothermal convection are considerably lower than those in purely conductive crust. This difference in average temperatures should result in crestal offsets at mid-ocean ridges. The observation that crestal offsets attributable to convective cooling are not larger than 50 m suggests that the depth of penetration of convection strong enough to produce conductive heat flow compatible with observed heat flow distributions is less than 5 km in crust younger than 2 m.y.

The downward flow of cold ocean water into the sea floor greatly reduces conductive heat flow. The magnitude of this inflow and hence the degree of heat flow reduction depends on the average permeability of the oceanic crust. Comparison of heat flow measurements from the FAMOUS area and from the Galapagos Spreading Center to heat flow over downwelling areas in our models indicates that the average permeability

[1] Originally published as: Fehn, U. and Cathles, L.M., 1979. Hydrothermal convection at the slow-spreading mid-ocean ridges. In: J. Francheteau (Editor), Processes at Mid-Ocean Ridges. Tectonophysics, 55: 239—260.
* Present address: Department of Geosciences, Pennsylvania State University, University Park, Pa. 16802 (U.S.A.)

(including faults or high permeability zones) in young oceanic crust is not larger than 2.5 millidarcy.

Finally, the integrated mass flux through sea floor between 0 and 2 m.y. old was found to be approximately $4 \cdot 10^6$ g yr^{-1} (cm of ridge). If this mass flux is considered representative for spreading centers and if older crust is included into the calculations, a total mass of $\sim 1 \cdot 10^{17}$ g yr^{-1} is convected through the worldwide system of mid-ocean ridges.

INTRODUCTION

Much evidence suggests that hydrothermal activity is associated with the emplacement of new oceanic crust along mid-ocean ridges. Numerous pieces of hydrothermally altered basaltic and ultramafic rocks have been dredged from ridge areas, and the metalliferous sediments from these regions, particullarly those in the vicinity of the East Pacific Rise, are best explained as the products of hydrothermal activity. Strong support for this hypothesis comes also from the recent discovery of active hydrothermal vents at the axis of the Galapagos Spreading Center (Von Herzen et al., 1977).

The presence of hydrothermal activity at spreading centers implies that convective as well as conductive heat transfer contribute to the cooling of newly formed oceanic crust. Consequently, the discrepancy observed between measured conductive heat flow and heat flow predicted by thermal models of mid-ocean ridges (e.g., Lister, 1972 and Morgan, 1975) as well as the great variability of closely-spaced heat flow measurements in these areas (Williams et al., 1974; Williams et al., 1977; Davis and Lister, 1977) are commonly attributed to hydrothermal convection in young oceanic crust. Estimates of the total difference between measured and predicted heat transfer at mid-ocean ridges range from $24 \cdot 10^{18}$ cal/yr to $65 \cdot 10^{18}$ cal/yr (Anderson et al., 1977; Wolery and Sleep, 1976; Williams and Von Herzen, 1974). The age of the ocean floor where measured and predicted heat flow begin to coincide varies between 5 m.y. for the Galapagos Spreading Center and 70 m.y. for the Atlantic Ocean (Anderson et al., 1977). This observation suggests that hydrothermal convection can occur in old crust far beyond the direct influence of hot intrusions at the axis of spreading centers.

While the presence of hydrothermal convection at mid-ocean ridges is commonly accepted, the precise manner in which this convection resolves the discrepancy between theoretical and observed heat flow is not well understood. In particular, since hydrothermal convection causes not only reduction of heat flow by the inflow of cold ocean water but also increase of heat flow by the exit of heated water, convection implies the existence of zones where heat flow is raised above the predicted values. The observation, that only a small percentage (<10%) of the heat flow values measured at mid-ocean ridges is higher than predicted, is an important constraint for models of hydrothermal convection at mid-ocean ridges.

A major obstacle to a better understanding of convection at mid-ocean ridges is the lack of knowledge about permeability and permeability distribution in oceanic crust. Estimates of the average permeability of oceanic crust range from 0.45 mD (1 millidarcy = 10^{-11} cm^2) (Ribando et al., 1976) to 10 mD (Anderson et al., 1977). The permeability in highly fractured parts of the oceanic crust was suggested to be as high as 10 darcy (Lister, 1974). Sediments are probably less permeable than fractured basement rocks; a range of 0.1 to 0.001 mD is given by Bryant et al. (1974), and a permeability of 0.25 mD was measured for a sediment sample from the Juan de Fuca Ridge (Pearson and Lister, 1973).

Qualitative explanations for patterns of hydrothermal convection associated with mid-ocean ridges have assumed either a uniform permeability of the oceanic crust (e.g., Lister, 1972) or flow through isolated fractures (Bodvarsson and Lowell, 1972). Similarly, two distinct types of quantitative models have been developed: Ribando et al. (1976) simulated the periodic heat flow distribution found by Williams et al. (1974) near the axis of the Galapagos Spreading Center by use of a continuum Darcy flow model with homogeneous permeability. Lowell (1975) investigated the effect of an isolated fracture loop on the heat flow distribution near a ridge system. Both of these models assumed a constant heat supply from below, and did not take into account the horizontal heat flow from narrow dike intrusions at the axis of spreading centers. The cooling of a single narrow dike intrusion by seawater convection was studied recently by Lowell and co-workers (Lowell et al., 1977, Lowell and Patterson, 1979).

We present here results of a Darcy flow model which can be used to compute fluid convection through oceanic crust of homogeneous permeability as well as through crust where flow is dominated by isolated fractures or shear zones. In this paper, we investigate hydrothermal convection in crust younger than 2 m.y. and without sediment cover at a ridge with half spreading rate of 1 cm/yr. The purpose of this investigation is to find permeability distributions and penetration depths for fluid flow which are compatible with heat flow distributions and topography observed at mid-ocean ridges. The results are then used to estimate mass fluxes and temperature distributions associated with hydrothermal convection at active spreading centers.

THE NUMERICAL MODEL

The numerical model we used for the calculation of hydrothermal convection at mid-ocean ridges is essentially the same model developed previously by Cathles (1977) for hydrothermal systems associated with the cooling of hot plutons. The model is based on equations for balance of momentum (Darcy's Law):

$$-\nabla p + \rho \underline{g} - \frac{\nu}{k}\, \underline{q} = 0 \tag{1}$$

balance of mass:

$$\nabla \cdot \underline{q} = 0 \tag{2}$$

and balance of heat:

$$\rho_m c_m \frac{\partial T}{\partial t} = -\nabla \cdot \underline{q} c_f T + K_m \nabla^2 T - \rho_m c_m \underline{v}_s \cdot \nabla T \tag{3}$$

where c_f = specific heat of fluid ($c_f = H/T$; H = enthalpy); c_m = specific heat of fluid saturated rock (0.2 cal/g dg); g = gravitational field strength (980 cm/sec^2); k = permeability; K_m = thermal conductivity of fluid saturated rock ($6 \cdot 10^{-3}$ cal/dg cm sec); p = pressure; q = mass flux; v_s = half spreading rate (1 cm/yr); ν = kinematic viscosity; ρ_m = density of fluid saturated rock (2.7 g/cm^3); ρ = density of fluid.

This system of equations is solved by standard finite difference techniques (Carnahan et al., 1969). An important feature of the model is that the pressure and temperature dependence of the kinematic viscosity, enthalpy and density of the circulating fluid (pure water) are taken into account (Cathles, 1977). The two-dimensional solution domains contain twenty grid points in the vertical z-direction and thirty points in the y-direction; the grid spacing of the model is variable and can be adjusted to the problem in question. All the calculations were carried out for a half-spreading rate of 1 cm/yr, an intrusion temperature of 1200°C, and a surface pressure of 200 bars corresponding to sea floor at 2000 m depth.

The above equations are applicable to cases of water convection through a homogeneous porous medium as well as through a system of interconnected fractures as long as the Reynolds numbers are smaller than 1 which was always the case in our calculations. In the case of flow through fractures, two approaches are possible: if flow takes place through fractures spaced less than a few hundred meters apart, the system can be modeled as flow through uniform porous media. When fracture spacings are in excess of a few hundred meters, the assumption that temperatures within matrix blocks between fractures are close to those calculated for a homogeneous porous medium is no longer valid. Convection can then be treated in terms of flow through individual zones of high permeability set in a matrix of low permeability.

BOUNDARY AND INITIAL CONDITIONS

The choice of boundary and initial conditions is of basic importance for the results of numerical models. The boundary conditions for fluid flow are obvious: no flow through the sides and through the base of the domain and free flow through the surface. The choice of temperature boundary conditions needs more discussion. Since our domain usually has a width of 20 km, horizontal heat flux through the right hand boundary due to a dike intrusion at the ridge axis (left hand boundary) is small and can be neglected. The right hand boundary of the domain was thus assumed to be insulating. The heat flow through the base of the domain was chosen to be constant at 6 HFU (1 HFU = 10^{-6} cal/cm^2 sec). This heat flow is increased by the heat flow from the ridge axis to a total average heat flow of approximately 15 HFU through the surface of the domain (see for example Fig. 3). The surface of the domain was kept at a constant temperature of 0°C. This boundary condition formally requires all heat flow at the surface to be conductive, although in a real situation this total heat flow may be partly conductive and partly convective. As an approximation to this real situation, we compute total and conductive heat flow at the first grid point which is less than 100 m below the surface. Total heat flow at this depth is a good estimate of the total heat flow through the surface, while the conductive contribution gives a useful measure of the minimum conductive heat flow in upwelling areas and of the maximum conductive heat flow in downwelling areas.

At the ridge axis, a boundary condition suggested by Oldenburg (1975) was chosen in most of our calculations:

$$-K_m \frac{\partial T}{\partial y} = \rho_m v_s (L + (T_m - T_f(t, z))c_m) \tag{4}$$

where L = latent heat (100 cal/g); T_m = intrusion temperature (1200°C); T_f = temperature at the left hand (ridge) boundary. The temperature T_f at a particular timestep t_n was chosen to be the temperature at the ridge boundary calculated for the previous timestep t_{n-1}. This boundary condition assumes a continuous spreading process at the ridge axis. Another possibility for simulating sea floor spreading at a rate of 1 cm/yr is the periodic intrusion of dikes of 200 m width at the ridge axis in time intervals of 20,000 years. Since evidence for such a periodicity has been found at the Mid-Atlantic Ridge (Moore et al., 1974, Ballard and Van Andel, 1977), this model was also tested.

To test the boundary conditions chosen and to evaluate the time necessary to reach steady state conditions, the evolution of fluid convection with time is compared for the following four cases, all computed for domains of 20 km width and 10 km depth and with a constant heat flow of 6 HFU from below:

(1) Intrusion of a dike of 200 m width and 9.5 km height at the ridge axis. The initial temperature in the domain of 0.25 mD uniform permeability was assumed to correspond to a heat flow of 6 HFU, i.e. 1000°C at the base, linearly decreasing to the surface. Intervals between intrusions were 20,000 years long.

(2) Heat flow from the ridge axis as given by equation (4) into a domain of 0.25 mD homogeneous permeability. Initial temperature of 1000°C at the base, linearly decreasing to the surface.

(3) Like case 2, but with initial temperature of 500°C at the base, linearly decreasing to the surface.

(4) Like case 2, but with uniform permeability of 0.5 mD in the domain.

The development with time of the average temperatures in the domain is demonstrated for these four cases in Fig. 1. Case 1, periodic intrusion at the ridge axis, begins at an average temperature of 500°C which is increased to

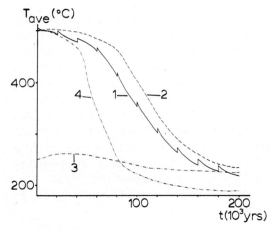

Fig. 1. Evolution of average temperatures with time in domains of 10 km depth and 20 km width. See text for description of cases *1—4*.

507°C due to the intrusion of a 200 m wide dike at 1200°C. Convection causes a decrease in average temperature to 493°C before the next episode of intrusion which increases the average temperature to 499°C, and so on. After about 50,000 years, the average temperature of the crust in the domain begins to decrease rapidly until a quasi-steady state temperature of about 225°C is reached.

The evolution with time of the average temperature in case *2*, continuous intrusion at the ridge axis, follows the trend of case *1* except for the somewhat later beginning of the decrease in average temperature. In case *3*, where an initial average temperature of 250°C was chosen, variation of the average temperature with time is small. These three cases, all of which have domains with uniform permeabilities of 0.25 mD, reach steady state average temperatures around 230°C. If a higher uniform permeability of 0.5 mD is assumed (case *4*) a sooner and steeper decline in average temperature results, and the steady state average temperature is reduced to 190°C.

The variation with time of total fluid mass efflux through the surface of the domain is compared in Fig. 2a for the four cases. The mass efflux is integrated over the width of the domain and given in 10^6 g/yr (cm of ridge axis). In a similar diagram (Fig. 2b), heat flow through the surface is plotted, also integrated over the width of the domain and given in cal/cm sec. Most of the mass flux shortly after the onset of convection is associated with the steep horizontal temperature gradients close to the ridge axis. After some time, several other convection cells develop in the domain which give rise to additional discharge areas. The increase of heat flow and mass flux caused by these additional convection cells coincides with the steep decline in average temperatures shown in Fig. 1. This decrease in average temperatures and the corresponding transient increase in mass flux and heat flow represents the fact that — because convective heat transfer is more efficient than conduc-

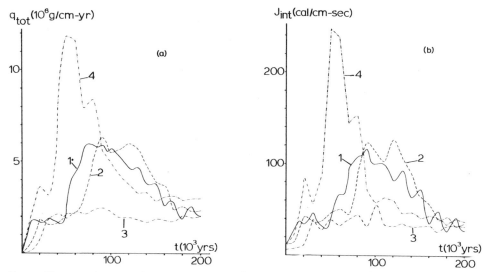

Fig. 2. Time evolution of integrated mass fluxes (a) and of integrated heat fluxes (b) through surface of 10 km deep and 20 km wide domain. See text for description of cases 1—4.

tive one — the heat stored and, correspondingly, the average temperature of the crust are higher in a purely conductive layer than in the same layer, if fluid convection takes place.

The comparison of these four cases suggests the following conclusions in regard to the choice of boundary and initial conditions for our models. First, the similarity of average temperatures, mass fluxes and heat fluxes found for the cases 1 and 2 indicates that the heat flow boundary condition (4) is a proper way to represent the intrusion of new oceanic crust at mid-ocean ridges. Since the boundary condition (4) is more flexible in choice of time steps and grid spacing than a periodic intrusion at the ridge axis, this boundary condition was chosen for the further calculations. Secondly, the observation that cases 1, 2 and 3 achieved very similar steady state temperatures, mass fluxes, and heat fluxes indicates that the solutions converge and reach steady state within ~200,000 yrs. This time span required to obtain steady state is short compared to the lifetime of spreading centers which usually have been spreading for millons of years. Steady state solutions should thus be typical for slow-spreading ocean ridges active today. Steady state in this context means that convection cells are fixed with respect to the ridge axis. This assumption implies that oceanic crust goes through zones of upwelling and downwelling convection and, correspondingly, through zones of high and low heat flow as it moves away from the ridge axis. Support for this assumption is offered by measurements made by Katz et al. (1977) who ascertained an oscillatory behavior of heat flow in drill core samples from the Mid-Atlantic Ridge.

FLUID CONVECTION FOR VARIOUS PERMEABILITY DISTRIBUTIONS IN THE OCEANIC CRUST

In Fig. 3, steady state temperature distribution, flow lines and heat flow are shown for crust with a uniform permeability of 0.25 mD (case 2). The ridge axis is at the left hand side of the domain of 10 km depth and 20 km width. Isotherms are shown in intervals of 100°C and stream lines in steps of 2.5 of the dimensionless stream function. A large convection cell driven by the horizontal temperature gradients at the ridge axis has developed. The convection cell has two discharge areas, one at the ridge axis with a width of

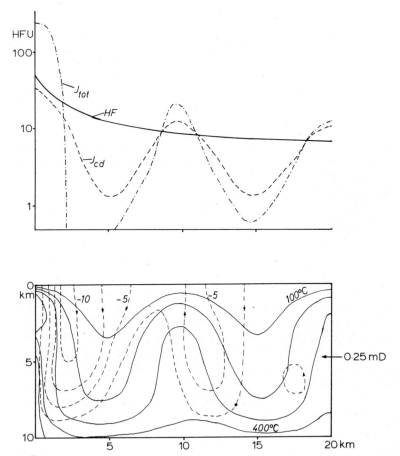

Fig. 3. Steady state temperatures and stream lines in crust of uniform 0.25 millidarcy permeability. Ridge axis is located at the left hand side of the domain. Total heat flow J_{tot} and conductive contribution J_{cd}, both calculated at a depth of 70 m, are compared to heat flow HF computed in absence of convection. Heat flow is given in heat flow units (1 HFU = 10^{-6} cal/cm^2 sec) on an exponential scale.

2.5 km and a second one at a distance of 10 km from the ridge axis. A weak counter cell has appeared at the right boundary of the domain. Temperatures in the domain are generally below 500°C, temperatures above 500°C are found only close to the ridge axis.

In the upper half of the diagram, heat flow through the surface is plotted against distance from the ridge axis. An exponential scale is used for the heat flow values which are given in HFU (1 HFU = 10^{-6} cal/cm^2 sec). Here and in the following diagrams, the total heat flow J_{tot} which is the sum of convective and conductive heat flow is plotted together with the conductive part J_{cd}, both calculated at a depth of 70 m. The conductive heat flow at the surface lies between these two curves. The exact breakdown between the conductive and convective portion of the total heat flow at the surface depends on fracture width and fracture spacing in a surface location. The curve *HF* indicates the heat flow distribution in absence of convection. In general, heat flow is depressed below the theoretical heat flow *HF* in areas of downwelling flow and is increased over discharge areas. Discharge areas cover approximately 30% of the surface of the domain in this case.

In Fig. 4, the flow pattern and isotherms are shown for a domain in which fluid flow is restricted to the upper five kilometers of the crust. The impermeable zone is separated by the heavy broken line from the permeable zone where a uniform permeability of 0.5 mD was assumed. Three separate convection cells can be distinguished, the main one convecting towards the ridge axis. The two weaker convection cells have discharge areas in distances of 10 and 20 km from the ridge axis. Temperatures in the impermeable part of the domain are very little affected by the convection and reach values of more than 1000°C at the ridge axis. In the permeable half of the domain, temperatures are below 500°C except for the area close to the ridge axis. The heat flow distribution is similar to the one of Fig. 3, but the discharge areas are somewhat narrower and cover only 25% of the surface area.

The assumption that oceanic crust is of uniform permeability is probably not a good representation of the situation at mid-ocean ridges. Since newly intruded magma is permeable to water circulation only after solidification and after contraction due to cooling has opened fractures, a temperature dependence of the permeability of oceanic crust is likely. The temperature at which cooling magma becomes permeable lies between 400°C and 800°C (Lister, 1974). Because temperatures in the convective part of the domain are generally lower than these cracking temperatures, it was found that models which took into account the impermeability of crust at temperatures above 500°C resulted in steady state distributions of temperatures and fluid flow very similar to those of models in which this temperature dependence of the permeability was neglected. This result assumes that the rate of propagation of a cracking front through newly intruded magma is faster than the average rate of intrusion, an assumption which is justified by calculations on the penetration of water into hot rock. Lister (1974, 1977) estimates that a cracking front propagates through cooling rock at rates between 2 and 60

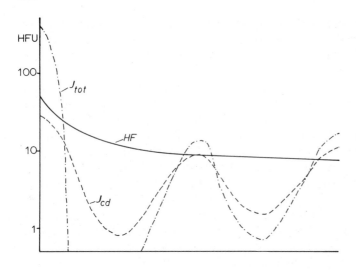

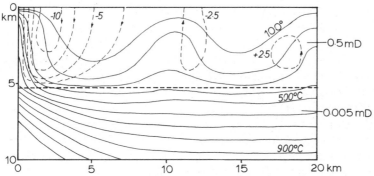

Fig. 4. Steady state temperatures, stream lines and heat flow distribution for domain of uniform permeability of 0.5 mD and penetration depth h_p = 5 km. Symbols in heat flow distribution as in Fig. 3.

m/yr. Thus, a dike of 200 m width which intrudes into an environment at temperatures steadily maintained by hydrothermal convection below the cracking temperature becomes permeable for fluid flow within a period of less than about 100 years. This time span is very short compared to the 10,000 to 20,000 years thought to ellapse between episodes of intrusion at mid-ocean ridges (Moore et al., 1974).

The increase of pressure with depth probably also affects the openness of fractures to fluid flow. Since the existence of fractures depends not on the hydrostatic pressure but on the lithostatic pressure which can be approximated by a linear increase with depth, pressure dependence can be modeled as depth dependence. In Fig. 5, the flow pattern is shown resulting from a

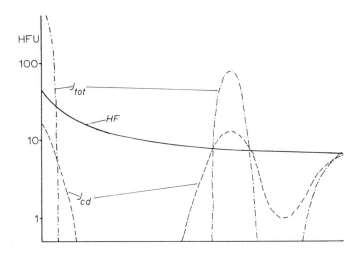

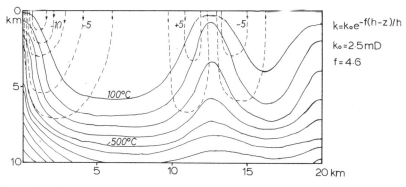

Fig. 5. Steady state temperatures, stream lines and heat flow distribution in crust where permeability decreases exponentially from surface permeability $k_0 = 2.5$ mD to base permeability $k_b = 0.025$ mD. Symbols in heat flow distribution as in Fig. 3.

permeability distribution which decreases exponentially with depth according to:

$$k(z) = k_0 \exp(-f(h - z)/h) \tag{5}$$

where k_0 = surface permeability; f = decay constant and h = depth of domain. The permeability in this case decreases from a surface permeability $k_0 = 2.5$ mD to a permeability of 0.025 mD at the base of the domain. A strong convection cell results at the ridge axis causing a large, narrow peak in the heat flow distribution. A second discharge area appears at a distance of 13 km from the ridge axis producing another sharp peak in the heat flow distribution. In the rest of the domain, isotherms are so much depressed that the heat flow over large parts of the domain is practically zero. As in the previous cases, temperatures within the convection cells are smaller than 500°C, although flow is concentrated closer to the surface than in cases with homogeneous permeability distribution in the crust.

In all the cases so far, it was assumed that the crust has a uniform permeability in horizontal direction. The presence of numerous fissures at mid-ocean ridges (e.g., Ballard and Van Andel, 1977) and the observation of temperature anomalies above fissures (Crane and Normark, 1977) suggests, however, that fractures or shear zones are of fundamental importance for flow regimes in oceanic crust. The next three examples illustrate the effect fractures can have on the flow distribution. We modeled the flow through frac-

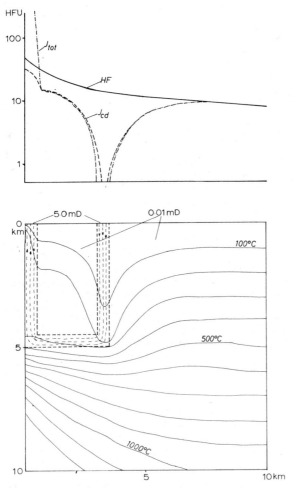

Fig. 6. Steady state temperatures, stream lines and heat flow distribution for loop of high permeability zones (k = 5.0 mD) in crust of low permeability (k = 0.01 mD). Symbols in heat flow distribution as in Fig. 3.

tures by introducing narrow zones of high permeability into a domain of low or intermediate permeability.

Figure 6 shows stream lines and isotherms for a fracture loop where 500 m wide zones of 5.0 mD permeability at the ridge axis and at a distance of 3 km from the ridge axis are connected by a similar zone of high permeability at a depth of 5 km. The rest of the domain has a permeability of 0.01 mD which practically prevents fluid flow. The fluid flow is thus restricted to the fracture loop; the flow direction is towards the ridge axis. Mainly the temperatures in the block between the fractures are affected by this kind of convection. Heat flow is depressed sharply over the downwelling branch of the fracture loop and reaches a high peak over the discharging branch. The influence of this kind of convection on the temperature distribution in the

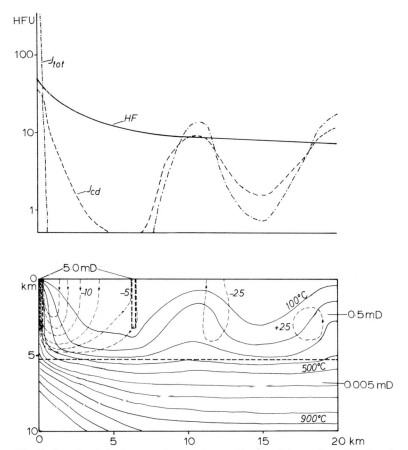

Fig. 7. Steady state temperatures, stream lines and heat flow distribution in crust of 0.5 mD permeability with 250 m wide zones of high permeability (k = 5.0 mD) at the ridge axis and in a distance of 6.25 km from the ridge axis. Symbols in heat flow distribution as in Fig. 3.

domain is negligible at distances greater than 5 km from the ridge axis.

A more realistic case is the situation where isolated zones of high permeability exist in a crust of otherwise low permeability. Fig. 7 and 8 illustrate the effect narrow zones of 5.0 mD permeability have on the flow distribution in the domain shown in Fig. 4, i.e., uniform permeability of 0.5 mD in the upper 5 km of the domain and very low permeability in the lower 5 km of the domain. In Fig. 7, two high permeability zones, each of which is 250 m wide and 3 km deep, are located at the the ridge axis and at a distance of 6.5 km from the ridge axis, respectively, i.e., one zone is situated in the upwelling limb, the other in the downwelling part of the convection cell flowing towards the ridge axis. Almost the entire upwelling flow of this con-

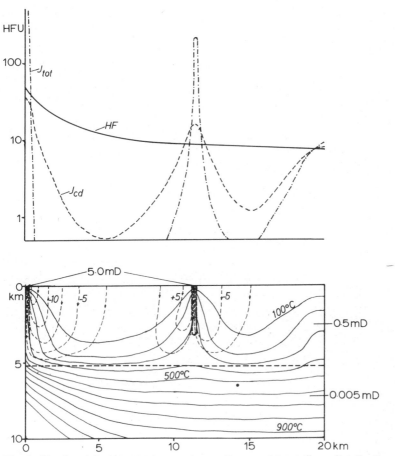

Fig. 8. Steady state temperatures, stream lines and heat flow distribution in crust of 0.5 mD permeability with 250 m wide zones of high permeability ($k = 5.0$ mD) at the ridge axis and in a distance of 11 km from the ridge axis. Symbols in heat flow distribution as in Fig. 3.

vection cell is concentrated in the zone of high permeability at the ridge axis. This concentration of fluid flow narrows the zone where heat flow is higher than *HF* to approximately 500 m as compared to 1.7 km in the case of uniform permeability (see Fig. 4). Although the high permeability zone in the downwelling part of the convection cell attracts some downwelling flow, it has only a small effect on fluid convection and heat flow distribution. Convection in the rest of the domain is not influenced by the presence of these two high permeability zones.

Fluid flow and heat flow distribution in the entire domain are strongly affected if the second zone of high permeability is located at a distance of 11 km from the ridge axis in the upwelling limbs of convection cells (Fig. 8). The two high permeability zones in this example attract practically the entire upwelling flow in the domain. Strong, narrow peaks in the heat flow distribution correspond to the discharge areas over the two high permeability zones. The percentage of the surface with heat flow higher than predicted is decreased to less than 10% of the domain as compared to 25% in the case of uniform permeability (see Fig. 4).

It is important to note that the concentration of upwelling flow into a fracture does not necessarily produce a similarly narrow zone of high heat flow. If the rock surrounding the fracture is impermeable an envelope of high conductive heat flow is caused by the high temperatures in the upwelling fluid (see Fig. 6). If, however, the surrounding rock is of a permeability which allows fluid flow, the steep horizontal temperature gradients around the fracture with upwelling flow induce downwelling flow very close to the fracture. This downwelling convection is what narrows the zone of high heat flow. The width of this zone depends on the width of the highly permeable zone and on the contrast in permeability between fracture and surrounding rock.

The presence of zones of high permeability in regions of upwelling flow influences not only the distribution of heat flow but also the mass flux through the surface. An increase of 23% in integrated mass flux was found for the model shown in Fig. 8 as compared to the mass flux through the surface of the same domain but without the high permeability zones (see Fig. 4). Interestingly, this increase in mass flux turns out to be proportional to the difference in average surface permeability between the two cases. Thus, it may be possible to estimate mass fluxes through a crust of non-uniform permeability by using the results obtained for domains with the same average, but uniform permeability.

The models shown in Figs. 6—8 are directly applicable only to cases where zones of high permeability are fixed with respect to the ridge axis. Examples for such cases are the area of the ridge axis itself and zones where plates go through episodes of high tectonic activity such as the uplift zones at the edges of a rift valley. In other cases, zones of high permeability may move with the moving plate. If the convection cells in permeable rock are indeed fixed with respect to the ridge axis, as suggested earlier, these fracture zones would go through areas of upwelling and downwelling flow. Fig. 7 and 8 can then be interpreted as providing two views of convection in a plate with a high-permeability zone moving along with the plate.

DISCUSSION

In Fig. 9, the average temperatures in domains of 10 km depth and 20 km width are plotted against permeability. If only conductive cooling occurs in these domains, an average temperature, T_{cd}, of 720°C results. Convection in the domain lowers the average temperature considerably; the degree of this decrease in average temperature depends primarily on the penetration depth, h_p, of the convection and secondarly on the permeability in the crust. Average temperatures range between 580°C for cases with a penetration depth of 3 km and 190°C for cases with a penetration depth of 10 km. The two circles in the diagram show cases calculated for domains in which the permeability decreases exponentially over two orders of magnitude over the depth of the domain. These two cases and the case with two high permeability zones (see Fig. 8) indicated by the closed triangle are plotted at their respective average surface permeability.

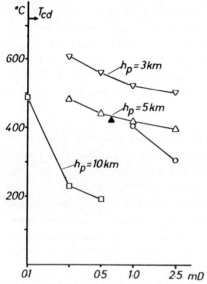

Fig. 9. Average steady state temperatures for oceanic crust between 0 and 20 km distance from the ridge axis and between 0 and 10 km depth below the sea floor for various permeabilities and fluid convection depths h_p. Open circles indicate cases where the permeability decreases exponentially with depth. The closed triangle shows the case (Fig. 8) where two zones of high permeability separated by 11 km exist in crust of 0.5 mD permeability. This case is plotted at its average surface permeability. The average temperature, T_{cd}, in absence of convection is indicated at the heat flow axis.

Convective cooling of the top 10 km of the oceanic crust has consequences for the relation between depth and age at mid-ocean ridges. The relation observed between depth and age of the sea floor can be explained by purely conductive cooling of oceanic crust as it moves away from a spreading center (e.g., Davis and Lister, 1974, and Parsons and Sclater, 1977). Since convective cooling of oceanic crust is not included in these models, the difference in average temperature between a purely conductive layer and layers with conductive and convective heat transfer (as illustrated in Fig. 9 for crust between 0 and 2 m.y. old) should result in a deviation from the depth-age relation close to spreading centers. The temperature difference between the conductive case and cases with convection can be converted into a crestal offset which — depending on the value of the thermal expansion coefficient of the crust — ranges between 26 m ($\alpha = 2.6 \cdot 10^{-5}$ dg^{-1}) and 40 m ($\alpha = 4 \cdot 10^{-5}$ dg^{-1}) per 100°C of temperature difference in the 10 km of the domain.

Hydrothermal convection not only causes a decrease in average temperatures near spreading centers but also alteration reactions in the crust some of which — as, for example, the serpentinization of peridotite — are accompanied by a considerable increase in volume (Coleman and Keith, 1971). The influence this volumetric increase may have on the elevation of mid-ocean ridges depends on the rate of hydrothermal alteration in the crust. If all the alteration occurs within the first 2 m.y., the volumetric increase due to alteration would probably result in an increase in height that would compensate the crestal offset due to contraction during convective cooling of the crust. In this case, only little deviation from the depth-age relation would be visible at mid-ocean ridge crests. If, however, hydrothermal alteration occurs over a longer period of time, expansion caused by this alteration produces a nearly constant offset of the elevation which does not affect the depth-age relationship. In the latter case, the temperature difference between conductive and convective regime would result in an observable crestal offset.

The crestal offset observed for the Mid-Atlantic Ridge is not larger than 250 m (Davis and Lister, 1974, Parsons and Sclater, 1977). Since about 200 m of this offset is attributable to horizontal heat conduction and changes in thermal parameters at the mantle-crust boundary (see discussion in Davis and Lister, 1974) contraction due to convective cooling in this example has to be less than about 50 m. We assume that hydrothermal alteration is not complete in 2 m.y. in this case because hydrothermal circulation is inferred to prevail in Atlantic crust up to 70 m.y. old (Anderson et al., 1977). Thus, because convection in regimes with penetration depths greater than 5 km result in contraction larger than 100 m, convection probably does not penetrate deeper than 5 km in crust younger than 2 m.y. in the Atlantic Ocean. Similar arguments can be used for other active spreading centers so that this limit of hydrothermal penetration is probably typical for most mid-ocean ridges.

Average temperatures show only a weak dependence on permeability and

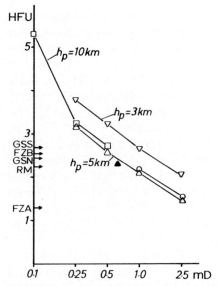

Fig. 10. Average steady state heat flow over areas with downwelling convection in crust of various permeabilities and penetration depths. Open circles and closed triangle as in Fig. 9. Average heat flow values in parts of the FAMOUS area and at the Galapagos Spreading Center are given as comparison; see text for explanation of symbols for these areas.

therefore do not usefully constrain the average permeability of oceanic crust. Heat flow distributions more directly reflect the permeability distribution of the crust. Because most of the heat flow values observed at mid-ocean ridges are lower than predicted by thermal models, they probably represent areas with downwelling convection and should be compared to heat flow values of areas with downwelling flow in our models. In Fig. 10, average values of conductive heat flow, J_{down}, over areas with downwelling convection in our models are plotted against permeability. The stronger convection associated with higher permeability in the crust causes a stronger decrease of temperatures in the downwelling parts of convection cells. Consequently, J_{down} decreases considerably with increasing permeability; the values of J_{down} range between 5.3 and 1.5 HFU.

In order to compare our results to observed heat flow values close to mid-ocean ridges, we used the heat flow distribution measured in the FAMOUS area of the Mid-Atlantic Ridge (Williams et al., 1977) and at the Galapagos Spreading Center (Williams et al., 1974, and K. Green, 1977, unpublished data). The results of the FAMOUS area are more applicable because the half-spreading rate of the Mid-Atlantic Ridge (0.9 cm/yr) is closer to that used in our calculations (1 cm/yr) than the half-spreading rate of the Galapagos Spreading Center (3.5 cm/yr). Heat flow values were measured in three separate parts of the FAMOUS area: fracture zone A (*FZA*), fracture zone B

(*FZB*) and in the western rift mountains (*RM*). The average heat flow values of these three areas and of the areas north (*GSN*) and south (*GSS*) of the Galapagos Spreading Center are indicated in Fig. 9. To insure that the average value calculated for these five areas represent heat flow over areas of downwelling convection, we included only heat flow values smaller than 5 HFU in the calculation of these averages. Since our models show that zones with weak downwelling convection can have heat flows significantly higher than 5 HFU, the restriction to values <5 HFU probably means that the averages calculated are minimum values for these areas.

Average heat flow J_{down} for fracture zone B, rift mountains and Galapagos Spreading Center fall into the range between 2.2 and 2.7 HFU while the value found for fracture zone A is considerably lower. Since 20 of the 35 stations in fracture zone A are given as "minimum value" only, the average value found for this area is probably underestimated. If the remaining four averages are considered as representative heat flow averages over downwelling areas close to mid-ocean ridges, they suggest that average permeabilities (including fractures) of young oceanic crust range between 0.25 and 2.5 mD.

Fig. 11. shows steady state mass fluxes through the surface of domains for crusts of various permeabilities. Mass fluxes are integrated over the first 20 km from the ridge axis and are given in 10^6 grams per year and centimeter of ridge axis. Increase in permeability causes stronger convection and thus higher mass fluxes. The calculated mass fluxes range between 1 and $5 \cdot 10^6$

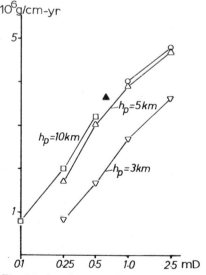

Fig. 11. Steady state fluid mass fluxes through the surface of domains of various permeabilities and penetration depths, integrated over the area between 0 and 20 km distance from the ridge axis. Open circles and closed triangle as in Fig. 9.

g/cm yr. The case with fractures in the upwelling limbs of convection cells (see Fig. 8) results in a mass flux of $3.7 \cdot 10^6$ g/cm yr.

If these mass fluxes are considered as being representative for hydrothermal convection at mid-ocean ridges, the total mass of sea water convecting through all ridges can be estimated. In order to convert our results into mass fluxes through the entire ridge system, the fluxes of Fig. 11 have to be multiplied by the length of the ridge systems (50,000 km) and by a factor of 2 for symmetry. If the model of Fig. 8 is used as an example, i.e. two fractures in a domain of 0.5 mD permeability to a depth of 5 km, a mass of $3.7 \cdot 10^{16}$ g/yr convects through crust younger than 2 m.y. old. Hydrothermal convection is, however, not restricted to crust of this age, but is inferred to occur in crust as old as 70 m.y. (Anderson et al., 1977). If older crust is included into the calculations, the total hydrothermal mass flux is increased by a factor between 1.5 and 4 depending on permeability distribution and sediment coverage of older crust (Fehn and Cathles, 1978). On this basis we estimate the total mass flux associated with hydrothermal convection at mid-ocean ridges to be smaller than $2 \cdot 10^{17}$ g/yr.

This mass flux is lower by at least a factor of 2 than the mass flux suggested by Wolery and Sleep (1976). This difference could be related to the occurrence of a more rapid convection at fast spreading ridges which are not considered in our calculations. Other reasons for this difference may be that the heat loss attributed by Wolery and Sleep to hydrothermal convection seems too high in view of new heat flow investigations (Anderson et al., 1977) and that their estimate of total mass flux is based on the assumption that the entire difference between observed and predicted heat flow is lost by the exit of heated water through the ocean floor. Any vent discharging heated water is, however, surrounded by a zone of high conductive heat flow, which can be responsible for an important part of the total heat transfer associated with hydrothermal convection. If these considerations are taken into account, the result obtained here is in good agreement with Wolery and Sleep's (1976) estimate.

CONCLUSIONS

Calculations were carried out for hydrothermal convection through oceanic crust of 10 km depth and 20 km width at a mid-ocean ridge spreading at a rate of 1 cm/yr. The following results were obtained:

(1) Heat flow over areas where downwelling convection occurs is compatible with observed heat flow values at active spreading centers provided the average permeability of the crust is less than 2.5 mD.

(2) Convection in crust of uniform permeability can depress heat flow values below the values predicted by conductive models over approximately 70% of the ocean floor. The rest of the area has heat flow higher than predicted by these models. If narrow zones of high permeability exist in domains of intermediate permeability, the percentage of zones with heat flow

lower than predicted is increased to more than 90% of the area. Since this latter result is in good agreement with the observed distribution of depressed and increased heat flow values near mid-ocean ridges, we suggest a model in which downwelling convection occurs in wide areas of the sea floor where the crust has permeabilities between 0.25 and 2.5 mD while upwelling flow is concentrated in narrow zones of considerably higher permeability.

(3) If the oceanic crust is not hydrothermally altered entirely or nearly entirely within the first two million years of its existence, the reasonably good agreement between observed ridge elevations and those expected from depth-age relations suggests that fluid circulation is restricted to depths smaller than 5 km in crust less than two million years old.

(4) Temperatures in the portions of the crust where hydrothermal convection occurs are generally lower than 400°C and reach values above 500°C in areas close to the ridge axis only.

(5) If circulation to a depth of 5 km is allowed, and the average crustal permeability is 1 mD, $4 \cdot 10^6$ g of water circulate per year through each cm wide and 20 km long strip of oceanic crust perpendicular to a ridge crest spreading at 1 cm/yr. If all ridges are similar to those spreading at 1 cm/yr, $4 \cdot 10^{16}$ g of water circulate through the oceanic crust within 20 km of all ridge crests each year. Up to $2 \cdot 10^{17}$ g/yr may circulate through the oceanic crust taking into account crust older than two million years. These mass fluxes are somewhat lower than those estimated by Wolery and Sleep (1976), but, for reasons cited in the text, can be considered in good agreement with Wolery and Sleep's estimate.

ACKNOWLEDGEMENTS

We are grateful to H.D. Holland for his support of the project and for many helpful discussions. We also thank K.E. Green for making available the unpublished heat flow data from the Galapagos Spreading Center. The second author wishes to thank Kennecott Copper Corporation for its support of the project. The research was funded by the National Science Foundation, Grant OCE 76-82188.

REFERENCES

Anderson, R.N., Langseth, M.G. and Sclater, J.G., 1977. The mechanism of heat transfer through the floor of the Indian Ocean. J. Geophys. Res., 82: 3391—3409.

Ballard, R.D. and Van Andel, T.H., 1977. Morphology and tectonics of the inner rift valley at lat 36°50'N on the Mid-Atlantic Ridge. Geol. Soc. Am. Bull., 88: 507—530.

Bodvarsson, G. and Lowell, R.P., 1972. Ocean-floor heat flow and the circulation of interstitial waters. J. Geophys. Res., 77: 4472—4475.

Bryant, W.R., Deflache, A.P. and Trabant, P.K., 1974. Consolidation of marine clays and carbonates. In: A.L. Interbitzen (Editor), Deep-Sea Sediments: Physical and Mechanical Properties. Plenum, New York, N.Y., pp. 209—244.

Carnahan, B., Luther, H.A. and Wilkes, J.O., 1969. Applied Numerical Methods. Wiley, New York, N.Y., pp. 604.

Cathles, L.M., 1977. An analysis of the cooling of intrusives by ground-water convection which includes boiling. Econ. Geol., 72: 804—826.

Coleman, R.G. and Keith, T.E., 1971. A chemical study of serpentinization — Burro Mountain, California. J. Petrol., 12: 311—328.

Crane, K. and Normark, W.R., 1977. Hydrothermal activity and crestal structure of the East Pacific Rise at 21°N. J. Geophys. Res., 82: 5336—5348.

Davis, E.E. and Lister, C.R.B., 1974. Fundamentals of ridge topography. Earth Planet. Sci. Lett., 21: 405—413.

Davis, E.E. and Lister, C.R.B., 1977. Heat flow measured over the Juan de Fuca Ridge: evidence for widespread hydrothermal circulation in a highly heat transportive crust. J. Geophys. Res., 82: 4845—4860.

Fehn, U. and Cathles, L.M., 1978. Hydrothermal convection through oceanic crust between 0 and 70 m.y. old. EOS, Trans., Am. Geophys. Union, Abstr., 59: 384.

Katz, B.J., Harrison, C.G.A. and Man, E.H., 1977. Organic geochemical evidence for oscillatory heat flow on the Mid-Atlantic Ridge (abstr.). EOS, Trans., Am. Geophys. Union, 58: 514.

Lister, C.R.B., 1972. On the thermal balance of a mid-ocean ridge. Geophys. J.R. Astron. Soc., 26: 515—535.

Lister, C.R.B., 1974. On the penetration of water into hot rock. Geophys. J.R. Astron. Soc., 39: 465—509.

Lister, C.R.B., 1977. Qualitative models of spreading-center processes including hydrothermal penetration. Tectonophysics, 37: 203—218.

Lowell, R.P., 1975. Circulation in fractures, hot springs, and convective heat transport on mid-ocean ridge crests. Geophys. J.R. Astron. Soc., 40: 351—365.

Lowell, R.P. and Patterson, P.L., 1979. Numerical models of hydrothermal circulation at an ocean ridge axis. In: F.T. Manheim and K. Fanning (Editors), The Dynamic Environment of the Sea Floor.

Lowell, R.P., Patterson, P.L. and Fulford, J.K., 1977. Numerical modeling of hydrothermal circulation at slow spreading ridges (abstr.). Geol. Soc. Am., Abstr. Progr., 9: 1075—1076.

Moore, J.G., Fleming, H.S. and Phillips, J.D., 1974. Preliminary model for the extrusion and rifting at the axis of the Mid-Atlantic Ridge, 36°48' North. Geology, 2: 437—450.

Morgan, W.J., 1975. Heat flow and vertical movement of the crust. In: A.G. Fischer and S. Judson (Editors), Petroleum and Global Tectonics. Princeton Univ. Press, Princeton, N.J., 23—43.

Oldenburg, D.W., 1975. A physical model for the creation of the lithosphere. Geophys. J.R. Astron. Soc., 43: 425—451.

Parsons, B. and Sclater, J.G., 1977. An analysis of the variation of ocean floor bathmetry and heat flow with age. J. Geophys. Res., 82: 803—827.

Pearson, W.C. and Lister, C.R.B., 1973. Permeability measurements on a deep sea core. J. Geophys. Res., 78: 7786—7787.

Ribando, R.J., Torrance, K.E. and Turcotte, D.L., 1976. Numerical models for hydrothermal circulation in the oceanic crust. J. Geophys. Res., 81: 3007—3012.

Von Herzen, R., Green, K.E. and Williams, D., 1977. Hydrothermal circulation at the Galapagos Spreading Center (abstr.). Geol. Soc. Am., Abstr. Progr., 9: 1212—1213.

Williams, D.L. and Von Herzen, R.P., 1974. Heat loss from the Earth: new estimate. Geology, 2: 327—328.

Williams, D.L., Von Herzen, R.P., Sclater, J.G. and Anderson, R.N., 1974. The Galapagos spreading center: lithospheric cooling and hydrothermal circulation. Geophys. J.R. Astron. Soc., 38: 587—608.

Williams, D.L., Lie, T-C., Von Herzen, R.P., Green, K.E. and Hobart, M.A., 1977. A geothermal study of the Mid-Atlantic Ridge near 37°N. Geol. Soc. Am. Bull., 88: 531—540.

Wolery, T.J. and Sleep, N.H., 1976. Hydrothermal circulation and geochemical flux at mid-ocean ridges. J. Geol., 84: 249—275.

Chapter 9

VERTICAL VARIATIONS IN THE EFFECTS OF HYDROTHERMAL
METAMORPHISM IN CHILEAN OPHIOLITES: THEIR IMPLICATIONS
FOR OCEAN FLOOR METAMORPHISM[1] *

CHARLES STERN and DON ELTHON **

*Lamont-Doherty Geological Observatory, Columbia Univeristy, Palisades,
N.Y. 10964 (U.S.A.)*

SUMMARY

The metamorphic overprint on the pseudostratigraphy of ophiolite complexes in
southern Chile exhibits a steep vertical metamorphic gradient passing from zeolite to
actinolite facies in 2 km followed by a transition to fresh gabbros. Metamorphic boundaries are irregular and disequilibrium "retrograde" effects are common. These observations are best explained by hydrothermal metamorphism associated with circulation of
seawater related to igneous and tectonic activity at a spreading center. Within the
gabbros, both the intensity of metamorphic recrystallization and the extent of retrograde
effects are low; epidote is sparse while secondary biotite and magnetite are common
accessory minerals. In the overlying dikes both the extent of metamorphic recrystallization and the degree of retrograde effects are intense; epidote is abundant while magnetite
and secondary biotite are absent. These petrologic and mineralogic differences are consistent with a greatly decreased circulation of water within the gabbro unit so that reactions
here have been buffered at the FMQ buffer of the igneous rock. At higher levels in the
ophiolites metamorphic reactions occurred within an oxidizing environment resulting
from the circulation of large volumes of water.
The pseudostratigraphic variations in the proportions of major metamorphic minerals,
even for rocks of the same facies, and the presence or absence of accessory phases such as
biotite and orthoclase has resulted in distinct metasomatic effects at different pseudostratigraphic levels within Chilean ophiolites, particularly with respect to K_2O and Rb.
Where the gabbro unit makes an igneous contact with the overlying dikes, the observed
petrochemical changes are abrupt and could produce a sharp seismic layer 2 to layer 3
transition.

[1] Originally published as: Stern, C. and Elthon, D., 1979. Vertical variations in the effects
of hydrothermal metamorphism in Chilean ophiolites; their implications for ocean floor
metamorphism. In: J. Francheteau (Editor), Processes at Mid-Ocean Ridges. Tectonophysics, 55: 179—213.
* Lamont-Doherty Geological Observatory Contribution No. 2772.
** And the Department of Geological Sciences of Columbia University, Palisades, New
York (U.S.A.).

INTRODUCTION

Petrochemical studies of meta-igneous rocks drilled and dredged from the ocean floor and exposed in ophiolite complexes, combined with the results of marine geophysical studies, suggest that ocean floor metamorphism is related to circulation of seawater and igneous and tectonic processes at spreading ridges (Cann, 1968; Lister, 1972, 1974; Gass and Smewing, 1973; Spooner and Fyfe, 1973; Williams et al., 1974; Spooner et al., 1974; Bonatti et al., 1975; Stern et. al., 1976; Muehlenbachs and Clayton, 1972; Wolery and Sleep, 1976). The type and extent of petrochemical modifications due to ocean floor metamorphism are important factors for understanding both the geophysical properties of the ocean crust and the chemistry of seawater. Petrochemical effects of basalt—seawater interactions have been shown to differ for glassy and crystalline rocks (Corliss, 1971; Coish, 1977) as well as for different temperatures and oxygen fugacities (Miyashiro et al., 1971; Spooner and Fyfe, 1973; Hart, 1973). The integrated effect of ocean floor metamorphism must depend upon a complicated interaction among the primary igneous pseudostratigraphy of the ocean floor, the extent and composition of circulating waters, the oceanic geothermal gradient, and the progress of the metamorphism itself which acts to change the chemistry of the circulating fluids, the pattern of circulation, and the geothermal gradient.

In this paper, we describe the overprint of metamorphism on the original igneous pseudostratigraphy of ophiolite complexes in southern Chile. Excellent vertical exposures have allowed us to closely correlate the effects of this metamorphism with the different pseudostratigraphic levels within the ophiolites. In general the grade of metamorphism increases from zeolite and greenschist facies in the pillow lavas to actinolite facies within the gabbro unit. The extent of metamorphic recrystallization is low in the pillow lavas, increases to a maximum within the sheeted dike complex, and then decreases again within the gabbro unit until the metamorphic effects terminate, leaving fresh gabbros (DeWit and Stern, 1976; Stern et al., 1976). Significantly, the petrochemical effects of greenschist facies metamorphism are different at different levels within the ophiolites. The differences suggest that within the gabbro unit, lack of permeability has limited the volume of seawater circulation so that metamorphic reactions were buffered at the fayalite—magnetite—quartz (FMQ) oxygen fugacity buffer of the igneous rocks, while within the dikes and pillow lavas, relatively extensive circulation of seawater produced a more oxidizing environment resulting in different metamorphic reactions. Such differences in extent and composition of circulating fluids at different levels within the oceanic crust may in part explain the different petrochemical modifications reported for oceanic metabasalts within the same metamorphic facies (Miyashiro et al., 1971).

CHILEAN OPHIOLITES

Field evidence indicates that the mafic complexes which outcrop from 49°S to Cape Horn in southern Chile represent the mafic portion of the floor of a Mesozoic extensional back-arc basin (Dalziel et al., 1974; DeWit, 1977). The early stages of evolution of the back-arc basin in southern Chile involved a change from subaerial silicic volcanism within a continental volcano-tectonic rift zone to submarine tholeiitic volcanism behind an active calc-alkaline volcanic arc (Bruhn et al., 1978). An active calc-alkaline volcanic arc, founded on a rifted sliver of continental crust, was separated from the adjacent continent by the formation of the basin. Sedimentary debris eroded from both the volcanic arc and the continental margin on the opposite side of the basin and was deposited on the basin floor. The lack of ophiolites north of 49°S and the southward widening of the outcrop belt of the sedimentary rocks which infilled the basin suggest that the width of the original basin increased from north to south (DeWit, 1977). Petrologic and geochemical data are presented in this paper for the Sarmiento and Tortuga complexes (Fig. 1) which are located at the extreme northern and southern ends respectively, of the outcrop belt of mafic complexes. DeWit (1977) has estimated the original basin width in the vicinity of the Sarmiento complex to be 25 km, and in the vicinity of the Tortuga complex to be 100 km.

A number of significant structural and geochemical differences are observed between the Sarmiento complex located at 51°S in the northern narrow part of the original basin, and the Tortuga complex located at 55°S in a more southern, wider part of the original basin. Chemically, the rocks of both complexes exhibit a tholeiitic differentiation, trend consistent with differentiation, within shallow level magma chambers involving crystallization and separation of olivine, pyroxenes, calcic plagioclase, and titanomagnetite; phases found in the cumulate gabbro unit of both mafic complexes (Stern, 1976; in press, a). However, the extent of differentiation was much greater within the Sarmiento complex where highly evolved ferro-basalts, icelandites and silicic dikes and lavas similar in chemistry to oceanic plagiogranites (Coleman and Peterman, 1975) are common. Dikes from among the dike swarms flanking both ophiolite complexes, representing the earliest manifestation of mafic igneous activity associated with the formation of the basin, and dikes and lavas from the Sarmiento complex in the narrowest part of the original basin, have normalized La/Yb ratios in the range $(La/Yb)_N$ = 1.9—2.8. Basalts from the Tortuga complex in the widest part of the original basin have $(La/Yb)_N$ = 0.6—0.8. These changes in $(La/Yb)_N$ ratio are interpreted as secular changes indicating either progressive tapping of distinct mantle source regions or depletion in large ion lithophile elements of an initially homogeneous mantle source region during evolution of the back-arc basin (Stern, 1976; in press, b).

The igneous pseudostratigraphy of both the Sarmiento and Tortuga complexes (Fig. 2) consists in part of a sheeted dike complex of 100% dikes

184

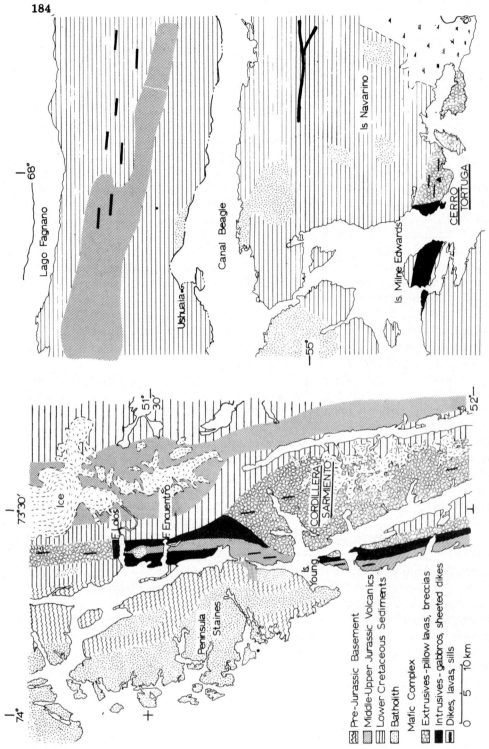

Fig. 1. Geologic maps of the regions around the Sarmiento and Tortuga complexes, southern Chile. Note that the Andean Cordillera strikes north–south in the vicinity of the Sarmiento complex and east–west in the vicinity of the Tortuga complex.

indicating 100% extension, consistent with their formation at a spreading ridge (Gass, 1968; Moores and Vine, 1971). Towards the top of the dike complexes screens of pillows are encountered between dikes, followed by regular sequences of horizontal pillow lavas and breccias as the proportion of dikes decreases. The extrusive unit of the Sarmiento complex contains large volumes of aquagene tuff and both intrusive and extrusive type pillow lavas (DeWit and Stern, 1978), while the extrusive unit of the Tortuga complex consists dominantly of massive flows and waterlain pillow lavas (Fig. 2).

Below the sheeted dike complexes are chemically layered plutonic rocks which grade downward into coarsely crystalline cumulate gabbros. In the Sarmiento complex the plutonic rocks exhibit the same large chemical diversity found in the lavas, with ferro-gabbros and silicic plagiogranites common just below the sheeted dike complex. The silicic plagiogranites intruded the sheeted dike complex by magmatic stoping, forming a chilled subhorizontal contact with the dike unit (Fig. 2). In the Tortuga complex plagiogranites are absent. Here the dike unit grades downward into a mixture of dikes and diabases in a continuous transition similar to that between the dike and pillow unit. Diabases first appear as large dikes 3—5 m across which grade downward into diabase masses 10—30 m across as the number of thinner fine-grained dikes decreases. These large diabase masses have fine-grained margins and are separated by 5—20 m of recognizable thin (<1 m) fine-grained dikes. The diabases seem to have intruded as very big slowly cooled

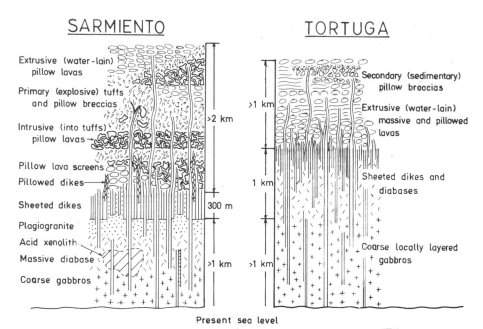

Fig. 2. Schematic pseudostratigraphy of the Sarmiento and Tortuga complexes. DeWit and Stern (1978) describe pillow lava types in detail.

dikes into the lower part of the sheeted dike unit (Fig. 2). Below this sheeted dike-diabase unit, coarse-grained, mineralogically layered cumulate gabbros become common.

The mafic complexes are considered to represent the upper parts of ophiolites (Dalziel et al., 1974). The apparent lack of exposed ultramafic rocks is considered to be due to the present level of exposure (Dalziel et al., 1974; Bruhn and Dalziel, 1977; Saunders et al., 1979). The complex sequence of processes of continued partial melting, intrusion, and extrusion which lead to the observed chemical variations and the formation of the pseudostratigraphy in the Chilean and other ophiolites are described in detail elsewhere (Moores and Vine, 1971; Greenbaum, 1972; Gass and Smewing, 1973; Church and Riccio, 1974; Stern, 1976; in press, a and b; DeWit and Stern, 1978). The differences in the chemistry and pseudostratigraphy observed in the Sarmiento and Tortuga complexes must relate to either more rapid spreading rate and/or the more prolonged development of the mafic portion of the floor of the basin towards the south which could also have caused the apparent widening of the southern part of the original basin (DeWit and Stern, 1978). As mentioned above, the change from basalts sligthly enriched to slightly depleted in light rare earth elements indicates either progressive tapping of chemically distinct mantle source regions or progressive depletion of a homogeneous mantle source region; in either case suggesting a more prolonged evolution of the magma source region. The development of highly differentiated rocks in the narrow part of the marginal basin suggests that here magma remained in magma chambers longer, which is consistent with their intrusion by magmatic stoping into the overlying sheeted dike unit such as is observed in the Sarmiento complex. In the wider part of the marginal basin fewer differentiates were developed and a greater proportion of magma was intruded in the form of extensional dikes and diabases, suggesting that here extension was more rapid relative to the rate of magma supply. Because the effects of ocean floor metamorphism, as outlined below, are closely related to ocean floor pseudostratigraphy they will depend in turn on rates of magma supply and spreading.

METAMORPHISM

The back-arc basin was closed and uplifted in the mid-Cretaceous (Bruhn and Dalziel, 1977). This event was accompanied by heterogeneously distributed penetrative deformation, concentrated predominantly in rock assemblages flanking and overlying the ophiolites. Associated regional metamorphism in the prehnite—pumpellyite facies is evident in the basin sediments (Watters, 1965) but not within the undeformed ophiolite complexes. The Sarmiento and Tortuga complexes have escaped both the regional penetrative deformation and metamorphism associated with the closure and uplift of the basin. Within these ophiolites, prehnite and/or pumpellyite are sometimes found infilling vesicles or secondary veins, but they occur as major phases only near fault contacts where deformation is apparent.

An earlier phase of metamorphism that was characterized by the development of secondary minerals without schistosity was not related to regional deformation and is restricted to the ophiolites, or is here best preserved. This type of metamorphism has been described for rocks from the ocean floor as well as other ophiolites and has been termed ocean floor metamorphism. It is clearly distinct from prograde regional metamorphism. Within the Chilean ophiolites four distinct metamorphic facies may be distinguished on the basis of what appear to be, in a number of lavas, dikes, and gabbros, *equilibrium* mineral assemblages. From the lowest to highest grade these are, in conventional metamorphic terminology:.

(1) Zeolite facies: various zeolite minerals, palagonatized glass, ± smectites ± calcite ± quartz ± sulfides ± sphene ± albite.

(2) Greenschist facies: chlorite, epidote, albite (<An 15), sphene, quartz ± calcite ± biotite ± sulfides.

(3) Lower actinolite facies: green fibrous actinolitic—tremolitic amphiboles (Al_2O_3 = 2.5 −5.0 wt%), calcic plagioclase (>An 50), sphene ± biotite ± calcite.

(4) Upper actinolite facies: brown blocky and green fibrous actinolitic—tremolitic amphiboles (Al_2O_3 = 5.0—8.0 wt%), calcic plagioclase (≥An 50), titanomagnetite ± ilmenite ± biotite.

The lower and upper actinolite facies as defined here corresponds to the actinolite—calcic plagioclase hornfels facies of Miyashiro (1961). The alumina content of amphiboles is higher within the upper actinolite facies (Table V and VI) but not as high as aluminous hornblendes (Al_2O_5> 10 wt%) which occur in hornblende amphibolites formed in regional metamorphic terrains. The differences between the actinolite facies and amphibolite facies of regional metamorphism are discussed in more detail below.

We re-emphasize that the facies divisions that we have employed have been defined by what appear to be *equilibrium* mineral assemblages. The mineral assemblage actinolite—chlorite—epidote—albite—sphene—quartz, which is very common, may represent a transitional Upper Greenschist facies, but in the majority of the rocks we have studied with this mineral assemblage, the actinolite is clearly being replaced by epidote and chlorite. The disequilibrium textures we have observed have not resulted from the assemblage actinolite—chlorite—epidote being stable over a range of T, P_{H_2O}, and F_{O_2} conditions, although such a multivariant stability field exists (Liou et al., 1974). They result instead from incomplete reaction, a kinetic failure to attain equilibrium during low temperature subsequent to high temperature metamorphism. These disequilibrium "retrograde" textures are *texturally* similar to retrograde textures formed during regional metamorphism, although ocean floor metamorphism involves a sequential series of metamorphic reactions occurring with decreasing temperature and no "prograde" event-senso stricto.

Figure 3 shows the relative distribution among the four facies divisions defined above of the metabasalts and metagabbros from different pseudostratigraphic levels of the Sarmiento complex. For each pseudostratigraphic level (pillow lavas, sheeted dikes, gabbros) all the available rock samples were classified into one of the facies groups according to the *lowest* facies mineral assemblages it contained. The relative distribution among the facies of the rocks at each pseudostratigraphic level is indicated in Fig. 3 by the horizontal thickness of the appropriate bars. The vertical dimensions of these bars correspond only to the vertical thickness of the appropriate pseudostratigraphic level. These are further subdivided into those rocks containing relict higher facies minerals, indicated by the diagonally ruled areas, and those rocks without relict higher facies minerals, indicated by the blackened portion of the bars. The average K_2O contents of all the rocks in each of these categories are shown within the appropriate bars. Gabbros and cross cutting dikes have been treated separately, at the left and right, respectively, of the schematic pseudostratigraphy of the Sarmiento complex. The figure was constructed without including silicic dikes or plagiogranites which are discussed separately below. Most of the data used to construct the figure are in Table I, II, III and IV. Only the pillow lava level incorporates untabulated data.

The figure illustrates the heterogeneous distribution of metamorphic

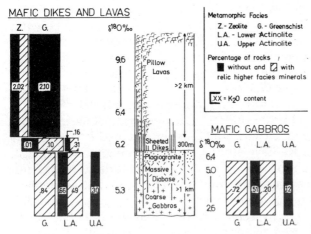

Fig. 3. Diagram illustrating the relative distribution of metabasalts and metagabbros from different pseudostratigraphic levels of the Sarmiento complex within the different metamorphic facies defined in the text. For each pseudostratigraphic level the relative thickness of each bar represents the relative abundance of rocks belonging to the appropriate facies. Solid bars represent the relative abundance of rocks without relict higher facies minerals, diagonally ruled bars are rocks with relict higher facies minerals. The numbers within each bar are the average K_2O contents of all rocks belonging to that category.

facies and the abundant presence of disequilibrium textures, which we interpret as "retrograde" effects of greenschist subsequent to actinolite facies metamorphism that characterize Chilean and other ophiolites (Bonatti et al., 1975; Stern et al., 1976). Figure 3 shows that despite the heterogeneous distribution of metamorphic facies and the abundance of disequilibrium effects, it is clear that the metamorphic grade increases downward from zeolite and greenschist facies within the pillow lavas to actinolite facies within the gabbro unit in a short vertical distance varying from 1—3 km in different areas.

Detailed petrographic examination of metabasalts from different pseudo-stratigraphic levels within the Sarmiento complex indicates that not only the grade of metamorphism, but also the extent of metamorphic replacement and recrystallization has varied in a relatively short vertical distance. In general, the extent of metamorphic recrystallization is low at the level of the pillow lavas, increases to a maximum within the sheeted dike unit, and then decreases again within the gabbro unit. Significantly, the extent of disequilibrium effects produced by greenschist subsequent to actinolite facies metamorphism follows a similar pattern.

TABLE I

Composition of pillow lavas and dikes cutting pillows in the Sarmiento Complex *

Components	Sample nr:	FL70A	PA56D	PA56E
	Facies **:	Z	G	G
	Relic facies **:	G		
SiO_2		52.99	53.34	55.31
TiO_2		0.87	1.29	1.84
Al_2O_3		16.16	15.41	14.92
FeO *		8.50	10.81	11.05
MnO		0.15	0.20	0.20
MgO		8.96	6.14	3.95
CaO		5.77	5.00	5.35
Na_2O		2.89	2.44	5.00
K_2O		3.65	4.34	1.27
P_2O_5		0.11	0.23	0.34
Total		100.05	99.20	99.23
FeO */MgO		0.95	1.66	2.80
Zr (ppm)		54	100	140
Y (ppm)		24	36	38
Rb (ppm)		95	95	24
Sr (ppm)		157	92	42

* Major element compositions determined by electron microprobe analysis of unfluxed fused glasses (Mazzulo and Bence, 1976). Total Fe expressed as FeO. Trace element compositions determined by X-ray fluorescence techniques (Norrish and Chappell, 1967).
** Abbreviations Z = zeolite facies; G = greenschist facies.

TABLE II

Composition of sheeted dikes in the Sarmiento Complex *

Compo- nents	Sample nr: Facies **: Relic facies **:	PA23T G L.A.	PA28C L.A. U.A.	PA28V G —	PA28B G L.A.	PA23D G U.A.	PA28T G —	PA23A G L.A.
SiO_2		53.93	51.72	50.19	50.97	50.52	52.50	53.29
TiO_2		0.92	0.97	1.00	1.04	1.04	1.09	1.13
Al_2O_3		15.00	16.54	17.15	16.67	16.39	16.34	15.38
FeO *		9.93	9.28	9.43	8.51	10.23	9.24	9.79
MnO		0.18	0.19	0.24	0.15	0.26	0.23	0.23
MgO		7.44	7.29	8.03	7.33	9.39	8.11	7.03
CaO		7.45	11.86	13.74	12.99	9.79	7.59	9.57
Na_2O		3.55	1.82	0.10	1.49	2.04	5.00	3.40
K_2O		0.03	0.20	0.01	0.01	0.11	0.02	0.03
P_2O_5		0.19	0.19	0.21	0.25	0.20	0.20	0.25
Total		98.62	100.06	100.10	99.41	99.97	100.32	100.10
FeO */MgO		1.33	1.27	1.17	1.16	1.10	1.14	1.39
Zr (ppm)		75	59	56	75	75	75	90
Y (ppm)		28	26	25	23	27	26	30
Rb (ppm)		1	6	<1	<1	2	<1	<1
Sr (ppm)		127	195	171	245	152	133	182

* See footnote for Table I.
** Abbreviations as in Table I, as well as L.A. = lower actinolite, U.A. = upper actinolite.

TABLE III

Composition of dikes cutting gabbros in the Sarmiento Complex *

Compo- nents	Sample nr: Facies **: Relic facies **:	PA32C L.A. U.A.	PA32A G L.A.	PA32I L.A. U.A.	FL-2 U.A. —	PA32H U.A. —
SiO_2		52.52	51.32	50.23	50.21	50.22
TiO_2		0.83	0.87	0.94	0.94	0.97
Al_2O_3		16.77	17.10	16.82	16.75	17.11
FeO *		8.56	6.81	8.26	9.00	8.45
MnO		0.15	0.14	0.16	0.18	0.14
MgO		8.17	8.73	8.65	8.23	8.96
CaO		9.86	11.59	11.05	12.21	11.54
Na_2O		2.37	1.94	2.29	1.98	1.90
K_2O		0.84	1.33	0.86	0.32	0.27
P_2O_5		0.19	0.22	0.20	0.12	0.21
Total		100.26	99.34	99.46	99.94	99.77
FeO */MgO		1.05	0.78	0.96	1.09	0.94
Zr (ppm)		70	56	65	64	59
Y (ppm)		30	21	24	21	21
Rb (ppm)		42	66	38	11	12
Sr (ppm)		164	161	179	132	160

* See footnote for Table I.
** Abbreviations as in Tables I and II.

PA23I G U.A.	PA23B G —	PA23H L.A. —	PA23R G U.A.	PA28P L.A. U.A.	PA23U L.A. U.A.	PA23Y G U.A.	PA28K G —	PA43A G —	PA23X G L.A.
51.44	51.57	49.77	51.63	53.34	52.29	53.09	54.28	54.79	53.37
1.16	1.17	1.20	1.21	1.24	1.31	1.66	1.67	1.87	2.09
16.44	16.06	16.54	16.15	15.99	15.77	14.71	14.54	15.66	13.99
9.70	10.15	10.10	11.80	10.64	11.41	11.41	13.32	11.89	13.38
0.22	0.21	0.18	0.29	0.15	0.22	0.12	0.26	0.15	0.20
7.88	8.13	6.45	7.12	5.86	6.79	5.51	3.83	3.02	4.24
10.66	8.92	10.45	9.36	8.73	9.00	9.96	11.95	11.56	8.27
2.12	2.74	2.50	2.39	3.42	3.04	3.07	0.25	0.30	3.21
0.13	0.01	0.16	0.10	0.58	0.15	0.24	0.01	0.01	0.21
0.23	0.28	0.22	0.25	0.22	0.20	0.29	0.31	0.46	0.31
99.98	99.24	97.57	100.28	100.17	100.18	100.06	100.43	99.71	99.27
1.23	1.25	1.57	1.66	1.82	1.68	2.07	3.48	3.94	3.16
87	69	75	65	93	99	93	115	137	187
20	26	27	27	35	27	39	41	59	45
6	<1	4	6	24	6	4	<1	<1	5
168	142	158	167	180	167	150	238	294	143

PA43H L.A. U.A.	PA32G G L.A.	PA32J G L.A.	PA43F L.A. —	PA51A L.A. —	PA32F G L.A.	FL-6 L.A. U.A.
52.29	51.09	51.77	52.54	52.08	52.90	51.14
1.00	1.09	1.11	1.18	1.32	1.49	1.71
16.29	16.61	16.55	15.76	15.80	15.34	14.87
10.20	8.85	8.93	9.74	11.05	12.98	12.14
0.14	0.17	0.16	0.15	0.17	0.20	0.18
7.48	8.14	7.34	5.46	5.68	5.35	6.14
10.91	10.76	11.32	10.06	8.70	8.43	9.95
1.59	2.65	2.10	2.37	2.93	3.27	3.11
0.13	0.54	0.56	0.29	1.03	0.92	0.12
0.24	0.24	0.25	0.29	0.24	0.29	0.29
100.27	100.14	100.09	97.84	99.00	101.17	99.65
1.36	1.09	1.22	1.78	1.95	2.43	1.98
62	82	72	118	103	109	109
22	26	27	27	39	33	40
6	13	16	9	33	42	3
164	172	293	183	160	143	145

TABLE IV

Composition of gabbros in the Sarmiento Complex *

Compo- nents	Sample nr: Facies **: Relic facies **:	PA31K L.A. —	PA31P L.A. U.A.	PA31N G U.A.	PA31A G L.A.	PA31B L.A. —
SiO$_2$		50.25	48.14	50.00	50.20	51.58
TiO$_2$		0.59	0.75	0.79	0.86	0.88
Al$_2$O$_3$		18.52	16.06	16.04	15.13	15.69
FeO *		6.89	7.34	8.27	9.90	9.33
MnO		0.14	0.13	0.16	0.14	0.16
MgO		7.38	10.28	7.94	7.43	6.87
CaO		13.95	15.09	12.89	11.88	11.63
Na$_2$O		2.12	1.25	2.21	2.33	2.52
K$_2$O		0.16	0.23	0.88	0.86	0.86
P$_2$O$_5$		0.07	0.20	0.22	0.27	0.20
Total		100.07	99.50	99.40	99.00	99.37
FeO */MgO		0.93	0.71	1.04	1.33	1.36
Zr (ppm)		23	13	19	35	38
Y (ppm)		17	10	21	21	24
Rb (ppm)		2	9	32	35	34
Sr (ppm)		131	200	107	150	137

* See footnote for Table I.
** Abbreviations as in Tables I and II.
*** Average of six samples.

PILLOW LAVAS AND DIKES CUTTING PILLOWS

Within the pillow lavas and cross cutting dikes both zeolite and greenschist facies mineral assemblages are found (Fig. 3). Neither completely fresh pillow lavas nor relict actinolite facies minerals have been found at this level. Metamorphic replacement of the original igneous minerals is incomplete; often cores of relict pyroxenes or calcic plagioclases are found. Metamorphic textures in pillow lavas and cross cutting dikes preserve or strongly reflect the texture and the degree of crystallinity of the original igneous rocks (Fig. 4). Albite and occasionally hydrothermally derived potassium feldspar (Or$_{99}$) replace calcic plagioclase, although the original plagioclase lath crystal forms are well preserved even when the original calcic plagioclase is completely replaced by albite. Mafic phases are replaced, often pseudomorphically, by either zeolites or mixtures of chlorite, epidote, and calcite depending on the metamorphic grade. Zeolites are often associated with greenschist facies rocks, but are usually restricted to vesicles and secondary veins. In such instances, they are interpreted as later, lower temperature cavity fillings. Hydrothermal sulfide mineralization, usually restricted to veins and vesicles, is common at this pseudostratigraphic level, a feature similar to other ophiolites (Upadhyay and Strong, 1973; Bonatti et al., 1975).

PA30F L.A. U.A.	PA31E L.A. U.A.	PA25B G L.A.	PA30E U.A. —	PA30H U.A. —	PA25A G L.A.	Plagiogranite *** —
51.00	48.61	50.72	46.86	43.47	51.57	74.59
1.39	1.44	1.70	2.24	2.97	2.36	0.30
15.73	17.59	11.64	13.60	13.00	13.83	12.60
10.36	10.30	12.07	16.12	22.03	14.28	2.73
0.18	0.16	0.23	0.21	0.29	0.19	0.03
5.91	5.78	7.74	6.22	7.71	4.50	0.26
11.84	11.89	12.87	11.75	6.86	9.36	5.00
2.64	2.77	1.76	2.18	2.65	2.73	3.55
0.10	0.26	0.55	0.16	0.08	0.62	0.15
0.22	0.22	0.35	0.28	0.16	0.27	0.06
99.37	98.72	99.63	99.62	99.22	99.74	99.27
1.75	1.78	1.56	2.59	2.85	3.17	
28	40	53	25	27	81	470
18	22	26	18	20	29	97
1	9	20	3	2	23	6
144	159	95	130	121	139	195

Fig. 4. Outline of olivine crystal pseudomorphed by calcite and chlorite within a pillow lava. Albite has replaced calcic plagioclase needles but original igneous texture is preserved. Photo is 2 mm across.

SHEETED DIKES

Within the sheeted dike unit, greenschist, lower and relict upper actinolite facies mineral assemblages are found (Fig. 3). Mafic minerals are replaced, by brown, blocky or green, fibrous amphiboles in the actinolite facies, or by chlorites and epidotes in the greenschist facies. When actinolitic amphibole, epidote, and chlorite occur together, the amphibole is usually clearly being replaced by the latter minerals. We interpret this as greenschist facies metamorphism subsequent to actinolite facies metamorphism which causes amphiboles to react, possibly in a complex reaction involving calcic plagioclase, to form chlorite, epidote, albite and quartz. Such disequilibrium effects are usually very heterogeneous; within a single thin section some areas will be completely replaced by greenschist facies minerals, others will preserve abundant relict higher facies minerals. Within the lower actinolite and greenschist facies, sphene is always the stable titanium bearing mineral. Sphene pseudomorphs titanomagnetite, even within the most highly recrystallized greenschist metabasalts. Within the sheeted dikes, titanomagnetite and ilmenite may persist within incompletely recrystallized metabasalts but are always enclosed by rims of sphene. Sulfides are very rare within the sheeted dikes, in contrast to the pillow lava unit. Biotite is uncommon at the level of sheeted dikes. The abundant presence of epidote indicates an oxidizing environment (Humphris, 1978; Coish, 1977) accounting for the absence of biotite, titanomagnetite and ilmenite (Miyashiro et al., 1971).

At this pseudostratigraphic level, metamorphic replacement and recrystallization is intense, no cores of primary mafic igneous silicates are found and primary igneous textures are only poorly or not at all preserved (Fig. 5). The degree of preservation of primary igneous texture depends upon the degree of recrystallization of plagioclase which is in turn related to metamorphic grade due to the changing stability of calcic plagioclase. Rocks preserving relict upper actinolite facies mineral assemblages, brown blocky amphiboles and titanomagnetite cores, may also preserve nearly pristine calcic plagioclase. Plagioclase within the lower actinolite facies often shows varying degrees of cloudy alteration to calcite and sericite, but not recrystallization to albite. At these metamorphic grades plagioclase retains its original igneous lath-like form and subophitic to intersertal textures are partially preserved. Within the greenschist facies plagioclase is extensively replaced by albite, carbonates, epidote and sericite. This replacement is accompanied by erosion and recrystallization of the original crystal boundaries. Extensive greenschist facies recrystallization at this pseudostratigraphic level has restricted the number of metabasalts which preserve igneous textures.

For rocks within the greenschist facies the great amount of recrystallization which occurs within the sheeted dike unit, rather than pseudomorphic replacement which occurs within the pillow lavas, may be related firstly to the higher temperature at this deeper level, evidenced by the close association with higher facies rock types, and secondly to the fact that many

Fig. 5. Highly recrystallized greenschist facies sheeted dike without relict igneous textures. Compare texture with Fig. 4. Photo is 2 mm across.

greenschist metabasalts within the sheeted dikes have formed by subsequent recrystallization of lower and upper actinolite facies metabasalts so that the more complicated recrystallization history has affected the final textures.

DIKES CUTTING GABBROS

Within the gabbro level of the Sarmiento complex, dikes cutting gabbros contain greenschist to upper actinolite facies mineral assemblages (Fig. 3). At this level all metabasalts within the greenschist facies contain relict higher facies assemblages, and some upper actinolite facies metabasalts without subsequent greenschist effects are preserved. Subophitic and intersertal textures are preserved because of the stability of calcic plagioclase within the actinolite facies (Fig. 6). Original mafic igneous minerals have been almost completely replaced by brown and green amphiboles but cores of pyroxenes are preserved at this level in contrast to their complete absence within the sheeted dike unit, indicating a decrease in the extent of metamorphic replacement and recrystallization. Consistent with this, titanomagnetite and ilmenite which are stable in upper actinolite facies metabasalts are rimmed but rarely completely replaced by sphene during lower grade metamorphism.

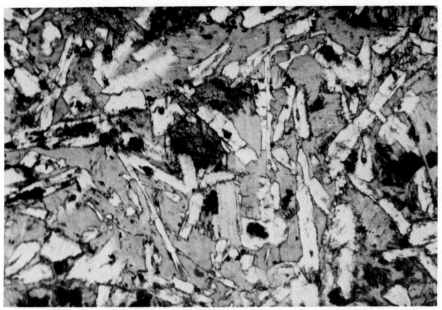

Fig. 6. Upper actinolite facies dike cutting gabbros with blocky brown as well as fibrous green amphibole and calcic plagioclase. Compare with Fig. 4 and 5. Photo is 2 mm across.

Biotite, absent within the sheeted dikes, is a common accessory at this level.

Although greenschist metamorphic effects are common (Fig. 3), the extent of greenschist replacement and recrystallization is much less than within the sheeted dike unit. Greenschist effects at this pseudostratigraphic level have, in all rocks, taken place subsequent to actinolite facies metamorphism and consist of the development of chlorites without significant epidote, in contrast to within the sheeted dike unit. Biotite in lower actinolite facies metabasalts does not seem to be affected by subsequent greenschist metamorphism. The paucity of epidote suggests a less oxidizing environment within the dikes cutting gabbros than the sheeted dike unit, consistent with the common presence of biotite, titanomagnetite and ilmenite.

GABBROS

For gabbros themselves, the distribution in the Sarmiento complex of metagabbros of different grades is very similar to the dikes cutting gabbros (Fig. 3). Gabbroic textures are well preserved, in part because of the high modal content (up to 50%) and stability within the actinolite facies of calcic plagioclase. Cores of pyroxenes are commonly preserved, sometimes with only very thin rims of amphibole developed. Increased grain size may be in part responsible for this low degree of metamorphic replacement and recrystallization.

Two-pyroxene gabbros which always exhibit some degree of actinolite grade metamorphism occur at the deepest level of exposure in the Sarmiento complex (Fig. 2). Equivalent two-pyroxene gabbros found below the sheeted dike-diabase unit in the Tortuga complex (Fig. 2) also exhibit the effects of actinolite grade metamorphism. Deeper in the gabbro unit of the Tortuga complex these effects become progressively more restricted to crystal boundaries. Olivine—pyroxene gabbros from the deepest level of the Tortuga complex (Fig. 2), as well as from less well studied ophiolite bodies lying between the Sarmiento and Tortuga complexes commonly contain completely pristine igneous minerals with no sign of hydrous alteration. Thus the extent of metamorphic replacement and recrystallization, which in the Sarmiento complex is lower in the gabbros and cross cutting dike than in the sheeted dikes, is seen, in the Tortuga complex where the level of exposure is deeper, to continue to decrease until metamorphic effects terminate leaving fresh gabbros.

Due to the preservation of igneous materials in all stages of metamorphic replacement within the gabbros, it is possible to observe the metamorphic reactions which transform pristine gabbros to actinolite grade metagabbros. At the onset of the lower stratigraphic boundary of hydrous reactions, olivine crystals containing internal veins of magnetite plus silica react to form serpentine, in some cases replacing all but the cores of the olivine crystals. At a higher level brown amphibole associated with quartz and magnetite appears along cracks within and as rims of olivine and pyroxene crystals. Serpentine is not observed in association with amphibole. The abundance of blocky brown amphibole characteristic of the upper actinolite facies increases upwards until only some cores of pyroxenes remain. The $\delta\ ^{18}O$ value of a brown amphibole separate has been determined to be 2.6‰, consistent with a metamorphic origin (Stern et al., 1976). These blocky brown amphiboles always contain microcrystalline magnetite scattered along cracks and cleavage planes. This microcrystalline magnetite is the result of metamorphic replacement and not simply relict igneous magnetite initially poikilitically enclosed within pyroxenes.

Lower actinolite grade metamorphism is characterized by the replacement of both pyroxenes and blocky brown amphiboles by fibrous green, slightly more iron-rich and alumina-poor (Table V, VI, and discussion below) amphibole without associated magnetite. Biotite is a common accessory in lower amphibolite metagabbros, just as in dikes cutting gabbros. In both upper and lower actinolite facies metagabbros, plagioclase appears to remain unaffected by metamorphism, but the $\delta\ ^{18}O$ value of a plagioclase separate from an upper actinolite facies metagabbro has been determined as 4.0‰, lower than primary igneous plagioclase (6.5‰) consistent with high temperature exchange with a reservoir of low $^{18}O/^{16}O$ (Stern et al., 1976).

A common feature of metagabbros from both the Sarmiento and Tortuga

TABLE V

Compositions of metamorphic minerals in the Sarmiento Complex *

Components	Chlorites				Epidotes		Amphiboles			
Rock FeO*/MgO	0.95	1.10	2.00	3.48	1.17	3.48	1.10	1.09	2.07	1.98
Facies:	Z	G	Z	G	G	G	G	U.A.	G	L.A.
Relic facies:	G	L.A.	G				L.A.		L.A.	U.A.
SiO$_2$	31.65	28.48	26.26	23.17	37.19	35.68	49.83	50.12	51.88	51.34
TiO$_2$	0.00	0.04	0.00	0.01	0.10	0.06	0.20	0.48	0.20	0.39
Al$_2$O$_3$	16.98	18.59	17.75	20.20	28.97	25.13	4.29	6.69	2.91	3.60
FeO *	21.84	22.10	28.46	32.54	5.14	10.77	15.87	14.81	19.50	18.84
MnO	0.20	0.27	0.46	0.45	0.19	0.06	0.36	0.26	0.23	0.27
MgO	19.11	18.24	15.27	11.54	0.02	0.02	13.97	14.00	11.44	12.42
CaO	0.17	0.09	0.04	0.10	22.34	23.01	11.63	11.76	11.73	11.20
Na$_2$O	0.00	0.00	0.00	0.00	0.00	0.00	0.49	0.80	0.33	0.63
K$_2$O	0.00	0.00	0.00	0.00	0.00	0.00	0.03	0.13	0.05	0.06
Total	89.95	87.81	88.24	88.01	93.95	94.73	96.67	99.05	98.27	98.75
FeO *	21.84	22.10	28.46	32.54	5.14	10.77	15.87	14.81	19.50	18.84
FeO */MgO	1.14	1.21	1.86	2.84	—	—	1.14	1.06	1.70	1.52

* Determined by microprobe analysis. Total Fe given as FeO *. Abbreviations as in Tables I and II.

TABLE VI

Variations in metamorphic mineral compositions

	Increasing T		Increasing rock FeO/MgO		
	chlorites	amphiboles	chlorites	epidotes	amphiboles
SiO$_2$	—	0	—	—	+
Al$_2$O$_3$	+	+	+	—	—
FeO *	0	—	+	+	+
FeO */MgO	0	—	+	+	+

+ = increase, — = decrease, 0 = no change.

complexes are "trellis" intergrowths of magnetite and ilmenite (Fig. 7) developed from both titanomagnetite and "sandwich" intergrowths of ilmenite and titanomagnetite (Fig. 8). While sandwich intergrowth may result from crystallization processes, trellis intergrowths can develop only by post-crystallization oxy-exsolution of ilmenite from Ti-rich magnetite (Haggerty, 1977). Up to four generations of ilmenite lamellae are found in trellis intergrowths of ilmenite and magnetite within Sarmiento and Tortuga gabbros, indicating continued re-equilibration of co-existing titanomagnetite and ilmenite during slow cooling. In upper actinolite facies metagabbros, the Ti-poor magnetites located along the ilmenite lamellae are often replaced by

Fig. 7. Reflected light, oil immersion photomicrograph of trellis intergrowth of ilmenite (dark grey) and magnetite (light grey) within a gabbro. Photo is 0.5 mm across.

Fig. 8. Partial pseudomorphic replacement by sphene of a skeletal ilmenite. Hydrothermal circulation has replaced the Ti-poor magnetite within a trellis intergrowth of ilmenite and magnetite (Fig. 7) with actinolite. Photo is 2 mm across.

amphibole (Fig. 8). The extent of replacement of magnetite is closely related to the extent of metamorphic replacement of the metagabbros, skeletal networks of ilmenite without associated magnetite only occur in highly recrystallized metagabbros without cores of primary mafic igneous phases. In lower actinolite facies metagabbros the skeletal networks of ilmenite are pseudomorphically replaced by sphene (Fig. 8). The pseudomorphic preservation of these skeletal networks attests to the relative immobility of titanium during metamorphism, as well as the lack of stress accompanying the metamorphism.

Within the Sarmiento complex, greenschist metamorphism in gabbros occurs subsequent to actinolite facies metamorphism, and is similar in its frequency, extent, and effects to greenschist metamorphism in dikes cutting gabbros. The development of chlorite occurs without significant epidote and biotite in lower amphibolite metagabbros is unaffected by subsequent greenschist metamorphism. As in the dikes cutting the gabbros, the paucity of epidote suggests a less oxidizing environment than within the sheeted dike unit, consistent with the common occurrence of biotite, ilmenite and magnetite. Moreover, the close association caused by reaction relations of olivines, pyroxenes and amphiboles with magnetite and minor quartz suggests that the oxygen fugacity has remained close to the value of the FMQ buffer, typical of oceanic basalts (Haggerty, 1977).

SILICIC DIKES AND PLAGIOGRANITES

Silicic dikes, both within the sheeted dike complex and cutting gabbros, and plagiogranites found just below the sheeted dike unit contain myrmekitic intergrowths of quartz and albite replacing twined plagioclase laths. Secondary albite is characteristic of greenschist facies metamorphism. However, although these textures are distinctly non-igneous, similar granophyric textures are found in the uppermost levels of layered basic intrusions due to crystallization under high P_{H_2O} caused by concentration of volatiles within late stage differentiation. Thus the highly evolved composition of the silicic plagiogranites prevents unequivocal assignment to the greenschist facies except when chlorites and epidotes are found. For this reason these rocks were not considered in the construction of Fig. 3.

CHEMICAL EFFECTS OF METAMORPHISM

The metabasalts and metagabbros of the Sarmiento complex show a large amount of chemical variation (Tables I—IV), in part due to primary igneous processes, in part due to secondary metamorphism. Metasomatic effects can only be evaluated if primary igneous effects are accounted for. The approach we use here is first to characterize the relevant igneous processes by considering the variations in the abundance of elements considered least mobile under a wide range of metamorphic conditions, such as TiO_2, Zr, Y, P_2O_5,

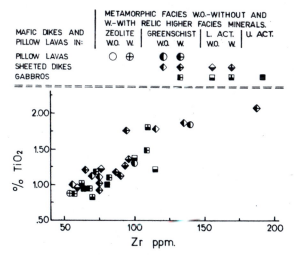

Fig. 9. Zr vs. TiO_2 for dikes and lavas of different metamorphic grades and from different pseudostratigraphic levels of the Sarmiento complex. Data are from Tables I—III. Coherency of chemical trends is consistent with low mobility of Ti and Zr under a wide range of metamorphic conditions.

and rare earth elements (Pearce and Cann, 1973; Frey et al., 1974; Kay and Senechal, 1976). For basic dikes and lavas from the Sarmiento complex, these elements exhibit a large range of variation (Fig. 12; Tables I—III; Stern, 1976; in press, a). These variations show a positive correlation with each other, illustrated for TiO_2 and Zr in Fig. 9. Also, as REE contents increase, a negative Eu anomaly develops but La/Yb does not vary. The large positive covariance of these elements is consistent with their progressive enrichment in residual liquids produced by crystal-liquid fractionation involving olivine, pyroxenes, and calcit plagioclase in low level magma chambers (Stern, in prep., a). This enrichment results from the low distribution coefficients that TiO_2, Zr, Y, P_2O_5, and rare earth elements have between these mafic silicates and silicate liquids. Fresh oceanic basalts exhibit similar covariance of these elements which has also been attributed to differentiation in shallow magma chambers (Kay et al., 1970; Miyashiro et al., 1971; Erlank and Kable, 1976; Clague and Bunch, 1976).

Significantly, the same sympathetic variation of these elements occurs in lavas and dikes within different metamorphic facies and from different pseudostratigraphic levels of the Sarmiento complex, again illustrated for TiO_2 and Zr in Fig. 9. This coherency, independent of pseudostratigraphic level or metamorphic grade, is consistent with the low mobility of these elements under a wide range of metamorphic conditions. The pseudomorphic replacement of titanomagnetite and skeletal ilmenite lamellae by sphene even in the most highly recrystallized metabasalts attests specifically to the low mobility of titanium (Fig. 8). Thus we conclude that the large covariance in the abundance of TiO_2, Zr, Y, P_2O_5 and rare earth elements found in lavas and dikes from the Sarmiento complex is a primary igneous effect, and the degree of enrichment of these elements is related to the extent of magmatic differentiation.

FeO^*, (FeO^* = total Fe as FeO), MgO and FeO^*/MgO also show large variations in dikes and lavas from the Sarmiento complex (Fig. 10; Tables I—III; Stern, 1976; in press, a). Variations of TiO_2, Zr, Y, P_2O_5, and rare earth elements correlate positively with both FeO^*/MgO (Fig. 10) and FeO^*, and negatively with MgO. This suggests that the variations of FeO^* and MgO are also related to the igneous process of differentiation in shallow level magma chambers. Similar covariances observed in fresh oceanic basalts are also explained by low pressure magmatic differentiation. For low and intermediate values (0.75—2.00) of FeO^*/MgO, the positive correlation with TiO_2 is less coherent. Highly recrystallized greenschist facies sheeted dikes have diverged from the trend of covariance towards higher values of FeO^*/MgO (Fig. 10). A small loss of MgO during greenschist facies metamorphism is the most likely explanation of this effect since a large addition of FeO^* would be needed to increase the FeO^*/MgO ratio of metabasalts with initially high FeO^*, and this effect would be more evident in metabasalts with lower values of FeO^* and FeO^*/MgO. The loss of MgO must be small since it is not observed in highly recrystallized greenschist facies sheeted dikes with ini-

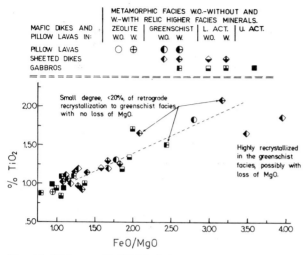

Fig. 10. TiO₂ vs. FeO*/MgO for lavas and dikes from the Sarmiento complex. Data are from Tables I—III.

tially high MgO, or in other dikes with initially low MgO that have been only partially recrystallized to greenschist facies (Fig. 10).

SiO₂ varies between 49.77 wt.% and 55.31 wt.% in lavas and dikes from the Sarmiento complex (Tables I—III), but this variation is independent of TiO₂, Zr, FeO, and FeO*/MgO. The lack of systematic increasing of SiO₂ with increasing FeO* and FeO*/MgO confirms that fractionation trends within the Sarmiento complex are tholeiitic, similar to oceanic basalts

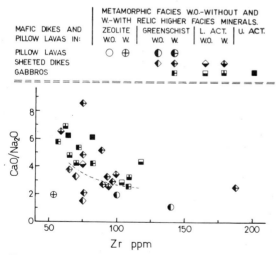

Fig. 11. Zr vs. CaO/Na₂O for lavas and dikes from the Sarmiento complex. Data are from Tables I—III.

(Miyashiro, 1973). The silica variations that do occur are related to meta-morphic grade and pseudostratigraphic position, and correlate with variations in CaO and Na_2O. Fig. 11 illustrates the variation of CaO/Na_2O with increasing Zr, or increasing differentiation. In fresh oceanic basalts the calcium—sodium ratio is commonly within the range of 4—6, decreasing slightly with increasing fractionation. Figure 11 shows that for dikes cutting gabbros in the Sarmiento complex, CaO/Na_2O follows this general pattern. However, pillow lavas and dikes cutting pillows have significantly lower CaO/Na_2O and a number of sheeted.dikes have distinctly lower CaO/Na_2O. Increase in SiO_2 is associated with decrease in CaO/Na_2O; metabasalts with CaO/Na_2O less than 3.0 average SiO_2 = 53.21 wt.% (Tables I—III). Increase of SiO_2 and Na_2O and decrease of CaO are common effects of spilitization. In the Sarmiento complex these effects are restricted to the upper levels of the ophiolite complex and are less extreme than reported for the Troodos ophiolite complex (Gass and Smewing, 1973; Kay and Senechal, 1976).

Not included in the construction of Fig. 11 are three dikes within the sheeted dike unit, PA28K, PA28V and PA43A which have anomalously low Na_2O and high CaO, but are coherent with other dikes with respect to TiO_2, Zr, Y, P_2O_5, FeO^*, and FeO^*/MgO (Table II). These dikes are located immediately above the intrusive contact of the plagiogranites, at the very bottom of the sheeted dike unit (Fig. 2). They are highly recrystallized greenschists without igneous textures or relict higher grade minerals, but with abundant secondary quartz and minor secondary anorthite although they exhibit only slight silica enrichment. Sodium may have been flushed out of these rocks due to circulation of silica-rich fluids related to the intrusion by magmatic stoping of the plagiogranites.

For the lavas and dikes from the Sarmiento complex K_2O and Rb exhibit the greatest amount of variation of any elements, from below detectability to 4.34 wt.% for K_2O, and to 95 ppm for Rb (Tables I—III). These variations show no relation with any parameter of degree of differentiation, as illustrated for Zr in Fig. 12. The variations depend instead on metamorphic grade and pseudostratigraphic level within the ophiolite. Fig. 12 shows that sheeted dikes have significantly lower K_2O contents than dikes cutting gabbros, and greenschist facies sheeted dikes without relict higher facies assemblages have less than 0.01 wt.% K_2O. The average values for K_2O for different pseudostratigraphic levels is shown in Fig. 3. Pillow lavas and dikes cutting pillows have high, >2.00 wt.% K_2O contents. Within the sheeted dike unit, average K_2O contents are low and decrease with decreasing metamorphic grade. In the dikes cutting gabbro, average K_2O contents are intermediate, and in contrast to within the sheeted dikes, increase with decreasing metamorphic grade. A similar pattern exists for the gabbros themselves as for the dikes cutting gabbros except that average K_2O contents are slightly lower for equivalent metamorphic facies. Corresponding relations occur for Rb (Tables I—III).

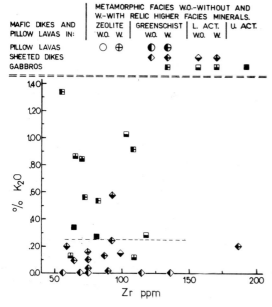

Fig. 12. K_2O vs. Zr for dikes from the sheeted dike unit and cutting gabbros in the Sarmiento complex. Data are from Tables II and III. Note the systematically lower values of K_2O in sheeted dikes.

No completely fresh basalts are available from the Sarmiento complex with which to determine unequivocally the K_2O content and Rb content before alteration. The average value of K_2O for unaltered mid-ocean ridge basalts is 0.26.% (Erlank and Kable, 1976), for Lau Basin basalts 0.18 wt.% (Hawkins, 1976), and for basalts from behind the Mariana arc 0.17 wt.% (Hart et al., 1972). Slight $[(La/Yb)_N \sim 1.9]$ relative enrichment of light rare earths indicates that Sarmiento basalts were derived from a source less depleted in large ion lithophile elements, such as K_2O and Rb, than MOR or Lau Basin basalts (Stern, in press, b). Fresh Sarmiento basalts more likely resembled Mariana Basin basalts which also exhibit a small enrichment of light relative to heavy REE. We conclude that metamorphic processes have caused pillow lavas and dikes cutting pillows to have gained K_2O, and dikes within the sheeted dike complex to have lost K_2O. Alteration of basaltic glass causes increase in K_2O (Hart et al., 1972), but is unlikely to have produced the extreme enrichment and variations found in the pillow lavas and dikes cutting pillows of the Sarmiento complex. These metabasalts often contain metamorphic orthoclase (Vallence, 1965; Montigny et al., 1973). The source of the potassium may have been either seawater or the sediments overlying the ophiolites, since it has been suggested that the effects of spilitization are related to and dependent upon the participation of marine sediments (Vallence, 1965; Montigny et al., 1973).

The progressive loss with decreasing metamorphic grade of K_2O from dikes within the sheeted dikes is consistent with the corresponding decrease in stability of primary igneous plagioclase in which K_2O is initially incorporated, as well as the lack of development of a metamorphic mineral which incorporates potassium. Completely recrystallized greenschist facies dikes preserving no relict higher facies minerals, in particular primary calcic plagioclase, contain the least amount of K_2O. For dikes cutting gabbros and gabbros, K_2O increases with decreasing grade of metamorphism, consistent with the presence of biotite as an accessory within the lower actinolite facies and its increased stability in the greenschist facies (Miyashiro, 1975). Within this pseudostratigraphic level, K_2O and Rb have clearly been metasomatically redistributed on a large scale, since their abundances no longer correlate with the degree of igneous evolution of the original rock (Fig. 12). Whether these metabasalts and metagabbros on the whole have lost or gained potassium is more problematic. The average value of K_2O for all dikes cutting gabbros is 0.60 wt.%, and for all gabbros 0.43 wt.%. These values are high in comparison to fresh basalts and cumulate gabbros from similar tectonic environments, as discussed above. On the other hand, the lower actinolite facies sheeted dikes and upper actinolite facies dikes cutting gabbros, preserving the greates amount of primary plagioclase, have similar K_2O contents (Fig. 3). This suggests that the highest grade metabasalts may most nearly preserve their original K_2O contents, 0.30 wt.%. Upper actinolite facies metagabbros contain lower K_2O, 0.12 wt.% consistent with their cumulate origin. Comparison of these values of K_2O with the average K_2O of all dikes cutting gabbros (0.60 wt.%) and all metagabbros (0.43 wt.%) suggests that this pseudostratigraphic level has gained postassium. The sheeted dikes may have been the source of this potassium, with freely mobilized potassium from this level being incorporated in biotite within the gabbros and dikes cutting gabbros. A mass balance calculation shows that 300 m of dikes, originally with 0.30 wt.% K_2O and now with an average 0.10 wt.% K_2O, can supply only 20% of the K_2O needed to increase a 1,000 meter section of 20% dikes and 80% gabbros from a combined initial K_2O content of 0.16 wt.% to a final K_2O content of 0.46 wt.%. However, the 100 m of plagiogranites lying directly below the sheeted dikes also suffered extensive loss of K_2O. These highly evolved differentiates have fivefold enrichments of REE, P_2O_5, Zr and Y relative to more primitive basalts (Stern, 1976 and in press, a), suggesting that their originial K_2O contents were as high as 1.50 wt.%. Also the upper actinolite facies metagabbros without "retrograde" effects are the deepest, most coarsely crystalline cumulates, and the overlying gabbro unit, containing a large volume of diabase (Fig. 2) may have had a higher average K_2O content than 0.12 wt.%. For a gabbro unit with an initial average K_2O = 0.20 wt.%, 400 m of dikes and plagiogranites, initially with a combined K_2O of 0.60 wt.% and now with 0.10 wt.%, would be sufficient to provide all the potassium needed to increase 900 meters of dikes and gabbros from a combined initial K_2O of 0.22 wt.% to a final value of 0.46 wt.%.

Metamorphic minerals also show chemical variations that correlate with both variations in rock chemistry and changing metamorphic grade. Table V lists chemical compositions of chlorites, epidotes and amphiboles in metabasalts of different metamorphic grade and with different FeO*/MgO. Table VI summarizes the major variations in mineral chemistry for changing temperature independent of rock composition and for changing rock composition (FeO*/MgO) independent of temperature. Of particular interest is the lower FeO* and FeO*/MgO for amphiboles within the upper actinolite facies compared to the lower actinolite facies independent of rock chemistry.

Finally, Fig. 3 shows the variations of δ ^{18}O found within the Sarmiento complex (Stern et al., 1976). δ ^{18}O decreases downward from 9.6‰ within pillow lavas to 2.6% within the lowest level actinolite bearing metagabbros. A completely pristine gabbro from the Tortuga comples has δ ^{18}O = 6.5% . This is within the range 5.5—6.5‰ of unaltered basaltic rocks (Muehlenbachs and Clayton, 1972). The vertical change from ^{18}O enriched to ^{18}O depleted basalts can be explained by re-equilibration, at successively higher temperatures, with a large reservoir of low δ ^{18}O water (Spooner et al., 1974; Stern et al., 1976).

METAMORPHISM WITHIN A SPREADING RIDGE

Hydrothermal metamorphism associated with circulation of seawater within the spreading center at which the ophiolites formed explains better than burial metamorphism the steep metamorphic gradient, the irregularity of metamorphic boundaries, and the decrease and termination of metamorphic effects observed in Chilean ophiolites (DeWit and Stern, 1976; Stern et al., 1976). Seawater rather than connate magmatic fluid is considered as the cause of the metamorphism, firstly because of δ ^{18}O variations which indicate reequilibration with non-magmatic water, and secondly because of the large scale (volume) of the alteration and the low estimates of weight percent of juvenile water in tholeiitic magmas (Moore, 1972). Tectonic activity within a spreading center would create a zone of structural weakness and high permeability along which seawater could penetrate until it is driven back by the heat from igneous intrusion.

In the tectonic setting of a spreading ridge, igneous activity occurs in a restricted region and igneous rocks move away from this region by continued spreading. The geothermal gradient will be, in general, high near the ridge crest and lower at a distance from the zone of igneous activity. In detail, the thermal history of any one rock will be very complicated, involving not only its own cooling, but occasional increases in temperature due to continued igneous activity. Away from the ridge such perturbations in the cooling history of any given igneous body will be less frequent. The detailed cooling histories of lavas, dikes and gabbros may differ due to the different spatial form and extent of these different igneous bodies. Superimposed on the thermal variations related to igneous processes will be thermal perturbations related to rising and descending hydrothermal circulation currents. Hydrothermal metamorphism will affect a pristine igneous rock through its complicated thermal history. As the rock moves away from the spreading ridge, the *general* decrease in nearby igneous activity and geothermal gradient will lead to the development of a sequence of increasingly lower grade minerals which, depending on the amount of fluid phase present and/or the residence time of the rock within a certain mineral stability field, may or may not completely replace the previously formed metamorphic phases or the original igneous minerals. The common presence in Chilean and other ophiolites, of disequilibrium textures involving relict igneous and high grade metamorphic minerals subsequently incompletely recrystallized at lower grade metamorphic conditions, is consistent with the complicated metamorphic history expected in the vicinity of a spreading ridge. The heterogeneous distribution of such disequilibrium effects in Chilean ophiolites, even on the scale of a thin section, suggests that hydrothermal circulation and the availability of a fluid phase to flux metamorphic reactions was very irregular even in the upper levels of the ophiolites.

The pseudostratigraphic variations in the metamorphic effects observed within Chilean ophiolites elucidates the nature of the interaction of the geothermal gradient and hydrothermal circulation that resulted in the observed metamorphic overprint. Firstly, the termination of metamorphic effects indicates that the circulation of water did not penetrate deeply into the cumulate gabbro layer, except possibly along very restricted fractures and/or dikes which continue to exhibit metamorphic effects at the lowest level of exposure of the Tortuga complex. Secondly, the lower frequency of occurrence and extent of retrograde metamorphism within the gabbros and dikes cutting gabbros implies a rapid decrease in local temperature, seawater circulation, or both. This suggests that the residence time of the metagabbro within the zone of metamorphism is short, the implication being that the deeper roots of this zone are narrowly restricted to near the spreading ridge. This restriction must be due to the dependence of the hydrothermal circulation on the igneous and tectonic activity associated with the spreading center. Away from the spreading center deep hydrothermal circulation acts to reduce itself by lowering the geothermal gradient by convective heat removal, and by effecting metasomatic redistribution of material which decreases the permeability of the rocks. The upper levels of the ophiolites may have undergone metamorphism for a more extended time as evidenced by the greater extent of metamorphic recrystallization and retrograde effects, as well as spilitization of the pillow lavas which may not have occurred until sediments were laid down over the ophiolites (Montigny et al., 1973). Hydrothermal circulation may exist within a wide area about a spreading center, but it is probably confined to shallower depths than near the spreading ridge.

Within the Sarmiento complex the plagiogranites which intrude the sheeted dike unit by magmatic stoping (Fig. 2) are the locus of a major discontinuity in the degree of metamorphic recrystallization, the extent of retrograde effects, the stability of epidote, biotite, titanomagnetite, ilmenite, and the chemical effects of metamorphism with respect particularly to K_2O and Rb. Just as the ultimate decrease and termination of metamorphic effects must be the result of decreasing volume and finally lack of hydrothermal circulation, the contrast in the extent of metamorphic replacement and recrystallization and "retrograde" effects between the sheeted dikes and gabbro unit is likely to be due to decreased volume of seawater circulation. Fracturing associated with the intrusion of the plagiogranites may have greatly increased the permeability of the overlying sheeted dikes as suggested by numerous silicified veins and fractures just above the sheeted dike-plagiogranite contact. At this level large volumes of water have caused extensive metamorphic recrystallization and retrograde effects, and produced an oxidizing environment as suggested by the abundance of epidote and the lack of biotite, ilmenite and titanomagnetite. At the level of the gabbro unit below the plagiogranites the paucity of epidote and the common occurrence of biotite, titanomagnetite, ilmenite and magnetite are consistent with a relatively small volume of water partaking in reactions that are buffered near the value of the FMQ buffer by the large volume of igneous rock. The different mineralogies within the two stratigraphic levels have resulted in differences in the geochemical effects of the metamorphism.

Metabasalts and metagabbros drilled and dredged from the ocean floor display many petrochemical features similar to those in the Chilean ophiolites (Cann, 1971). $^{18}O/^{16}O$ studies indicate extensive high-temperature interaction of these rocks with ocean water, as well as isotopic disequilibrium of plagioclase and pyroxenes from metagabbros (Muehlenbachs and Clayton, 1972), implying that oceanic metagabbros, like Chilean metagabbros, have not remained long within the zone of effective metamorphism. Bonatti et al., (1975) dredged metagabbros (amphibolites) from the axial valley wall of part of the Mid-Atlantic Ridge which apparently had never been buried more than 1.5 km. They concluded that such a steep metamorphic gradient could best be explained by hydrothermal and contact metamorphism within the axial valley of the spreading ridge. Extensive hydrothermal systems within active spreading axes are supported by several lines of evidence, in particular the need for convective heat transfer to explain the lower than predicted heat flow along oceanic ridges (Lister, 1972; Williams et al., 1974; Hyndman et al., 1976).

Figure 13 illustrates our model of metamorphism in the vicinity of a spreading ridge based on the above discussion and evidence from the Chilean ophiolites. The figure illustrates the restriction of the zone of deep hydrothermal circulation and high grade metamorphism to near the zone of igneous intrusion. It also shows a decrease in the volume of circulating fluid within the gabbro unit indicated by the decreased thickness of the schematic circulation paths. Due to the decreased volume of water circulating through

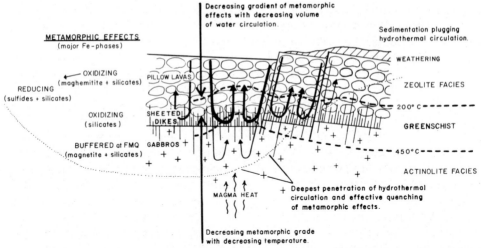

Fig. 13. Model of metamorphism in the vicinity of a ridge crest. Note decreasing volume of circulating water with depth, represented by thinner arrows. See text for detailed description of the construction of the diagram.

the deeper levels of the ocean crust, metamorphic reactions within the gabbro unit are buffered at the FMQ buffer of the igneous rocks. Within the dikes and probably the pillow lavas, as suggested by the development of titanomaghemites in lavas within the medium valley of the Mid-Atlantic Ridge (Irving et al., 1970), oxidation of the basalts occurs. Away from the spreading ridge the deeper metamorphic levels are no longer affected by hydrothermal circulation and lower grade metamorphism, while the pillow lavas and dikes continue to undergo low grade alteration. Secondary veins of sulfides within the pillow lavas of the Sarmiento complex suggest that these low grade effects are reducing, a condition which, like spilitization, may depend on sedimentation. The model is consistent with metamorphic effects found in the Troodos, South Oman, Apennine, and Bay of Islands ophiolite complexes (Cann, 1971; Christensen and Salisbury, 1975; Bonatti et al., 1975; Spooner and Fyfe, 1973; Coish, 1977).

Schematic isotherms (Fig. 13) are drawn so that the dike unit is within the field of actinolite grade metamorphism near the zone of igneous intrusion while some gabbros as well as the sheeted dikes pass through the field of greenschist metamorphism before they leave the region of hydrothermal circulation and metamorphism. The isotherm separating conditions of greenschist and actinolite facies metamorphism has been estimated as approximately 450°C, based on the limited experimental data (Liou et al., 1974) suggesting that the temperature stability conditions of the actinolite facies is broadly correspondent to the amphibolite facies of regional prograde metamorphism although the Al_2O_3 contents of amphibole in the upper actinolite metabasalts from the Sarmiento complex are not as high as reported from hornblende amphibolites occurring in regional metamorphic terrains ($Al_2O_3 > 10$ wt.%). Grapes et al., (1977) have demonstrated that strain recrystallization in a metagabbro—amphibolite sequence increases the alumina content of the amphiboles. Besides the lack of stress associated with hydrothermal metamorphism within a spreading ridge, the low alumina content of the amphiboles in the actinolite facies metabasalts and metagabbros from the Sarmiento complex may be related to the apparent stability of calcic plagioclase ($>An_{50}$) which acts as the major alumina reservoir in these rocks. Even when the extent of metamorphism has been intense enough to result in the complete replacement of mafic phases, calcic plagioclase in actinolite facies rocks retains the various twinning characteristics of the original igneous plagioclase and remains morphologically unaltered although oxygen isotope studies of separates of calcic plagioclase from lower and upper actinolite facies metagabbros indicate isotopic re-equilibration with non-magmatic water, suggesting that these plagioclases are in fact stable mineral phases under the metamorphic conditions at which the rock was altered (Stern et al., 1976). The stability of the mineral assemblage calcic plagioclase ($>An_{50}$)-low alumina amphibole ($Al_2O_3 < 8$ wt.%) within ocean floor rocks hydrothermally altered at actinolite facies temperature conditions in the vicinity of a spreading center may result from both the lack of associated stress, the highly vapor saturated nature of the metamorphism, and the composition of the igneous rocks being altered.

DISCUSSION

DeWit and Stern (1976) have discussed the implications of a similar model of hydrothermal metamorphism in the vicinity of a ridge crest for the seismic layering and remnant magnetism of the ocean floor. They suggested that the transition from seismic layer 2 to layer 3 (2B to 3A) corresponds to the greenschist to actinolite (amphibolite) transition, consistent with compressional velocity data (Christensen and Salisbury, 1975). Figure 3 indicates that in the Sarmiento complex greenschist facies mineral assemblages occur together with actinolite facies minerals over a large vertical range. In the Sarmiento complex, the transition from greenschist to actinolite (amphibolite) facies is found in the sub-hand specimen scale due to extensive retrograde metamorphic effects. Nevertheless, the sheeted dike-gabbro transition is an abrupt discontinuity in the primary igneous pseudostratigraphy. Thus, in Sarmiento, the most likely cause of a seismic transition from layer 2B to 3A would be the primary igneous structure of the crust.

In the Tortuga complex plagiogranites are absent and the sheeted dike complex grades downward in a continuous transition to gabbros (Fig. 2). Petrochemical studies are not yet sufficiently detailed to determine if the Tortuga complex is characterized by similar metamorphic effects as the Sarmiento complex, although a decrease in extent of metamorphic replacement within the gabbroic unit is apparent. Within the oceanic crust plagiogranites are a common minor component of dredge samples. Both types of igneous pseudostratigraphy probably occur due to variations in the rate of magma supply and spreading rate between and at different times within spreading ridges. This may account for the suitability of the uniform layer model of the oceanic crust in some regions, and a continously varying velocity structure in others (Kennett and Orcutt, 1976).

Both greenschist and actinolite facies metamorphism have resulted in different chemical modifications of the original basalts at different levels of the Sarmiento complex. Greenschist facies metabasalts within the pillow lavas have gained K_2O, Na_2O, SiO_2 and lost CaO, common effects of spilitization. These effects could be related to the participation of marine sediments (Montigny et al., 1973). At deeper pseudostratigraphic levels the chemical effects of metamorphism have been more nearly isochemical, the only significant chemical modification being loss of K_2O and Rb from within the sheeted dike unit and their increase within the underlying gabbro level. This is directly related to the lack of biotite at the level of the sheeted dikes and its common occurrence at the level of the gabbros. The changing stability of biotite has resulted from variations in oxygen fugacity, which in turn has resulted from differences in the ratio of the volume of water and rock at different pseudostratigraphic levels of the ophiolite complex. Metabasalts dredged from the ocean floor and exposed in other ophiolite complexes have also been found to exhibit different chemical modifications even when whitin the same metamorphic facies (Miyashiro et al., 1971). Wolery and Sleep (1976) suggest that some of these differences may be due to the participation of seawater on the one hand, juvenile water on the other. Our data indicate that differences in oxygen fugacity related to the ratio of the volume of water and igneous rock at different levels of the ocean crust may affect the chemical modifications caused by hydrothermal metamorphism.

ACKNOWLEDGEMENTS

We are grateful to V. Alvarez, E. Bonatti, M. Dobbs, D. Fornari, P.J. Fox, E. Godoy, F. Herve, L. Oviedo, W.I. Ridley, R. Schweickert, M.A. Skewes and W.S. Snyder for advice and assistance, and to R.N. Anderson for a critical review of the manuscript. We would like to thank L. Vodanovic, R. Pochapsky and J. Latham for typing the manuscript. The work was supported by National Science Foundation grants DES 75-04076, EAR 76-82456 and OPP 7421415.

REFERENCES

Bonatti, E., Honnorez, J., Kirst, P. and Radicati, F., 1975. Metagabbros from the Mid-Atlantic Ridge at 06°N: contact-hydrothermal-dynamic metamorphism beneath the axial valley. J. Geol., 83: 61—78.
Bruhn, R.L., and Dalziel, J.W.D., 1977. Destruction of the Early Cretaceous Marginal

Basin in the Andes of Tierra del Fuego. In: M. Talwani and W. Pitman (Editors), Island Arcs, Deep Sea Trenches and Back-arc Basins. pp. 395—406.

Bruhn, R.L., Stern, C.R. and DeWit, M.J., 1978. The bearing of new field and geochemical data on the origin and development of a Mesozoic volcano-tectonic rift zone and back-arc basin in southernmost South America. Earth Planet. Sci. Lett., 41: 32—46.

Cann, J.R., 1968. Geological processes at mid-ocean ridge crests. Geophys. J. R. Astron. Soc., 15: 331—341.

Cann, J.R., 1971. Petrology of basement rocks from Palmer Ridge, N.E. Atlantic. Philos. Trans. R. Soc. London, Ser. A., 268: 605—617.

Christensen, N.I. and Salisbury, M.H., 1975. Structure and constitution of the lower oceanic crust. Rev. Geophys. Space Phys., 13: 57—86.

Church, W.R. and Riccio, L., 1974. The sheeted dike layer of the Betts Cove ophiolite complex does not represent spreading: discussion. Can. J. Earth Sci., 11: 1491—1502.

Clague, D.A. and Bunch, T.E., 1976. Formation of ferrobasalt at east Pacific midocean spreading centers. J. Geophys. Res., 81: 4247—4256.

Coish, R.A., 1977. Ocean floor metamorphism in the Betts Cove ophiolite, Newfoundland. Contrib. Mineral. Petrol., 60: 255—270.

Coleman, R.G. and Peterman, Z.E., 1975. Oceanic plagiogranite. J. Geophys. Res., 80: 1099—1108.

Corliss, J.G., 1971. The origin of metal bearing submarine hydrothermal solution. J. Geophys. Res., 76: 8128—8138.

Dalziel, I.W.D., DeWit, M.J. and Palmer, K.F., 1974. Fossil marginal basin in the southern Andes. Nature, 250: 291—294.

DeWit, M.J., 1977. The evolution of the Scotia Arc as a key to the reconstruction of southwestern Gondwanaland. Tectonophysics, 37: 53—81.

DeWit, M.J. and Stern, C.R., 1976. Ocean floor metamorphism, seismic layering and magnetism. Nature, 264: 615—619.

DeWit, M.J. and Stern, C.R., 1978. Pillow talk. J. Volcanol. Geotherm. Res., 4: 55—80.

Erlank, A.J. and Kable, E.J.D., 1976. The significance of incompatible elements in Mid-Atlantic Ridge basalts from 45°N with particular reference to Zr/Nb. Contrib. Mineral. Petrol., 54: 281—291.

Frey, F.E., Bryan, W.B. and Thompson, G., 1974. Atlantic ocean floor: geochemistry and petrology of basalts from legs 2 and 3 of the Deep-Sea Drilling Project. J. Geophys. Res., 79: 5507—5527.

Gass, I.G., 1968. Is the Troodos Massif of Cyprus a fragment of Mesozoic ocean floor? Nature, 220: 39—42.

Gass, I.G. and Smewing, J.D., 1973. Intrusion, extrusion and metamorphism at constructive margins: evidence from the Troodos Massif, Cyprus. Nature, 242: 26—29.

Grapes, R.H., Hashimoto, S. and Miyashita, M., 1977. Amphiboles of a metagabbro—amphibolite sequence, Hikaka metamorphic belt, Hokkaido. J. Petrol., 18: 285—318.

Greenbaum, D., 1972. Magmatic processes at oceanic ridges: evidence from the Troodos Massif, Cyprus. Nature, 238: 18—21.

Haggerty, S.E., 1977. Oxidation of opaque mineral oxides in basalts. In: D. Rumble (Editor), Mineralogical Society of America Short Course Notes, Vol. 3. Oxide Minerals, HG 1—100.

Hart, R., 1973. A model for the chemical exchange in the basalt—seawater system of oceanic layer 2. Can. J. Earth Sci., 10: 799—816.

Hart, S.R., Glassley, W.E. and Karig, D.E., 1972. Basalts and sea floor spreading behind the Mariana Island Arc. Earth Planet. Sci. Lett., 15: 12—18.

Hawkins, J.W., 1978. Petrology and geochemistry of basaltic rocks of the Lau Basin. Earth Planet. Sci. Lett., 28: 283—298.

Humphris, S.E., 1978. The Hydrothermal Alteration of Oceanic Basalts by Seawater. Thesis, Mass. Inst. Technol., Mass.

Hyndman, R.D., Von Herzen, R.P., Erickson, A.J. and Jolivet, J., 1976. Heat flow measurements in deep crustal holes on the Mid-Atlantic Ridge. J. Geophys. Res., 81: 4053—4060.

Irving, E., 1970. The Mid-Atlantic Ridge at 45°N. Oxidation and magnetic properties of basalts, review and discussion. Can. J. Earth Sci., 7: 1528—1538.

Kay, R. and Senechal, R.G., 1976. The rare earth geochemistry of the Troodos ophiolite complex. J. Geophys. Res., 81: 964—970.

Kay, R., Hubbard, H.J. and Gast, P.W., 1970. Chemical characteristics and origin of ocean ridge volcanic rocks. J. Geophys. Res., 75: 1585—1613.

Kennett, B.L.N. and Orcutt, J.A., 1976. A comparison of travel time inversions for marine refraction profiles. J. Geophys. Res., 81: 4061—4070.

Liou, J.G., Kuniyoshi, S. and Ito, K., 1974. Experimental studies of phase relations between greenschist and amphibolite in a basaltic system. Am. J. Sci., 274: 613—632.

Lister, C.R.B., 1972. On the thermal balance of a mid-ocean ridge. Geophys. J. R. Astron. Soc., 26: 515—535.

Lister, C.R.B., 1974. On the penetration of water into hot rock. Geophys. J. R. Astron. Soc., 39: 465—509.

Mazzulo, L.J. and Bence, A.E., 1976. Abyssal tholeiites from DSDP Leg 34: the Nazca plate. J. Geophys. Res., 81: 4327—4352.

Miyashiro, A., 1961. Evolution of metamorphic belts. J. Petrol., 2: 277—311.

Miyashiro, A., 1973. The Troodos ophiolite complex was probably formed in an island arc. Earth Planet. Sci. Lett., 19: 218—224.

Miyashiro, A., 1975. Classification, characteristics and origin of ophiolites. J. Geol., 83: 249—281.

Miyashiro, A., Shido, F. and Ewing, M., 1971. Metamorphism in the Mid-Atlantic Ridge near 24°N and 30°N. Philos. Trans. R. Soc. London, Ser. A, 268: 589—603.

Montigny, R., Bougault, H., Bottinga, Y. and Allegre, C.J., 1973. Trace element geochemistry and genesis of the Pindos ophiolite suite. Geochim. Cosmochim. Acta, 37: 2135—2147.

Moore, J.G., 1972. Water content of basalts erupted on the ocean floor. Contrib. Mineral. Petrol., 28: 272—279.

Moores, E.M. and Vine, F.J., 1971. The Troodos Massif, Cyprus, and other ophiolites as oceanic crust: evaluation and implications. Philos. Trans. R. Soc. London, Ser. A, 268: 443—466.

Muehlenbachs, K. and Clayton, R.N., 1972. Oxygen isotope geochemistry of the oceanic crust and its bearing on seawater. J. Geophys. Res., 81: 4365—4369.

Norrish, K. and Chappell, B.W., 1967. X-ray fluorescence spectrography. In: J. Zussman (Editor), Physical Methods in Determinative Mineralogy, Academic Press, New York, N.Y. pp. 161—214.

Pearce, J.A. and Cann, J., 1973. Tectonic setting of basic volcanic rocks using trace element analyses. Earth Planet. Sci. Lett., 19: 290—300.

Saunders, A.D., Tarney, J., Stern, C.R. and Dalziel, I.W.D., 1979. Geochemistry of Mesozoic marginal basin floor igneous rocks from southern Chile. Geol. Soc. Am. Bull. (in press).

Spooner, E.T.C. and Fyfe, W.S., 1973. Sub sea-floor metamorphism, heat and mass transfer. Contrib. Mineral. Petrol., 42: 287—304.

Spooner, E.T.C., Beckinsale, R.D., Fyfe, W.S. and Smewing, J.D., 1974. O^{18} enriched ophiolite metabasic rocks from E. Liguria (Italy), Pintos (Greece) and Troodos (Cyprus). Contrib. Mineral. Petrol., 47: 41—62.

Stern, C.R., 1976. Geochemistry of ophiolite complexes formed in an ensialic marginal basin. Geol. Soc. Am., Abstr. Progr., 1976. Annu. Meet., 8: 1121.

Stern, C.R., in press, a. Chemical differences between two Chilean ophiolites: Their relation to igneous fractionation within open periodically refilled and closed magma chambers and their tectonic implications. Contrib. Mineral. Petrol.

Stern, C.R., in press, b. Geochemistry of Chilean ophiolites: Evidence for the compositional evolution of the mantle source of back-arc basin basalts. J. Geophys. Res.

Stern, C.R., de Wit, M.J. and Lawrence, J.R., 1976. Igneous and metamorphic processes associated with the formation of Chilean ophiolites and their implication for ocean floor metamorphism, seismic layering and magnetism. J. Geophys. Res., 81: 4370—4380.

Upadhyay, H.D. and Strong, D.F., 1973. Geologic setting of the Betts Cove Copper Deposits, Newfoundland: an example of ophiolite sulfide mineralization. Econ. Geol., 68: 161—167.

Vallence, T.G., 1965. Spilitic degradation of a tholeiitic basalt. J. Petrol., 15: 79—96.

Watters, W.A., 1965. Prehnitization in the Yahgan formation of Navarino Island, southernmost Chile. Mineral. Mag., Tilley Volume: 517—527.

Williams, D.L., Von Herzen, R.P., Sclater, J.G. and Anderson, R.N., 1974. The Galapagos spreading center: lithospheric cooling and hydrothermal circulation. Geophys. J. R. Astron. Soc., 38: 587—608.

Wolery, T.J. and Sleep, N.H., 1976. Hydrothermal circulation and geochemical flux at mid-ocean ridges. J. Geol., 84: 249—275.

Chapter 10

ZONE REFINING AT THE BASE OF LITHOSPHERIC PLATES: A MODEL FOR A STEADY-STATE ASTHENOSPHERE*

ROBERT W. KAY

Department of Geological Sciences, Cornell University, Ithaca, N.Y. 14853 (U.S.A.)

SUMMARY

Isotopic heterogeneity of oceanic basalt types provides evidence of long-term chemical heterogeneity in the earth's upper mantle. Processes operating at oceanic ridges and beneath cooling lithospheric plates are capable of generating heterogeneities that, when allowed to age for billions of years in the mantle, resemble heterogeneities now present in the mantle. The processes involve migration of melt relative to solid, both at the ridge and, in a model developed in this paper, under thickening oceanic plates. A steady-state asthenosphere, enriched in chemically "incompatible" elements, is possible under the following conditons: "incompatible" element — enriched peridotite from the base of thickened lithospheric plates is added to the asthenosphere at a rate equal to the rate of asthenospheric mantle processed during basalt magma generation.

INTRODUCTION

Large scale mantle heterogeneity extending over a billion-year time scale is indicated by Pb, Sr and Nd isotope differences among mantle-derived oceanic basalt types. The most abundant oceanic basalt type, the normal ocean ridge basalts (NORB), are generally lower in the radiogenic isotopes of Sr, Pb and higher in radiogenic Nd than "Icelandic type" (IT) ridge basalt and intraplate (IP) basalt from oceanic seamounts and islands (Church and Tatsumoto, 1975; De Paolo and Wasserburg, 1976; O'Nions et al., 1977; Richard et al., 1976; Sun and Hanson, 1975a, b, 1976). Details of this isotopic variation may be complex even within one locality (e.g., range in Sr isotope ratios in Iceland, O'Nions et al., 1976; range in Pb isotopes in Hawaii, Tatsumoto, 1978). Source mixing along mid-oceanic ridges and within oceanic islands (Sun et al., 1975; Vollmer, 1976; Tatsumoto, 1978), and fractionation of elements during magma genesis at the ocean ridges (Treuil and Juron, 1975; Langmuir et al., 1977) are two processes that pro-

---

*Originally published as: Kay, R.W., 1979. Zone refining at the base of lithospheric plates: a model for a steady-state asthenosphere. In: J. Francheteau (Editor), Processes at Mid-Ocean Ridges. Tectonophysics, 55: 1—9.

bably occur, and result in magmas with hybrid sources. The first order difference between NORB and IT—IP mantle sources is clear despite complication introduced by these processes. In addition to longevity, mantle heterogeneities have other important features:

(1) The NORB-type mantle does not appear to be the source for any of the intraplate magmas. Thus, in addition to being the mantle most commonly melted to yield basalt, it also upwells in only one tectonic setting: the oceanic ridge.

(2) The bulk composition of the two source types differs. In particular, large ion lithophile (LIL) elements are lower in the NORB sources. The basalt types derived from the two source types differ even more than the sources themselves, due to element fractionation during partial melting and crystal fractionation (e.g., Gast, 1968; Kay and Gast, 1973; O'Nions and Pankhurst, 1974; Flower et al., 1976; Treuil and Joron, 1975; O'Nions et al., 1976).

The arrangement of the two end-member oceanic mantle regions is a fundamental problem in geodynamics. Among several proposed mantle geometries that recognize and isolate the two types of mantle are the deep mantle plume hypothesis (Anderson, 1975; Deffeys, 1972; Morgan, 1972; Schilling, 1973; Sun and Hanson, 1975a) and various asthenospheric melting spot hypotheses (Green and Lieberman, 1976; Ringwood, 1975). Tatsumoto (1978) has reviewed the bearing of isotopic and geochemical data on the nature of large scale mantle heterogeneity and has proposed a model that is similar to the one used in this paper. A novel aspect of Tatsumoto's (1978) mantle geometry is that the depleted material comes up from the mesosphere (not the asthenosphere) to oceanic ridges. I will focus on the evaluation of mass flux through the asthenospheric mantle reservoir and illustrate that (1) the asthenosphere is an inadequate long-term reservoir for the derivation of NORB, mainly due to the small asthenospheric volume relative to the rate of NORB volcanism and lack of adequate mechanisms for regenerating NORB-type asthenosphere, and (2) the asthenosphere is an adequate long-term source reservoir for IT ridge basalts and IP basalt, which have a lower production rate than NORB. One mechanism for replenishing the asthenosphere involves addition of segregated, but unerupted LIL-element enriched magma pockets that develop during melting of mantle under the MOR (Bottinga and Allègre, 1976; Langmuir et al., 1977; Turcotte and Ahern, 1978). A second mechanism, the one explored here, involves addition of a downward zone refined layer at the base of thickened oceanic lithospheric plates (Bottinga and Allègre, 1973; Bottinga, 1974; Forsyth, 1975, 1977; Kay, 1975; Leeds, 1975; Parker and Oldenburg, 1973; Schubert et al., 1976; Yoshii, 1975; Yoshii et al., 1977). The zone refined layer is enriched in LIL elements. Replacement time of the steady state asthenosphere is about 2 b.y., in agreement with isotopic pseudoisochrons (Brooks et al., 1977; Sun and Hanson, 1975a; Church and Tatsumoto, 1975).

THE ASTHENOSPHERE AS A RESERVOIR PRODUCING BASALTIC VOLCANISM

The earth's asthenosphere is the mechanically weak region between the rigid lithospheric plates above and the mesosphere below. The chemical characteristics of asthenospheric mantle are a subject of current debate: the two extreme choices are that asthenospheric mantle yields NORB magmas or intraplate magmas. One boundary condition bearing on this question is the volumes and rates involved (Armstrong, 1968; Armstrong and Hein, 1973; Deffeys, 1972; Dickinson and Luth, 1971). The total volume of the astheno-sphere is not well-defined, but is perhaps no larger than a quarter of the volume of the upper mantle, or $5 \cdot 10^{10}$ km^3. NORB at mid-oceanic ridges erupt at a rate of about 12.5 km^3/yr (2.5 km^2/yr times 5 km thickness: Deffeys, 1972); however, a larger volume of mantle peridotite must be processed at the ridges, because it is likely that the entire volume of thickened lithosphere to a depth of 80—100 km upwells at the ridge axis, even if it does not yield basalt (Forsyth, 1977). In this case the flux is 200 km^3/yr (2.5 km^2/yr times 80 km) and the turnover time of a steady-state asthenosphere source would be 250 m.y. The asthenospheric reservoir would require replacement with NORB-type mantle at the same rate from below because, once processed, the plate is not likely to yield basalt at mid-oceanic ridges again. The eruption rate of IP basalts and IT ridge basalts is perhaps 1/8 the rate of ridge volcanism. This figure follows from the observation (Sun et al., 1975; White et al., 1975) that the oceanic ridge system is esti-mated to be 75% normal ridge and 25% IT ridge plus transitional zones (assumed to be 1/2 the NORB component and 1/2 IT component). IP mag-mas are slightly less abundant than IT basalts. The turnover time of a steady-state asthenosphere yielding IT and IP basalts would be about 2 b.y.

THICKENING OF OCEANIC LITHOSPHERE AND DEVELOPMENT OF A CHEMICALLY ENRICHED LAYER

Oceanic lithosphere is created at oceanic ridges by intrusion and extrusion of basalt derived from upper mantle at the ridge axis. The base of the lithosphere is commonly assumed to coincide with the peridotite solidus (see Fig. 1), and is also presumed to be a shear boundary. As the newly formed thin lithospheric plate moves away from the ridge, it thickens by underplating as underlying asthenosphere cools below its solidus. The crys-tallization of partly molten asthenosphere can follow several paths (Yoshii et al., 1977). If any vertical mass transfer during crystallization is controlled by molecular diffusion, the crystallization will follow an almost closed system behavior. This follows from the observation that thickening of about 50 km occurs over a time period of about 10 m.y. and diffusion rates of 10^{-6} cm^2/sec (Sr in basalt at 1400°C: Hofmann and Margaritz, 1978) are too slow for significant material transfer. Thickening rates become progressively slower, but are still faster than molecular diffusion rates, even for the oldest

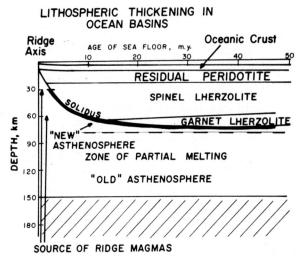

Fig. 1. Schematic depiction of lithospheric thickening in the ocean basins, after Forsyth (1975). Oceanic asthenosphere is interpreted to be a zone of partly molten peridotite under a solid oceanic lithospheric peridotite (garnet-, spinel-, or plagioclase-bearing at successively shallower depth). The uppermost mantle layer is residual peridotite complementary to the basaltic oceanic crust. The NORB-type mantle that wells up under oceanic ridge axes is shown to originate from the mesosphere (diagonally-lined region beneath the asthenosphere). Asthenospheric volume is continually decreased by removal of lithospheric plate-puncturing intraplate magmas, with sources centered at the top of convection cells in the asthenosphere (IP-type), as well as entrainment of asthenosphere in upwelling NORB-type mantle (IT-type). Asthenospheric volume is continuously increased at about the same rate as it is decreased by addition of a downward zone-refined layer at the base of the thickened lithospheric plate.

oceanic lithosphere. However, if eddy diffusion operates at the shear interface the effective diffusion coefficient may be higher than the simple ion mobility, and elements concentrated in interstitial melt may diffuse downward from the solidus interface. The net result of the latter process would be the inefficient purging of crystallizing lithosphere in elements that do not crystallize in the phases stable at the solidus. In particular, water may be excluded from the lithosphere near the oceanic ridges, since the thermal stability of hydrous phases may be exceeded (see review of Hofmann and Hart, 1978). A layer just under the base of the growing lithosphere may accumulate melt rich in water and other incompatible elements — the same elements that are in high concentrations in IT and IP basalts.

Forsyth (1977) shows that extensive melting to a depth of 60 km below the active ridge (corresponding to the bottom of Turcotte and Ahern's, 1978, melting zone) will result in a rapid increase in plate thickness to 60 km during the first 10 m.y. Further thickening of the plate follows the wet solidus, as shown on Fig. 1. I have assumed that the downward zone refining process operates between 30 and 80 km in the thickening lithosphere, with mantle from 30 km to 5 km being the depleted peridotite complimentary to the NORB.

Quantitative evaluation of the evolving composition of a layer at the base of the oceanic lithosphere is possible using a downward zone-refining model. Consider a partly molten sheared zone of thickness compatible with the effective eddy diffusion distance, with its upper surface the base of the lithosphere. Near the oceanic ridge, the composition of melt within the zone is the same as the composition of primitive ocean ridge basalt (see Fig. 2). As the lithosphere thickens, a thickness of peridotite crystallizes with mineralogy appropriate to the ambient pressure and temperature, and a thickness of

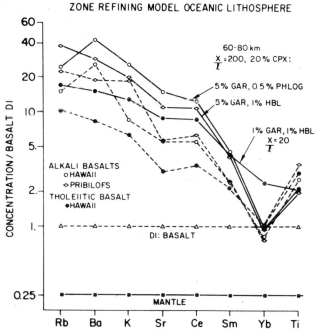

Fig. 2. The concentration of eight elements in four natural basalts (dashed lines) and three calculated basalts (solid lines), expressed as multiples of the concentration in a primative low Fe/Mg normal ocean ridge basalt (sample *D1*, East Pacific Rise, Kay and Hubbard, 1978). The mantle compositions were calculated by starting with a mantle column of composition that yields basalt *D1* on 25% melting. Zone refining at the base of the thickening lithosphere was calculated in three steps, since mineralogy, and therefore the bulk distribution coefficient, changes with depth in the mantle. Only the third step (garnet peridotite stage) is shown on the figure. The first two steps are: zone refining from 30 km to 40 km with x/l = 10 and mantle mineralogy with 20% clinopyroxene and 5% plagioclase; zone refining from 40 to 60 km with x/l = 50 and mantle mineralogy with 20% clinopyroxene and 1% hornblende. Distribution coefficients were estimated from various literature sources and are listed in Table I. The results show that high concentrations of the largest cations can be expected in zone-refined layers, and that fractionation of the elements shown resembles fractionation in natural basalts.

TABLE I

Distribution coefficients* (K)

	Ol	Opx	Cpx	Am	Ga	Phlog	Plag
Rb	0.0002	0.001	0.002	<u>0.41</u>	0.001	<u>3.1</u>	0.03
Ba	0.0001	0.001	0.001	<u>0.73</u>	0.001	<u>1.1</u>	<u>0.15</u>
K	0.0002	0.001	0.002	<u>1.15</u>	0.001	<u>2.7</u>	<u>0.16</u>
Sr	0.0002	0.001	<u>0.078</u>	<u>0.64</u>	0.01	<u>0.08</u>	<u>1.4</u>
Ce	0.001	0.0015	<u>0.098</u>	<u>0.26</u>	0.003	0.034	0.04
Sm	0.001	0.0018	<u>0.26</u>	<u>0.40</u>	<u>0.11</u>	0.031	0.02
Yb	0.001	0.024	<u>0.28</u>	<u>0.29</u>	<u>4.0</u>	0.042	0.015
Ti	0.001	0.05	<u>0.5</u>	<u>2.1</u>	0.1*	0.1*	0.01*

Ol = olivine; Opx = orthopyroxene (ratio of Ol = Opx assumed to be 3.5 in all calculations); Cpx = clinopyroxene; Am = Amphibole; Ga = garnet; Phlog = phlogopite; Plag = plagioclase. * : values estimated, no data. All coefficients are between mineral phase and basalt melt at liquidus temperatures. For each element, underlined coefficients are significantly higher than the others.

partly molten mantle is admitted to diffusive interchange at the base of the layer.

Focusing on the melt phase, zone refining theory (Harris, 1957; Pfann, 1952) yields the relation:

$$\frac{C}{C_0} = \frac{1}{K} + \frac{1}{K}(KC_i/C_0 - 1) \exp(K \cdot \frac{x}{l})$$

where C = concentration of element in molten zone at end of its movement over length l; C_0 = concentration of element in mantle; K = distribution coefficient: concentration of element in solid divided by its concentration in coexiting liquid; C_i = concentration of element in molten zone at start of pass; x = thickness of molten zone; l = length traversed by molten zone.

Figure 2 shows the results of several calculations of zone refining on the thickening oceanic lithosphere (see Table I for the distribution coefficients used). Good matches can be made between the trace element patterns of some oceanic intraplate basalts and the patterns in some calculated mantle zones enriched in the largest cations (Rb, Ba) relative to the smallest ones (Yb, Ti). Thus, the zone-refining calculation shows that it is possible to convert NORB-type mantle (mantle that melts to form NORB) into intraplate basalt-type mantle by a continuous process operating now in the oceanic crust. Note that many elements of high K value (Ni, Cr) and major elements (Fe, Ca, Al, Mn) remain almost constant in the mantle undergoing the downward zone refining.

FATE OF THE ENRICHED LAYER

The enriched layer always remains in the mechanically weak asthenosphere; it is not subducted at ocean trenches. It can serve as the source of intraplate basaltic magmas, but only after a long storage period, since isotopic differences between ocean ridge and intraplate magmas must be allowed to develop. Evidence from the basalts themselves indicates that a mean storage time of about 2 b.y. may be required (Sun and Hanson, 1975; Brooks et al., 1977; Tatsumoto, 1978). The layer is possibly scraped off the bottom of the lithospheric plate as the plate is subducted, with the scraped-off enriched zone being added to the asthenosphere. For an asthenospheric volume of $5 \cdot 10^{10}$ km^3, addition of a 10 km thick peridotite layer at the bottom of all plates being subducted at their present-day rates (about 2.5 km^2/yr) would yield a replacement time of 2 b.y., in agreement with the isotopic data. The enriched zones shown in Fig. 2 are much narrower than 10 km (for the phlogopite peridotite the melt is 100 m thick) implying that a heterogeneous mixture of enriched mantle and NORB-type mantle may be recirculated into the asthenosphere. The concentration levels in the enriched zones are sufficient that approximately equal amounts of elements like Rb, Ba, K and Ce will be contributed by the two mixing components, whereas other elements (Yb, Ti) will come almost entirely from the NORB component. The resulting asthenosphere closely resembles the inferred source regions of intraplate magmas.

ACKNOWLEDGEMENTS

I thank D. Turcotte and S. Hart for discussions. Research support through NSF grants EAR77-13682 and OCE76-23826 is acknowledged.

REFERENCES

Anderson, D., 1975. Chemical plumes in the mantle. Geol. Soc. Am. Bull., 86: 1593—1600.
Armstrong, R.L., 1968. A model for the evolution of strontium and lead isotopes in a dynamic earth. Rev. Geophys., 6: 175—199.
Armstrong, R.L. and Hein, S.M., 1973. Computer simulation of Pb and Sr isotope evolution of the earth's crust and upper mantle. Geochim. Cosmochim. Acta, 37: 1—18.
Bottinga, Y., 1974. Thermal aspects of sea floor spreading and the nature of the suboceanic lithosphere. Tectonophysics, 21: 15—38.
Bottinga, Y. and Allègre, C.J., 1973. Thermal aspects of sea floor spreading and the nature of the oceanic crust. Tectonophysics, 18: 1—17.
Bottinga, Y. and Allègre, C.J., 1976. Geophysical, petrological and geochemical models of the oceanic lithosphere. Tectonophysics, 32: 9—59.
Brooks, C., Hart, S.R., Hofmann, A. and James, D.E., 1977. Rb—Sr mantle isochrons from oceanic regions. Earth Planet. Sci. Lett., 32: 51—61.
Church, S.E. and Tatsumoto, M., 1975. Lead isotope relations in oceanic ridge basalts from the Juan de Fuca—Gorda Ridge area, N.E. Pacific Ocean. Contrib. Miner. Petrol., 53: 253—279.

Deffeys, K.S., 1972. Plume convection with an upper-mantle temperature inversion. Nature, 240: 539—544.

De Paolo, D.J. and Wasserburg, G.J., 1976. Inferences about magma sources and mantle structure from variations of $^{143}Nd/^{144}Nd$. Geophys. Res. Lett., 3: 743—746.

Dickinson, W.R. and Luth, W.C., 1971. A model for plate tectonic evolution of mantle layers. Science, 174: 400—404.

Flower, M.F.J., Schmincke, H.U. and Thompson, R.N., 1975. Phlogopite stability and the $^{87}Sr/^{86}Sr$ step in basalts along the Reykjanes ridge. Nature 254: 404—406.

Forsyth, D., 1975. The early structural evolution and anisotropy of the oceanic upper mantle. Geophys. J. R. Astron. Soc., 43: 103—162.

Forsyth, D., 1977. The evolution of the upper mantle beneath mid-ocean ridges. Tectonophysics, 38: 89—118.

Gast, P.W., 1968. Trace element fractionation and the origin of tholeiitic and alkaline magma types. Geochim. Cosmochim. Acta, 32: 1057—1086.

Green, D.H. and Lieberman, R.C., 1976. Phase equilibria and elastic properties of a pyrolite model for the oceanic upper meantle. Tectonophysics, 32: 61—92.

Harris, P.G. 1957. Zone refining and the origin of potassic basalts. Geochim. Cosmochim. Acta, 12: 195—208.

Hofmann, A.W. and Hart, S.R., 1978. An assessment of local and regional isotopic equilibrium in the mantle. Earth Planet. Sci. Lett., 38: 44—62.

Hofmann, A. and Magaritz, M., 1977. Diffusion of Ca, Sr, Ba and Co in a basalt melt: implications for the geochemistry of the mantle. J. Geophys. Res., 82: 5432—5440.

Kay, R.W., 1975. Chemical zonation of the oceanic mantle. Eos, Trans. Am. Geophys. Union, 56: 1077.

Kay, R.W. and Gast, P.W., 1973. The rare earth content and origin of alkali-rich basalts. J. Geol., 81: 653—682.

Kay, R.W. and Hubbard, N.J., 1978. Trace elements in ocean ridge basalts. Earth Planet. Sci. Lett., 38: 95—116.

Langmuir, C., Bender, J., Bence, A. and Hanson, G., 1977. Petrogenesis of basalts from the FAMOUS area: Mid-Atlantic Ridge. Earth Planet. Sci. Lett., 36: 135—156.

Leeds, A.R., 1975. Lithospheric thickness in the western Pacific. Phys. Earth Planet. Inter. 17: 61—64.

Morgan, W.J., 1972. Deep mantle convection plumes and plate motions. Am. Assoc. Petrol. Geol. Bull., 56: 203—213.

O'Nions, R.K. and Pankhurst, R.J., 1974. Petrogenetic significance of isotope and trace element variations in volcanic rocks from the Mid-Atlantic. J. Petrol., 15: 605—654.

O'Nions, R.K., Pankhurst, R.J. and Gronvold, K., 1976. Nature and development of basalt magma sources beneath Iceland and the Reykjanes Ridge. J. Petrol., 17: 291.

O'Nions, R.K., Hamilton, P.J. and Evensen, N.M., 1977. Variations in $^{143}Nd/^{144}Nd$ and $^{87}Sr/^{86}Sr$ in oceanic basalts. Earth Planet. Sci. Lett., 34: 13—22.

Parker, R.L. and Oldenburg, D.W., 1973. Thermal model of oceanic ridges. Nature, 242: 137—139.

Pfann, W.G., 1952. Principles of zone-melting. J. Met., 4: 747—753.

Richard, P., Shimizu, N. and Allègre, C.J., 1976. $^{143}Nd/^{144}Nd$, a natural tracer; an application to oceanic basalts. Earth Planet. Sci. Lett., 31: 209—278.

Ringwood, A.E., 1975. Composition and Petrology of the Earth's Mantle. McGraw-Hill, New York, N.Y., 618 pp.

Schilling, J.-G., 1973. Iceland mantle plume: geochemical study of the Reykjanes Ridge. Nature, 242: 55.

Schubert, G., Froidevaux, C. and Yuen, D.A., 1976. Oceanic lithosphere and asthenosphere: thermal and mechanical structure. J. Geophys. Res., 81: 3525—3540.

Sun, S.-S. and Hanson, G.N., 1975a. Evolution of the mantle: geochemical evidence from alkali basalt. Geology, 3: 297—302.

Sun, S.-S. and Hanson, G.N., 1975b. Origin of Ross Island basanitoids and limitations upon the heterogeneity of sources for alkali basalts. Contrib. Mineral. Petrol., 52: 77—106.

Sun, S.-S. and Hanson, G.N., 1976. Evolution of the mantle: geochemical evidence from alkali basalt (reply to J.B. Gill's comment). Geology, 4: 626—631.

Sun, S.-S., Tatsumoto, M. and Schilling, J.G., 1975. Mantle mixing along the Reykjanes Ridge axis: lead isotope evidence. Science, 190: 143—146.

Tatsumoto, M., 1978. Isotopic composition of lead in oceanic basalt and its implication to mantle evolution. Earth Planet. Sci. Lett., 38: 44—62.

Treuil, M. and Joron, J.M., 1975. Utilisation des éléments hygromagmatophiles pour la simplification de la modelisation quantitative des processus magmatiques. Exemples de l'Afar et de la dorsale médio-atlantique. Soc. Ital. Mineral. Petrol., 31: 125—174.

Turcotte, C. and Ahern, J., 1978. A porous flow model for magma migration in the asthenosphere. J. Geophys. Res., 83: 767—772.

Vollmer, R., 1976. Rb—Sr and U—Th—Pb systematics of alkaline rocks: the alkaline rocks from Italy. Geochim. Cosmochim. Acta, 40: 283—296.

White, W., Hart, S.R. and Schilling, J.G., 1975. Geochemistry of the Azores and the Mid-Atlantic Ridge, 29° to 60°N. Carnegie Inst. Washington Yearb., 74: 224—234.

Yoshii, T., 1975. Regionality of group velocities of Rayleigh waves in the Pacific and thickening of the plate. Earth Planet. Sci. Lett., 25: 305—312.

Yoshii, T., Kono, Y and Ito, K., 1977. Thickening of the oceanic lithosphere. Trans. Am. Geophys. Union, Monogr. 19: 423—430.

Chapter 11

EVIDENCE FOR VARIABILITY OF MAGMATIC PROCESSES AND
UPPER MANTLE HETEROGENEITY IN THE AXIAL REGION OF THE
MID-ATLANTIC RIDGE NEAR 22° AND 36°N*

H. BOUGAULT [1], P. CAMBON [1], O. CORRE [1], J.L. JORON [2] and M. TREUIL [2]

[1] *Centre Océanologique de Bretagne, 29273 Brest (France)*

[2] *Groupe des Sciences de la Terre, Laboratoire Pierre Süe, C.E.N. Saclay, 91190 Gif-sur-Yvette (France) and Institut de Physique du Globe de Paris (France)*

SUMMARY

Basalt samples at 36°N (FAMOUS area) and at 22°N (leg 45 and leg 46 of D.S.D.P.) are compared with particular reference to their trace element concentrations. The results are interpreted in terms of variability of magmatic processes and compositional heterogeneity of the upper mantle material. It is shown that the concentrations of elements such as nickel, chromium and cobalt, which have high partition coefficients, do not vary significantly either in the residue or in the liquid as a result of varying degrees of partial melting. The large variations of Ni and Cr, together with the narrow range of variation of Co in oceanic basalts provide valuable constraints for models of fractional crystallization. The concentrations of hygromagmaphile elements (low partition coefficient elements) in basalts and the ratios of two of these elements are a function of the initial values in the source material, partial melting and to a lesser extent crystallization processes. When a liquid is generated through partial melting, it is likely that the solid should be treated as a partial residue (previous melting) and not as fertile mantle material.

The mantle sources at 22°N are the same both for sites 395 and 396 which are almost symmetric in relation to the ridge. The mantle source at 36°N is compositionally different. These sources (at 22°N and 36°N) can be considered either as original different sources constituting the upper mantle in both areas or as different residues of an initial homogeneous mantle.

INTRODUCTION

It is generally admitted that there are two basic processes controlling the compositions of ocean floor basalts: partial melting of the mantle source

*Originally published as: Bougault, H., Cambon, P., Corré, O., Joron, J.L. and Treuil, M., 1979. Evidence for variability of magmatic processes and upper mantle heterogeneity in the axial region of the Mid-Atlantic Ridge near 22° and 36°N. In: J. Francheteau (Editor), Processes at Mid-Ocean Ridges. Tectonophysics, 55: 11—34.
Contribution no. 579 of the Département Scientifique, Centre Océanologique de Bretagne.

followed by fractional crystallization of the liquid. The extent of these two processes is variable (Kay et al., 1970; Bryan et al., 1976; O'Hara, 1977). In addition to these fundamental processes, there may be some degree of magma mixing (O'Hara, 1977; Rhodes et al., in press; Bryan, in press). The observed wide variation of basalt compositions is more in agreement with discontinuous processes occurring at a mid-oceanic ridge, than a steady-state magma chamber.

In general, spreading rates at mid-ocean ridges are defined on the basis of geophysical observations, and usually over a very large time scale compared with magmatic events. On the one hand, an average rate for sea-floor spreading is obtained over a large time interval but, on the other, the residence time of a magmatic chamber is associated with discrete tectonic events. It is important to bear in mind that spreading rates may be very variable on a short time scale when considering a suitable time scale for magmatic events. Thus a magma chamber might undergo fractional crystallization; but following a tectonic event the size of magma chamber can be modified — it can be refilled by a new liquid with the possibility of mixing — or it can disappear and a new chamber can be created.

Variable spreading rates and tectonic events probably imply variable partial melting; in turn this variable partial melting implies different compositions (at least for trace elements) not only of the liquid, but of the residual source too. When a new liquid is generated through partial melting, it is unlikely that melting would involve only fertile mantle; it is more likely that the solid has already undergone some partial melting and may need to be treated as a partial residue (Langmuir et al., 1977).

Because of the discontinuous nature of processes at the ridge, it is necessary, therefore, when considering the suitable time scale for magmatic processes, to take into account: (a) partial melting and fractional crystallization as fundamental processes; (b) a variable extent of these processes in the same area; (c) the possibility of magma mixing.

As a consequence, the explanation of trace element behavior should take into account the following points: (a) the initial solid may have undergone some previous partial melting; (b) different degrees of partial melting; (c) variable extent of fractional crystallization; and (d) magma mixing.

Furthermore, in order to establish that the source areas are compositionally different (i.e., homogeneity or heterogeneity of the upper mantle) the use of trace elements relies on the recognition of parameters independent of these four items.

DESCRIPTION OF RESULTS

Basaltic rocks were sampled at 36°N on the Mid-Atlantic Ridge during the French—American operation "FAMOUS" in 1974. Descriptions of the area, of the samples, and the major element analyses of these samples were given by Arcyana (1977). Three parts of the area were sampled using the submers-

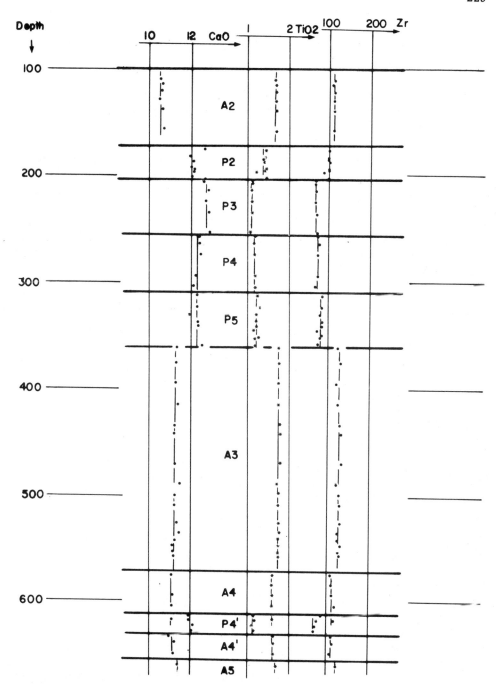

Fig. 1. Example of definition of geochemical basalt units: Hole 395A D.S.D.P. (Leg 45). Three elements (from elements analysed on board) are chosen and plotted versus depth: CaO (%), TiO$_2$ (%), Zr (ppm).

ible "Archimède" and the diving saucer "Cyana": the inner floor of the Rift valley; fracture zone A; and the intersection of the Rift valley and fracture zone A. All of the samples are tholeiitic aphyric to phyric basalts. Phenocrysts in "plagioclase phyric basalts" consist largely of plagioclases (some of them up to 1 cm diameter) together with small amounts of olivine and occasional clinopyroxenes. Picrites, with olivine phenocrysts up to 5 mm diameter, have been recovered from Mont de Vénus, the central high on the inner floor of the Rift valley.

Two sites were drilled at 22°N (sites 395 and 396) during Legs 45 and 46 of the Deep-Sea Drilling Project, almost symmetrically placed in relation to the ridge and about 7 m.y. and 10 m.y. old respectively (Melson and Rabinowitz, in press; Heirtzler and Dmitriev, in press). From shipboard studies (Bougault et al., in press), several chemical units were defined on the basis of major elements (SiO_2, Al_2O_3, MgO, CaO, K_2O, TiO_2) and four trace elements (Sr, Zr, Ni and Cr). Figure 1 shows the downhole variations for three selected elements in Hole 395A.

TABLE I

Trace element concentrations (c) found in unit A3 of Hole 395A of D.S.D.P. (Leg 45) *

Element	Method	c (ppm)	σ
Sc	NA	38.7	0.7
Ti	FX	10,350	150
V	FX	306	3.6
Cr	FX	251	6
Co	FX	42.7	1.5
Co	NA	43.4	0.7
Ni	FX	117	4.4
Ni	NA	118	3.8
Cu	AA	57.8	2.1
Zn	AA	84.6	5.6
Sr	FX	131.2	2
Y	FX	41.5	1.3
Zr	FX	125.6	4.8
Zr	NA	131.8	8.7
Nb	FX	4.3	0.5
La	NA	4.1	0.1
Eu	NA	1.52	0.04
Tb	NA	0.89	0.03
Hf	NA	3.16	0.09
Ta	NA	0.23	0.007
Th	NA	0.164	0.017

* A3 is an aphyric unit 210 m thick; fifteen samples were analysed in this unit. Standard deviation σ related to a single determination has been calculated from these measurements and has to be regarded as the precision of the analytical procedure.
Key for all tables: NA = neutron activation; FX (or XRF) = X-ray fluorescence spectrometry; AA = atomic absorption.

TABLE II

Average values for the different basaltic units in Holes 395 and 395A of D.S.D.P. (Leg 45) (concentration expressed in ppm) *

		A1 (A) 395	A2 (B) 395 6	A2 (B) 395A 6	A3 (E) 395A 15	A4 (G) 395A 4	A4' (G') 395A	A5 (H)
Sc	NA	38.4	37.8 / 1.4	36.6 / 1.2	38.7 / 0.7	36.8 / 0.4	36.8 / 0.4	37.8
Ti	FX	10 200	9710 / 59	9820 / 62	10 337 / 152	9528 / 26	9800 / 34	10 140
V	FX	301	268 / 3.7	270 / 2.3	306 / 3.6	281 / 5.8	273 / 5	297
Cr	FX	279	271 / 5	270 / 4	251 / 6	303 / 8	296 / 2	263
Co	FX	48	48.6 / 1	49.5 / 0.8	42.7 / 1.5	46.8 / 1.3	46.3 / 1.1	44
Co	NA	49	49.7 / 0.5	49.6 / 0.5	43.4 / 0.7	47.3 / 0.6	47.3 / 0.6	44
Ni	FX	161	176 / 3.9	178 / 4.8	117 / 4.4	172 / 6.5	169 / 11	129
Ni	NA	160	185 / 3.3	182 / 9.5	118 / 3.8	176 / 5.5	176 / 5	131
Cu	AA	69	67.8 / 2	72.8 / 2.3	57.8 / 2.1	65.5 / 3.5	60.3 / 2.9	54
Zn	AA	92	85.5 / 2.8	84.2 / 2.1	84.6 / 5.6	81.5 / 1.9	84.5 / 2.4	83
Sr	FX	127	119 / 1	121 / 2.4	131.2 / 2	127.5 / 6.5	129 / 2.4	132
Y	FX		39.5 / 1	39.5 / 1	41.5 / 1.3	37.1 / 1.9	38.5 / 1.0	39.9
Zr	FX	121	116 / 3.3	110 / ~2.3	125.6 / 4.8	111.2 / 5.6	109 / 4.1	124
Zr	NA	112	118 / 11	115 / 15	131.8 / 8.7	114.7 / 4.2	116 / 10	112

TABLE II (continued)

		A1 (A) 395	A2 (B) 395 6	A2 (B) 395A 6	A3 (E) 395A 15	A4 (G) 395A 4	A4′ (G′) 395A	A5 (H)
Nb	FX			4.2 / 0.6	4.3 / 0.5	5.0 / 1.1	5.1 / 0.5	4.9
La	NA	3.25	3.39 / 0.3	3.25 / 0.1	4.1 / 0.1	3.5 / 0.1	3.6 / 0.1	4.0
Eu	NA	1.5	1.39 / 0.04	1.39 / 0.07	1.52 / 0.04	1.41 / 0.03	1.4 / 0.03	1.46
Tb	NA	0.87	0.83 / 0.01	0.82 / 0.06	0.89 / 0.03	0.81 / 0.01	0.82 / 0.01	0.87
Hf	NA	2.84	2.82 / 0.06	2.79 / 0.15	3.16 / 0.09	2.8 / 0.04	2.93 / 0.12	3.07
Ta	NA	0.2	0.190 / 0.004	0.189 / 0.008	0.228 / 0.007	0.193 / 0.002	0.198 / 0.002	0.225
Th	NA	0.137	0.134 / 0.004	0.130 / 0.013	0.164 / 0.017	0.144 / 0.014	0.147 / 0.010	0.142

		P2 (D) 395A	P1 (D) 395	P3 (C1) 395	P3 (C1) 395A	P4 (C2) 395A	P4′ (C2′)	P5 (C3)
Sc	NA	32.6 / 0.4	30.6	30.6	31.5 / 0.9	32.1 / 0.9	31.7 / 1.5	30.12 / 8
Ti	FX	8177 / 137	7680	6090	6300 / 114	6694 / 129	6750 / 216	7045 / 245
V	FX	235 / 4	219	212	221 / 1.7	226 / 6.3	229 / 18	206 / 8
Cr	FX	232 / 4	222	350	351 / 8	see text		295 / 8
Co	FX	37.6 / 1.5	34	37.5	38.8 / 0.7	39.4 / 1.4	38.5 / 0.6	37 / 1.1
Co	NA	38.1 / 0.7	35	39.5	40.5 / 1.4	40.8 / 1.6	40.2 / 1.5	38.2 / 1.0

TABLE II (continued)

		P2 (D) 395A	P1 (D) 395	P3 (C1) 395	P3 (C1) 395A	P4 (C2) 395A	P4' (C2')	P5 (C3)
Ni	FX	94.8 (8)	86	135	140 (4.6)	see text		136.1 (7.6)
Ni	NA	101 (4.6)	90	145	143.6 (8.2)			145.3 (11)
Cu	AA	50.6 (1.3)	55	64	65.8 (3.2)	69 (2.9)	64.7 (4.3)	59.5 (2.5)
Zn	AA	64.3 (2.5)	58	55	59 (3.7)	63.8 (3.3)	65 (2.6)	64 (3.3)
Sr	FX	162 (5)	162	113	117 (2.8)	135.6 (2.1)	132.2 (4.7)	163.8 (5.6)
Y	FX	31.9 (0.5)			26.4 (1.5)	27.2 (1.1)	26.1 (0.4)	28.1 (1.5)
Zr	FX	97.6 (1.6)	94	67	67 (2.2)	72.1 (4.9)	73.5 (9)	78.2 (5)
Zr	NA	104.8 (12)	98	65	73 (6.2)	82.3 (7.1)	74.2 (7.6)	88.6 (7.2)
Nb	FX	3.5 (0.3)			2.85 (0.2)	3.1 (0.7)	3.5 (0.3)	2.8 (0.3)
La	NA	3.0 (0.1)	2.8	2.0	2.1 (0.1)	2.2 (0.1)	2.2 (0.1)	2.45- (0.2)
Eu	NA	1.24 (0.03)	1.18	0.96	0.97 (0.04)	1.02 (0.03)	1.03 (0.07)	1.07 (0.04)
Tb	NA	0.63 (0.02)	0.64	0.53 (0.01)	0.55 (0.02)	0.56 (0.01)	0.57 (0.04)	0.58 (0.03)
Hf	NA	2.37 (0.05)	2.21	1.65	1.73 (0.06)	1.88 (0.08)	1.83 (0.14)	1.97 (0.10)
Ta	NA	0.161 (0.005)	0.149	0.106	0.112 (0.004)	0.110 (0.004)	0.106 (0.006)	0.125 (0.007)
Th	NA	0.120 (0.012)	0.113	0.081	0.084 (0.013)	0.083 (0.014)	0.074 (0.009)	0.096 (0.013)

* For explanation of symbols, see note below Table I.

TABLE III

Average values for the different basaltic units in Hole 396 (Leg 45) (concentrations in ppm) *

		a	b	c
Sc	NA	36.8	35	34.7
Ti	FX	9000	7600	7740
V	FX	312	264	255
Cr	FX	288	393	290
Co	FX	40	40	40.5
Co	NA	40.7	41	42
Ni	FX	111	142	121
Ni	NA	114	143	124
Cu	AA	58	66	65
Zn	AA	73	66	72
Sr	FX	148	156	131
Y	FX	34.6	29.4	30.9
Zr	FX	109	87	86
Zr	NA	118	87	97
Nb	FX	4	4.6	3.2
La	NA	3.5	2.9	2.7
Eu	NA	1.36	1.2	1.22
Tb	NA	0.79	0.65	0.68
Hf	NA	2.64	2.2	2.2
Ta	NA	0.21	0.17	0.156
Th	NA	0.16	0.13	0.114

* For explanation of symbols see note below Table I.

TABLE IV

Average values for the different basaltic units in Hole 396B (Leg 46) (concentrations in ppm) *

Unit		Ti XRF	V XRF	Cr XRF	Mn XRF	Fe XRF	Co			Ni	
							XRF	AA	NA	XRF	NA
A1	$\overline{x}$	8473	266	298	1380	70 920	40.6	42.3	43.7	136	139
	s	153	19	18	58	2 168	1.1	2.1	1.7	7.5	6.7
A2	$\overline{x}$	9160	293	282	1428	73 741	43.0	41.0	43.5	129	133
	s	99	5	12	87	1 537	1.6	1.3	1.3	8	5.5
A3	$\overline{x}$	9822	282	265	1338	77 010	43.0	42.0	44.0	118	134
	s	113	5	7.7	38	1 661	1	0.7	0.8	1.7	3.2
B1	$\overline{x}$	7360	257	309	1316	66 161	41.0	39.0	42.0	133	139
	s	305	22	7.5	54	2 025	2	2	1.2	14	15
B2	$\overline{x}$	6360	214	337	1223	61 096	41.0	40.0	41.0	141	150
	s	417	21	15	34	1 541	2	4	0.4	15	15
C	$\overline{x}$	9050	274	291	1432	730 566	40.0	40.0	41.0	133	140
	s	272	13	14	42	1 806	1.3	1.6	0.8	7	7

* For explanation of symbols see note below Table I.

All the chemical data presented in this paper were obtained by X-ray fluorescence analysis (Bougault et al., 1977) or neutron activation analysis (Treuil, 1973). To enable assessment of analytical precision, average values and sigma values (related to one determination) are given in Table I for unit A3 (210 m thick) of hole 395A, of which fifteen samples were analysed. Three elements have been measured both by X-ray fluorescence and neutron activation techniques: both results are given.

Shore-based trace element analyses have confirmed the chemical units which were defined by shipboard data. However, some elements have been shown to be much more variable due to alteration and sea water effects: K_2O, Cs, Rb and Sb (Bougault et al., in press).

The concentration of all elements investigated (major and trace) at 22°N and 36°N falls within the normal range of values for tholeiites. Tables II, III, IV and V concern, respectively, Site 395, 396 (Leg 45), 396 (Leg 46) and the FAMOUS area. Plagioclase phyric basalts (with some olivine phenocrysts and occasional clinopyroxene phenocrysts) tend to have lower trace element concentrations than aphyric basalts (except Sr). This feature could be due partly to the dilution effect of plagioclase phenocrysts. Ta and Nb concentrations are lower in the 22°N samples than for those from 36°N. The contrary is observed for Ti, Y, Zr, Tb and Hf. These differences will be considered in more detail later.

Cu AA	Zn AA	Sr XRF	Zr		Eu NA	Tb NA	Hf NA	Ta NA	Th NA
			XRF	NA					
65	79.4	126	94	103	1.24	0.73	2.47	0.174	0.124
4	1.2	10	4	7.3	0.03	0.01	0.05	0.004	0.01
64	81.0	132	101	104	1.35	0.76	2.65	0.182	0.136
2.2	3.8	6	6	7	0.04	0.02	0.006	0.005	0.016
63	82	145	120	123	1.45	0.86	2.96	0.195	0.149
0.8	4.3	5.3	6.3	13	0.03	0.017	0.10	0.006	0.013
67	74	133	80	91	1.10	0.63	2.14	0.14	0.1
2.5	4	5	2.2	4	0.05	0.038	0.09	0.008	0.014
70	67	135	69	71	0.96	0.52	1.8	0.12	0.089
3	3.6	12	4.6	11.7	0.04	0.03	0.14	0.008	0.015
65	77	156	108	125	1.30	0.76	2.6	0.295	0.02
2.3	3.3	4.7	4.4	9	0.06	0.02	0.14	0.03	0.04

TABLE V

(a) Concentrations (in ppm) of samples from the inner floor of FAMOUS area

	Sc NA	Ti FX	V FX	Cr FX	Co		Ni		Cu AA	Zn AA	Rb FX
					AA	NA	FX	NA			
ARP 74 7-5	32.5	8580	270	263		39	115	121			5
ARP 74 7-6	33.4	8340	271	264	46	39	125	123	64	84	
ARP 74 7-6 glass		8580		267	46		122		65	78	
ARP 74 7-7	32.6	8520	265	265		39	118	120			
ARP 74 7-8	33.1	7500	275	362		39	144	139			6
ARP 74 7-9	32.2	7660	270	366		38	142	141			5
ARP 74 8-10	34.6	6960	277	170	38	36	87	85	64	74	
ARP 74 9-12	27	5280	188	612	38	37	238	248	72	59	2.5
ARP 74 9-13	31	7380		250			93	92	59	74	2
ARP 74 10-14 glass		5040	189	570			255		75	65	
ARP 74 10-15		6120	199	548	43		226		72	66	
ARP 74 10-15 glass		5940	199	549	46		229		72	66	
ARP 74 10-16	32.6	4920	192	625	41	37	240	245	74	60	
ARP 74 10-16 glass		4920		626			247		74	61	
ARP 74 11-17	28.6	6840	217	473	40	37	196	199	72	64	
ARP 74 11-18	28.4	6420	215	402		36	159	161	63	63	2
ARP 74 12-19	33.2	6000	239	412	37	36	153	167	73	63	
ARP 74 12-19 glass		6000		410	46		141		73	69	
CYP 30-32		3840	187	1333			381				
CYP 30-33 xx glass	29.8	7020	225	453		33	159	154			2
CYP 30-34 xx	30.9	3780	182	1321		47	392	371			0.5
CYP 31-35	27	5160		645		34	253	234	73	64	
CYP 31-35 glass	27.3	5100	189	628		35	255	297			
CYP 31-36 x	20.7	3300	153	338		24	91	98			
CYP 31-37x	20.5	3300				23.1		99			
CYP 31-38	33.4	5880	251	83		36	61	69			6.5
CYP 31-39	29.9	6840	237	370		35	115	122			
CYP 31-40	30	6940		360		34	110	115			
ARP 73(1 and 4)	25	5280	185	600	40	40	260	260	73	60	
ARP 73-2	31.4	9480	300	130	42	40	78	81	54	86	
ARP 73-3	18.2	2700	140	1640	72	76	900	1155	85	50	

Y FX	Zr NA	Nb FX	Sb NA	Cs NA	La NA	Ce NA	Eu NA	Tb NA	Hf NA	Ta NA	Th NA
32	82	14	0.133	0.08	8.7	15	1.08	0.72	2.42	0.83	0.82
	73		0.023	0.06	8.6	16	1.15	0.75	2.40	0.86	0.91
	110		0.09		8.8	16	1.12	0.71	2.37	0.85	0.86
27	93	14	0.032	0.12	8.0	14	1.0	0.59	2.03	0.85	0.86
27		15	0.022	0.08	8.3	13	0.95	0.61	2.02	0.83	0.83
	80				7.4	13	0.95	0.55	1.62	0.64	0.68
19	51	10	0.014	0.045	5	10	0.68	0.42	1.38	0.49	0.51
28	69	10	0.016	0.06	6.2	11	0.97	0.6	1.76	0.57	0.6
	54		0.016	0.03	5.6		0.85	0.43	1.32	0.48	
	70		0.028	0.06	6.5	13	0.92	0.55	1.9	0.66	0.69
26	57	12			6.1	10	0.9	0.51	1.6	0.55	0.62
	52				4.8		0.81	0.5	1.74	0.4	0.44
20		5			4.1	7	0.5	0.34	1	0.3	0.45
25.3	67	10	0.015	0.041	5.8	11	0.94	0.56	1.76	0.6	0.66
17.5	48	5			3.1	6.8	0.54	0.36	0.9	0.27	0.32
	45		0.033	0.06	4.8	9	0.76	0.40	1.31	0.49	0.57
	57		0.034	0.16	5.4	9	0.76	0.42	1.32	0.52	0.58
13.4	40	6	0.05	0.081	3.1	5.2	0.58	0.29	0.94	0.28	0.32
	31		0.028	0.084	2.4	4.4	0.51	0.266	0.8	0.27	0.28
22	48	9.5	0.08		5.6	11	0.85	0.5	1.43	0.6	0.63
	41		0.026	0.045	7.0	11	1	0.56	1.91	0.68	0.74
			0.046	0.038	7.5	12	0.98	0.56	1.8	0.69	0.71
	75				4.5		0.79	0.42	1.43	0.52	0.58
	147				8.7		1.18	0.73	2.5	1.03	1.07
	33				1.41		0.34	0.2	0.62	0.15	0.14

TABLE V (continued)

(b) Concentrations (in ppm) of samples from fracture zone A of FAMOUS area

	Sc	Ti	V	Cr	Co NA	Ni FX	Ni NA	Zr NA
CYP 74 19-6	—	6300	237	212	—	87	—	—
CYP 74 22-9	32.1	5460	197	466	47	242	260	19
CYP 74 22-10	28.9	3360	149	658	49	336	381	19
CYP 74 23-11	28.6	5100	193	505	56	258	258	27
CYP 74 27-18	27.6	6660	226	535	40	212	222	75
CYP 74 27-20	30.7	7080	237	541	38	—	193	65
CYP 74 27-21	30.1	6720	233	536	39	210	210	58
CYP 74 27-26	31	6960	245	435	40	195	196	72
CYP 74 29-30	30.7	8880	278	304	38	127	136	—
CYP 74 29-31 C	30.9	6600	246	399	37	157	163	67
CYP 74 29-31 A	33.1	6780	242	404	37	164	176	—

(c) Concentrations (in ppm) of samples from the intersection of the rift and fracture zone A in FAMOUS area

	Sc NA	Ti FX	V FX	Cr FX	Co AA	Co NA	Ni FX	Ni NA	Cu AA	Zn AA	Zr NA
ARP 74 13-22	32.9	6330	250	167	41	38	79	72	70	66	52
ARP 74 13-23	33.4	6330	253	170		40	77	76			
ARP 74 13-24	30.8	6300	250	167		37	82	73			64
ARP 74 14-26	33.9	6240	242	123		39	65	61			
ARP 74 14-27	33.7	6180	243	122		36	68	58			43
ARP 74 14-28 F	32.7	6300	245	127	41	39	72	66	86	67	54
ARP 74 14-28 A	33.9	6120	235	122	41	39	73	66	86	64	50
ARP 74 14-29	33.4	6180	240	119		35	66	57			69
ARP 74 14-31 x	26.9	2580	142	401		36	176	171			18
ARP 74 14-32 x	29.7	2880		430		39	183	183			19
ARP 74 14-33 x	27	2700	143	412		35	177	172			25
ARP 74 16-36		6540	237	511			160				
ARP 74 16-37 xx		6240		1407			718				
ARP 74 17-40		6660	248	131			77				
ARP 74 17-41	39.6	7260	265	89		35	64	65			85
ARP 74 18-44 B	45.1	6300	257	80		37	65	63			64

HIGH PARTITION COEFFICIENT ELEMENTS: Co, Ni, Cr: FRACTIONAL CRYSTALLIZATION

To show that the behavior of Co, Ni and Cr is related mainly to the fractional crystallization process, it is necessary to demonstrate that their initial magma concentrations are not greatly dependent upon the first process

Sb	Cs	La	Ce	Eu	Tb	Hf	Ta	Th
—	—	—	—	—	—	—	—	—
—	0.14	3.7	10	0.8	0.46	1.3	0.36	0.41
—	—	1.3	2	0.5	0.29	0.7	0.08	0.17
0.146	0.15	4.8	9	0.71	0.47	1.19	—	0.51
0.066	0.19	6.7	13	0.93	0.60	1.82	0.65	0.71
0.02	0.04	6.7	14	0.99	0.55	1.88	0.66	0.65
0.063	0.11	6.8	12	0.93	0.55	1.82	0.66	0.66
0.036	0.1	7.2	15	0.95	0.55	1.77	0.69	0.74
0.024	—	7	13	1.2	0.75	2.35	0.65	0.73
0.015	0.07	6.4	11	0.91	0.57	1.73	0.62	0.68
—	—	6.0	15	0.95	0.58	1.85	0,48	0.49

Sb	Cs	La	Ce	Eu	Tb	Hf	Ta	Th
NA	NA	NA	NA	NA	NA	NA	NA	NA
0.03	0.09	5.8	10	0.87	0.53	1.59	0.57	0.59
	0.09	5.5	9	0.8	0.53	1.6	0.57	0.61
	0.09	5.2	10	0.8	0.5	1.5	0.55	0.55
0.021	0.67	6.1	12	0.88	0.39	1.6	0.6	0.63
0.045	0.15	6.6	11	0.90	0.5	1.58	0.61	0.62
	0.06	6.8	11	0.88	0.51	1.57	0.61	0.62
0.026	0.13	5.8	10	0.89	0.51	1.62	0.61	0.63
0.055	0.59	6.6	12	0.86	0.50	1.51	0.61	0.64
	0.07	1.8	2	0.41	0.29	0.51	0.13	0.17
		1.7		0.46	0.30	0.66	0.15	0.17
0.015		1.6	6.4	0.44	0.28	0.53	0.14	0.16
0.024	0.09	9.2	14	1.16	0.57	2.09	0.89	0.94
0.016	0.06	6.4	10	1.15	0.54	1.68	0.54	0.65

— partial melting (Bougault, 1977). Using the known partition coefficients (mineral/liquid) and the proportions of minerals in the peridotite source to calculate the bulk partition coefficients, DO, and P in the equation of Shaw (1970) (see Table VI):

$$\frac{CLO}{CSO} = \frac{1}{DO + F(1 - P)}$$

TABLE VI

Partition coefficients of Cr, Ni and Co and proportions of minerals (in the initial solid: X_i and melting phases: XF_i) used to compute the possible variations of Cr, Ni and Co as partial melting occurs. Two models are considered A and B with different proportions of garnet and spinel (Allègre et al., 1973)

		OL	OPX	CPX	PLAG	GRT	SP
		partition coefficient					
Cr		1	7.3	13	0.02	7	220
Ni		12.6	6.1	4.4	0.06	0.1	19.6
Co		3	2.08	1.32	0.1	2.2	5.1
		proportion of minerals					
MODEL A	X_i	0.5	0.25	0.2	0.00	0.00	0.05
	XF_i	0.05	0.05	0.8	–	–	0.10
MODEL B step 1	X_i	0.55	0.30	0.12	0.00	0.01	0.02
	XF_i	0.05	0.05	0.7	–	0.1	0.1
		F = 0.1 : garnet disappears					
step 2	X_i	0.54	0.29	0.05	0.00	0.00	0.01
	XF_i	0.055	0.055	0.77	–	–	0.11

CLO = concentration in the liquid; *CSO* = concentration in the initial solid; *F* = degree of melting; *DO* = bulk partition coefficient of the initial assemblage; *P* = bulk partition coefficient for melting phases; it can be shown that the ratios CLO/CSO and CS/CSO remain nearly constant for cobalt, nickel and chromium (Fig. 2). This is because, for the initial solid, olivine is a major constituent and melts only in low proportions compared to clinopyroxene.

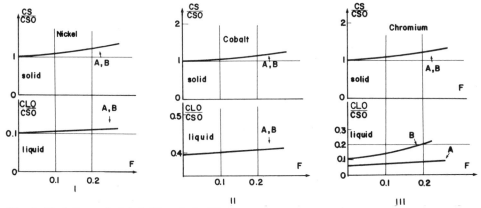

Fig. 2. Variation of nickel (I), cobalt (II) and chromium (III) versus partial melting (F) in the solid phase (residue). CS/CSO and in the liquid phase CLO/CSO do not vary significantly for nickel and cobalt. For chromium, CS/CSO does not vary but CLO/CSO can vary, depending on the proportion of spinel (model A or B).

For nickel and cobalt, olivine acts as a buffer during the melting process. For chromium, this statement is not true because bulk partition coefficients vary with spinel contents: the chromium concentration both in liquids and in residues can be considered to show limited variation with melting but only within a restricted degree of melting for a given composition of the initial solid.

Figure 2 shows the limited variation in the ratio CS/CSO (concentration in the residue/concentration in the initial solid) for Ni, Co and Cr. This theoretical result is confirmed by Sato (1977) and by our own data obtained for different peridotites (Lhertzolite, alkali basalt nodules, kimberlite nodules). Nickel concentrations mostly lie between 2000 and 2,200 ppm and cobalt concentrations do not depart much from 115 ppm. Chromium values show more dispersion, because of their dependence on the spinel content.

In the liquids, the ratio CLO/CSO is found to be about 0.4 for cobalt (nearly independent of partial melting). Taking the value of 115 ppm for the source, CLO (concentration in the liquid) is equal to 47 ppm which is in good agreement with observed values in tholeiites (40—50 ppm). For nickel, CLO/CSO is about 0.115 (nearly independent of partial melting) and using 2,100 ppm as the value in the initial solid, the liquid concentration CLO is 241 ppm. The highest Ni value recorded so far in tholeiites (aphyric samples or glasses, undifferentiated by crystallization process, with approximately 11% MgO is 250 ppm which is in good agreement with the theoretical value (241 ppm). For the same tholeiite samples (Ni = 250 ppm), the Cr content is 600 ppm, which is also in good agreement with the theoretical estimate; but this result has less significance because of the possible influence of spinel.

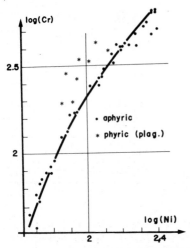

Fig. 3. Log (Cr) versus log (Ni). Points plotted include the FAMOUS area (data in this paper) and Leg 37 results. This pattern confirms previous results obtained from dredged material in the FAMOUS area (Bougault and Hékinian, 1974).

The fit obtained between predicted variations of cobalt, nickel and chromium and observed values, both in solids (peridotites) and undifferentiated liquids (glasses or aphyric samples with MgO ~11%), confirms that the observed variations of these elements in tholeiites is not due to partial melting.

The large variations observed for Ni and Cr in tholeiites (Fig. 3) and the limited variation of Co (40—50 ppm) place some constraints on the relative proportions of the crystallizing minerals (Bougault et al., 1977). As an example, a large variation in nickel concentrations implies crystallization of olivine and/or spinel. However, if only olivine and/or spinel are involved, cobalt would be depleted at the same time (partition coefficients of Co for olivine and spinel, respectively, are: 3 and 5): this is not the case since cobalt varies mostly between 40 and 50 ppm in tholeiites. To reconcile the theoretical and observed results, the bulk partition coefficient for Co should be about one and hence another mineral, the partition coefficient of which is lower than one (e.g., plagioclase), has to be removed along with olivine and/ or spinel. Deducing the proportion of minerals crystallizing to account for the observed levels of Co, Ni and Cr does not always give a unique solution; this point will not be discussed here. Nevertheless we can state that cobalt, nickel and chromium concentrations do not vary significantly with partial melting. The large variations of nickel and chromium contents in tholeiites are mainly due to fractional crystallization. These observations together with the limited variation of cobalt provide constraints for any model of crystallization processes.

PARTIAL MELTING, INITIAL SOLIDS—HYGRO MAGMAPHILE ELEMENTS

Using equations 15 and 16 of Shaw (see also Treuil and Joron, 1975), we can deduce the ratio of concentrations of two hygromagmaphile elements (i.e., those with very low partition coefficients) in a solid considered as a residue of melting (f):

$$\frac{CS1}{CS2} = \frac{CSO1}{CSO2} \frac{DO1}{DO2} \frac{DO2 + f(1-P)}{DO1 + f(1-P)} \tag{1}$$

(1 and 2 are related to element 1 and 2)
in which CS = concentration in the solid residue; CSO = concentration in the initial solid; DO = bulk partition coefficient of the initial mineral assemblage; P = bulk partition coefficient of melting phases.

For hygromagmaphile elements, P can be neglected, hence eq. 1 simplifies to:

$$\frac{CS1}{CS2} = \frac{CSO1}{CSO2} \frac{DO1}{DO2} \frac{DO2 + f}{DO1 + f} \tag{2}$$

It is apparent that for low values of f (0.01—0.05) the ratio $CS1/CS2$ can vary significantly when $DO2$ and $DO1$ differ by an order of magnitude (e.g., 0.1 and 0.01). Figure 4 shows the variations of CS/CSO related to two different bulk partition coefficients of 0.01 (element 1) and 0.1 (element 2). $CS1/CS2$ varies from 1 to 0.2 when f varies from 0 to 5%. With the same simplification made above for P, the relationship between two hygromagmaphile elements in the liquid phase (basalt) derived from melting of the residue can

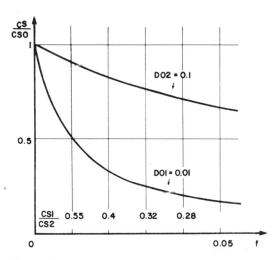

Fig. 4. Hygromagmaphile elements: variation of concentration in the solid phase (residue) CS/CSO versus partial melting f. Two examples with bulk partition coefficient (DO) chosen as 0.1 and 0.01, respectively.

244

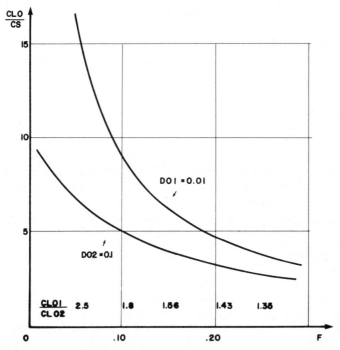

Fig. 5. Hygromagmaphile elements: variation of concentration in the liquid phase CLO/CS versus partial melting. Two examples: bulk partition coefficient DO chosen 0.1 and 0.01.

be written as:

$$\frac{CLO1}{CLO2} = \frac{CS1}{CS2} \frac{DO2 + F}{DO1 + F} \tag{3}$$

F is the degree of partial melting; O in $CLO1$ means that the concentration in the liquid phase is the initial concentration with respect to subsequent processes (fractional crystallization). Variation of CLO/CS is presented in fig. 5 for two elements: element 1 $DO1 = 0.01$ and element 2 $DO2 = 0.1$. The ratio $CLO1/CLO2$ varies from 2.4 ($F = 5\%$) to 1.35 ($F = 25\%$) for a given solid ($CS1/CS2$), assuming that basaltic liquids are produced in this range of melting. The range of variation of $CLO1/CLO2$ could be higher if basaltic liquids are produced through melting lower than 5%. In any event, simply considering that tholeiites are produced by melting a given solid between 5 to 25% and that the solid itself may previously have undergone some partial melting (0 to 5%), combining equation 2 and 3, we obtain:

$$\frac{CLO1}{CLO2} = \frac{CSO1}{CSO2} \frac{DO1}{DO2} \frac{DO2 + f}{DO1 + f} \frac{DO2 + F}{DO1 + F} \tag{4}$$

From what has previously been found, the quantity:

$$\frac{CLO1/CSO1}{CLO2/CSO2} = \frac{DO1}{DO2} \cdot \frac{DO2+f}{DO1+f} \cdot \frac{DO2+F}{DO1+F}$$

may vary from 0.34 (f = 5%, F = 25%) to 2.5 (f = 0%), F = 5%).

Postulating only that basalts can be produced through variable partial melting and further that the solid undergoing melting can itself be considered as a residue of previous melting, we observe that the quantity ($CLO1/CSO1$)/($CLO2/CSO2$) is variable simply because of these two processes. As a consequence, we should be cautious in using two hydromagmaphile elements whose partition coefficients differ by an order of magnitude, to deduce information about the chemistry of the mantle in a given area (Schilling, 1975).

If $DO1$ and $DO2$ are both of the order of 0.01 (as is the case with Th, Ta, La, Nb (Treuil, 1973) the term ($DO2+F$)/($DO1+F$) remains invariable with F. Then $CLO1/CLO2$, or $CL1/CL2$ (concentrations after crystallization processes, as $CL1/CL2 = CLO1/CLO2$ for hygromagmaphile elements) is only function of $CSO1/CSO2$ and f.

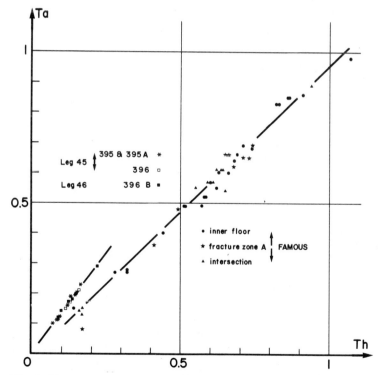

Fig. 6. Plot of two hygromagmaphile elements whose bulk partition coefficient is ~0.01: Ta versus Th.

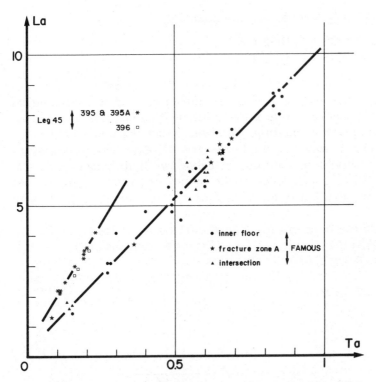

Fig. 7. Plot of two hygromagmaphile elements whose bulk partition coefficient is ~0.01: La versus Ta.

Plots of Ta versus Th and La versus Ta are shown in Figs. 6 and 7, using data for Sites 395 and 396 on one hand and for the FAMOUS area on the other. The first interesting feature is the different element ratios seen in these two areas. The second feature is that Sites 395 and 396 plot exactly on the same line. Since $(DO2 + F)/(DO1 + F)$ is invariable for such elements, we conclude that:

$$\frac{CS1}{CS2} = \frac{CSO1}{CSO2} \frac{DO1}{DO2} \frac{DO2 + f}{DO1 + f}$$

is constant in each zone but has a different value. This means that, even if solids melting in each zone are already residues of previous melting, these solids are compositionally different with respect to these elements in each area.

As a first step, it can be inferred that $CSO1/CSO2$ and f are constant within each area. However, in order to explain the differences between 22°N and 36°N, we are left with two possibilities: either the ratios $CSO1/CSO2$ are different for each zone, or they are the same, the difference between ratios $CL1/CL2$ in each zone being due to two different values of

f (same initial mantle material previously differentiated through low partial melting, f). The choice between these two possibilities will have to be made when additional data are available on North Atlantic basalts; it is necessary to demonstrate distinct variations of f and of $CSO1/CSO2$. This last point can be investigated by considering a pair of elements whose ratio is not variable through all magmatic processes; these two elements probably have to be chosen among very low partition coefficient elements such as thorium, tantalum, niobium or lanthanum.

Considering a group of samples for which it has been concluded from Ta/Th or La/Ta ratios that they are derived from the same solid (even if a residue) it is possible to see evidence of different degrees of partial melting, F, by observing the behaviour of a pair of elements whose partition coefficients are lower than one, but different ($DO1 = 0.01$; $DO2 = 0.1$). For such samples, it is easy to see that:

$$\frac{CL1}{CL2} = \frac{CLO1}{CLO2} \sim K\frac{DO2 + F}{F}$$

(K = constant: same initial solid demonstrated by Ta/Th, Nb/Th or La/Th

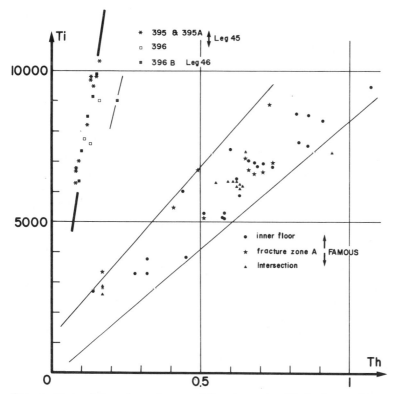

Fig. 8. Plot of two hygromagmaphile elements: Ti (bulk partition coefficient ~ 0.1) versus Th (bulk partition coefficient ~ 0.01).

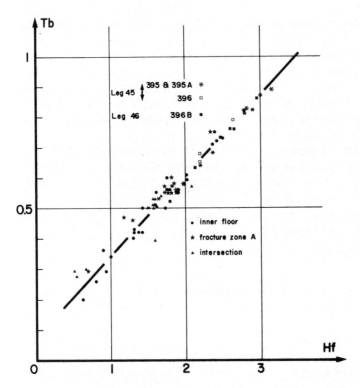

Fig. 9. Plot of two hygromagmaphile elements whose bulk partition coefficients are around 0.1 : Tb versus Hf.

ratios.) With the above condition (same initial solid) different slopes for $CL1$ versus $CL2$ or dispersed points will account for different partial melting F. An example is given in Fig. 8, where titanium is plotted against thorium. Similar relationships can be obtained by plotting Hf, Zr, Tb . . . versus Th, Ta, La or Nb. For FAMOUS area basalts, it is clear that points are more dispersed in the Ti/Th diagram than in the Ta/Th diagram. In addition, the highest Ti/Th ratios should correspond to the lowest Ti and Th concentrations. This is, in fact, observed for plagioclase phyric samples in FAMOUS area. At site 396, one group of samples has to be attributed to a lower degree of partial melting than other groups (one point on the right side of 395, 396 trend. In Ta/Th or La/Th diagram this unit is on the 395—396 trend).

If we now consider two elements whose partition coefficients are of the order of ca 0.1, it is easy to deduce that:

$$\frac{DO1}{DO2} \simeq \frac{DO2 + f}{DO1 + f} \simeq \frac{DO2 + F}{DO1 + F}$$

cannot vary significantly. As an example, a plot of Tb versus Hf is given in Fig. 9; a similar relationship is observed both for sites 395, 396 and the

FAMOUS area. This suggests similar initial ratios ($CSO1/CSO2$) for such elements in the mantle. Similar results are obtained when Tb or Hf are plotted against Ti, Zr and Y.

CONCLUSION

Concentrations of trace elements with high partition coefficients do not vary (Co, Ni) or vary little (Cr) both in solid residues and liquids when partial melting occurs. Aphyric basaltic samples or glasses undifferentiated through crystal fractionation (MgO: 11%) show values for Ni, Co and Cr in agreement with computed values obtained from partition coefficients and their concentrations in peridotites. Fractional crystallization is responsible for the large variation of these elements in basaltic samples.

Hygromagmaphile elements show a large range of variation due to both the partial melting process and the fractional crystallization process. Choosing pairs of these elements among Th, La, Nb and Ta it is shown that the mantle sources (before melting) at 22°N are the same both for Site 395 and 396, almost symmetric in relation to the ridge. The mantle source at 36°N is compositionally different. These solids (at 22°N and at 36°N) can be considered either as: (1) original different solids constituting the upper mantle in both areas; or, (2) as different residues of an initial homogeneous mantle.

Magma mixing, one of the processes which can occur, is not apparent from our data. If this process occurs, it can be inferred that those magmas which have been derived by mixing, have nevertheless been derived from one source characteristic of the area.

ACKNOWLEDGEMENTS

We thank J. Francheteau, H.D. Needham, J. Erlank and J. Tarney for their comments and for reading the manuscript.

REFERENCES

Allègre, C.J., Montigny, R. and Bottinga, Y., 1973. Cortège ophiolitique et cortège océanique, géochimie comparée et mode de genèse. Bull. Soc. Géol. Fr. (7), XV, 5-6: 461—477.

Arcyana, 1977. Rocks collected in the FAMOUS area by bathyscaphe and diving saucer from the Rift valley of the Mid-Atlantic Ridge: petrological diversity and structural setting. Deep-Sea Res., 24: 565—569.

Bougault, H., 1977. Evidence de la cristallisation fractionnée au niveau d'une ride médio-océanique: Co, Ni, Cr — FAMOUS — leg 37 du D.S.D.P. Bull. Soc. Géol. Fr., XIX, no. 6: 1207—1212.

Bougault, H. and Hékinian, R., 1974. Rift valley in the Atlantic ocean near 36°50'N: petrology and geochemistry of basaltic rocks. Earth Planet. Sci. Lett., 24: 249—261.

Bougault, H., Cambon, P. and Toulhoat, H., 1977. X-ray spectrometric analysis of trace elements in rocks. Correction for instrumental interferences. X-Ray Spectrom., 6 (2) 2: 66—72.

Bougault, H., Treuil, M. and Joron, J.L., 1978. Trace elements from 22°N and 36°N in the Atlantic Ocean: fractional crystallization, partial melting and heterogeneity of the upper mantle. Leg 45 D.S.D.P. In: Initial Reports of the Deep-Sea Drilling Project, Vol. 45. U.S. Government Printing Office, Washington, in press.

Bryan, W.B., 1978. Regional variation and petrogenesis of basalt glasses from the FAMOUS area, Mid-Atlantic Ridge. J. Petrol., in press.

Bryan, W.B., Thompson, G., Frey, F.A. and Dickey, J.S., 1976. Inferred settings and differentiation in basalts from Deep-Sea Drilling Project. J. Geophys. Res., 81: 4285—4304.

Heirtzler, J. and Dmitriev, L., 1978. Leg 46 DSDP in Initial Reports of the Deep-Sea Drilling Project, Vol. 46. U.S. Government Printing Office, Washington, in press.

Kay, R., Hubbart, N.J. and Gast, P.W., 1970. Chemical characteristics and origin of oceanic ridge volcanic rocks. J. Geophys. Res., 75: 1585—1613.

Langmuir, C.H., Bender, J.F., Bence, A.E., Hanson, G.N. and Taylor, S.R., 1977. Petrogenis of basalts from the FAMOUS area, Mid-Atlantic Ridge. Earth Planet. Sci. Lett., 36: 133—156.

Melson, W.G. and Rabinowitz, P.D., 1978. Leg 45 DSDP in Initial Reports of the Deep-Sea Drilling Project, Vol. 45. U.S. Government Printing Office, Washington, in press.

O'Hara, M.J., 1977. Geochemical evolution during fractional crystallization of a periodically refilled magma chamber. Nature, 266: 503—507.

Rhodes, J.M., Blanchard, D.P., Dungan, M.A., Rodgers, K.W. and Brannon, J.C., 1978. Chemistry of basalts from Leg 45 of the Deep-Sea Drilling Project, D.S.D.P. In: Initial Reports of the Deep-Sea Drilling Project, Vol. 45. U.S. Government Printing Office, Washington, in press.

Sato, H., 1977. Nickel content of basaltic magmas: identification of primary magmas and a measure of the degree of olivine fractionation. Lithos, 10: 113—120.

Shaw, D.M., 1970. Trace elements fractionation during anatexis. Geochim. Cosmochim. Acta, 34: 237—243.

Schilling, J.G., 1975. Rare earth variations across "normal segments" of the Reykjanes ridge, 60—53°N, Mid-Atlantic Ridge, 29°S, and East Pacific Rise 2—19°S, and evidence on the composition of the underlying low velocity layer. J. Geophys. Res., 80: 1459—1473.

Treuil, M., 1973. Critères pétrologiques géochimiques et Structuraux de la Genèse et de la Différenciation des Magmas basaltiques: Exemple de l'AFAR. Thèse, Université d'Orleans.

Treuil, M. and Joron, J.L., 1975. Utilisation des éléments hygromagmatophiles pour la simplification de la modélisation quantitative des processus magmatiques: exemple de l'AFAR et de la dorsale médio-atlantique. Soc. Ital. Mineral. Petrol., 31: 125—174.

Chapter 12

MAGMA MIXING AT MID-OCEAN RIDGES: EVIDENCE FROM BASALTS DRILLED NEAR 22°N ON THE MID-ATLANTIC RIDGE[a]

J.M. RHODES [1]*, M.A. DUNGAN [2]*, D.P. BLANCHARD [3] and P.E. LONG [3]*

[1] *Lockheed Electronics Co., Inc., Houston, Texas (U.S.A.)*
[2] *Lunar Science Institute, Houston, Texas (U.S.A.)*
[3] *NASA Johnson Space Center, Houston, Texas (U.S.A.)*

SUMMARY

This chapter summarizes the results of a detailed petrological, geochemical and experimental study of basalts recovered during Legs 45 and 46 of the Deep Sea Drilling Project, and presents evidence that magma mixing has played an important role, along with crystal fractionation, in the evolution of these and other sea-floor basalts. Three major lines of evidence are considered, including phenocryst mineralogy, melt inclusion compositions, and whole rock major and trace element chemistry. Taken together, this evidence indicates that "primitive" magmas and their attendant phenocrysts are episodically injected into fractionating magma chambers containing more evolved cogenetic magmas. The dominant products of this steady-state process of repeated magma injection, mixing and fractionation are moderately evolved ocean floor basalts. Compositional extremes, including "primitive", mantle-derived primary magmas and highly evolved differentiation products, will be rare. This model accounts for several enigmatic petrological characteristics of ocean-floor basalts previously inadequately explained by partial melting and crystal fractionation processes. Consequently, it is suggested that magma mixing is a fundamental process that is intrinsically interrelated with crystal fractionation in ocean-floor basalt petrogenesis.

INTRODUCTION

The petrogenesis of ocean-floor basaltic rocks has been commonly modeled in terms of partial melting of a mantle source, with subsequent modification by crystal fractionation processes (e.g., Kay et al., 1970; Schilling, 1971; Frey et al., 1974; Bryan et al., 1976). However, it has become increasingly apparent that magmatic processes along mid-ocean ridges are more complex than these models would suggest, and that there are inconsistencies that need additional explanation. For example, it is found that crystal fractionation models frequently fail to provide an entirely satisfactory

---

[a]Originally published as: Rhodes, J.M., Dungan, M.A., Blanchard, D.P. and Long, P.E., 1979. Magma mixing at mid-ocean ridges: evidence from basalts drilled near 22°N on the Mid-Atlantic Ridge. In: J. Francheteau (Editor), Processes at Mid-Ocean Ridges. Tectonophysics, 55: 35—61.
* For present address, see p. 277.

relationship for basalts that are closely associated in space and time, can be presumed to have shared the same plumbing systems, and appear to have been derived from the same or similar mantle source. An alternative explanation, which is equally perplexing, requires that each compositionally distinct basaltic unit is independently derived from the mantle and has undergone a separate and distinct fractionation history (e.g., Blanchard et al., 1976; Bryan et al., 1976; Rhodes et al., 1976; Bryan and Moore, 1977).

Recently, O'Hara (1977) has introduced a model involving the fractionation of a repeated filled magma chamber in an attempt to account for large differences in trace element abundances in otherwise rather similar magmas. This conceptual model of magma mixing and fractionation may provide an answer to the inconsistencies in simple fractionation models providing it can be shown that magma mixing is indeed a common and prevailing process associated with mid-ocean ridge volcanism.

Anderson (1976) states that on the basis of volcanological studies one should expect magma mixing to occur and recur over and over with each eruption, and he argues convincingly that magma mixing is a widespread phenomenon that plays a particularly significant role in circum-Pacific volcanism. That this process is confined not only to andesitic and other intermediate rocks is evident from the work of Wright and Fiske (1971) on Kilauea, and from the suggestion of Irvine (1977) that magma mixing has occurred in the large stratiform intrusions, such as the Muskox intrusion, and that such mixing episodes may initiate precipitation of monomineralic layers.

In view of the tectonic regime of mid-ocean ridge volcanism, where magma is constantly being generated and replenished along a narrow zone of volcanic activity, it appears likely that new batches of magma will be intruded prior to the solidification of previous magma batches, and therefore that magma mixing is an inherently plausible process. Indeed, continual replenishment of magma chambers is required by two models for the generation of the oceanic crust recently proposed by Sleep (1975) and Dewey and Kidd (1977).

It is the purpose of this paper, therefore, to summarize results of a detailed, integrated petrological, geochemical and experimental study of basalts recovered during Legs 45 and 46 of the Deep Sea Drilling Project, and to present evidence that magma mixing has played an important role, along with crystal fractionation, in the evolution of these and other ocean-floor basalts, suggesting that magma mixing is indeed a fundamental process in the petrogenesis of ocean-floor basalts.

OUTLINE OF BASALT PETROLOGY AND CHEMISTRY

A total of 1019 m of volcanic basement was sampled at Sites 395 and 396 during Legs 45 and 46 of the Deep Sea Drilling Project, the majority of which consisted of basaltic pillow lavas, thin flows, and occasional more massive cooling units. The two sites are located in young oceanic crust (Miocene) about 300 km apart astride the Mid-Atlantic Ridge at about 22°N, to the south of the Kane Fracture zone. The basement at Site 395 is inferred, on the basis of magnetic anomalies, to be about 7 m.y. old, whereas the inferred age at Site 396 is about 10 m.y. The basaltic basement at these sites is thought to have been generated at different times along roughly the same segment of the ridge crest, and subsequently separated by sea floor spreading.

Two holes were drilled at both sites, a shallow, single-bit exploratory hole, followed by a deeper, multiple re-entry hole. At Site 395, 668 m of basaltic basement was penetrated, and 351 m at Site 396.

The following section summarizes the results of Dungan et al. (1978a), Rhodes et al. (1978) and Dungan et al. (1978b) on the petrology, geochemistry and melting experiments on these basalts. The major basalt types and their stratigraphic relationships are shown in Fig. 1. At both sites, two broad basalt types, differing fundamentally in both petrography and composition are prevalent. These are, respectively, the aphyric and phyric basalts. The

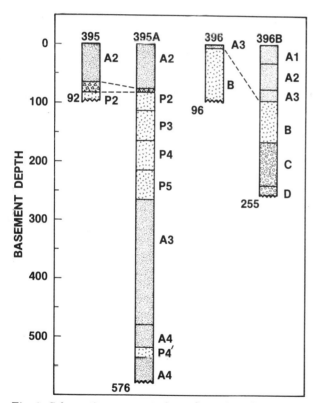

Fig. 1. Schematic representation of stratigraphic relations at Sites 395 and 396. Depth of penetration is shown from top of volcanic basement. The dense stipple and open symbols designate the aphyric and phyric basalt types, respectively. References in the text to individual units follow the numbering scheme shown here.

TABLE I

Average compositions of basaltic units from D.S.D.P. sites 395 and 396 (Data from Rhodes et al., 1978; Dungan et al., 1978)

	Site 395				
	Aphyric basalts			Phyric basalts	
	A2	A3	A4	P2	P3
Major elements (wt%)					
SiO_2	49.86	49.28	48.07	49.16	49.27
TiO_2	1.61	1.71	1.58	1.31	1.07
Al_2O_3	14.90	15.07	14.98	17.98	17.54
FeO^*	10.81	9.84	10.01	8.11	7.85
MnO	0.19	0.18	0.18	0.14	0.14
MgO	8.27	7.48	8.36	6.75	7.27
CaO	10.28	11.03	10.71	11.68	12.50
Na_2O	2.89	2.85	2.82	2.78	2.40
K_2O	0.15	0.23	0.16	0.10	0.13
P_2O_5	0.13	0.16	0.13	0.11	0.09
S	0.11	0.06	0.04	0.09	0.02
Mg'-value	0.602	0.601	0.623	0.622	0.647
CaO/Al_2O_3	0.69	0.73	0.71	0.65	0.71
Trace elements (ppm)					
Rb	1.4	2.3	1.6	0.9	1.9
Sr	116	130	125	154	114
Y	34.9	36.5	34.4	27.7	23.0
Zr	105	119	108	88	63
Nb	1.9	2.2	2.1	1.5	1.2
La	3.05	3.94	3.30	2.70	2.05
Ce	10.1	12.2	10.4	8.9	6.7
Sm	3.55	4.20	3.63	2.99	2.42
Eu	1.33	1.44	1.35	1.11	0.92
Tb	0.94	1.04	0.95	0.73	0.64
Yb	3.5	3.8	3.4	2.7	2.4
Lu	0.51	0.57	0.50	0.41	0.36
Hf	2.9	3.2	3.1	2.3	1.9
Sc	36.9	38.4	36.7	30.4	32.5
Cr	295	280	330	262	379
Ni	150	98	150	90	120
La/Sm	0.86	0.94	0.91	0.90	0.85
Zr/Y	3.0	3.3	3.1	3.2	2.7

FeO^* total iron expressed as FeO: Mg'-value = mol. prop. $Mg/(Mg + Fe)$ after adjusting $Fe^{3+}/FeO^* = 0.1$.

aphyric basalts lack megascopic phenocrysts, but contain microphenocrysts of olivine and plagioclase, whereas the phyric basalts contain abundant, large phenocrysts of plagioclase, and olivine. Clinopyroxene is an occasional phenocryst phase at Site 395 only, whereas minor chromium spinel occurs in

		Site 396				
		Aphyric basalts			Phyric basalts	
P4	P5	A1	A2	A3	B	C
49.22	49.13	49.28	49.65	49.11	48.70	49.08
1.16	1.14	1.40	1.52	1.61	1.08	1.51
17.77	18.08	15.44	15.24	15.11	17.31	16.15
7.71	7.74	9.16	9.32	9.75	7.84	9.04
0.15	0.14	0.18	0.19	0.19	0.17	0.18
7.20	7.03	7.86	7.63	7.52	7.40	7.19
12.22	12.07	11.49	11.11	10.64	11.92	11.33
2.70	2.75	2.68	2.78	2.95	2.59	2.83
0.15	0.16	0.22	0.20	0.20	0.16	0.24
0.11	0.10	0.11	0.12	0.14	0.08	0.13
0.01	0.02	0.07	0.08	0.09	0.05	0.06
0.649	0.643	0.629	0.615	0.605	0.651	0.611
0.69	0.67	0.74	0.73	0.70	0.69	0.70
2.0	2.5	1.7	1.6	1.9	1.9	2.0
139	154	119	127	137	123	158
24.8	24.5	30.6	32.4	34.8	23	32.5
74	75	93	103	116	67	109
1.2	1.1	1.8	2.0	1.9	1.1	3.5
2.37	2.48	2.99	3.23	3.73	2.24	4.12
7.7	8.1	10.2	10.6	11.9	7.6	12.5
2.75	2.82	3.34	3.52	3.99	2.61	3.63
1.05	1.06	1.24	1.26	1.43	1.00	1.27
0.76	0.67	0.87	0.92	0.97	0.70	0.92
2.6	2.7	3.2	3.4	3.7	2.5	3.4
0.40	0.41	0.50	0.52	0.58	0.40	0.52
2.3	2.2	2.8	2.7	3.1	2.0	2.8
33.7	32.3	38.5	38.8	39.2	33.2	37.1
340	310	330	305	295	335	320
140	140	125	105	120	115	115
0.86	0.88	0.90	0.92	0.93	0.86	1.13
3.0	3.1	3.0	3.2	3.3	2.9	3.35

the phyric rocks at both sites. Averages for the major basalt units are given in Table I.

At site 395, seven compositionally distinct magmatic units are recognized within the two holes, and there is good stratigraphic correlation between

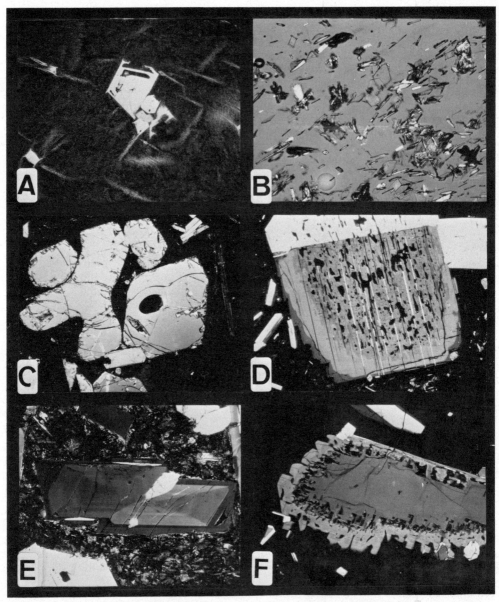

Fig. 2. A. Olivine microphenocrysts in aphyric basalt (Unit *A2—395A*). This texture is characteristic of basalts in which olivine occurs alone on the liquidus. B. Plagioclase and olivine microphenocrysts in the glassy rind of a multiply saturated basalt (Unit *A2—396B*). Width of field for A, B, E and F = 1.4 mm. C. Anhedral olivine phenocryst (Fo_{89}) in a phyric basalt (*395A*). Unlike many of the olivine phenocrysts that do have euhedral morphologies, this crystal is representative of those which exhibit evidence of resorption. D. Complexly zoned plagioclase phenocryst in a phyric basalt (*395A*). The core of this grain is An_{75-78} and is riddled with melt inclusions produced by resorption. The core is mantled by an inclusion free intermediate zone that is more calcic (An_{85}). The outer rim is normally zoned from An_{73} to An_{68}. Width of field for C and D = 2.8 mm. E. Typical calcic plagioclase phenocryst in a phyric basalt (*395A*). Note the sharp hiatus between the core (An_{86-83}) and the rim (An_{73-65}). This pattern is characteristic of the majority of the plagioclase phenocrysts in these rocks. F. Reverse zoning in a plagioclase phenocryst in a phyric basalt (*395A*). A core of An_{68} is mantled by a rim of An_{93-68}. Note the vermicular resorption zone between the two.

them. The majority of the basalts are aphyric (60%), with a thick sequence of phyric basalts sandwiched between the upper and lower aphyric units. Units A3 and A4 comprise rapidly chilled pillow basalts containing skeletal microphenocrysts of olivine (Fig. 2A). The other aphyric unit (A3) is more evolved and contains microphenocrysts of both olivine and plagioclase (Fig. 2B). The phyric basalts (P2—P5) contain large phenocrysts of plagioclase (10—30%), olivine (<10%), and occasional clinopyroxene. Compositional variation within the phyric basalt units is due in part to the discrete, chemically-distinct eruptive units and in part to the heterogeneous distribution of phenocryst phases. At Site 396 there are six distinct compositional units: a 95 m sequence of three aphyric basalts (A1—A3) that are chemically similar, but are progressively enriched in magmaphile elements with depth, overlays a thick sequence of phyric (B), and sparsely phyric basalts (C—D). The aphyric basalts contain both olivine and plagioclase microphenocrysts, and are similar to the evolved A3 basalt type at Site 395. The phyric basalts (B) are also similar to those at Site 395, containing abundant plagioclase phenocrysts with lesser amounts of olivine and minor chromian spinel, but unlike the Site 395 phyric basalts, clinopyroxene phenocrysts are not observed. The lower part of the stratigraphic column (C—D) is composed of clastic breccias and hyaloclastites, with fragments of sparsely-phyric and phyric basalt within a basaltic sand and gravel matrix. At both sites, the individual basaltic units are chemically homogeneous, with sharp compositional breaks between them. We attribute these features to discrete magma batches that attained their present compositions by differentiation within shallow magma chambers prior to episodic eruption. Differentiation is used here in its broadest sense to include any process that may modify a magma composition.

Figure 3 illustrates the normative olivine—plagioclase—pyroxene relationships for these basalts relative to the inferred olivine—plagioclase cotectic of Shido et al. (1971). The aphyric basalts from both sites are close to liquid compositions and plot in the olivine tholeiite field of the diagram, along a band sub-parallel to the inferred cotectic. Since the more evolved of the aphyric basalts contain both olivine and plagioclase microphenocrysts, the olivine—plagioclase cotectic must lie at, or close to, these basalt compositions. One atmosphere, controlled oxygen fugacity melting experiments on typical examples of these basalt units confirm this assumption, since both phases crystallize within a few degrees of each other. It is important to note that the "less-evolved" of the aphyric basalts, containing only micropheno-crysts of olivine, are not displaced into the olivine field along an olivine control line, but also fall close to the inferred cotectic. This is somewhat surprising since experimental results support the petrographic inference that olivine is the first phase to crystallize, and show that it is joined by plagioclase about 25°C below the liquidus. We believe this to be an important clue, indicating that processes other than crystal fractionation have played an important role in the genesis of these basalts. In contrast to the aphyric basalts, the phyric basalts from both sites are widely scattered in the plagioclase tholeiite

258

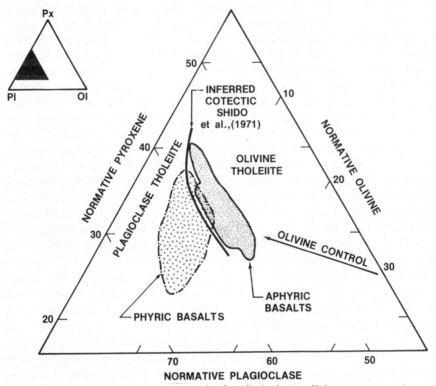

Fig. 3. Plot of normative components in the plagioclase—olivine—pyroxene ternary.

field (Fig. 3). There are no apparent systematic relationships between the various phyric basalt types, with the exception that the less aluminous, sparsely-phyric variants from each type plot closer to the inferred olivine—plagioclase cotectic. Glassy selvages from phyric basalt pillow margins plot within the field of aphyric basalts. Several lines of evidence indicate that the phyric basalts do not reflect liquid compositions, and that their varied compositions result from incorporation of varying amounts of plagioclase and olivine phenocrysts into melts that were initially at the olivine—plagioclase cotectic and not substantially different from the aphyric basalts. These lines of evidence are:

(a) The compilation of Melson et al. (1976) of over 600 basaltic glass analyses from spreading ocean ridges fails to identify highly aluminous basaltic liquids that clearly plot within the plagioclase tholeiite field.

(b) First-formed plagioclase and olivine crystals in controlled oxygen fugacity melting experiments on phyric basalts are, respectively, less anorthitic and less forsteritic than the observed phenocryst cores, an indication of plagioclase and olivine accumulation.

(c) Compositional variation within and between the various phyric basalt groups broadly follow trends dominated by plagioclase control.

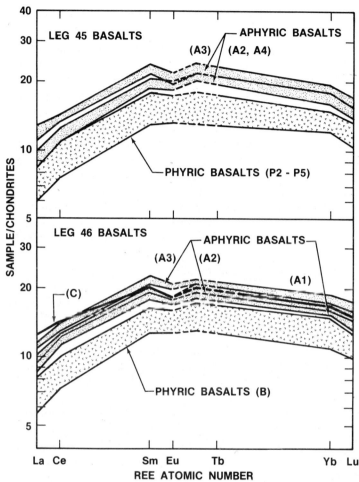

Fig. 4. Summary plot of chondrite-normalized REE-contents in both types of basalts from both sites.

The Site 395 and 396 basalts have the overall compositional and petrographic characteristics typical of mid-ocean ridge tholeiites (e.g., Engel et al., 1965; Melson et al., 1976). Mg'-values, [mol. $Mg'/(Mg + Fe)$], in these rocks are variable, ranging from 0.57 to 0.67 (Fig. 4), which is within the range common for mid-ocean ridge basalts, but is lower than the values found in the most "primitive" basalts identified to date (Frey et al., 1974; Bryan and Moore, 1977; Rhodes and Dungan, 1977). These values, together with low Ni concentrations, moderately high magmaphile element abundances, and the presence of multiple phenocryst phases in most of the basalt types, are indicative of the evolved nature of these basalts. The minor and trace element ratios are similarly within the range of typical, but evolved, mid-ocean ridge tholeiites (Kay et al., 1970; Schilling, 1971; Pearce and Cann, 1973;

Bryan et al., 1976). This is illustrated by the chondrite-normalized rare earth patterns (Fig. 4), which display the light rare-earth depletion characteristic of "normal" or type I ocean ridge basalts (Bryan et al., 1976). In keeping with their evolved nature, the aphyric basalts possess a small but significant europium anomaly. This is particularly pronounced in the plagioclase and olivine saturated cotectic basalts from both sites, but is also evident in the olivine saturated aphyric basalts from Site 395 ($A2$, $A4$). Since this observation is critical to arguments developed later that these basalts have undergone a complex history involving magma mixing, we have carefully analyzed a representative sample of the $A2$ basalt type from Site 395 by both instrumental neutron activation and isotope dilution analysis, confirming that the europium anomally, although small, is indeed significant. The phyric basalts are notably lower in rare-earths and other magmaphile elements relative to the aphyric basalts. This results partly from dilution due to accumulation of plagioclase and olivine phenocrysts, and partly from lower initial concentrations in the host melt.

An important feature of all the basalts from both sites is the overall similarity of the rare-earth abundance patterns (Fig. 4), and of the magmaphile element ratios. Since the characteristic rare-earth patterns and many magmaphile element ratios are not substantially changed by fractional crystallization (Kay et al., 1970; Schilling, 1971; Frey et al., 1974; Schilling, 1975), we suggest that these ratios reflect the characteristics of the mantle source, and that all of these basalts were derived from an essentially homogeneous source similarly depleted in magmaphile elements. The implied homogeneity of this source, both spatially and temporally, contrasts with the variable and complex mantle sources, or processes, necessary to account for the range in magmaphile element ratios observed in the Leg 37 and FAMOUS area basalts (Blanchard et al., 1976; Langmuir et al., 1977).

EVIDENCE SUPPORTING MAGMA MIXING

Recent studies of ocean floor basalts have benefited from improved spatial and temporal control achieved either by drilling into the oceanic basement, or by detailed sampling along segments of a mid-ocean ridge (Initial Reports of the Deep Sea Drilling Project, Vol. 34, 37, 45, 46; Hekinian et al., 1976; Bryan and Moore, 1977). A common problem encountered by these studies is the inadequacy of crystal fractionation models to account satisfactorily for the derivation of spatially associated basalts from a common parental magma (e.g., Rhodes et al., 1976; Blanchard et al., 1976; Bryan and Moore, 1977; Bryan, 1979). Although these models, by using observed phenocryst phases, may provide satisfactory solutions for most of the major elements, they frequently fail to match the higher TiO_2 or magmaphile element abundances in the derivative basalts. This discrepancy can be partially overcome by resorting to fractionation dominated by pyroxene, even though it may be absent or a minor component relative to plagioclase and olivine, in the phenocryst assemblage (Clague and Bunch, 1976; Thompson et al., 1976; Bryan, 1979). The other alternative, that each distinct basaltic unit is independently derived from a specific parental magma, and is unrelated to associated basaltic units (Rhodes et al., 1976; Blanchard et al., 1976), is hardly more tenable. The conceptual magma mixing model of O'Hara (1977) offers a solution to this dilemma. Repeated mixing of evolved and more primitive basaltic magmas provides a mechanism for increasing magmaphile element abundances, simultaneously buffering the major element constituents of the magma. Continued or episodic replenishment of small magma chambers beneath ridge crests is also an inherent feature of two recent geophysical models for the development of the oceanic crust (Sleep, 1975; Dewey and Kidd, 1977). It is the purpose of this section of the paper, therefore, to present specific evidence for magma mixing resulting from our studies of the Leg 45 and 46 basalts. Three major lines of evidence are considered, including (a) phenocryst assemblages; (b) melt inclusions; and (c) bulk chemistry.

PHENOCRYST ASSEMBLAGES

Phenocryst assemblages that are clearly not in equilibrium with their host liquids are a common feature of andesites and other intermediate volcanic rocks. Mixing of basaltic and more acidic magmas has been cited as a possible mode of origin for such rocks (e.g., Kuno, 1936; Larsen et al., 1938; Turner and Verhoogen, 1960; Wager et al., 1965; Bowman et al., 1973; Anderson, 1976). Similarly, ocean floor basalts commonly contain large phenocrysts of both olivine and plagioclase that are, respectively, too forsteritic and too anorthitic to be in equilibrium with a melt corresponding to the host rock composition (e.g., Muir and Tilley, 1964; Muir and Tilley, 1966; Melson et al., 1968; Melson and Thompson, 1971; Hekinian et al., 1976; Donaldson et al., 1976; Donaldson and Brown, 1977). In these studies, such phenocrysts have been variously interpreted as (a) accidental xenocrysts incorporated from the mantle or oceanic crust; (b) cognate phenocrysts, in the sense of relict crystals formed at an earlier stage in the closed system fractionation of a magma body; and (c) phenocrysts crystallizing in a primitive basaltic melt which was subsequently mixed with another magma at a different stage of differentiation. The olivine and plagioclase phenocrysts from Sites 395 and 396 are clearly not xenocrysts since they lack strain lamellae, and some contain abundant melt inclusions (Fig. 2C, D). Although some are resorbed and complexly zoned, attesting to a multi-stage crystallization history, others are idiomorphic (Fig. 2C–F).

Fig. 5 is a plot of observed olivine compositions versus compositions calculated from whole-rock data using iron—magnesium partitioning relationships between olivine and liquid (Roeder and Emslie, 1970) and a K_D of 0.27. The partition coefficient of 0.27 was determined empirically using glass-microphenocryst data from samples from Sites 395 and 396. Melting experiments on selected samples yielded slightly higher K_D values ranging

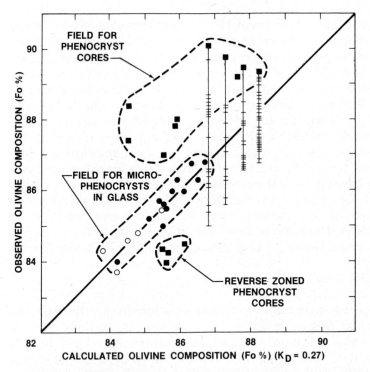

Fig. 5. Olivine-liquid relations in aphyric and phyric basalts. Open and closed circles represent coexisting microphenocrysts and residual glasses in aphyric and phyric basalts respectively. Filled square symbols above the 1 : 1 line are the maximum Fo-contents of phenocryst cores in several phyric basalts from both sites 395 and 396. The four vertical lines with horizontal tick marks are the bulk compositions of four rocks from a single phyric unit at 396. The tick marks are the compositions of olivine phenocrysts and microphenocrysts.

between 0.27 and 0.30. Bryan and Moore (1977) similarly report K_D values of about 0.28 for natural olivine-glass pairs from the FAMOUS area. In contrast to the close coherence between observed and predicted compositions for olivine microphenocrysts and phenocryst rims, the cores of the large phenocrysts are too forsteritic to have been in equilibrium with melts corresponding to the host rock compositions. Applying the K_D-value of 0.27, found empirically for microphenocryst-glass pairs, it follows that these phenocrysts (Fo_{85-90}) must at one time have been in equilibrium with more "primitive" melts with Mg'-values of about 0.70—0.72. Such values compare closely with those of the most "primitive" ocean floor basalts identified to date (Frey et al., 1974; Bryan and Moore, 1977; Rhodes and Dungan, 1977), and are thought to be characteristic of primary magmas derived directly from the mantle. Since these forsteritic olivine phenocrysts have been incorporated into the phyric basalt magma, it follows that the Mg'-value for the melt must have been lower than the present whole-rock Mg' values. Addition of only 5% Fo_{90} will change an evolved basalt with an Mg'-value of 0.56 to one with an Mg'-value of 0.64, similar to many of the phyric basalts. Evidence for the existence of such evolved precursors to the phyric basalts is to be found in occasional reverse zoned phenocrysts with cores of Fo_{82-83} and rims of Fo_{84-86} associated in the same sample with normally zoned phenocrysts.

A similar plot for the plagioclase data is shown in Fig. 6. Observed compositions are plotted against compositions calculated from the bulk rock data using the expression developed by Drake (1976) for plagioclase-liquid equilibria, and crystallization temperatures extrapolated from melting experiments on typical samples. As with the olivine data, there is close correspondence with the observed and calculated compositions for microphenocrysts in aphyric basalts, for rims of phenocrysts in contact with residual glass, and for plagioclase-melt relationships in melting experiments. On the other hand, the cores of many of the large phenocrysts are too anorthitic (An_{83-86}) to be in equilibrium with melts appropriate to the whole-rock composition. We infer that, like the olivine phenocrysts, they crystallized from more "primitive", calcic melts prior to incorporation in the phyric basalt magmas.

Complex zoning patterns and resorption features are ubiquitous character-

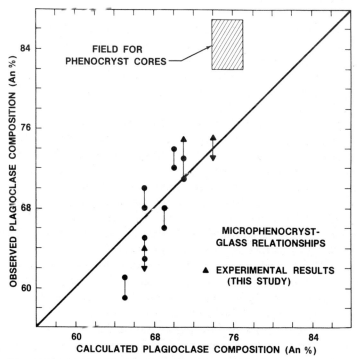

Fig. 6. Plagioclase-liquid relations in aphyric and phyric basalts. The equilibrium compositions has been calculated by the method of Drake (1976).

istics of these plagioclase phenocrysts. The most common variety (Fig. 2E) have a broad, unzoned core of An_{83-86}, separated by a compositional hiatus from the normally zoned rim of An_{73-68}. The rim compositions compare closely in range with the compositions of the smaller phenocrysts and with values calculated from whole-rock data. Other, less common, large phenocrysts are reverse zoned, some with resorbed cores of An_{78-75}, surrounded by a more calcic rim of An_{85-82}, which in turn is surrounded by a normal rim of An_{73-68} (Fig. 2D), and others with resorbed cores of An_{68} mantled by a more calcic, normally zoned rim of An_{73-68} (Fig. 2F).

The presence of both plagioclase and olivine phenocrysts which have initially crystallized from melts that were more "primitive" than their host rocks, together with complex and reversed zoning patterns, attests to a mixed parentage for these phenocrysts. In particular, the sharp breaks in the zoning patterns, accompanied by distinct compositional hiatuses and resorbed cores is indicative of episodic events, rather than continuous evolution of a single magmatic body. We interpret these disequilibrium assemblages and complex zoning patterns as evidence for injection of "primitive" magmas and their associated phenocrysts into fractionating magma bodies containing more evolved basalts.

COMPOSITION OF MELT INCLUSIONS

It is evident from the preceding discussion that many of the olivine and plagioclase phenocrysts must have crystallized in a more "primitive" melt prior to incorporation in the phyric basalt magmas. The high forsterite (Fo_{85-90}) and anorthite (An_{83-86}) content of the cores implies that this "primitive" magma must have high Mg'-values (0.70—0.72) and low $Na_2O/$ CaO ratios. Melt inclusions within both olivine and plagioclase phenocrysts (Fig. 2C, D, F) provide additional insight into the composition of these "primitive" magmas, and lend further support to the magma-mixing hypothesis (argued more fully in a follow-up paper: Dungan and Rhodes, 1978).

Fig. 7 illustrates the relationships between Mg'-value and TiO_2 content of melt inclusions in phenocrysts of olivine and plagioclase in phyric basalts from both sites. The compositions of these inclusions are highly variable and will have been modified by a variety of processes, including entrapment of melt at various stages of magmatic evolution, and reaction with the enclosing host crystal. Nonetheless, several useful parameters can be deduced concerning the most "primitive" parental magmas from which these phenocrysts crystallized. Reaction of entrapped melt with plagioclase will not alter the Mg'-value, whereas it will be rapidly reduced by small amounts of olivine crystallization. Consequently, the highest Mg'-values recorded for these melts, 0.68—0.70, place a lower limit on the Mg'-value of the parental magma. Another important characteristic shown by Fig. 7 is the low TiO_2 content of many of the inclusions relative to the host rocks. Since both olivine and plagioclase crystallization will increase the TiO_2 content of the melt, it

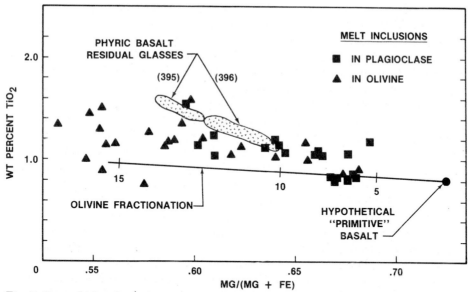

Fig. 7. Compositional relations of glass melt inclusions and residual glasses in the phyric basalts, Sites 395 and 396.

follows that the lowest value recorded (0.77) is an upper limit for the "primitive" magma. Since crystallization of all phenocryst phases increases the TiO_2 and K_2O content of the melt, it is possible to identify specific melt inclusions that have undergone minimal amounts of fractionation from the ratio of their TiO_2 and K_2O contents to the lowest observed values (Anderson and Greenland, 1969). Therefore, since the least fractionated melts should have an Mg'-value greater than 0.70, it is possible to allow for olivine and plagioclase fractionation and to arrive at an estimate of the "primitive" melt composition. The following values have been deduced on this basis adopting an arbitrary Mg'-value of 0.70: SiO_2 = 49.8, TiO_2 = 0.68, Al_2O_3 = 15.0, FeO = 8.4, MgO = 10.0, CaO = 13.4, Na_2O = 1.7, K_2O = 0.03. Further refinement may slightly modify these values, but the basic characteristics will remain the same. A melt with these characteristics should crystallize olivine of Fo_{89-91} (using K_D = 0.27) and plagioclase of about An_{84} (Drake, 1976), values which, within experimental error, are close to the observed olivine (Fo_{85-90}) and plagioclase (An_{83-86}) phenocryst cores. Consequently, such a liquid is a suitable parent for the most refractory of the phenocryst phases. It is important to note that the composition of this melt shares many similarities with the most "primitive" basaltic liquids identified to date from the ocean floor (Frey et al., 1974; Bryan and Moore, 1977; Rhodes and Dungan, 1977; Bryan, 1978), and with melt inclusions from Bouvet Island (Watson, 1976) and Leg 37 DSDP samples (Donaldson and Brown, 1977). All are typified by high Mg'-values (0.70—0.72), MgO contents of about 10%, high

CaO (>12%) and CaO/Al$_2$O$_3$ (>0.81), and low abundances of TiO$_2$ (0.7–0.9%) and Na$_2$O (<2.0%), and are good candidates for primary, mantle-derived melts.

Anderson (1976) has argued that evidence of magma mixing is obtained if the composition of an identified liquid lies outside the range defined by the whole-rock on the one-hand and residual glass on the other. The "primitive" melt inclusions, characterized above, fall well beyond the compositional range for several components for the phyric basalts and their residual glassy selvages. In particular they differ in Mg′-values, TiO$_2$ content and CaO/Al$_2$O$_3$ ratios. Anderson's argument is particularly strong if the melt inclusion is more evolved than either the rock or residual glass. If it is more primitive, as is the case here, one might argue, as an alternative to the mixing hypothesis, that the melt inclusions represent parental magma entrapped in early pheno-crysts, and that these phenocrysts persisted as relicts, out of equilibrium with surrounding magma, as it fractionated within a closed system to more evolved compositions. As noted earlier, the sharp discontinuities and the complexities in phenocryst zoning patterns do not support such a model. Furthermore, since the "primitive" parental melt composition is character-ized by a high CaO/Al$_2$O$_3$ ratio (0.89), it would be necessary for the fractio-nation process to lower this ratio to that of the derivative basalts (0.67–0.80). Only clinopyroxene fractionation is capable of effecting this reduc-tion (Clague and Bunch, 1976). It cannot be produced by any combination of olivine and plagioclase fractionation, olivine will have no effect on the ratio, and plagioclase fractionation will cause it to increase. Since plagioclase and olivine are the only observed phenocryst phases in the Site 396 basalts, and clinopyroxene phenocrysts are sparse and erratically distributed in the Site 395 basalts, we conclude that the fractionation alternative is unlikely. The reduction in CaO/Al$_2$O$_3$ ratios can be brought about by mixing the prim-itive high CaO/Al$_2$O$_3$ magma and its attendant phenocrysts with an evolved basalt that is low in CaO/Al$_2$O$_3$ as a consequence of prior extensive pyroxene fractionation. The resulting mixed magma would no longer have pyroxene on the liquidus and would crystallize only plagioclase and olivine.

BASALT CHEMISTRY

Before evaluating the chemical evidence supporting magma mixing, it is necessary to consider briefly the nature of the primary, mantle-derived melts from which these basalts might be derived. Fig. 8 shows the TiO$_2$ content versus Mg′-value for the Site 395 and 396 basalts relative to the composi-tional field over 600 abyssal volcanic glasses (Melson et al., 1976).

These data, supplemented by other recent glass analyses (Bryan and Moore, 1977) provide the best information currently available for ocean-floor basaltic liquid compositions. It is important to note that the most mafic glasses observed have Mg′-values of around 0.70, and that none exceed 0.72. Such values are compatible with the most forsteritic olivine cores

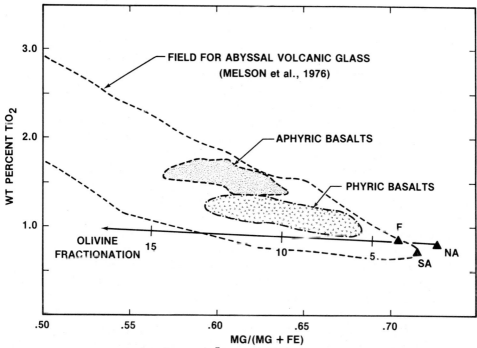

Fig. 8. Major element relations among Legs 45 and 46 as exemplified by TiO$_2$ vs Mg/ (Mg + Fe). Shown for comparison are the field of MORB glasses compiled by Melson et al. (1976) and three rock compositions that are believed to approximate primitive unfractionated MORB (*F* = FAMOUS area, Hekinian et al. (1976); *SA* = South Atlantic, Frey et al. (1974) and *NA* = 45°N on MAR, J.M. Rhodes, unpublished data).

(Fo$_{90}$) occurring in the phyric basalts, and in ocean floor basalts in general, and with the "primitive" melt inclusions discussed in the preceding section. They also correspond closely with Mg′-values predicted for primary magmas derived by partial melting of model mantle compositions (Green, 1970; O'Hara et al., 1975). Fig. 8 also shows that the TiO$_2$ content of these glasses has a limited range (0.65—0.90%) at these high Mg′-values. Similarly, it can be shown that the contents of other components have a restricted range, and that it therefore appears possible to place relatively tight constraints on such "primitive" ocean basalt magma compositions (Rhodes and Dungan, 1977). Those include, in summary, high Mg′-values (0.70—0.72) and low TiO$_2$ abundances (0.65—0.90), high CaO contents (12—13.5%) and CaO/Al$_2$O$_3$ ratios (>0.81) and low Na$_2$O abundances (<2.0%). These values correspond closely with the characteristics deduced earlier from melt inclusion data and from the anorthitic and forsteritic phenocryst cores in the Site 395 and 396 phyric samples. Basalts with these compositions have been identified at several localities along the Mid-Atlantic Ridge (Frey et al., 1974; Bougault and Hekinian, 1974; Bryan and Moore, 1977; J.M. Rhodes et al., unpublished data), typical examples of which are indicated in Fig. 8.

The evidence advanced thus far in support of magma mixing has emphasized the importance of the phyric basalts. Chemical arguments favour the aphyric basalts, since they are close to liquid compositions, and the microphenocryst assemblages and melting experiments place strong constraints on their magmatic history. The aphyric units $A2$ and $A4$ from Site 395 are particularly important in this respect. The only liquidus phase is olivine, which according to our melting experiments is not joined by plagioclase and clinopyroxene until 25° and 50°C below the liquidus, respectively. Consequently, only olivine can have been involved in the low pressure fractionation history of these basalts. However, several serious problems are encountered if one attempts to derive these basalts solely by olivine fractionation from more "primitive" parental basalts, of the type discussed above. These are as follows:

(a) Fractionation of olivine from "primitive" melts with Mg'-values between 0.70 and 0.72 to produce basalts with Mg'-values of about 0.60 similar to the $A2$ and $A4$ aphyric basalts, requires the removal of about 12% olivine (Fo_{90-86}). The effect of this would be to reduce drastically the MgO content from 10 to 5% maintaining the FeO content between 7 and 8%. Similarly, since the Ni content of "primitive" ocean floor basalt melts is between 250 and 300 ppm (Kay et al., 1970; Sato, 1977), it too, would be drastically reduced by 12% olivine fractionation to only about 60 ppm. As both the $A2$ and $A4$ aphyric basalts contain over 8% MgO, 10—11% FeO, and about 150 ppm Ni (Table I), simple olivine fractionation is clearly implausible.

(b) The TiO_2 content of these aphyric basalts ($\simeq 1.6\%$) is much too high for them to have been derived solely by olivine fractionation from the low TiO_2 "primitive" basalt melts. This is illustrated in Fig. 8 by an olivine fractionation trend from one such composition. It could be argued that the parental magmas for these basalts were in fact a good deal higher in TiO_2 than our proposed "primitive" parental compositions, yet if this is the case such "primitive" high TiO_2 basalts have not been identified as is clearly shown by Fig. 8.

(c) The Al_2O_3 and CaO contents of these two aphyric basalt units also pose problems for an olivine fractionation model. Both would be increased by 12% olivine fractionation from our preferred parental magma composition to values well above those found in the actual basalts (Table I). The CaO content is particularly problematic (10.3—10.7%), since it is too low to be derived from any realistic parental magma composition by olivine fractionation. The ratio CaO/Al_2O_3 introduces a further complication. Inspection of ocean floor basaltic glass data (Frey et al., 1974; Melson et al., 1976; Bryan and Moore, 1977; Bryan, 1979) indicates that there is a positive relationship between Mg'-values and CaO/Al_2O_3 ratios, and that basaltic liquids with Mg'-values greater than 0.67 have CaO/Al_2O_3 ratios above 0.75. As noted earlier, we infer ratios between 0.81 and 0.90 for "primitive" mantle-derived melts. Olivine fractionation on its own will not change this ratio, making it extremely difficult to account for CaO/Al_2O_3 ratios in the $A2$ and $A4$ aphyric basalts of about 0.70 solely by olivine fractionation from a less evolved composition. Only pyroxene fractionation can effect a reduction in the ratio, and this is why it figures prominently in ocean floor basalt fractionation models (as noted by Clague and Bunch, 1976). The dilemma in the case of these aphyric basalts is that olivine is clearly the liquidus phase and clinopyroxene does not crystallize until 50°C below the liquidus.

(d) We have drawn attention earlier to the small, but significant Eu anomalies in all of the aphyric basalt units. These anomalies are commonly ascribed to plagioclase fractionation (e.g., Kay et al., 1970), although between 20 to 30% plagioclase must crystallize to produce anomalies comparable to those in Fig. 4. This is an acceptable hypothesis for the more evolved of the aphyric basalts which have both plagioclase and olivine on the liquidus, but is unsatisfactory for Site 395 units $A2$ and $A4$, where olivine is the sole microphenocryst phase. What, then, is the explanation for these small, but significant anomalies? We have already ruled out experimental error by analyzing the same sample by instrumental neutron activation and mass spectrometric isotope dilution analyses. Three other explanations are possible:

(1) The Eu anomaly is characteristic of the source and is inherited by the basalts in an analogous manner to lunar basalts. We have no means to evaluate this possibility.

(2) The basalts are derived by partial melting of plagioclase lherzolite, with plagioclase remaining in the refractory residuum. Adopting the pyrolite model of Ringwood (1975) which will contain about 11% modal plagioclase at shallow depths, melts with Eu anomalies similar to those in the Site 395 basalts (Eu/Eu*-0.95) can be produced by about 15% partial melting. Larger amounts of melting will remove all of the plagioclase with the consequence that the resultant basalts will lack a Eu anomaly. Such a model is clearly plausible providing melting occurs at shallow depths within the stability field of plagioclase pyrolite, and providing that the extent of melting is about 15% or less. On the other hand, segregation of magma from a mobilized mantle diapir at such shallow depths (<30 kms) will tend to favour large amounts of melting (Ringwood, 1975; Green and Liebermann, 1976), with the consequence that it is unlikely that plagioclase will remain in the residue.

(3) Mixing of a "primitive" magma, with olivine on the liquidus, with an evolved, multiply-saturated magma that has undergone extensive plagioclase fractionation and therefore will have a Eu anomaly, will result in a mixed magma which inherits a smaller Eu anomaly, and yet retains olivine as the only liquidus phase. In view of the other evidence assembled for magma mixing this alternative appears the more plausible, appeals less to special pleading, and is in fact a necessary consequence of the mixing process. It may be that the Eu anomalies described in other basalts have originated in a similar manner, but that this process has escaped detection due to a lack of petrographic control, or to the presence of both olivine and plagioclase phenocryst.

All the problems of derivation of the Site 395 *A2* and *A4* aphyric basalts from a "primitive" magma discussed above have as a common denominator, an apparent necessity to fractionate phases other than olivine. Most require the participation of clinopyroxene, except the Eu anomaly which requires plagioclase. Yet petrographic evidence, corroborated by experimental studies, clearly indicates that olivine is the liquidus phase and the only component that can be involved in the fractionation process. These problems can be overcome if these aphyric basalts resulted from mixing of cogenetic magmas prior to eruption. Consider an evolved magma that is multiply-saturated with olivine, plagioclase and pyroxene, and which has undergone extensive crystallization. If this magma body is intruded by a more "primitive" magma with olivine as the sole liquidus phase, the resultant mixed magma will bear the compositional imprints of both pyroxene and plagioclase fractionation even though olivine may be the only liquidus phase. Furthermore, the TiO_2 and magmaphile element content of the mixed magma will be higher than could have been achieved solely by olivine fractionation.

The preceding dilemma is not specific to the Site 395 aphyric basalts; it applies also to the other basalts from both sites and to ocean-floor basalts in general. For example, although the basalts at each of the two sites are spatially associated, have probably been derived from the same mantle source, and presumably shared the same plumbing system, it is not feasible to relate them to a common parental magma by crystal fractionation processes using the observed phenocrysts. It is only when clinopyroxene is introduced as the major component along with plagioclase and olivine that mathematically satisfying solutions are obtained. As noted earlier, this is not an uncommon problem in ocean-floor basalt petrogenesis. It is frequently found that substantial pyroxene fractionation is required in order to relate spatially associated basalts, even though pyroxene may not be present, or is subordinate in the phenocryst assemblage (e.g., Thompson et al., 1976; Bryan, 1979). This results from the dual necessity of decreasing the CaO/Al_2O_3 ratio with increasing fractionation, and increasing the TiO_2 and magmaphile element content of the melt. Clinopyroxene fractionation in consort with olivine and plagioclase is a useful device for buffering the major element composition, while crystallizing a sufficient quantity of the melt to effect substantial increases in the magmaphile element content. In many instances, even extensive pyroxene fractionation is insufficient to account for the high TiO_2 and other magmaphile element abundances in the derivative basalts (e.g., Bryan et al., 1976; Bryan and Moore, 1977; Bryan, 1978). This is partly a consequence of the models used. TiO_2, K_2O etc. behave as magmaphile elements in these rocks. Consequently, their abundances will be exponentially related to fractionation, and will tend to be higher than indicated by the linear mixing calculations commonly used for estimating fractionation processes. We believe it is also in part, a consequence of magma mixing, since the mixing process will always result in a higher magmaphile element content for a magma than could have been achieved solely by crystal fractionation processes.

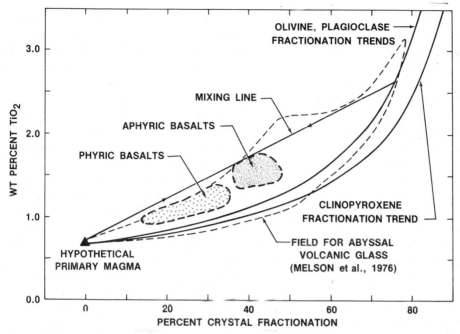

Fig. 9. Effects of crystal fractionation and magma mixing on the TiO$_2$ content of ocean floor basalts (see text for discussion).

This is illustrated in Fig. 9, where TiO$_2$ content is plotted against the amount of crystallization necessary for a hypothetical "primitive" magma (SiO$_2$ = 49.9; TiO$_2$ = 0.7; Al$_2$O$_3$ = 16.1; FeO* = 7.7; MgO = 10.1; CaO = 13.1; Na$_2$O = 1.9; K$_2$O = 0.09) to produce a derivative magma composition. The amount of crystal fractionation is estimated by linear least squares mixing calculations (Lawson and Hanson, 1974) using only the major mineral-forming oxides. TiO$_2$ and other minor elements (e.g., K$_2$O, P$_2$O$_5$) are omitted from these calculations because as magmaphile components their behavior will be exponential and not linear with the consequence that their inclusion in the mixing calculations would lead to overestimation of the extent of fractionation in an attempt to match the higher content of these components. Crystal fraction curves are shown for olivine, plagioclase and clinopyroxene, using TiO$_2$ distribution coefficients of Bougault and Hekinian (1974). In this diagram, magmas that are derived by crystal fractionation from a hypothetical "primitive" magma, such as the one shown, should fall close to the band defined by the fractionation curves. Magmas that result from mixing of any two magmas along this fractionation trend will plot above these curves, and will have higher TiO$_2$ concentrations than would be predicted solely from a crystal fractionation model. Fig. 9 shows that the phyric and aphyric basalts from both sites plot well above the fractionation curves, and can therefore be regarded as mixed magmas.

The simplest example of this process would be to mix the "primitive" primary magma with a magma that is highly fractionated (70—80%), such as the FeTi basalts commonly found in the Pacific (Clague and Bunch, 1976). This is shown schematically in Fig. 9. In reality, a single-step process such as this seems improbable. It is more likely to be a continuum of processes, bounded by the crystal fractionation curves on the one hand, and by the single-stage mixing of extreme magma compositions on the other. Repeated mixing of small volumes of magma, followed by fractionation prior to the next mixing episode, as envisioned by O'Hara (1977), will ultimately produce magma compositions, such as the aphyric basalts, that plot well above the fractionation curves. This argument is clearly model-dependent, hinging upon the parental magma composition chosen. In this case we have used our best estimate from published glass analyses (Frey et al., 1974; Melson et al., 1976; Bryan and Moore, 1977; Rhodes and Dungan, 1977), which corresponds closely with our estimates of the "primitive" parental magma for these basalts based on melt inclusion data and phenocryst compositions.

Also shown in this diagram is the field for abyssal volcanic basaltic glasses (Melson et al., 1976; Bryan and Moore, 1977). Several points deserve attention: (1) with decreasing fractionation the composition field for these glasses narrows and converges towards our hypothetical "primitive" melt composition, thus providing support for the composition used; (2) the lower boundary of the field closely mimics the two crystal fractionation curves, a necessary condition if this model is realistic; (3) a large number of basaltic glass analyses plot above the two curves, indicating either that they also are of mixed origin, or that their parental magmas were much higher in TiO_2 than the value used in this model. As noted earlier, there is no evidence for "primitive" liquid compositions with TiO_2 values greater than 0.9 percent, as would be required if these basalts have attained their compositions solely by crystal fractionation processes.

We conclude therefore that magma mixing is a common process in ocean-floor basalts in general, as well as in the basalts sampled at Sites 395 and 396.

DISCUSSION

In the preceding sections, we have argued from several lines of evidence including petrography, phenocryst compositions, melt inclusion data, and whole-rock major and trace element chemistry that mixing of cogenetic basaltic magmas was an important process in the origin of the basalts from Sites 395 and 396. Probably no one aspect of this evidence is sufficient in itself, but taken together there can be little doubt concerning the significance of the magma mixing process. We have also shown that these basalts are not unique, but typical of ocean-floor basalts in general, with which they share common problems of petrogenesis. Consequently, we believe that magma mixing is a fundamental process, which is intrinsically interactive with crystal fractionation in ocean-floor basalt petrogenesis.

We envision that mixing takes place episodically in small, shallow magma chambers immediately beneath the spreading ridge axis. "Primitive" magma, derived by partial melting of upwelling mantle, is repeatedly injected into the magma chamber and mixed with a more evolved, consanguineous basalt that has fractionated from a prior episode of injection and mixing. The injected magma may be a primary melt with olivine on the liquidus, or it may have fractionated somewhat during ascent and be saturated with both olivine and plagioclase. The compositional characteristics of this magma are deduced from melt inclusion data, refractory phenocryst assemblages, and the few "primitive" basalts that have been sampled to date. Compositional arguments indicate that the magma within the chamber is evolved, and multiply saturated: simultaneously crystallizing clinopyroxene, plagioclase and olivine. Extreme examples of this type would be the FeTi-basalts that are common along the Galapagos and Juan de Fuca ridges (Clague and Bunch, 1976). Both magmas, although at different stages of evolution, may well have been derived from the same source, and will therefore have comparable magmaphile element ratios and relative rare-earth element patterns. Temperature differences between the evolved magma and the more "primitive" incoming magma should be sufficiently large ($>50°$C) to initiate turbulent convection, thereby ensuring efficient, almost instantaneous mixing. Magmas that are displaced from the chamber at this stage of the cycle, in response to injection of a fresh batch of "primitive" magma, will tend to be the least evolved in the stratigraphic sequence, and will include refractory phenocrysts derived from the incoming "primitive" melt that are out of equilibrium within the resultant mixed melt composition. These are the phyric basalts. Why these magmas preferentially accumulate plagioclase relative to other phenocryst phases is not clear.

Following each episode of injection and mixing, there must be a quiescent period during which the mixed magma fractionates to more evolved compositions. Crystals formed during this period of the evolutionary cycle are presumably plated onto the roof of the magma chamber in the manner envisioned by Dewey and Kidd (1977). Passive tapping of the magma chamber during this stage of the evolutionary cycle, through fissures created in response to the spreading process, results in the eruption of the evolved, phenocryst-free, aphyric basalts.

The model outlined above has several implications that are of importance to ocean-floor basalt petrogenesis and volcanological processes occurring along mid-ocean ridges.

It explains why spatially associated magmas frequently cannot be inter-related by crystal fractionation processes. First, differences in residence time of the fractionating magma in the magma chamber prior to the next episode of injection and mixing will result in magmas that are evolved to varying degrees. Mixing of these magmas with the incoming "primitive" magma will produce mixed magmas that cannot be inter-related simply by crystal fractionation models. Second, the mixed magmas will always be higher in TiO_2 and other magmaphile elements than compositionally equivalent magmas that have evolved solely by crystal fractionation (Fig. 9). Third, the necessity to postulate a crystal fractionation model dominated by clinopyroxene, when pyroxene is rarely an important phenocryst phase, is readily explained. Pyroxene fractionation has in fact actually occurred but in the evolved magma prior to mixing with the more primitive one.

A further consequence of this magma mixing model is that the dominant products are moderately evolved basalts, buffered by the steady state mixing process. This means that the magma chamber is probably quite small. There is no necessity to postulate large fractionating magma batches in order to arrive at these moderately evolved compositions. This also explains why "primitive" basalts and highly evolved intermediate derivatives are rare amongst ocean floor basalts. Only when the cyclical process of magma injection, mixing, and fractionation ceases or magmatic activity shifts laterally, resulting in solidification of the magma body, will the probability of erupting "primitive" magmas or highly evolved derivatives be high. It has frequently been observed that basalts erupted along the East Pacific Rise tend to be more evolved than those erupted along the Mid-Atlantic Ridge (e.g., Melson et al., 1976). Seen in the context of our model, it is not that the East Pacific Rise basalts tend to be more evolved, but that they are in fact less mixed than basalts from the Atlantic. This readily understandable in view of the differences in spreading rate of the two ridge systems. Fast spreading rates will result in more frequent tapping of a magma chamber, through fracturing and fissure eruptions. Consequently, the incidence of fracturing relative to the frequency of injection and mixing of "primitive" magma should be greater along fast spreading ridges. This will favour a higher probability of sampling the magma chamber during the quiescent fractionating part of the injection—mixing—fractionation cycle along the East Pacific Rise.

CONCLUSIONS

We have presented evidence indicating that mixing of consanguineous evolved and "primitive" magmas and their attendant phenocrysts was an important process in the generation of the Site 395 and 396 basalts. We believe that magma mixing is not unique to this part of the Mid-Atlantic Ridge and that it is a fundamental process intrinsically interactive with crystal fractionation in ocean basalt petrogenesis.

ACKNOWLEDGEMENTS

We thank Henry Wiesmann for careful confirmation of our INAA rare earth determinations by isotope dilution. Lina Romero typed the manuscript. This paper was prepared in partial fulfillment of NASA contract NAS-9-15200 to Lockheed Electronics Company, Inc. The participation of

M.A.D. and P.E.L. was partially supported by the National Research Council while they were Research Associates at the NASA-Johnson Space Center, and (M.A.D.) by the Lunar and Planetary Institute which is operated by the Universities Space Research Association under contract No. NSR 09-051-001 with the National Aeronautics and Space Administration. This paper constitutes the Lunar and Planetary Institute Contribution No. 327.

REFERENCES

Anderson, A.T., 1976. Magma mixing: petrological process and volcanological tool. J. Volc. Geotherm. Res., 1: 3—33.

Anderson, A.T. and Greenland, L.P., 1969. Phosphorus fractionation diagram as a quantitative indicator of crystallization differentiation of basaltic liquids. Geochim. Cosmochim. Acta, 33: 493—505.

Blanchard, D.P., Rhodes, J.M., Dungan, M.A., Rodgers, K.V., Donaldson, C.H., Brannon, J.C., Jacobs, J.W. and Gibson, E.K., 1976. The chemistry and petrology of basalts from Leg 37 of the Deep Sea Drilling Project. J. Geophys. Res., 81: 4231—4246.

Bougault, H. and Hekinian, R., 1974. Rift valley in the Atlantic Ocean near 36°50'N: petrology and geochemistry of basaltic rocks. Earth Planet. Sci. Lett., 24: 249—261.

Bowman, H.R., Asaro, F. and Perlman, I., 1973. On the uniformity of compositions in obsidions and evidence for magmatic mixing. J. Geol., 81: 312—327.

Bryan, W.B., 1979. Regional variation and petrogenesis of basalt glasses from the FAMOUS area, Mid-Atlantic Ridge. J. Petrol. (in press).

Bryan, W.B. and Moore, J.G., 1977. Compositional variations of young basalts in the Mid-Atlantic Ridge rift valley near lat 36°49'N. Geol. Soc. Am. Bull., 88: 556—570.

Bryan, W.B., Thompson, G., Frey, F.A. and Dickey, J.S., 1976. Inferred settings and differentiation in basalts from the Deep Sea Drilling Project. J. Geophys. Res., 81: 4285—4304.

Clague, D.A. and Bunch, T.E., 1976. Formation of ferrobasalt at East Pacific mid-ocean spreading centers. J. Geophys. Res., 81: 4247—4256.

Dewey, J.F. and Kidd, W.S.F., 1977. Geometry of plate accretion. Geol. Soc. Am. Bull., 88: 960—968.

Donaldson, C.H. and Brown, R.W., 1977. Refractory megacrysts and magnesium-rich melt inclusions within spinel in oceanic tholeiites: indicators of magma mixing and parental magma composition. Earth Planet. Sci. Lett. 37: 81—89.

Donaldson, C.H., Brown, R.W. and Reid, A.M., 1976. Petrology and chemistry of basalts from the Nazca Plate, Part 1. Petrography and mineral chemistry. In: Initial Reports of the Deep Sea Drilling Project. Vol. 34. U.S. Government Printing Office, Washington, D.C., pp. 227—238.

Drake, M.J., 1976. Plagioclase-melt equilibria. Geochim. Cosmochim. Acta, 40: 457—466.

Dungan, M.A. and Rhodes, J.M., 1978. Residual glasses and melt inclusions in basalts from DSDP Legs 45 and 46: Evidence for magma mixing. Contrib. Mineral. Petrol., 67: 417—431.

Dungan, M.A., Long, P.E. and Rhodes, J.M., 1978a. The petrography, mineral chemistry and one-atmosphere phase relations of basalts from Site 395 — Leg 45 D.S.D.P. In: Initial Reports of the Deep Sea Drilling Project, Vol. 45. U.S. Government Printing Office, Washington, D.C., pp. 461—472.

Dungan, M.A., Rhodes, J.M., Long, P.E., Blanchard, D.P., Brannon, J.C. and Rodgers, K.V., 1978b. The petrology and geochemistry of basalts from Site 396 — Legs 45 and 46 of the Deep Sea Drilling Project. In: Initial Reports of the Deep Sea Drilling Project, V. 46. U.S. Government Printing Office, Washington, D.C., pp. 89—113.

Engel, A.E., Engel, C.G. and Havens, R.G., 1965. Chemical characteristics of oceanic basalts and the upper mantle. Geol. Soc. Am. Bull., 76: 719—734.

Frey, F.A., Bryan, W.B. and Thompson, G., 1974. Atlantic Ocean floor: Geochemistry and petrology of basalts from Legs 2 and 3 of the Deep Sea Drilling Project. J. Geophys. Res., 79: 5507—5527.

Green, D.H., 1970. The origin of basaltic and nephelenetic magmas. Trans. Leicester Lit. Philos. Soc., 64: 28—54.

Green, D.H. and Liebermann, R.C., 1976. Phase equilibria and elastic properties of a pyrolite model for the oceanic upper mantle. Tectonophysics, 32: 61—92.

Hekinian, R., Moore, J.G. and Bryan, W.B., 1976. Volcanic rocks and processes of the Mid-Atlantic Ridge rift valley near 36°49'N. Contrib. Mineral. Petrol., 58: 83—110.

Irvine, T.N., 1977. Origin of chromitite layers in the Muskox intrusion and other stratiform intrusions: a new interpretation. Geology, 5: 273—277.

Kay, R., Hubbard, N.J. and Gast, P.W., 1970. Chemical characteristics and origin of oceanic ridge volcanic rocks. J. Geophys. Res., 75: 1585—1613.

Kuno, H., 1936. Petrological notes on some pyroxene andesites from Hakone volcano, with special reference to some types with pigeonite phenocrysts. Jap. J. Geol. Geogr., 13: 107—140.

Langmuir, C.H., Bender, J.F., Bence, A.E. and Hanson, G.N., 1977. Petrogenesis of basalts from the FAMOUS area: Mid-Atlantic Ridge. Earth Planet. Sci. Lett., 36: 133—156.

Larsen, E.S., Irving, J. and Bonyer, F.A., 1938. Petrological results of a study of the minerals from the Tertiary volcanic rocks of the San Juan region, Colorado. Am. Mineral., 23: 227—257.

Lawson, C.L. and Hanson, R.J., 1974. Solving Least Squares Problems. Prentice Hall, New Jersey, 340 pp.

Melson, W.G. and Thompson, G., 1971. Petrology of a transform fault zone and adjacent ridge segments. Philos. Trans. R. Soc. London, Ser. A, 268: 423—441.

Melson, W.G., Thompson, G. and Van Andel, T.H., 1968. Volcanism and metamorphism in the Mid-Atlantic Ridge, 22°N latitude. J. Geophys. Res., 73: 5925—5941.

Melson, W.G., Vallier, T.L., Wright, T.L., Byerly, G. and Nelen, J., 1976. Chemical diversity of abyssal volcanic glass erupted along Pacific, Atlantic, and Indian ocean sea-floor spreading centers. In: G.H. Sutton, M.H. Manghani and R. Moberly (Editors), The Geophysics of the Pacific Ocean basin and its margin. American Geophysical Union, Washington, D.C., pp. 351—367.

Muir, I.D. and Tilley, C.E., 1964. Basalts from the northern part of the rift zone of the mid-Atlantic Ridge. J. Petrol., 5: 409—434.

Muir, I.D. and Tilley, C.E., 1966. Basalts from the northern part of the mid-Atlantic Ridge. II The Atlantis collection near 30°N. J. Petrol., 7: 193—201.

O'Hara, M.J., 1977. Geochemical evolution during fractional crystallisation of a periodically refilled magma chamber. Nature, 266: 503—507.

O'Hara, M.J., Saunders, M.J. and Mercy, E.L.P., 1975. Garnet-peridotite, primary ultrabasic magma and eclogite; interpretation of upper mantle processes in kimberlite. Phys. Chem. Earth, 9: 571—604.

Pearce, J.A. and Cann, J.R., 1973. Tectonic setting of basic volcanic rocks determined using trace element analyses. Earth Planet. Sci. Lett., 19: 290—300.

Rhodes, J.M. and Dungan, M.A., 1977. The nature of primary ocean-floor basalts. In: Papers Presented to the Second Inter-Team Meeting, Basaltic Volcanism Study Project, Lunar Science Institute, Houston, pp. 50—52.

Rhodes, J.M., Blanchard, D.P., Rodgers, K.V., Jacobs, J.W. and Brannon, J.C., 1976. Petrology and chemistry of basalts from the Nazca plate. Part 2. Major and trace element chemistry. In: Initial Reports of the Deep Sea Drilling Project, V. 34. (U.S. Government Printing Office) Washington, D.C., pp. 239—244.

Rhodes, J.M., Blanchard, D.P., Dungan, M.A., Rodgers, K.V. and Brannon, J.C., 1978. Chemistry of basalts from Leg 45 of the Deep Sea Drilling Project. In: Initial Reports of the Deep Sea Drilling Project, V. 45. U.S. Government Printing Office, Washington, D.C., pp. 447—459.

Ringwood, A.E., 1975. Composition and Petrology of the Earth's Mantle. McGraw Hill, New York, 618 pp.

Roeder, P.L. and Emslie, P.F., 1970. Olivine-liquid equilibrium. Contrib. Mineral. Petrol., 29: 275—289.

Sato, H., 1977. Nickel content of basaltic magmas: identification of primary magmas and a measure of the degree of olivine fractionation. Lithos, 10: 113—120.

Schilling, J.-G., 1971. Sea-floor evolution: Rare-earth evidence. Philos. Trans. R. Soc. London, Ser. A., 268: 663—706.

Schilling, J.-G., 1975. Rare-earth variations across "normal segments" of the Reykjanes Ridge, 60°—53°N, Mid-Atlantic Ridge, 29°S, and East Pacific rise 2°—19°5, and evidence on the composition of the underlying low velocity layer. J. Geophys. Res., 80: 1459—1473.

Shido, F.A., Miyashiro, A. and Ewing, M., 1971. Crystallization of abyssal tholeiites. Contrib. Mineral. Petrol., 31: 251—266.

Sleep, N.H., 1975. Formation of oceanic crust: some thermal constraints. J. Geophys. Res., 80: 4037—4042.

Thompson, G., Bryan, W.B., Frey, F.A., Dickey, J.S. and Suen, C.J., 1976. Petrology and geochemistry of basalts from DSDP Leg 34, Nazca plate. In: Initial Reports of the Deep Sea Drilling Project, V. 34. Washington (U.S. Government Printing Office), pp. 215—226.

Turner, F.J. and Verhoogen, J., 1960. Igneous and Metamorphic Petrology. McGraw-Hill, New York, N.Y., 694 pp.

Wager, L.R., Vincent, E.A., Brown and Bell, J.D., 1965. Marscoite and related rocks of the Western Red Hills Complex, Isle of Skye. Philos. Trans. R. Soc. London, Ser. A, 257: 273—307.

Watson, E.B., 1976. Glass inclusions as samples of early magmatic liquid: determinative method and application to a South Atlantic basalt. J. Volcanol. Geotherm. Res., 1: 73—84.

Wright, T.L. and Fiske, R.S., 1971. Origin of the differentiated and hybrid lavas of Kilauea Volcano, Hawaii. J. Petrol., 12: 1—65.

NOTE ADDED IN PROOF

Present addresses:

J.M. Rhodes. *Department of Geology, University of Massachusetts, Amherst, MA 01003 (U.S.A.).*

M.A. Dungan. *Department of Geological Sciences, Southern Methodist University, Dallas, TX 75275 (U.S.A.).*

P.E. Long. *Rockwell International, Richland, WA 99352 (U.S.A.).*

Chapter 13

LATE STAGE DEVELOPMENT OF MATURE ATLANTIC-TYPE CONTINENTAL MARGINS*

H.J. NEUGEBAUER and T. SPOHN

Institute of Meteorology and Geophysics, J.W. Goethe University, Frankfurt/Main (West Germany)

SUMMARY

Elongated sedimentary basins are found along continental margins of the Atlantic-type. The evolution of these basins is seen to be related to continental break-up and the associated fault system, cooling of the spreading lithosphere, sedimentary loading, and metamorphic processes.

In this study we assume that the transformation of metastable phases of the lower crust beneath the continental margin will dominate the mature phase of the development of the margins. Further on this process is assumed to be involved in the possible development of lithospheric instabilities at continental margins of the Atlantic type.

By means of the finite element technique, assuming non-linear rheology, we calculated a number of models on the dynamics of passive continental margins. We studied in detail the effect of metastable phase transition, sedimentary loading, active faulting at the continental oceanic transition as well as horizontal loading of the lithosphere on the subsidence of Atlantic-type continental margins. These mechanisms may be responsible for the late development of sedimentary basins along passive continental margins. The numerical results support the idea that the late development of a passive continental margin will be dominated by loading at the continental rise.

INTRODUCTION

Atlantic continental margins are formed by a sequence of tectonic processes like doming, rifting, and sea-floor spreading (Dewey and Bird, 1970). This type of continental margin marks the transition between continental and oceanic lithosphere (Rabinowitz, 1974). In terms of plate-tectonic theory they do not form plate boundaries and so do not serve as concentrators of seismic activity (Oliver et al., 1974). Both tectonic and sedimentary processes will contribute to the development of such a passive margin. The continental margins along the Atlantic Ocean are mainly in a stage of development where a three-fold division of shelf, slope and rise are to be observed

*Originally published as: Neugebauer, H.J. and Spohn, T., 1978. Late stage development of mature Atlantic-type continental margins. In: M.N. Toksöz (Editor), Numerical Modeling in Geodynamics. Tectonophysics, 50: 275—305.

(Drake et al., 1959; Sheridan, 1974; Heezen, 1974). The development of these divisions seems to be interrelated.

Besides most parts of the Atlantic coast, continental margin structures of the Atlantic type will be found along the eastern coast of Africa and along the coast of India (Mitchell and Reading, 1969). The most important aspect in the history of passive continental margins is the substantial vertical movements which were heterogeneous in both space and time (Whitten, 1976).

In order to explain the inferred subsidence, different mechanisms have been cited. Thermal effects like the cooling of the lithosphere after the continental break-up will cause differential subsidence on the oceanic (Sclater and Francheteau, 1970; Parker and Oldenburg, 1973; Davis and Lister, 1974) as well as on the continental side of the continental margin (Sleep, 1971), as indicated by sediments.

Subsidence of the shelf and slope region has been discussed by Drake and Nafe (1968), Sheridan (1969), Emergy and Uchupi (1972), and Falvey (1974) in the context of transition of the underlying crust from continental to oceanic. Bott (1971) attributed shelf subsidence near continental margins to thinning of the crust by creep in response to stresses caused by differential loading.

The sedimentary deposits will be a contributing factor to subsidence of the shelf (Walcott, 1972; Watts and Ryan, 1976). Dietz and Holden (1966) explain subsidence of rifted aseismic continental shelves by the sedimentary loading of the deep ocean floor. Finally, Turcotte et al. (1977) proposed that the evolution of Atlantic-type margins is controlled dominantly by the vertical displacement of the oceanic—continental lithosphere interface.

The life span of a seismically inactive continental margin can be assumed to be a few hundred million years. During this period the influence of thermal effects inherited from the rift stage will cease. Following the concept of plate theory, continental margins of the Atlantic-type might become convergent plate boundaries and/or form orogens (Kay, 1951; Dewey and Bird, 1970; Bird and Dewey, 1970). In this case the substantial sedimentary deposits at the continental shelf, slope and rise would become involved in tectonic processes which will be conducted by mechanisms different than lithospheric cooling.

The distribution and the age of the sediments accumulated at the continental margin structures provide some information about their subsidence histories in space and time. Figure 1 exhibits structural cross-sections of the continental margin off eastern North America after Emery et al. (1970). The lower section shows the continental margin structure off Nova Scotia and can be taken to be characteristic of the general Atlantic-type margin structure, while the structure above, taken off Cape Fear in the NE—SW direction, exhibits the main characteristics upon which our hypothetical mechanism of mature stage subsidence is based. The upper part of each cross-section I and II presents the "basement" of the structure. The basement has

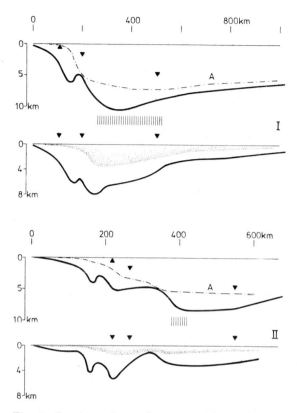

Fig. 1. Continental margin cross-sections off eastern North America. Cape Fear section above (I), Nova Scotia section (Halifax) below (II) after Emery et al. (1970). The upper part of each section represents the depth to basement (solid line) and the horizon A (dash-dot line). Lined pattern indicates extent of intermediate seismic velocities beneath the continental rise. Lower part of each section shows the total thickness of sediments (solid line) while the shaded area indicates the amount of Cenozoic sediments. Triangles indicate bathymetry of the cross-sections, from left to right 200 m, 2000 m and 5000 m position.

been indicated as the top of layer 2 of the oceanic crust and is defined by compressional velocities of 4.2—5.7 km/s. Under the coastal plain and the continental shelf and slope it indicates pre-Triassic rocks. The age of the oceanic basement is suggested to be late Jurassic. The layer boundary above the basement is known as horizon A. Its age has been dated between late Cretaceous to middle Eocene time. Thus the area between the two horizons shown represents deposits of a time span of 130—150 m.y. On the other side the lower curves at cross-sections I and II show corresponding thicknesses of all sediments, while the thickness of Cenozoic sediments is indicated additionally.

The amount of Cretaceous sediments, as indicated by the area between the basement and horizon A along each cross-section, appears to be similar on both the continental shelf and continental rise. This period of development might have been dominantly affected by gradual thermal contraction of the lithosphere. The increasing weight of the sediments would have dragged down the shelf and the rise. At the young continental margin, shelf basins may have developed by oceanward creep of lower continental crustal material accompanied by wedge subsidence in the upper crust.

During the Tertiary period the rate of loading of the lithosphere by thermal contraction would be less, because cooling of the lithospheric plate appears to be represented by a decreasing exponential law. This seems to be supported by the thickness and the spatial distribution of Cenozoic sediments at section II.

On the other hand, the Cenozoic sediments at section I indicate a pronounced maximum at the continental rise, which is at least three times the amount of the sediment thickness at the shelf and the abyssal plain of the same section and the shelf and rise of section II as well. This maximum is accompanied by a zone of intermediate crust mantle boundary, where compressional velocities are 7.2—7.7 km/s instead of 7.8 km/s.

According to Emery et al. (1970) this zone shows a small relative maximum of free-air gravity. Mass calculations based on an assumed isostatic equilibrium, the inferred crustal structure and a 40 km depth of compensation indicate an excess mass with respect to the average for the continental rise at section I. As the accumulation of sediments is supposed to reflect subsidence of the corresponding basement, the significantly higher accumulation of Cenozoic sediments at the continental rise indicates a higher rate of subsidence there. This view is supported by a down-bending of horizon A, which generally is found to be nearly horizontal. These features can be attributed to a mechanism causing subsidence, which is related to the continental rise region only.

Mature stage subsidence along Atlantic-type margins and their possible development due at loading of the continental rise are the main concern of this paper. The response of dynamical models of an Atlantic-type continental margin structure to both vertical loading at the rise region as well as horizontal loading of the lithosphere will be investigated.

STRUCTURE

In our study we will consider structural data of the continental margin bordering eastern North America. It includes all the features common to Atlantic-type margins and it is one of the most extensively studied margins.

The margin is characterized by a broad, flat continental shelf which may vary from 30 to over 300 km, a gentle continental slope with a width of 20—100 km and an even more gently sloping continental rise which spreads out to more than 500 km (Drake and Burk, 1974; Heezen, 1974).

Drake et al. (1959) have shown that two major sedimentary troughs, separated by a ridge, occur near the shelf break. The inner trough is located beneath the continental shelf while the outer one is found under the continental rise and slope.

Recent seismic studies (Sheridan, 1974, 1976; where references to most of the work may be found) provide basement depths of from 7 to 14 km for individual basins. In particular, for the continental rise, great thicknesses of stratified sediments have been inferred. Deep crustal and mantle structures have been derived from refraction profiles and gravity measurements (Drake and Nafe, 1968; Worzel, 1974; Sheridan, 1974; Mayhew, 1974).

The crust-mantle boundary depths below sea level are consistently about 30—35 km under the shelf, about 20 km under the slope and between 12 and 15 km under the outer rise. Deep-crustal refraction measurements reveal a gentle dip of the oceanic crust towards the continent near the continental slope.

The crust beneath the continental rise appears to show significant zones of intermediate compressional seismic velocities between 7.2 and 7.7 km/s with respect to standard velocities of the crust and mantle. These intermediate seismic velocities might indicate a dense nature of the corresponding crustal sections beneath the continental rise and thus could serve to explain its subsidence, which is well documented by the sedimentary record of the basins.

A number of geological and geophysical lineaments are found bordering the east coast of North America. They give rise to interpretations of the location of the edge of the continent. Subsurface basement ridges have been derived from seismic data and appear to be common in continental margins (Burk, 1968). A continuous free-air gravity high located near the shelf break is observed (Emery et al., 1970; Rabinowitz, 1974) suggesting intrabasement density highs. Taylor et al. (1968) mapped the east coast magnetic anomaly, which parallels the slope.

While the edge of the eastern North American continent is thought to be

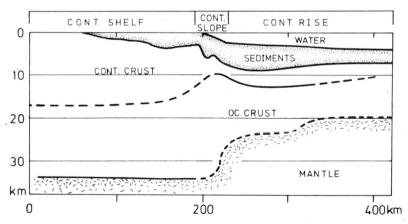

Fig. 2. Crustal cross-section off Portland, Maine, after Drake et al. (1959, fig. 28).

under the slope, Talwani and Eldholm (1973) have demonstrated that the boundary between oceanic and continental basement is not necessarily associated with bathymetric contours.

From a mapping of major basement faults it appears that the basins of the Atlantic margin are marked by boundary faults (Sheridan, 1974, 1976). Thus it can be assumed that basement faulting affected the basin's boundaries and the overlying sediments, as the major faults will have formed during the initial rifting of the Atlantic ocean.

In order to model a cross-section of a typical Atlantic-type continental margin structure we adopted a crustal structure (Fig. 2) after Drake et al. (1959, fig. 28) in our model of the upper mantle. To study the influence of deep faulting on the dynamics of the modelled marginal structure we introduced a fault zone at the transition from oceanic to continental lithosphere.

MECHANISM OF LOADING

Broad drastic subsidence at the continental margin has occurred since the Jurassic, carrying shallow water sediments to depths of 8—12 km. Restoring the basement under the shelf to sea level makes the basement of the adjacent continental rise much shallower, otherwise large-scale vertical faulting between the shelf and the rise is required to explain the present structure. The latter aspect has been emphasized by Turcotte et al. (1977).

The sediments accumulated at the shelf and rise have been dated to possible Triassic, Jurassic, Cretaceous and Tertiary. Phases of rapid accumulation of sediments occurred in geographically discrete basins. Sedimentation rates are locally sometimes an order of magnitude faster than during the immediately-earlier or subsequent times. During the periods of rapid subsidence, the inferred rate reaches 200 m/m.y., as shown by Whitten (1976) for west Florida.

Heterogeneous subsidence in space and time and any development of a plate boundary at continental margins of the Atlantic-type requires a mechanism different from cooling of the lithosphere.

Our preferred mechanism governing the mature and final stage of the development of a passive continental margin is the transformation of metastable lower crustal material to its denser stable phase. The mechanism has been discussed by Ringwood (1975). We investigated the transformation of metastable phases at lower crustal conditions by means of combined numerical models including temperature changes and diffusion for the reacting grains (Spohn and Neugebauer, 1978). The reaction has been presumed to be induced by an increase of crustal temperature beneath sedimentary deposits due to the effect of thermal blanketing of the sediments. As the wavelength of the fundamental horizontal mode of the temperature field is much longer than that of the vertical mode, mainly a vertical diffusion of heat can be concluded. This mechanism would require some minimum thickness of sediments. Providing reasonable crustal conditions the reaction rate of the transformation will be a function of the rate of sedimentation. In this study it has been shown that a finite layer of deposited sediments will induce a proportional layer of transformed material at the lower crust. The transformation rate appears to be geologically significant, with a time delay between sedimentation and lower crustal material facies formation varying between 60 and 200 m.y., depending on the prescribed conditions. The modelled reaction zone of the lower crust covers a depth of several kilometers.

Apart from a particular transformation, this process could provide a reasonable increase in density and thus could serve as a mechanism causing subsidence at existing sedimentary basins along the passive continental margins. Figure 3 relates the increasing density ρ of possible lower crustal material to the layer thickness of transformed material and corresponding body forces (vertical axis). The marked area encloses an example of the variation of density:

$3 \leqslant \rho \leqslant 3.9$ (g/cm^3)

and an inferred layer thickness:

$0.9 \leqslant d \leqslant 5$ (km)

The body forces then cover a range between: $0.34 \leqslant F/l \leqslant 1.72$ ($\cdot 10^{11}$ N/m)

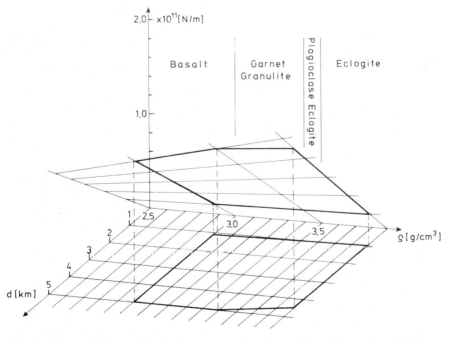

Fig. 3. The variation of body forces as a function of density ρ of possible lower crust material and the inferred layer thickness of transformed material. Marked surface indicates presumptions made for this study.

To determine the forces affecting subsidence due to the transformation of metastable material one has to take the excess of density with respect to a fixed value, for example, of basalt at $\rho = 3.0$ g/cm^3. This would imply vertical body forces up to $1.19 \cdot 10^{11}$ (N/m) as a function of the above range of d and density increase between $0 < \Delta\rho \leqslant 0.9$ (g/cm^3). According to the metastable phase transition models by Spohn and Neugebauer (1978), these changes could be caused by sedimentary deposits of 3—10 km thickness.

A spreading ocean like the Atlantic Ocean provides horizontal forces acting on the moving plates (Solomon et al., 1975; Forsyth and Uyeda, 1975; Von Braunmuehl, 1977). Gravity sliding due to an increase of the thickness of the lithosphere with time has been discussed by Jacoby (1970), Lister (1975) and Neugebauer and Jacoby (1975).

Williams and Poehls (1975) suggested an increasing lithospheric thickness up to 80 m.y. remaining constant after this time. Parker and Oldenburg (1973) and Sclater et al. (1975) on the contrary proposed a continuous thickening of the lithosphere with time. Therefore, the structure of a passive continental margin might be affected by horizontal loading as well. This aspect appears to be of importance to the mature development of the continental margin, i.e., in the context of possible formation of a plate boundary. For this purpose we considered a horizontal push against the fixed continent. For simplification the push was assumed to be perpendicular to the continental margin. Following Neugebauer and Jacoby (1975), we applied horizontal forces of $0.4 \cdot 10^{10}$ N/m at the oceanic lithosphere.

In order to represent the effect of the sedimentary fill of the subsiding region, we assumed a density contrast of 1.2 g/cm^3 for the loads, due to a mean density of 2.3 g/cm^3 and 1.1 g/cm^3 for the sediments and water, respectively.

Thus a system of body forces due to metastable phase transition applied at the lower crust beneath the continental rise and a set of horizontal forces acting on the continental lithosphere as well as sedimentary loading at the top will affect the dynamics of our model structure.

RHEOLOGY

The present knowledge of the upper mantle rheology is based on experimental and theoretical studies of the deformation of rocks under appropriate conditions. Further on, models on the bending response of the lithosphere and asthenosphere to loading and unloading processes have been used to infer the rheology of the upper mantle.

The thickness of the lithosphere is generally defined thermally by isotherms indicating incipient melting. The depth range of observed intraplate seismicity appears to be direct evidence for a brittle-elastic rheology of rocks under deviatoric stresses. This seismic activity is observed at continental tectonic structures to depths of 20—35 km (Wohlenberg, 1975; Bonjer and Gelbke, 1976; Smith, 1977).

These observations are supported by investigations on the bending of the lithosphere in response to loads (Walcott, 1970, 1972; Watts and Cochran, 1974; Haxby et al., 1976). Turcotte and Oxburgh (1976) have shown that stress might accumulate in the lithosphere for geologically relevant times.

In terms of these investigations, the elastic lithosphere is characterized by the flexural rigidity D:

$$D = \frac{Eh^3}{12(1 - v^2)} \tag{1}$$

with Young's modulus E, the Poisson's ratio v, and the thickness h of the "elastic" lithosphere. A typical value for h is 30 km.

Beneath the elastic lithosphere the material will reach conditions where diffusion and dislocation deformation mechanisms will imply extensive steady state plastic flow. This is consistent with laboratory studies (most of the references will be cited by Carter, 1976) and theoretical investigations (Weertman, 1970; Stocker and Ashby, 1973; Weertman and Weertman, 1975).

In this study, we distinguished two mechanisms performing steady deformations which will yield additional contributions: linear diffusional flow and nonlinear dislocation creep.

Dislocation creep appears to be the dominant mechanism in the upper mantle at deviatoric stresses greater than about 1 bar. While the rate of dislocation creep is given by:

$$\dot{\epsilon} = \frac{D\mu b}{kT} \frac{\sigma^n}{\mu} \tag{2}$$

where μ is the shear modulus, b is Burger's vector, k is Boltzmann's constant, σ shear stress, T absolute temperature; the rate due to diffusional flow is:

$$\dot{\epsilon} = \frac{D\Omega}{kTL^2} \sigma \tag{3}$$

with Ω the atomic volume, L grain diameter. Both mechanisms depend on the diffusion coefficient D which is represented either by:

$$D = D_0 \exp - \frac{Q + pV}{RT} \tag{4}$$

where Q is activation energy, V is activation volume, p is pressure, and R gas constant, or by the empirical relation (Weertman, 1970):

$$D = D_0 \exp - g \frac{T_m}{T} \tag{5}$$

which assumes that the effect of pressure on the diffusion coefficient is given approximately by its effect on the melting temperature T_m. D_0 is the diffusion constant, g is proportional to the activation energy at melting.

STATEMENT OF THE MODEL

According to the discussed structure (Fig. 2), and the rheology, we assumed an elastic lithosphere (Fig. 4, dark and intermediate zone in the uppermost structure) and plastic flow governed by eqs. 2 and 3 for the whole structure below the elastic section.

The elastic lithosphere exhibits different flexural rigidities due to regional variations of lithospheric thickness and elastic constants (Table I). With respect to the structural subdivision we distinguished two continental and three oceanic layers of the elastic lithosphere beside the sediments. Youngs' modulus varies from $0.5 \cdot 10^{12}$ dn/cm^2 to $1.43 \cdot 10^{12}$ dn/cm^2, these data were adopted from Haddon and Bullen (1969) and Forsyth and Press (1971). Our flexural rigidities, given in Table I, are based on the calculated models which provide the best fit with the shape of the basement and the total thickness of sediment curves. This will be discussed later.

The transition stress for the applied deformation mechanisms, (2), (3) has been taken to be 1 bar and the power index n of shear stresses in equation (2) was taken to be equal to 3 in accordance with the observations. The grain size was set at $L = 1$ mm (Stocker and Ashby, 1973). The diffusion coefficient, (equation 5), and thus the rate equations (2) and (3) exhibit a strong dependence on the thermal regime of the material. In this study we used the temperature field of the continent—ocean transition presented by MacDonald (1965), model C, and the melting temperature of pyrolite after Ringwood (1969).

Laboratory experiments require generally a ratio of ambient to the melting temperature $T/T_m > 0.5$ for material undergoing creep. On the base of the involved thermal conditions T/T_m varies from 0.53 to 0.94 across our

TABLE I

Elastic lithosphere

Region	Thickness elastic lithosphere (km)	Flexural rigidity (dyn. cm)	Reference
contin. plate	34	$4.9 \cdot 10^{30}$	this study
contin. shelf	34	$3.9 \cdot 10^{30}$	this study
contin. rise	21	$0.8 \cdot 10^{30}$	this study
oceanic plate	16	$0.4 \cdot 10^{30}$	this study
contin. platform	—	$4.0 \cdot 10^{30}$	Walcott (1970)
Michigan Basin	45	$4.8 \cdot 10^{30}$	Haxby et al. (1976)
contin. shelf	—	$1.0 \cdot 10^{30}$	Watts and Ryan (1976)
contin. shelf	60	$1.15 \cdot 10^{31}$	Turcotte et al. (1977)
contin. rise	30	$1.44 \cdot 10^{30}$	Turcotte et al. (1977)

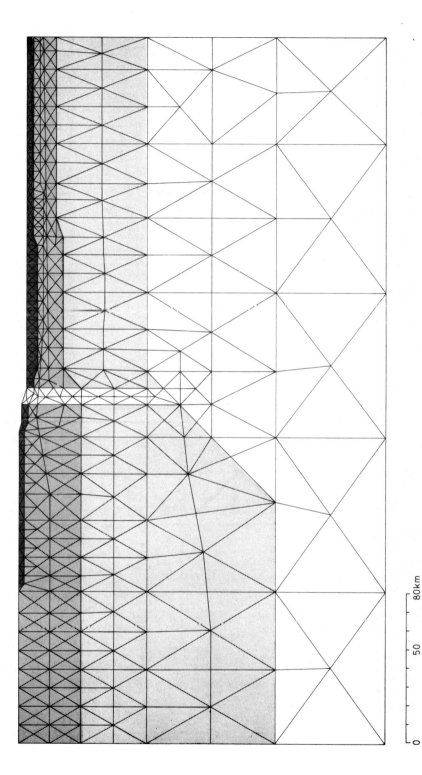

Fig. 4. Structural approximation of a two-dimensional cross-section of a continental margin of the Atlantic type. Lateral extension 380 km, depth 200 km. Shaded areas indicate: sedimentary basins (dark), crustal layers (intermediate), after Drake et al. (1959). For the definition of the lithosphere (light area) see text. The interface between continental and oceanic lithosphere shows an assumed fault zone. The structure comprises 24 areas of different material parameters.

0   50   80km

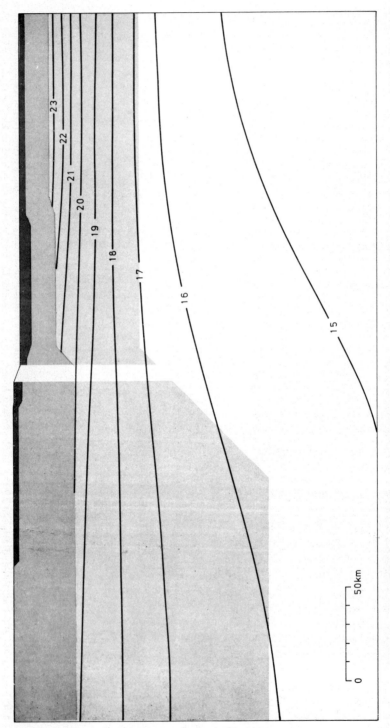

Fig. 5. Deformation map of the model structure of a continental margin of the Atlantic type. The layers on top exhibit the elastic lithosphere. Contour lines are related to $\dot{\epsilon} = 10^{-n}$ (s^{-1}), power index n is given to each line.

model which therefore appears to be related to the high-temperature creep regime. In order to present the rheology across the finite element structure, Fig. 4, which is due to the different assumed parameters, we have chosen a strain rate representation because the strain rate summarizes the effect of all creep parameters at an element at an assumed constant stress level for the whole structure.

The deformation map, shown in Fig. 5, is both statement and result because the strain rates contoured as well as the thickness of the elastic lithosphere are presumptious of our successful models. Based on diffusion flow and dislocation creep mechanism at 1 bar shear stress, the strain rates per second indicate the manner in which the continental and oceanic thermal lithosphere is described. To emphasize the substantial effect of the dislocation creep mechanisms on the shown strain rates above 1 bar shear stress, the strain rates for a deformed model related to shear stresses in Fig. 7 will increase by two to four orders of magnitudes due to the predicted local shear stresses. Although shear stresses increase only about one order of magnitude, the strain level and its distribution appear to be modified significantly.

To study the influence of densification of crustal rocks due to metastable phase transition we have chosen a simple load model. That is, these forces have been related to a homogeneous layer of constant thickness with horizontal extension of 80 km and they have been applied to our model at the base of the crust beneath the continental rise basin. Loading of the model by sediments or by horizontal forces due to a pushing plate are taken to be boundary forces. In the model the base of the structure is only allowed to move horizontally, the top of the structure is taken to be free or partially loaded by sediments. In order to simplify the finite element analysis we took only the excess of density due to metastable transition. That means this study does not include deformations and stresses caused by changes in crustal thickness as investigated by Bott and Dean (1972).

A number of elements have been assigned to approximate a main marginal fault zone at the transition of continental to oceanic lithosphere with variable fault activity. For this purpose we modified the elastic and creep parameters respectively, as stated at the models shown.

METHOD

In order to model a time-dependent non-linear rheology governed by equations like (2) and (3) we applied the initial strain approach described elsewhere (Zienkiewicz, 1971). Thus only the basic idea of the technique should be given here. The procedure is an incremental strain approach to non-linear rheology. The essential assembled equations of the structure:

$$[K] \{\delta\} - \{R\} = 0 \qquad (6)$$

relate the displacements $\{\delta\}$ of the nodal points of the structure to forces

{R} like external loads, body forces, or others. The so-called system stiffness matrix [K] comprises all information about geometrical, physical and mechanical properties of the individual elements and their interrelation.

The element stresses obtained after the initial solution of (6) are applied to appropriate rate equations, like (2) and (3), to determine creep strains associated with a finite time interval. In a second solution of the system equations (6) the incremental creep strains will be applied as initial strains, i.e. as initial forces. In order to obtain a successful approximation to the nonlinear rheology, constraints have to be made on the size of a time step and the allowable stress change during a time interval.

On the basis of this technique we determined the rates of subsidence from our model after thirty solutions of the system equations (6). In this case the calculated rates of deformation as well as the stresses became constant.

NUMERICAL CALCULATIONS

The modelled continental margin structure is assumed to be in a stage of development which is characterized by the existence of large sedimentary basins, in particular at the continental rise. To study the development of such a marginal zone in reaction to densification of the lower crust and in the context of possible plate boundary formation we will consider four specific groups of models.

Continuous transition models

Our first models are based on a continuous lithosphere transition between continent and ocean. The densification due to a metastable phase transition has been assumed beneath the continental rise. The amount of loads correspond to various thicknesses of layers of transformed material. Providing that the increase in density is $\Delta\rho = 0.52$ (g/cm^3) the curve parameters in Fig. 6 indicate the thickness of layers of the transformed material in km.

This model exhibits the rates of coupled subsidence of the continental rise and the continental shelf initiated by loads at the rise. While wavelength of the subsiding zone is affected by the extent of the loads, the differential flexural rigidity influences the non-symmetric shape of the rate distribution across the continental margin model. For the shelf region an increasing rate of subsidence towards the continental slope is calculated. Here the gradient of the rate increases by a factor 1.35 over 125 km. At the continental rise basin the maximum rate observed indicating 14, 105, 350 and 815 m/y with respect to corresponding loads between $0.51 \cdot 10^{10}$ N/m to $2.04 \cdot 10^{10}$ N/m at the rise. The outer rise shows a higher gradient of the rate by a factor 1.6 in comparison with the shelf values. Adjacent to the subsiding marginal zone fairly low rates of uplift at the continental lithosphere appear landward of the fall-line.

293

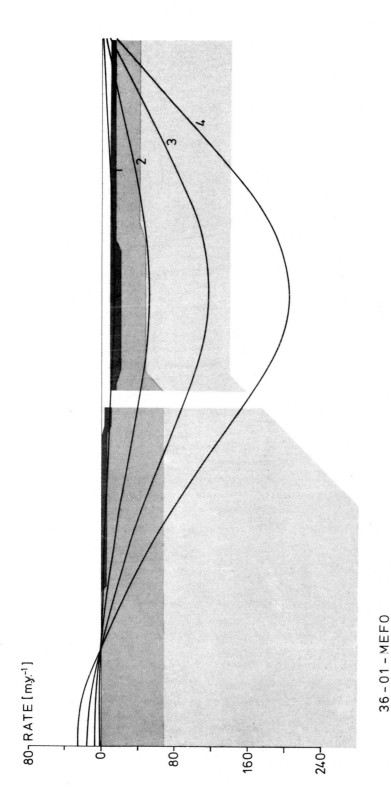

Fig. 6. Coupled subsidence of continental margin structure caused by densification beneath the continental rise basin. Model *A*. Subsidence is given in m/m.y. by multiplying the scale by the shown curve parameters. Curve parameters indicate thickness of transformed lower crustal material in km at the same time.

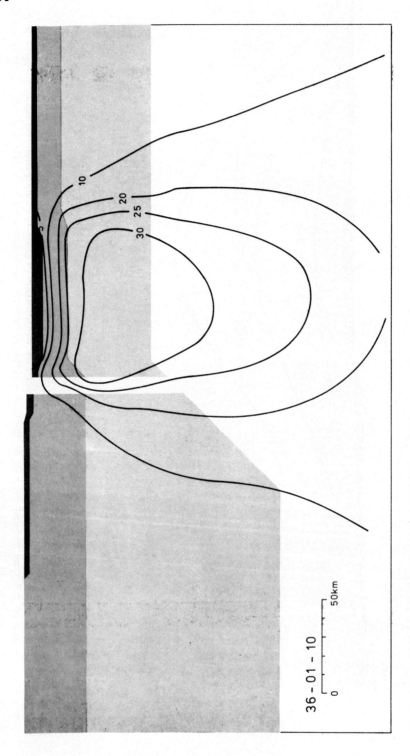

Fig. 7. Shear stresses corresponding to model 2, Fig. 6, contoured in bars.

As we computed a steady rate of subsidence for our models the total amount of bending is small compared with the observed basement depth. Therefore the corresponding stresses appear to be also on a low level but their spatial distribution will be representative for the applied loads (Fig. 7). The contoured maximum shear stresses are concentrated at the crust and upper mantle beneath the continental rise. The amount of shear stresses will reach several kbars at the elastic lithosphere in reaction to large deflections (Neugebauer, 1978; Neugebauer and Braner, 1978). Different approaches to lithospheric bending support these results (Turcotte and Oxburgh, 1976). Further on the shown stresses are only due to the applied loads and thus will not include thermal stresses or stresses caused by changes of crustal thickness.

Active fault models

Seismic exploration gives evidence for significant tensional basement faulting which appears to control sedimentary basin boundaries. This continental margin fault system is assumed to be formed during the initial rifting of the Atlantic margin in the Jurassic or perhaps Triassic. The sedimentary record at both the shelf and rise basins supports the idea that the continental margin fault system has been active during the development of the marginal structure. Thus differential subsidence between the continental and the oceanic lithosphere is expected to affect those zones.

In order to study the influences of a fault zone on the structural development of the continental margin we modified our model using an approximation of a major fault zone between the continental and oceanic lithosphere as stated in Fig. 4. For this purpose we reduced the elastic constants and the creep rate corresponding to a decrease of T_m of about $200°$ K in the fault zone of our model which causes the material to deform on a higher rate there than the surroundings. That is, in terms of flexured rigidity D, a decrease from $D = 3.5 \cdot 10^{30}$ to $D = 0.35 \cdot 10^{30}$ dyn. cm at the fault zone for the models shown in Fig. 8 and Fig. 9. Both sets of models show an offset for the calculated rates at the continental slope region. While the calculated offset of the rate varies from 2 to 110 m/m.y. for model B and from 4 to 210 m/m.y. for model C, the increase of the rate of subsidence appears to be concentrated to the slope and rise of the margin structure. This is obviously the effect of the reduced flexural and creep parameters at the fault zone combined with the lower flexural rigidity at the rise compared with the continental values.

The shallow-water facies sediments found at the shelf and rise give strong support to the idea that the offset between shelf and rise basement (up to 1 : 3) is active during the development of the sedimentary deposits.

On the basis of the implied model and the inferred kinetics we think that the development of continental margin basins, especially existing sedimentary basins, could be severely affected by subcrustal densification and marginal tectonics.

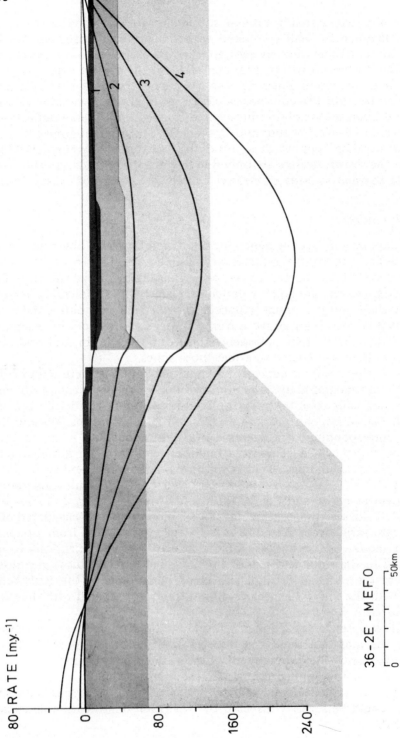

Fig. 8. Active fault model. *B*. The flexural parameter of the elastic lithosphere is reduced by a factor 0.5 at the fault, reducing *E*. The shown rates are scaled as in Fig. 6.

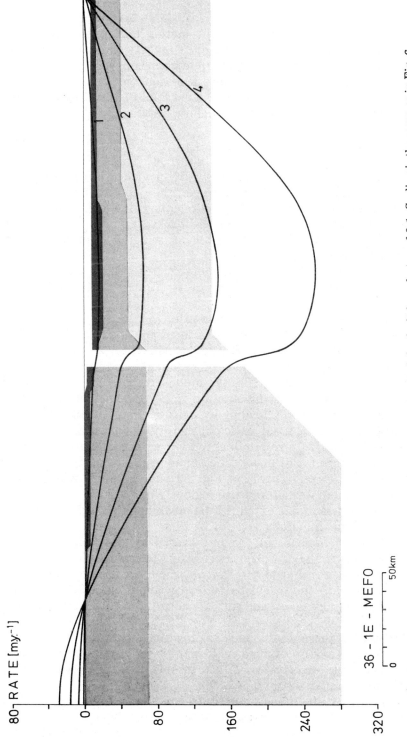

Fig. 9. Active fault model. C. Reduction of the flexural parameters of the fault by a factor of 0.1. Scaling is the same as in Fig. 6.

Whitten (1976) emphasized that the marine transgressions were heterogeneous in both space and time. Phases of rapid accumulation of shelf subsidence occurred in geographically discrete basins on shelf areas.

The rate of sedimentation provides a reasonable measure of the rate of down-buckling of the crust, because virtually all of the Cretaceous rocks of the emerged shelf were deposited in shallow water.

Sedimentary loading

The deposit of sediments requires subsidence caused by a different mechanism than sedimentary loading. However, the deposit of sediments will be involved in driving subsidence by sedimentary loading. Watts and Ryan (1976) have shown by the method of "backstripping" sediments that the effect of sediment loading and the effect of remaining "driving forces" can be quantitatively determined.

The weight of accumulating sediments is believed to be responsible for half or less of the subsidence of the crust (Rona, 1974; Watts and Ryan, 1976).

In order to determine the contribution of the weight of sediments to the calculated rate of subsidence we considered the load case 2 of the three families of model *A*, *B* and *C*. Besides densification at the lower crust as stated before we applied the loads of sediments corresponding to the calculated rates at the surface across the modelled structure. We took the rate of subsidence and that of sedimentation to be equal. Figure 10 shows three types of models comprising loading by densification and the weight of sediments assuming an excess of sediment density against that of water of 1.2 g/cm^3 (dashed lines). Corresponding models without sedimentary loads are shown for reference (solid lines). The apparent sedimentary loading has the wavelength of the solid curves in Fig. 10 and therefore will considerably affect the continental shelf and rise region. While the fall line at the continent remains constant, the extent of the down-buckled region at the oceanic lithosphere shown an increase.

The calculated increase of the rate caused by the weight of sediments varies between a factor of 1.6—1.9 according to the weight of the different models, which means that in the context of our model the sediment loading could contribute to the calculated rates by about 38—48% providing that both rates of sedimentation and subsidence are equal. This result agrees with other studies.

On the basis of these results one can draw the conclusion that the required degree of transformation of metastable material or the thickness of the transformed layer could be even lower than assumed for the previous models to cause the shown rates of subsidence.

The concept of our models applies to existing sedimentary basins at the continental rise along continental margins of the Atlantic-type. Providing that the lower crust along passive continental margins is in a metastable

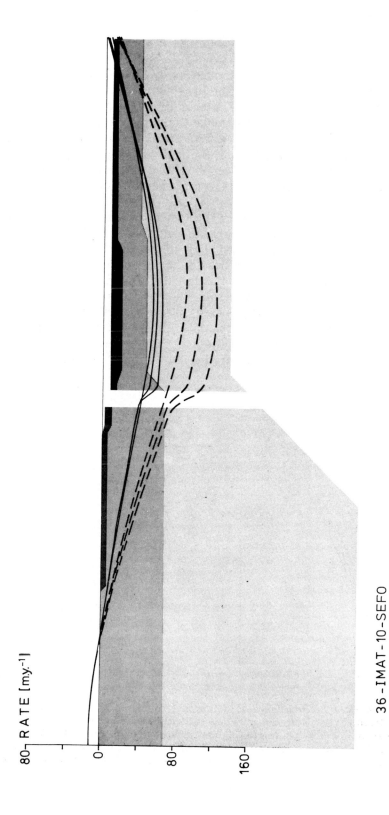

Fig. 10. The rate of subsidence related to subcrustal deformation (solid lines) and additional sedimentary loads (dashed lines) providing rates of sedimentation are given by solid lines. The shown models refer to *A*, *B* and *C*, load case 2. Scaling factor for the rates is 2.

state, a finite layer of sediments can induce a corresponding proportional layer of transformed dense material in the lower crust for reasonable conditions. This mechanism might be able to explain both episodic as well as continuous subsidence of a sedimentary basin like those at continental margins.

Figure 11 gives a summary of results inferred from the models considered so far. Heavy lines show the maximum rates of subsidence calculated for the continental rise basin as a function of applied loads beneath the rise. The indicated maximum rates are related to the sequence of models A, B and C, respectively.

These curves present an upper limit of the rate of subsidence: first, they represent the maximum rate at the rise within each model. Second, isostatic adjustment will reduce the initial values. This mechanism has an insignificant influence on the shown curves because of the short time-span and the small amount of subsidence modelled. Finally, the weight of the accumulating sediments will contribute to the total rate of subsidence. However, the most outstanding feature, which is based on the prescribed rheology, is the nonlinear, accelerated increase of the rate, shown for the continental rise, in response to a linear increase of the loads at the lower crust. In the same context, the change in the rate at the fault zone is demonstrated in

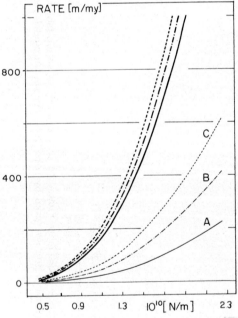

Fig. 11. Calculated rates of subsidence as a function of loading at the lower crust beneath the continental rise. Note the rates are related to loads across 80 km. Heavy lines indicate maximum rates at the rise. Model A: locked fault zone (straight line), model B: intermediate fault zone (dash-dot line), model C: active fault (dotted line). Curves at the lower right exhibit the corresponding offset at the rates across the continental slope.

Fig. 11, lower right. These curves represent the change of the rate of subsidence across the 23 km of the modelled continental slope due to increasing loads beneath the rise. While A corresponds to the continuous models (Fig. 6), curves B and C indicate an increasing offset of the rate across the fault zone (Figs. 8, 9) as a function of the applied loads and the chosen approximation of major faults.

According to our calculations the conversion of lower crustal material provides a suitable mechanism to govern the late development of sedimentary basins along passive continental margins. A proceeding densification beneath the rise will be able to explain the rates of subsidence inferred from sedimentary records.

Associated with the subsidence significant differential vertical displacements at major marginal fault zones seem to be appropriate to explain observed basement topographies. On the basis of our numerical results we suggest that the possible formation of a plate boundary at passive continental margins will be considerably affected by both an active marginal fault system as well as pronounced subsidence of the continental rise basin.

Horizontal loading

One might speculate whether an additional boundary condition might contribute to the development of a continental margin structure of the Atlantic-type. The driving mechanism for plate tectonics is expected to provide horizontal loading of the lithosphere. These horizontal forces will imply compressions to the passive continental margin structure.

To model such a situation at an Atlantic-type continental margin we fixed the continental lithosphere horizontally at the upper left of our structure, and applied horizontal forces at the oceanic lithosphere. This condition assumed a plate push perpendicular to the continental margin in addition to the vertical loads stated before.

To study the significance of the applied horizontal loads we calculated a number of models with various ratios of vertical to horizontal loads. On the basis of the active fault model C, Fig. 12 illustrates the generated creep deformation of the model surface due to five different applied load ratios. The shown numbers indicate the excess of vertical to horizontal loads which has been taken to be the unit. Hence, the dominance of vertical loads decreases with lower numbers and, in the same way, the deformation of the modelled structure appears to be dominated by the horizontal component. Following this sequence it can be recognized that the size of the subsiding rise basin shrinks from 170 km, model 5, to 110 km, model 1; the average slope of basin subsidence turns from 90° to about 50° with respect to the horizontal axis. The subsidence at the continental shelf appears to turn to the horizontal movements with increasing influence of horizontal forces, while the continental crust landward will be lifted up.

In connection with the shrinking size of the subsiding continental rise

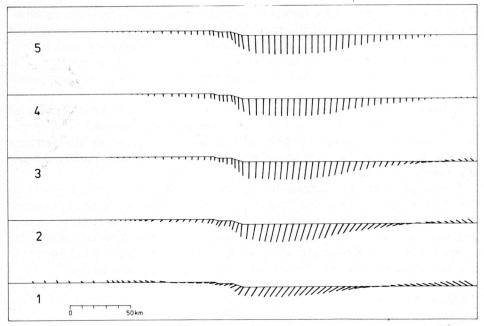

Fig. 12. Creep deformation of the model surface due to vertical and horizontal loads. Arbitrary scale. Numbers indicate excess vertical forces with respect to horizontal load unit.

basin a pronounced uplift of the oceanic lithosphere appears seaward of the rise. This type of uplift is observed at most subduction zones and is called an outer rise or forebulge seaward of the trench. It has been modelled by thin elastic plate theory as an effect of bending at the lithosphere (Watts and Talwani, 1974; Caldwell et al., 1976). McAdoo et al. (1978), however, showed that the deflection of the lithosphere at trenches like that at Kuril Island can be explained by applying vertical as well as horizontal forces. Thus, the numerical results support the view that compressive horizontal loading of the oceanic lithosphere is able to cause significant influence on the subsiding continental margin structure. This, however, requires horizontal loads which approach the amount of the vertical and probably even higher amounts. In this case the models predict a general trend of surface deformation like compression at the shelf zone, a shrinking width of the subsiding rise and the up-bending of the lithosphere oceanward of the rise. As the modelled time span represents only a "snap-shot" of the structural deformation due to the applied conditions a relation of these main features to an Anden-type continental margin structure must remain speculative. However, we would like to point out that the general trend of the predicted deformations does not contradict the general features at a convergent plate boundary of the Anden-type. So one might speculate that a possible development of a plate boundary at a continental margin of the Atlantic-type would be significantly affected by vertical loads at the rise, the existence of major faults as well as a sufficient horizontal compressive loading of the marginal structure.

COMPARISON WITH OBSERVATIONS

Now we will compare the wavelength and shape of the rate of subsidence predicted by our models with basement topography and the total amount of accumulated sediments. As the amount of sediments of known age presents a measure for the average rate of sedimentation, and as far as it reflects basement subsidence, it appears to be a measure for local subsidence as well.

The observations for the Portland section are given in Fig. 13 after Drake et al. (1959). In Fig. 13a and c, we compare the rate predicted by continuous transition models with the topography of the basement and the total thickness of the sediments, respectively. Reasonably good agreement of the predicted shape of the rate distribution with the total thickness of sediments is obtained, keeping in mind that the influence of shelf basins, shown as maxima at the left side of the observations, is not included in the models. However, the predicted rate for the continental rise is much too low and

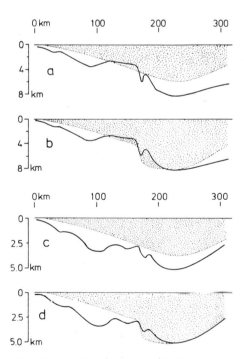

Fig. 13. Portland section; solid line represents depth to the basement at a and b; thickness of sediments at c and d, after Drake et al. (1959). Shaded area exhibits the rate of subsidence distribution on an arbitrary scale to the continuous transition model of a and c, and for the active fault model, of b and d.

304

does not reflect the obvious offset between continental shelf and rise. The nature of the offset is very well demonstrated by the shown accumulation of Cenozoic sediments in Fig. 1.

In Fig. 13b and d, we compare the observations with the distribution of the rate along the active fault model. Good agreement is obtained with the basement topography as well as with the total thickness of sediments and the predicted shape of the rate. The curvature of the basement topography and the predicted rate seaward of the continental rise differ from each other. This can be attributed to the insufficient representation of an extended oceanic lithosphere seaward of the continental margin by the chosen model structure. To model this specific feature a much larger extension of the

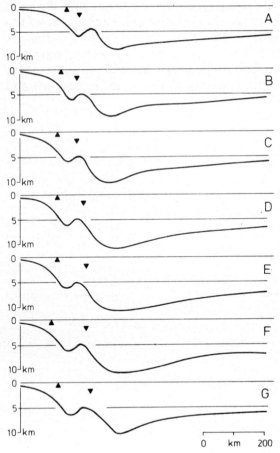

Fig. 14. Depth to the basement (km) off Cape Fear after Emery et al. (1970). Curves indicate pre-Triassic rocks under continental shelf and slope and oceanic basement beneath continental rise and abyssal plains. Triangles indicate bathymetric lines, 200 m and 2000 m from left to right, respectively. For further explanation see text.

model structure, Fig. 4, to the right would be required. However, both basement topography and total sediment accumulation strongly favour the active fault models.

Further on we conclude that mature development of continental margins of the Atlantic-type might be dominated by loading of the continental rise. The associated subsidence can be explained by lower crust densification at the continental rise, loading by sediments and active faults. The calculated flexure of the lithosphere in response to these loads can explain the accumulation of sediments on the oceanic as well as on the continental rise of the continental margin during its mature development.

The good agreement of the predicted rates with the shown continental margin section provides constraints on the inferred flexural rigidity at the elastic lithosphere and the rheology of the structure. The latter affects the flexure of the elastic lithosphere significantly because of its stress dependence.

It should be emphasized that the discussed mature development of a passive continental margin is assumed to be associated with a required minimum amount of accumulated sediments at the continental rise. So only a few sections along continental margins might show the proposed mechanism of loading as indicated by Cenozoic sediments.

To demonstrate the effects of the proposed loading and probable fault motion at a specific continental margin zone we show a set of basement topography profiles off Cape Fear, Fig. 14, after Emery et al. (1970). All the cross-sections meet at the same point at their left end. While profile A strikes E—W along 34°N, the following turn southward with the final section G in SSE direction. The shape of the profiles shows the regional character of the pronounced rise trough. Sections D, E and F exhibit good agreement with the general shape of the rate distribution predicted by our models.

CONCLUSIONS

The basement topography and even more significant, the distribution and thickness of Cenozoic sediments at particular continental margin structures of the Atlantic type indicate a distinct subsidence at the continental rise. These structures are associated with high amounts of total sediment accumulations and intermediate compressional seismic velocities at the crust-mantle boundary. The mature development, indicated by Cenozoic sediments, requires another mechanism than thermal contraction due to lithospheric cooling. Additionally, the nature of the rise sediments indicates significant differential subsidence between the continental shelf and rise. As a consequence of active sea floor-spreading in the Atlantic horizontal loading of the lithosphere will be expected at the continental margins.

These features will be observed at the Atlantic coast of North America, off Cape Fear. To combine both the incipient evolution of the mentioned observations with a reasonable horizontal extent of the cross-section, we modelled a structural section off Portland, Maine.

By means of the finite-element technique we tried to apply current knowledge of crustal and upper mantle rheology like variable flexural parameters of the elastic lithosphere and non-linear plastic flow in this study.

On the basis of the numerical calculations one can draw the following conclusions:

(1) The predicted relative distribution of the rate of subsidence is in good agreement with the general shape of the basement topography and the total thickness of sediments. Thus the proposed loading at the continental rise might dominate mature development of passive continental margin structures in question.

(2) The generation and the location of the loads are able to explain episodic as well as continuous subsidence at the observed rates for both the shelf and continental rise at their mature stage of development.

(3) The development of the investigated structures is controlled dominantly by the vertical displacement of the major fault zone at the oceanic—continental transition.

(4) The regional flexural response of the lithosphere can be attributed entirely to regional variations in lithospheric flexural rigidities associated with non-linear plastic flow beneath the elastic lithosphere.

(5) The strain rate at the non-elastic lithosphere varies between 10^{-20} s^{-1} and 10^{-15} s^{-1} at constant shear stress of 1 bar.

(6) Horizontal compressive loading of the lithosphere appears to be of tectonic significance at loads which equal or exceed the applied vertical loads.

ACKNOWLEDGEMENTS

This research was supported by the Deutsche Forschungsgemeinschaft.

REFERENCES

Bird, J.M. and Dewey, J.F., 1970. Lithosphere plate-continental margin tectonics and the evolution of the Appalachian orogen. Geol. Soc. Am. Bull., 81: 1031—1060.
Bonjer, K.P. and Gelbke, C., 1976. Seismizität und Dynamik im Bereich des Oberrheingrabens. (Abstract.) 36th Annual Meeting, DGG, Boduum.
Bott, M.H.P., 1971. Evolution of young continental margins and formation of shelf basins. Tectonophysics, 11: 319—327.
Bott, M.H.P. and Dean, D.S., 1972. Stress at young continental margins. Nature, 235: 23—25.
Burk, C.A., 1968. Buried ridges within continental margins. Trans. N.Y. Acad. Sci., 30: 397—409.
Caldwell, J.G., Haxby, W.F., Karig, D.E. and Turcotte, D.L., 1976. On the applicability of a universal elastic trench profile. Earth Planet. Sci. Lett., 31: 239—246.
Carter, N.L., 1976. Steady state flow of rocks. Rev. Geophys. Space Phys., 14: 301—360.

Davis, E.E. and Lister, C.R.B., 1974. Fundamentals of ridge crest topography. Earth Planet. Sci. Lett., 21: 405—413.

Dewey, J.F. and Bird, J.M., 1970. Mountain belts and the new global tectonics. J. Geophys. Res., 75: 2625—2647.

Dietz, R. and Holden, J.C., 1966. Miogeoclines in space and time. Geology, 75: 566—583.

Drake, C.L. and Nafe, J.E., 1968. The transition from ocean to continent from seismic refraction data. In: L. Knopoff, C.L. Drake and P. Hart (Editors), The Crust and Upper Mantle of the Pacific Area. Geophys. Monogr., Am. Geophys. Union, 12: 174—186.

Drake, C.L. and Burk, C.A., 1974. Geological significance of continental margins. In: C.A. Burk and D.L. Drake (Editors), Geology of Continental Margins. Springer, Berlin, pp. 3—10.

Drake, C.L., Ewing, M. and Sutton, G.H., 1959. Continental margins and geosynclines: the eastern coast of North America north of Cape Hatteras. In: L.H. Ahrens, F. Press, F. Rankama and S.K. Runcorn (Editors), Physics and Chemistry of the Earth. Pergamon, Elmsford, pp. 110—198.

Emery, K.O. and Uchupi, E., 1972. Western North Atlantic Ocean: topography, rocks structure, water, life and sediments. Am. Assoc. Petrol. Geol. Mem, 17: 532 pp.

Emery, K.O., Uchupi, E., Phillips, J.D., Bowin, C.O., Bunce, E.T. and Knott, S.T., 1970. Continental rise off eastern North America. Am. Assoc. Petrol. Geol., Bull., 54: 44—108.

Falvey, D.A., 1974. The development of continental margins in plate tectonic theory. Aust. Petrol. Explor. Assoc. J., 14: 95—106.

Forsyth, D.W. and Press, F., 1971. Geophysical tests of petrological models of the spreading lithosphere. J. Geophys. Res., 76: 7963—7979.

Forsyth, D.W. and Uyeda, S., 1975. On the relative importance of driving forces of plate motion. Geophys. J.R. Astron. Soc. 43: 163—200.

Haddon, R.A.W. and Bullen, K.E., 1969. An earth model incorporating free earth oscillation data. Phys. Earth Planet. Inter., 2: 35—49.

Haxby, W.F., Turcotte, D.L. and Bird, J.M., 1976. Thermal and mechanical evolution of the Michigan Basin. Tectonophysics, 56: 57—75.

Heezen, B.C., 1974. Atlantic-type continental margins. In: C.A. Burk and D.L. Drake (Editors), Geology of Continental Margins. Springer, Berlin, pp. 13—24.

Jacoby, W.R., 1970. Instability in the upper mantle and global plate movements. J. Geophys. Res., 75: 5671—5680.

Kay, M., 1951. North American geosynclines. Geol. Soc. Am. Mem., 48: 143 p.

Lister, C.R.B., 1975. Gravitational drive on oceanic plates caused by thermal contraction. Nature, 257: 663—665.

MacDonald, G.L.F., 1965. Geophysical deductions from observations of heat flow. In: W.H.K. Lee (Editor), Terrestrial Heat Flow. Am. Geophys. Union. pp. 191—210.

Mayhew, M.A., 1974. Geophysics of Atlantic North America, In: C.A. Burk and C.L. Drake (Editors), Geology of Continental Margins, Springer, Berlin, pp. 409—427.

McAdoo, D.C., Caldwell, J.G. and Turcotte, D.L., 1978. On the elastic-perfectly plastic bending of the lithosphere under generalized loading with application to the Kuril Trench. Geophys. J. R. Astron. Soc., 54: 11—26.

Mitchell, A.H. and Reading, H.G., 1969. Continental margins, geosynclines, and ocean floor spreading. J. Geol., 77: 629—646.

Neugebauer, H.J., 1977. Crustal doming and the mechanism of rifting, Part I: Rift formation. Tectonophysics, 45: 159—186.

Neugebauer, H.J. and Braner, B., 1977. Crustal doming and the mechanism of rifting. Part II: Rift development of the Upper Rhine Graben. Tectonophysics, 46: 1—20.

Neugebauer, H.J. and Jacoby, W.R., 1975. Dynamic numerical models of ocean ridges with nonlinear rheology (abstract). 16th General Assembly, IUGG, Grenoble.

Oliver, J.E., Isacks, B.L. and Barazangi, M., 1974. Seismicity at continental margins. In: C.A. Burk and D.L. Drake (Editors), Geology of Continental Margins. Springer, Berlin, pp. 85—92.

Parker, R.L. and Oldenburg, D.W., 1973. A thermal model of ocean ridges. Nature, 242: 137—139.

Rabinowitz, P.D., 1974. The boundary between oceanic and continental crust in the Western North Atlantic. In: C.A. Burk and D.L. Drake (Editors), Geology of Continental Margins. Springer, Berlin, pp. 67—84.

Ringwood, A.E., 1969. Composition and evolution of the upper mantle. In: P.J. Hart (Editor), The Earth's Crust and Upper Mantle. Geophys. Monogr., 13, Am. Geophys. Union, Washington, D.C., pp. 1—18.

Ringwood, A.E., 1975. Composition and Petrology of the Earth's Mantle. McGraw-Hill, New York, 618 pp.

Rona, P.A., 1974. Subsidence of Atlantic continental margins. Tectonophysics, 22: 283—299.

Sclater, J.C. and Francheteau, J., 1970. The implications of terrestrial heat flow observations on current tectonic and geochemical models of the crust and upper mantle of the earth. Geophys. J.R. Astron. Soc., 20: 509—542.

Sclater, J.G., Lawrence, A.L. and Parsons, B., 1975. Comparison of long-wavelength residual elevation and free-air gravity anomalies in the North Atlantic and possible implications for the thickness of the lithospheric plate. J. Geophys. Res., 80: 1031—1052.

Sheridan, R.E., 1969. Subsidence of continental margins. Tectonophysics, 7: 219—229.

Sheridan, R.E., 1974. Atlantic continental margin of North America. In: C.A. Burk and C.L. Drake (Editors), Geology of Continental Margins. Springer, Berlin, pp. 409—427.

Sheridan, R.E., 1976. Sedimentary basins of the Atlantic margin of North America. In: M.H.P. Bott (Editor), Sedimentary basins of continental margins and grabens. Tectonophysics, 36 (1—3): 113—132.

Sleep, N.H., 1971. Thermal effects of the formation of Atlantic continental margins by continental break-up. Geophys. J.R. Astron. Soc., 24: 325—350.

Smith, R.B., 1977. Seismicity, crustal structure and intraplate tectonics of the Western North American plate. IASPEI meeting, Durham (abstract).

Solomon, S.C., Sleep, N.H. and Richardson, M.R., 1975. On the forces driving plate tectonics: inferences from absolute plate velocities and intraplate stresses. Geophys. J. R. Astron. Soc., 42: 769—801.

Spohn, T. and Neugebauer, H.J., 1978. Metastable phase transition models and their bearing on the development of Atlantic-type geosynclines. Tectonophysics, 50: 387—412.

Stocker, R.L. and Ashby, M.F., 1973. On the rheology of the upper mantle. Rev. Geophys. Space Phys., 11: 391—426.

Talwani, M. and Eldholm, O., 1973. The boundary between continental and oceanic basement at the margin of the rifted continents. Nature, 241: 325—330.

Taylor, P.J., Zietz, I. and Dennis, L.S., 1968. Geologic implications of aeromagnetic data for the eastern continental margin of the United States. Geophysics, 33: 755—780.

Turcotte, D.L. and Oxburgh, E.R., 1976. Stress accumulation in the lithosphere. Tectonophysics, 35: 183—199.

Turcotte, D.L., Ahern, J.L. and Bird, J.M., 1977. The state of stress at continental margins. Tectonophysics, 42: 1—28.

Von Braunmuehl, W., 1977. Modelle zur Dynamik Kontinentaler und Ozeanischer Lithosphaerenplatten. Thesis, Inst. Met. Geophys. Frankfurt/Main.

Walcott, R.J., 1970. Flexural rigidity, thickness and viscosity of the lithosphere. J. Geophys. Res., 75: 3941—3954.

Walcott, R.J., 1972. Gravity, flexure and the growth of sedimentary basins at a continental edge. Geol. Soc. Am. Bull., 83: 1845—1848.

Watts, A.B. and Cochran, J.R., 1974. Gravity anomalies and flexure of the lithosphere along the Hawaiian—Emperor seamount chain. Geophys. J.R. Astron. Soc., 38: 119—141.

Watts, A.B. and Talwani, M., 1974. Gravity anomalies seaward of deep-sea trenches and their tectonic implications. Geophys. J.R. Astron. Soc. 36: 57—90.

Watts, A.B. and Ryan, W.B.F., 1976. Flexure of the lithosphere and continental margin basins. In: M.H.P. Bott (Editor), Sedimentary Basins of Continental Margins and Grabens. Tectonophysics, 56 (1—3): 25—44.

Weertman, J., 1970. The creep strength of the earth's mantle. Rev. Geophys. Space Phys., 14: 301—360.

Weertman, J. and Weertman, J.R., 1975. High temperature creep of rock and mantle viscosity. Annu. Rev. Earth Planet. Sci., 3: 293—315.

Whitten, E.H.T., 1976. Geodynamic significance of spasmodic, Cretaceous, rapid subsidence rates, continental shelf, U.S.A. In: M.H.P. Bott (Editor), Sedimentary Basins of Continental Margins and Grabens. Tectonophysics, 36 (1—3): 133—142.

Williams, D.L. and Poehls, K.A., 1975. On the thermal evolution of oceanic lithosphere. Geophys. Res. Lett., 2: 321.

Wohlenberg, J., 1975. Geophysikalische Aspekte der ostafrikanischen Grabenzonen. Geol. Jahrb., Reihe E, 4, 84 pp.

Worzel, J.L., 1974. Standard oceanic and continental structure. In: C.A. Burk and D.L. Drake (Editors), Geology of Continental Margins. Springer, Berlin, pp. 59—66.

Zienkiewicz, O.C., 1971. The Finite Element Method in Engineering Sciences. McGraw-Hill, London, 521 pp.

Chapter 14

AN ELASTIC-PERFECTLY PLASTIC ANALYSIS OF THE BENDING OF THE LITHOSPHERE AT A TRENCH*

D.L. TURCOTTE, D.C. McADOO and J.G. CALDWELL

Department of Geological Sciences, Cornell University, Ithaca, N.Y. 14853 (U.S.A.)

SUMMARY

A number of authors have modelled the flexure of the lithosphere at an oceanic trench using a thin elastic plate with a hydrostatic restoring force. In some cases good agreement with observed topography is obtained but in other cases the slope of the lithosphere within the trench is greater than that predicted by the elastic theory. In this chapter the bending of a thin plate is considered using an elastic-perfectly plastic rheology. It is found that the lithosphere behaves elastically seaward of the trench, but that plasticity decreases the radius of curvature within the trench. The results are compared with a number of observed trench profiles. The elastic-perfectly plastic profiles are in excellent agreement with those profiles that deviate from elastic behavior.

INTRODUCTION

At oceanic trenches the lithosphere appears to bend and descend into the mantle as a coherent plate. The deformation of the plate seaward of the trench can be studied directly by observations of topography. In most cases an outer rise is observed. Measurements of surface gravity anomalies provide further information on the plate deformation.

In order to explain the observed flexure of the lithosphere a number of authors have studied the elastic bending of a plate with a hydrostatic restoring force (Gunn, 1937, 1947; Walcott, 1970; Hanks, 1971; Le Pichon et al., 1973; Watts and Talwani, 1974; Caldwell et al., 1976; Parsons and Molnar, 1976). With given plate properties (flexural rigidity) and restoring force the deformation of the plate depends upon specified values of the vertical force, horizontal force, and bending moment at any point. This is illustrated in Fig. 1.

It was shown by Caldwell et al. (1976) that reasonable values of the horizontal force have only a small effect on the deformation of the lithosphere.

Originally published as: Turcotte, D.L., McAdoo, D.C. and Caldwell, J.G., 1978. An elastic-perfectly plastic analysis of the bending of the lithosphere at a trench. In: K.L. Burns and R.W.R. Rutland (Editors), Structural Characteristics of Tectonic Zones. Tectonophysics, 47: 193—205.

312

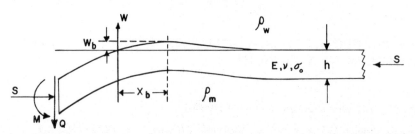

Fig. 1. Illustration of the bending of an elastic-perfectly plastic plate with a hydrostatic restoring force. Bending is caused by an applied moment M, vertical force Q, and horizontal force S.

This is consistent with the much earlier studies (Smoluchowski, 1909; Goldstein, 1926) which showed that it was not possible to buckle the lithosphere using horizontal forces.

In the absence of horizontal forces, Caldwell et al. showed that the deforming lithosphere at trenches fits an universal elastic trench profile. They also showed a number of cases in which there was a good fit with observed topography. In order to obtain a good fit the plate thickness was between 20 and 30 km. This is the thickness of the elastic lithosphere. Only at low temperatures (below about $400°C$) does the thermal lithosphere behave elastically on geological time scales. The maximum elastic bending stresses (deviatoric) approached 10 kb.

Although the elastic bending of the lithosphere at oceanic trenches gives results that are in good agreement with observations in some cases, there are other cases in which the discrepancies are considerable. It is not surprising that the lithosphere at trenches deviates from elastic behavior since the derived elastic stresses approach 10 kb. These stresses are certainly of the order of the yield strength of rock.

Deviations from elastic behavior can take the form of fractures or plastic flow. In general, rock will fracture at confining pressures small compared with the yield stress and will flow plastically at higher confining pressures. We certainly agree that the upper part of the lithosphere seaward of ocean trenches fractures under the tensional bending stresses. Kanamori (1971) has argued that the entire lithosphere seaward of a trench fractures. His conclusion is based primarily on the Sanriku earthquake of 1933. We do not believe that either the seismic evidence or the topography favors lithospheric fracturing seaward of trenches except in localized regions. We propose that the oceanic lithosphere at trenches behaves as either an elastic or plastic material at depth.

RHEOLOGY MODEL

In order to model the inelastic bending of the oceanic lithosphere seaward of a trench we will use an elastic-perfectly plastic rheology (Prager and

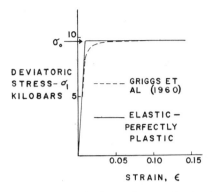

Fig. 2. The elastic-perfectly plastic rheology is compared with the experimental results of Griggs et al. (1960) for Dun Mountain dunite at a confining pressure of 5 kb and a temperature of 300°C.

Hodge, 1951). The uniaxial stress—strain dependence of an elastic-perfectly plastic material is shown in Fig. 2. At stresses below the yield stress, σ_0, this material is linearly elastic with a Young's modulus E and a Poisson's ratio ν. At the yield stress σ_0 the material flows plastically without strain hardening. When the material is unloaded by reducing the stress below the yield stress, it behaves elastically in a manner unaffected by the plastic flow. Experimental studies of rock deformation at high confining pressures and strain rates produce uniaxial, deviatoric stress—strain curves similar to this idealized behavior. The stress—strain curve for Dun Mountain dunite obtained by Griggs et al. (1960) at a confining pressure of 5 kb and a temperature of 300°C is given in Fig. 2 for comparison.

MODEL

We next consider the steady-state bending of an elastic-perfectly plastic plate with a hydrostatic restoring force. The elastic-perfectly plastic bending of a plate has been considered by Prager and Hodge (1951) and by Kachanov (1974). Based on the results of Caldwell et al. (1976) we will neglect the horizontal force, i.e., take $S = 0$.

As long as the bending moment M at any point along the plate is less than the value M_0 the entire cross-section of the plate is elastic. This is region I in Fig. 3a. The stress and strain are linearly proportional to the distance from the center of the plate and the maximum stresses at the top and bottom of the plate are less than the yield stress, as shown in Fig. 3d. In this elastic region the deflection of the plate satisfies:

$$D \frac{d^4 w}{dx^4} + (\rho_m - \rho_w) g w = 0 \tag{1}$$

where $D = Eh^3/12(1 - \nu^2)$ is the flexural rigidity of the plate, ρ_m is the mantle density, ρ_w is the water density, g is the gravitational acceleration, and h

314

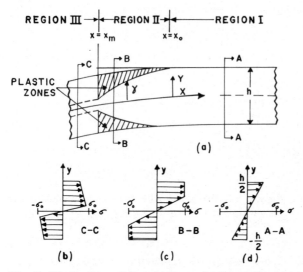

Fig. 3. Illustration of the elastic-perfectly plastic bending of the lithosphere at a trench (a). Region I is elastic and the linear distribution of stress at cross-section A—A is given in (d). Region II has an elastic core with a linear distribution of stress and outer plastic zones at the yield stress as shown for cross-section B—B in (c). In Region III the plate is unbending elastically because of the reduction in bending moment. However, because of prior plastic deformation, the stress is not a linear function of y. The stress distribution at cross-section C—C is given in (b).

is the thickness of the elastic plate.

It is convenient to choose the origin, $x = 0$, as the point nearest the trench axis where the deflection, w, is zero. With $w \to 0$ and $dw/dx \to 0$ as $x \to +\infty$, the solution of (1) is:

$$w = A \sin x/\alpha \; e^{-x/\alpha} \tag{2}$$

where $\alpha = [4\,D/(\rho_m - \rho_w)g]^{1/4}$ is the flexural parameter and A is a constant to be determined.

It was shown by Caldwell et al. (1976) that it is convenient to introduce the height of the forebulge (maximum positive deflection) w_b and the coordinate position of the forebulge x_b (see Fig. 1). From (2):

$$x_b = \frac{\pi\alpha}{4} = \frac{\pi}{4}\left[\frac{4\,D}{(\rho_m - \rho_w)g}\right]^{1/4} \tag{3}$$

Introducing $\overline{w} = w/w_b$ and $\overline{x} = x/x_b$, (2) can be written:

$$\overline{w} = 2^{1/2} \sin \frac{\pi\overline{x}}{4} \exp \frac{\pi}{4}(1-\overline{x}) \tag{4}$$

This is a universal trench profile which is valid for elastic behavior. It is plotted in Fig. 4a.

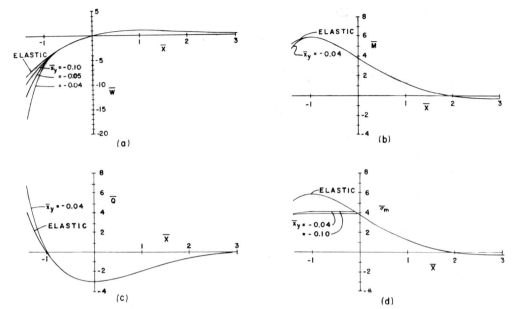

Fig. 4. Dependence of the non-dimensional deflection (a), bending moment (b), shear force (c), and maximum bending stress (d) on the horizontal coordinate for the elastic bending of the lithosphere and the onset of plastic deformation at $\bar{x} = -0.10$, -0.05 and -0.04.

It is also convenient to introduce a non-dimensional bending moment:

$$\overline{M} = \frac{x_b^2 M}{w_b D} = -\frac{x_b^2}{w_b}\frac{d^2 w}{dx^2} = -\frac{d^2 \overline{w}}{d\overline{x}^2} \tag{5}$$

and a non-dimensional shear force:

$$\overline{Q} = \frac{x_b^3 Q}{w_b D} = -\frac{x_b^3}{w_b}\frac{d^3 w}{dx^3} = -\frac{d^3 \overline{w}}{d\overline{x}^3} \tag{6}$$

in terms of the local bending moment M and shear force. Q. From (4) the non-dimensional bending moment is given by:

$$\overline{M} = \frac{2^{1/2}\pi^2}{8}\cos\frac{\pi\overline{x}}{4}\exp\frac{\pi}{4}(1-\overline{x}) \tag{7}$$

and the non-dimensional shear force by:

$$\overline{Q} = -\frac{2^{1/2}\pi^3}{32}\left(\cos\frac{\pi\overline{x}}{4}+\sin\frac{\pi\overline{x}}{4}\right)\exp\frac{\pi}{4}(1-\overline{x}) \tag{8}$$

The solutions for $\overline{M}$ and $\overline{Q}$ as functions of $\overline{x}$ are plotted in Figs. 4b and 4c.

The maximum bending stress (deviatoric) in the elastic region σ_{max} occurs at the top (bottom) of the plate and is given by:

$$\sigma_{max} = \frac{6\,M}{h^2} = \frac{h}{2}\frac{Ew_b}{(1-v^2)x_b^2}\,\overline{M}$$ (9)

A non-dimensional bending stress is defined by:

$$\overline{\sigma} = \frac{2(1-v^2)x_b^2}{Ehw_b}\,\sigma_x$$ (10)

In the elastic region the maximum non-dimensional bending stress σ_{max} occurs at the top (bottom) of the plate and is related to the non-dimensional bending moment by:

$$\overline{\sigma}_{max} = \overline{M}$$ (11)

The maximum non-dimensional elastic bending stress is given in Fig. 4d.
   The average shear stress is given by:

$$\sigma_s = \frac{Q}{h} = \frac{w_b D}{x_b^3 h}\,\overline{Q}$$ (12)

The ratio of the maximum bending stress to the shear stress is given by:

$$\frac{\sigma_{max}}{\sigma_s} = \frac{6\,x_b}{h}\frac{\overline{M}}{\overline{Q}}$$ (13)

Since $|\overline{M}/\overline{Q}|$ is near unity and $x_b/h \simeq 2$, the maximum bending stresses are an order of magnitude larger than the shear stressed. For the bending of thin beams shear stresses are small compared with bending stresses. Therefore, in determining the onset of plasticity, it is sufficient to consider only the bending stresses.
   The problem we are considering is one of plane strain, i.e., $\epsilon_z = 0$. We assume that plastic yielding occurs when the deviatoric bending stress attains the yield stress:

$$|\sigma_x| = \sigma_0$$ (14)

If we accept Tresca's yield condition, then σ_0 is the yield stress in a state of uniaxial stress. If the Von Mises yield condition is considered, then the yield stress σ_0 defined in (14) is slightly greater than the uniaxial yield stress (Hill, 1950).
   When the maximum stress in the plate reaches the yield stress, the outer portions of the plate behave plastically; see Region II in Fig. 3a. Setting $\sigma = \sigma_0$ in (9), the maximum non-dimensional moment for elastic bending M_0 is given by:

$$\overline{M}_0 = \frac{2(1-v^2)x_b^2}{Ehw_b}\,\sigma_0$$ (15)

When $\overline{M} \geqslant \overline{M}_0$ the solution given in (4), (7), and (8) is no longer valid; $\overline{M} = \overline{M}_0$ at $\overline{x} = \overline{x}_0$. The elastic solution is valid for $\overline{x} > \overline{x}_0$.

In the partially plastic section we make the usual assumption that transverse sections remain plane while bending. Therefore, the strain remains a linear function of the vertical distance $|y|$ from the center line of the plate. With $|y| > \gamma$ defining the plastic regions of the plate, the deviatoric stress within the plate is given by Kachanov (1974):

$$\left.\begin{array}{ll} \sigma_x = \sigma_0 & , \quad \gamma \leqslant y \leqslant \dfrac{h}{2} \\[2ex] \sigma_x = \sigma_0 \dfrac{y}{\gamma}, & -\gamma \leqslant y \leqslant \gamma \\[2ex] \sigma_x = -\sigma_0, & -\dfrac{h}{2} \leqslant y \leqslant -\gamma. \end{array}\right\} \tag{16}$$

This distribution of stress is illustrated in Fig. 3(C). It should be noted that the stress component σ_z and the strain component ϵ_y in the plastic zone can be obtained only by specifying a plastic stress—strain increment law (Hill, 1950; Kachanov, 1974). However, the bending moment obtained from (16) is independent of this law.

When the bending plate is partially plastic, the bending moment from the distribution of stress given in (16) is:

$$M = M_c \left[1 - \frac{4}{3}\left(\frac{\gamma}{h}\right)^2\right] \tag{17}$$

where $M_c = \sigma_0 h^2/4$ is the maximum bending moment that the plate can support when it is entirely plastic, i.e., when $\gamma = 0$. Note that the maximum (or critical) bending moment M_c is related to the bending moment at the onset of plasticity M_0, i.e., when $\gamma = h/2$, by:

$$M_0 = \tfrac{2}{3} M_c \tag{18}$$

The curvature of the plate is related to the stress distribution in the elastic core by:

$$\frac{d^2 w}{dx^2} = -\frac{\sigma_0 (1 - \nu^2)}{\gamma E}. \tag{19}$$

Eliminating the width of the elastic region 2γ from (17) and (19) we obtain:

$$\frac{d^2 w}{dx^2} = -\frac{2\sigma_0 (1 - \nu^2)}{Eh\sqrt{3}\left(1 - \dfrac{2}{3}\dfrac{M}{M_0}\right)^{1/2}} \tag{20}$$

which relates the moment to the curvature of the plate. We note that a singularity (infinite curvature) occurs when $M = M_c$. This singularity represents a plastic hinge, and the curvature of the plates will increase until the bending moment drops below the critical value. The equilibrium force balance between the bending moment and the hydrostatic restoring force requires that:

$$\frac{d^2 M}{dx^2} = (\rho_m - \rho_w)\, gw \qquad (21)$$

In the elastic-perfectly plastic region of bending (Region II in Fig. 3), (20) and (21) replace (1).

Introducing the non-dimensional variables which we have previously defined, (20) and (21) can be written, respectively, as:

$$\frac{d^2\overline{w}}{d\overline{x}^2} = -\frac{\overline{M}_0}{\sqrt{3}\left(1 - \frac{2}{3}\frac{M}{M_0}\right)^{1/2}} \qquad (22)$$

$$\frac{d^2\overline{M}}{d\overline{x}^2} = \frac{\pi^4}{64}\,\overline{w}. \qquad (23)$$

These two equations have been integrated numerically using a fourth-order Runge-Kutta algorithm to determine $\overline{w}$ and $\overline{M}$ as functions of $\overline{x}$ for various values of $\overline{M}_0$. The results are given in Figs. 4a and 4b. We found that whenever $\overline{x}_0 > 0.038$, a plastic hinge develops. Since this singular behavior must be avoided, the bending lithosphere *must* behave elastically for $\overline{x} > -0.038$.

The solution to (22) and (23) is valid as long as the bending moment is increasing. The bending moment is a maximum at $\overline{x} = \overline{x}_m$. For $\overline{x} > \overline{x}_m$, the fact that the lithospheric plate is actually moving has not affected the solution. In the region $\overline{x}_0 > x > \overline{x}_m$, the elastic-perfectly plastic solution is valid. However, in the region where the bending moment decreases (Region III in Fig. 3a) the prior history of plastic deformation influences the bending response of the plate (Kachanov, 1974). This reduction of curvature, or unbending, is an elastic process. The change in stress is proportional to y. However, only in the region that has not experienced plastic deformation, $|y| < \gamma_{min}$, is the stress proportional to y. A typical distribution of stress in the unbending region is shown in Fig. 3b. In this region (Region III of Fig. 3a) (23) is valid but (22) is replaced by:

$$\frac{d^3\overline{w}}{d\overline{x}^3} = -\frac{d\overline{M}}{d\overline{x}}. \qquad (24)$$

The solution valid for $\overline{x} > \overline{x}_m$ has been extended into Region III by the numerical integration of (23) and (24), again using a Runge-Kutta algorithm. The initial conditions for this integration are obtained from the continuity of $\overline{w}$, $d\overline{w}/d\overline{x}$, $d^2\overline{w}/d\overline{x}^2$, $\overline{M}$, and $\overline{Q}$ at $\overline{x} = \overline{x}_m$. Note that these unloading equations apply only as long as the magnitude of the stresses remains less than the yield stress σ_0.

RESULTS

The non-dimensional deflection, bending moment, shear force, and maximum bending stress for elastic-perfectly plastic bending are compared with the elastic results in Figs. 4(a–d). Results are given for various values of x_y. In Fig. 4(a) it is seen that inclusion of plasticity in the model affects the deflection significantly only for $\bar{x} < -1$; this is well within the trench. Plasticity cannot affect the location of the forebulge although it can certainly affect its height.

Plasticity has only a small effect on the bending moment — Fig. 4b — and the shear force: Fig. 4c. Plasticity reduces the maximum bending stress by about a third. This reduction is insensitive to the extent of the plastic region as can be seen in Fig. 4d.

COMPARISON WITH OBSERVATIONS

We compare the derived deflection curves with observed topographic profiles for four trench systems: Mariana (Fig. 5), Kuril (Fig. 6), Tonga (Fig. 7), and central Aleutian (Fig. 8). Profiles have been selected which have the least extraneous topography, i.e., seamounts, vertical offsets, etc. Where necessary corrections have been applied for variations in sediment thickness and the subsidence of the lithosphere due to its cooling; the methods given by Caldwell et al. (1976) have been used.

Values of w_b and x_b were obtained by fitting the observed topography to the elastic deflection curve — Fig. 4a — for $\bar{x} > 0$. The resulting values are given in Table I. The flexural rigidities are then obtained from (3) and lithospheric thicknesses from the definition of the flexural rigidity. In making

TABLE I

Trench parameters

		Mariana	Kuril	Tonga	Aleutian
x_b	(km)	55	40	60	50
w_b	(km)	0.5	0.25	0.20	0.35
D	(dyne cm)	$1.38 \cdot 10^{30}$	$3.87 \cdot 10^{29}$	$1.96 \cdot 10^{30}$	$9.44 \cdot 10^{29}$
h	(km)	28.8	18.8	32.3	25.4
x_t	(km)	−68	−64	−84	−50
M_t	(dyne)	$1.30 \cdot 10^{22}$	$2.49 \cdot 10^{21}$	$5.08 \cdot 10^{21}$	$7.80 \cdot 10^{21}$
Q_t	(dyne cm^{-1})	$8.57 \cdot 10^{14}$	$1.02 \cdot 10^{15}$	$1.37 \cdot 10^{15}$	$5.07 \cdot 10^{13}$
x_m	(km)	−55	−40	−59	−49
M_m	(dyne)	$1.35 \cdot 10^{22}$	$3.56 \cdot 10^{21}$	$6.45 \cdot 10^{21}$	$7.81 \cdot 10^{21}$
σ_m	(kb)	9.77	6.04	2.47	4.85
x_o	(km)	—	—	−2.34	−1.95
M_o	(dyne)	—	—	$4.30 \cdot 10^{21}$	$5.21 \cdot 10^{21}$
σ_o	(kb)	—	—	2.47	4.85

these calculations we have taken $E = 6.5 \cdot 10^{11}$ dyn/cm^2, $\nu = 0.25$, and $(\rho_m - \rho_w) g = 2.3 \cdot 10^3$ dyn/cm^3.

For the Mariana and Kuril profiles we find that the best fits are with the fully elastic profile. For the Tonga and the Aleutian profiles we fit the topography with the elastic-perfectly plastic profiles for $\bar{x}_0 = -0.039$ ($\overline{M}_0 = 3.944$). The position of the trench axis x_t, the bending moment at the trench axis M_t from Fig. 4(b) and eq. 5, and the shear force at the trench axis Q_t from Fig. 4(c) and eq. 6 are tabulated in Table I. Also given are the position of the maximum moment x_m, the value of the maximum moment M_m from Fig. 4(b) and eq. 5, and the maximum bending stress σ_m from Fig. 4(d) and eq. 9. For the Tonga and Aleutian profiles the maximum bending stress is equal to the yield stress σ_0.

Mariana profile

The Mariana profile given in Fig. 5 is from the Scan 5 cruise. A sediment correction has been applied using the data of Karig (1971). This profile has the highest forebulge (0.5 km), indicating strong bending. However, the profile is in good agreement with the elastic theory. The maximum bending stress is very high, 9.77 kb. This value is nearly equal to the yield strength of dunite obtained at high strain rates in the laboratory (see Fig. 2).

Kuril profile

The Kuril profile given in Fig. 6 is from the Zetes 2 cruise. A sediment correction has been applied using the data of Wertenbaker (1974). The forebulge is relatively close to the trench indicating a relatively thin lithosphere (18.8 km). Good agreement is obtained with the elastic profile.

Tonga profile

The Tonga profile given in Fig. 7 is from the 1970 cruise of the NOAA ship Oceanographer. No sediment correction was required for this profile.

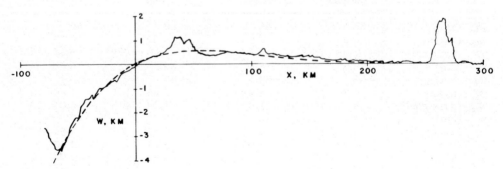

Fig. 5. Mariana profile from the Scan 5 cruise compared with the elastic deformation profile.

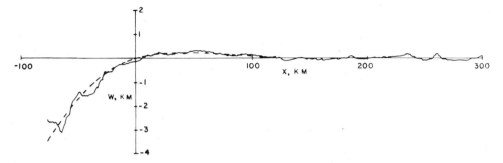

Fig. 6. Kuril profile from the Zetes 2 cruise compared with the elastic deformation profile.

Comparison with the elastic profile shows poor agreement. The elastic-perfectly plastic profile gives good agreement with $\bar{x}_0 = -0.039$. The deviation of the Tonga profile from the elastic profile is very close to the point predicted by the elastic-perfectly plastic analysis. The required yield stress is quite low (2.47 kb).

Central Aleutian profile

The central Aleutian profile given in Fig. 8 is from the Seamap 13 cruise. A sediment correction has been applied using the data of Hayes and Ewing (1968). An age correction has also been applied using the data of Sclater et al. (1974). The profile appears to be in better agreement with the elastic-perfectly plastic deformation curve ($\bar{x}_0 = -0.039$) than with the elastic deformation curve. The differences are not as striking as in the Tonga case because the Aleutian Trench is considerably shallower due to the large sediment wedge.

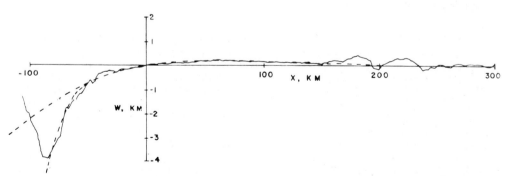

Fig. 7. Tonga profile from the 1970 cruise of the NOAA ship Oceanographer compared with the elastic deformation profile and the elastic-perfectly plastic profile with $\bar{x}_0 = -0.039$.

322

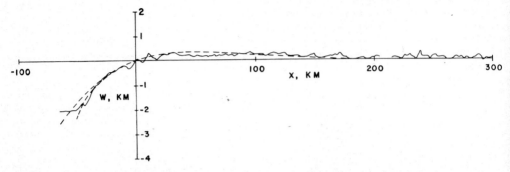

Fig. 8. Central Aleutian profile from the Seamap 13 cruise compared with the elastic deformation profile and the elastic-perfectly plastic profile with $\bar{x}_0 = -0.039$.

CONCLUSIONS

Observed trench profiles which do not fit the elastic theory appear to be in good agreement with the elastic-perfectly plastic analysis. The deviation from elastic behavior occurs near the predicted point on the profile.

Derived lithospheric thicknesses vary from 19 to 32 km. There does not appear to be a systematic dependence on lithospheric age. The maximum bending stresses vary from 2.5 to 9.8 kb. The maximum stress does not appear to correlate with lithospheric age or with lithospheric thickness.

ACKNOWLEDGEMENTS

This research was supported by the National Science Foundation Grants EAR 74-03259 and DES 75-0415 and by the National Aeronautics and Space Administration NASA Grant NSG 5060. We would like to thank D.E. Karig for his contributions to this work.

REFERENCES

Caldwell, J.G., Haxby, W.F., Karig, D.E. and Turcotte, D.L., 1976. On the applicability of a universal elastic trench profile. Earth Planet. Sci. Lett., 31: 239—246.
Goldstein, S., 1926. The stability of a strut under thrust when buckling is resisted by a force proportional to displacement. Proc. Cambridge Philos. Soc., 23: 120—129.
Griggs, D.T., Turner, F.J. and Heard, H.C., 1960. Deformation of rocks at 500° to 800°C. Geol. Soc. Am. Mem., 79: 39—104.
Gunn, R., 1937. A quantitative study of mountain building on an unsymmetrical earth. J. Franklin Inst., 224: 19—53.
Gunn, R., 1947. Quantitative aspects of juxtaposed ocean deeps, mountain chains, and volcanic ranges. Geophysics, 12: 238—255.
Hanks, T.C., 1971. The Kuril Trench-Hokkaido rise system: Large shallow earthquakes and simple models of deformation. Geophys. J.R. Astron. Soc., 23: 173—189.
Hayes, D.E. and Ewing, M., 1968. Pacific boundary structure. In: A.E. Maxwell (Editor), The Sea, 4: 29—72.

Hill, R., 1950. The Mathematical Theory of Plasticity. Oxford University Press, London, 355 pp.

Kachanov, L.M., 1974. Fundamentals of the Theory of Plasticity. Mir, Moscow, 445 pp.

Karig, D.E., 1971. Site surveys in the Mariana area (Scan IV). In: Initial Reports of the Deep-Sea Drilling Project, 6: 681—689.

Le Pichon, X., Francheteau, J. and Bonnin, J., 1973. Plate Tectonics. Elsevier, Amsterdam, 300 pp.

Parsons, B. and Molnar, P., 1976. The origin of outer topographic rises associated with trenches. Geophys. J. R. Astron. Soc., 45: 707—712.

Prager, W. and Hodge, P.G., 1951. Theory of Perfectly Plastic Solids. Wiley, New York, N.Y., 264 pp.

Sclater, J.G., Jarrard, R.D., McGowran, B. and Garther, S., 1974. Comparison of the magnetic and biostratigraphic time scales since the Late Cretaceous. In: Initial Reports of the Deep-Sea Drilling Project, 22: 381—386.

Smoluchowski, M., 1909. Über ein gewisses Stabilitätsproblem der Elastizitätslehre und dessen Beziehung zur Entstehung von Faltengebirgen. Bull. Int. Acad. Sci. Cracovie, pp. 3—20.

Walcott, R.I., 1970. Flexural rigidity, thickness, and viscosity of the lithosphere. J. Geophys. Res., 75: 3941—3954.

Watts, A.B. and Talwani, M., 1974. Gravity anomalies seaward of deep sea trenches and their tectonic implications, Geophys. J. R. Astron. Soc., 36: 57—90.

Wertenbaker, W., 1974. The Floor of the Sea. Little, Brown, Boston, Mass., 275 pp.

Chapter 15

THE EVOLUTION OF THERMAL STRUCTURES BENEATH A SUBDUCTION ZONE*

ALBERT T. HSUI and M. NAFI TOKSÖZ

Department of Earth and Planetary Sciences, Massachusetts Institute of Technology, Cambridge, Mass. 02139 (U.S.A.)

SUMMARY

Thermal models of the subducting lithosphere and the surrounding mantle are investigated simultaneously. In the model, slab subduction is treated as a kinematic process. Subduction angle and subduction velocity are prescribed. The mantle is permitted to respond dynamically to subduction. Mantle flow is driven by both mechanical shearing and thermal buoyancy. Finite difference numerical techniques are utilized to model the subduction process. Results indicate that the descending lithosphere and the induced flow in the wedge area dictate the thermal structure beneath plate convergent regions. The temperature of a slab is found to be insensitive to the mantle condition prior to slab subduction. Viscous shear heating effects within the dynamic mantle are found to be important in maintaining a high temperature region immediately beneath the inter-arc basins.

INTRODUCTION

The subduction regions are among the most complex tectonic provinces on Earth. Many processes taking place in these regions have not yet been understood satisfactorily (Uyeda, 1977). The thermal regime of the region is undoubtedly a piece of important information which enable us to better understand many geophysical and geological processes taking place in the converging plate boundaries. The thermal structure of a downgoing slab has been studied by a number of investigators (McKenzie, 1969; Turcotte and Oxburgh, 1969; Minear and Toksöz, 1970a, b; Hasebe et al., 1970; Toksöz et al., 1971, 1973; Turcotte and Schubert, 1971, 1973; Griggs, 1972; Schubert et al., 1975). These models were calculated without taking into account the effects of the dynamic mantle. Using these slab models as boundary conditions, the thermal structure of the surrounding mantle has also been investigated (Andrews and Sleep, 1974; Jones, 1977; Toksöz and Bird, 1977; Toksöz and Hsui, 1978). These two families of models generally ignored the continuous interaction between the slab and the mantle. In this

*Originally published as: Hsui, A.T. and Toksöz, M.N., 1979. The evolution of thermal structures beneath a subduction zone. Tectonophysics, 60: 43—60.

chapter we construct a more realistic model beneath the subduction region by calculating the temperature within a slab and that of the surrounding dynamic mantle simultaneously.

Another problem we investigate is why the temperature above the slab is relatively high. High temperature in the wedge area is inferred by the high surface heat flow measurements (for a good summary, see Watanabe et al., 1977) and the high attenuation of seismic waves (both P and S) in these regions (Oliver and Isacks, 1967; Molnar and Oliver, 1969; Utsu, 1971; Barazangi and Isacks, 1971; Barazangi et al., 1975). Beneath the converging plate boundaries, the mantle is cooled by the subducting oceanic lithosphere. If the subduction rate is 10 cm/yr and the plate thickness is 100 km, a volume of about 10^4 km^3 of cold lithospheric material will be injected into the mantle per million years per km width of the subduction zone. Taking an average temperature difference of about 500°C between the lithosphere and the mantle beneath, it follows that the same volume of the mantle will be replaced by the 500-degree cooler lithospheric material in a million years. Within the same period of time, radiogenic heat production (assuming that heat production is at equilibrium with the surface heat fluxes and the heat sources are uniformly distributed within the mantle) is only able to increase the temperature by about 1°C. Then, what is the mechanism that keeps the wedge region from cooling? This chapter looks at this problem in some detail.

MODEL DESCRIPTION

Ideally, it would be desirable to treat the subduction and mantle-convection problem as a closed, continuous and dynamic system with appropriate mantle properties. However, this presents extreme difficulties, both analytically and numerically. In this paper, the thermal structure of the subduction region is studied by modeling a slab subducting *kinematically* into a deformable mantle. In other words, the subduction velocity and the angle of subduction are prescribed. Instead of posing the problem as a continuously evolving system, we calculate the thermal regime at discrete time intervals. The basic approach is very similar in philosophy to that described by Minear and Toksöz (1970a, b) such that we are in essence studying the heating of a discretely descending slab by the dynamic mantle. The slab is moved downward by one grid space at a time. Calculations are then carried out to determine how the slab is heated up by the surrounding mantle within the time interval necessary for the slab to reach that depth. The convection and the thermal regime of the mantle are also determined simultaneously. Our model is an extension of most of the published slab models.

The model consists of a 100 km thick conducting layer (lithosphere) lying on top of a convecting 700 km thick upper mantle. Calculations for our model are separated into two parts. One part is to calculate the thermal structure within the lithosphere and the subducting slab, while the other is to compute the thermal regime within the dynamic mantle that lies beneath the conducting lithosphere and surrounds the descending slab. The two calculations interact with each other through their common boundaries. When the slab subducts, its temperature is translated to a new position, the mantle is then allowed to respond to this disturbance within the time interval necessary for the slab to travel to its new position. The effects of slab shearing, thermal buoyancy, and adiabatic compression are incorporated in the calculations. Computations of the slab temperatures and the mantle temperatures are self-consistent at each time step through the matching of boundary conditions. An explicit, forward time march, finite difference numerical scheme is used to simulate the thermal consequences of the subduction process. Details of the mathematical formulation, the numerical approximations, and the computational procedures are very similar to those given in Toksöz and Hsui (1978).

Boundary conditions. The boundary conditions are difficult to specify since neither the mantle flow structure nor the nature of the bottom of the mantle convection layer is known with great confidence. However, if the vertical boundaries are sufficiently far away from the subduction region, the effects of the vertical boundary conditions will be small on the thermal structure of the slab and its immediate environment. In our models, the horizontal dimension spans about 2800 km. The location of subduction at the surface is chosen at least 800 km away from either end, which is more than the vertical length scale of the convecting mantle. After the slab has been translated to its new position, the thermomechanical structure of the mantle is calculated assuming the vertical boundaries are shear stress free, mass exchange free, and thermally insulated. At the top of the dynamic mantle, a prescribed velocity equivalent to the slab velocity is imposed on the oceanic side of the slab while a no-slip condition is imposed on the continental side. The outward heat flux from the mantle at this boundary is determined from the thermal structure of the overlying lithosphere. At the bottom, a prescribed inward heat flux is imposed and a no-slip mechanical condition is used. We also calculated models with a shear-stress-free bottom boundary condition. The thermal structures are not very different from those of a no-slip bottom boundary condition. This is in agreement with the conclusions of Hsui (1978) that thermal convection solutions are not sensitive to the mechanical boundary conditions at the bottom. The most difficult boundary condition to choose is probably that at the tip of the slab. In this paper, we choose a no flow condition at this boundary, so that it is consistent with the no mass flux condition imposed on the two vertical boundaries. Mass flow induced by the tip strongly depends on the shape of the tip. Thus the choice of this boundary condition may introduce different effects locally. Richter (1977) studied the effects of tip push and found that it could be significant. However, the effects of variable mantle viscosity and different boundary geometries are not investigated. Furthermore, when one introduces a mass flux at the tip of a slab (mass input) one also has to specify the distriburion of outward mass flux at other boundaries. Different distributions of outward mass flux will affect the solutions in their own way. Therefore, this alternative will also introduce its own uncertainties to the problem. It should be stated that the "tip" problem goes away when the slab penetrates the bottom.

Initial conditions. Since the problem is cast as a transient problem, initial conditions are required to complete the mathematical formulation. Again, neither the mechanical nor the thermal conditions within the mantle prior to slab subduction are known. Therefore, we choose to treat the initial condition as a parameter and study the sensitivity of the thermal regime of a slab and its immediate environment to the initial conditions chosen. Even though very little data exists which can be used to constrain the initial conditions, we assume that the mantle was probably at equilibrium with the heat inputs to the system prior to slab subduction. Therefore, a steady state solution for the dynamic mantle in the absence of a slab is obtained and used as the initial condition for the mantle. Generally, any solution of thermal convection will yield upwelling as well as downgoing currents. In order to study the local effects upon a subducting slab, subduction has been initiated at three different locations: above the upwelling current, above the downgoing current, and above the center of a convection cell. With these three modes of subduction, the regional effects of different initial conditions on slab temperatures can be analyzed.

Heat sources. The mantle not only deforms in response to the mechanical shearing of a descending slab, but motions can also occur within the mantle due to thermal buoyancy generated by internal heat release. Therefore, the specification of heat source strength within the system is necessary. However, the distributions of U, Th, and K heat sources within the Earth are not known in detail. In this paper, the strength of the heat production rate is estimated by assuming that it is at equilibrium with the surface heat flux. The average heat flow values from the ocean floor vary from $50-80$ erg/cm^2-s (Sclater et al., 1979). We further assume that the lithosphere and the lower mantle each contribute about 25% of this heat budget while the rest is generated internally within the upper mantle. (i.e., The heat production rate within the lithosphere is about $1.5 \cdot 10^{-6}$ erg/cm^3-s. The heat production rate within the upper mantle is $4.5 \cdot 10^{-7}$ erg/cm^3-s and the heat flux from the lower mantle is about 13.5 erg/cm^2-s.) The choice of heat source distribution does not appear to affect the results very much since the time duration is short for the radiogenic isotopes to generate significant heating. For the same reason, the heat production rate is taken to be constant in time in our calculation.

Viscosity. The relationship between the rate of deformation and the shear stress within the mantle remains an obscure subject mainly because of the absence of relevant laboratory data at the high pressures and low strain rates appropriate to the mantle conditions. However, numerical experiments (Turcotte et al., 1973; Parmentier et al., 1976) indicate that the thermal structure of a convective system is very sensitive to the choice of a constant or a strongly temperature and pressure dependent viscosity, while relatively insensitive to the assumption of a newtonian or a non-newtonian behavior. Therefore, for mathematical simplicity, our model assumes a newtonian medium with a strongly temperature and pressure dependent viscosity. The viscosity ν is given by: $\nu = A \exp(BT_m/T)$, where T_m is the absolute melting temperature which is taken from the 0.1% hydrous pyrolite mantle model of Ringwood (1975). T is the local absolute temperature. A and B are two constants so adjusted that the resultant viscosities are in agreement with those determined from the study of glacial uplift data (Cathles, 1975; Peltier and Andrews, 1976; Peltier, 1976). In this paper, $A = 1.8 \cdot 10^{16}$ cm^2/s and $B = 9.0$, are used.

Phase boundaries. The upper mantle phase changes (basalt-eclogite, pyrolite-garnet, olivine-spinel, spinel-post spinel, etc.) could affect both the convection and thermal regime by density changes and heats of reaction. The effects of phase transitions on mantle dynamics have been studied by Schubert and Turcotte (1971) and Schubert et al. (1975). Their results indicate that the olivine-spinel phase transition is unlikely to inhibit mantle convection. Whether the spinel to post-spinel phase transition will inhibit mantle convection depends on the composition of the post-spinel phase. Since this transition is near the bottom of the convection zone, it will have little effect on this problem.

Besides the deformable mantle, phase transitions also take place within the descending slab. Their effects on the thermal regime of a subducting lithosphere have been studied by Toksöz et al. (1971). They concluded that the contribution of phase transitions to the heating of the interior of a downgoing slab is small when compared with other heating mechanisms such as conductive and convective heat transport, adiabatic heating and shear-strain heating. Based on the latent heat associated with these phase transitions (Verhoogen, 1965; Akimoto and Fujisawa, 1968; and Ringwood, 1970), the temperature increase due to phase changes within a slab is less than 50°C locally. Therefore, in this study, we neglect the effects of phase transitions.

THERMAL MODELS

The numerical models presented here are obtained using a 57 × 15 (50 km grid spacing) *x-z* grid system for the convecting mantle and a 10 × 10 km grid system for calculating the temperature within the slab. A test case using a 141 × 36 (20 km grid spacing) grid system for the mantle was carried out. Results are very similar to those obtained using the coarse grid. They differ by no more than 10 percent. Considering the uncertainties on many of the

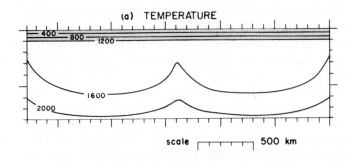

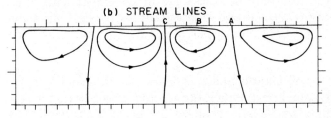

Fig. 1. The initial thermal and mechanical structure of the model calculations presented in this paper. a. Thermal profiles: the solid lines are the isotherms. The temperature in °C is given by the number associated with each isotherm. The shaded area represents the lithosphere. b. The stream lines plot: the maximum absolute stream function $|\psi|_{max}$ = 1.24. Stream lines plotted represent $|\psi|/|\psi|_{max}$ = 1/3, 2/3. Points A, B, and C indicate the locations where slab subduction is initiated.

parameters used in the models, the 50 km mantle grid system is adequate for calculating the gross thermal structure.

Figure 1 shows the initial conditions we used for the models. The diagram on the top is a plot of isotherms (constant temperature lines), while the one at the bottom is a plot of stream lines. In the isotherm plot, the lithosphere (shaded layer) is superimposed on top of the convecting mantle. However, in the stream line plot, the lithosphere has been removed. Points A, B, and C in the stream line plot indicate the locations where the initiation of subduction takes place for the three cases. In all the models, the slab subducts at an angle of 45° counterclockwise with respect to the horizon and at a speed of 8 cm/yr.

The following sequence of figures (Figs. 2—4) shows the thermal evolution history of a subducting slab and its surrounding mantle, for subduction initiating at point B. The shaded area represents the lithosphere and the subducting slab. As in Fig. 1, the top diagrams in this sequence are plots of isotherms while the bottom ones are plots of stream lines. Figure 2 shows the thermal and flow structure at 5.0 m.y. after the initiation of subduction. The thermal field has not been disturbed greatly except near the descending slab where the geotherms are being depressed downward. However, the flow field has been altered significantly. It is mainly because mechanical disturbances nor-

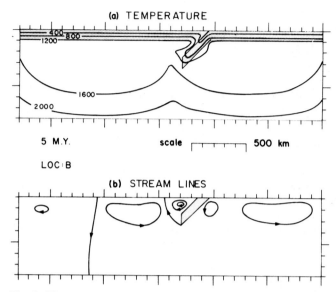

Fig. 2. Thermal-mechanical structure of the subduction zone at 5.0 m.y. after the initiation of slab subduction. $|\psi|_{max}$ = 2.31 and stream lines represent $|\psi|/|\psi|_{max}$ = 1/3, 2/3. Other notations are the same as in Fig. 1.

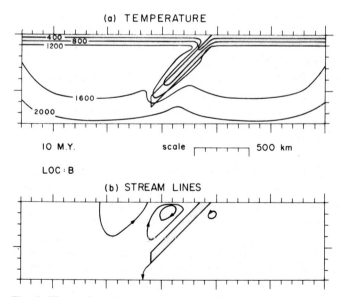

Fig. 3. Thermal-mechanical structure of the subduction zone at 10.0 m.y. after the initiation of slab subduction. $|\psi|_{max}$ = 4.00 and stream lines represent $|\psi|/|\psi|_{max}$ = 1/3, 2/3. Other notations are the same as in Fig. 1.

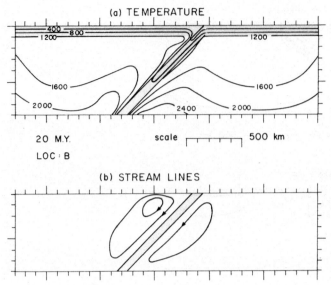

(a) TEMPERATURE

400
800
1200
1200

1600
1600

2000
2400
2000

20 M.Y.
LOC : B

scale ⊢———⊣ 500 km

(b) STREAM LINES

Fig. 4. Thermal-mechanical structure of the subduction zone at 20.0 m.y. after the initiation of slab subduction. $|\psi|_{max}$ = 5.21 and stream lines represent $|\psi|/|\psi|_{max}$ = 1/3, 2/3. Other notations are the same as in Fig. 1.

mally propagate much faster than thermal disturbances. Therefore, the whole region feels the existence of the descending slab and adjusts the flow pattern accordingly even though the slab has penetrated only a small distance into the mantle. As to the thermal structure, the low temperature geotherms are being carried downward into the mantle by the slab. In essence, this demonstrates the cooling of the mantle by the subducting oceanic lithosphere. As the lithosphere subducts, induced convection currents are generated on both sides of the slab. Our models indicate that the induced flow above the slab (on the continental side and within the wedge area) has a higher intensity than that generated on the oceanic side of the slab.

This phenomenon is not observed if the flow is driven only by mechanical shearing (Richter, 1977; Tovish et al., 1978). Flow patterns driven by both mechanical shear and thermal buoyancy have been studied by Torrance et al. (1972). Even though the specific problems they studied are not related to the Earth's mantle, the physical consequences they derived are directly applicable to our study. In their study, Torrance et al. (1972) concluded that buoyancy can significantly change or even dominate the flow driven by shear only, resulting in higher flow velocities. This is in agreement with our results. Due to the combined effects of the large thermal gradient introduced by the descending slab, the strongly temperature dependent viscosity and a small flow cross section, the wedge flow turns out to be a strong current. Our calculation shows that the flow velocity in the wedge can reach as high as two or three times the slab velocity.

At 10 m.y. (Fig. 3), the downgoing oceanic lithosphere has reached a depth of about 550 km. The mantle is cooled on both sides of the slab and the induced flow in the wedge area remains dominant. At 20 m.y. (Fig. 4), the slab has penetrated the bottom of our model. Even though the choice of a bottom at 800 km depth is somewhat arbitrary, comparison between Fig. 4 and Fig. 3 indicates that the thermal structure in the immediate environment of the slab above 300 km appears not to be affected significantly by the presence of this boundary. Since most of the geophysical and geological constraints lie above 300 km depth, the assumption of a bottom at 800 km in our models will not affect very much the near surface results derived from this study.

In the stream line plots of Figs. 1—4, stream functions are plotted at equal spacing for each diagram. Since the maximum absolute stream functions vary from one diagram to the other as indicated in the figure captions, the diminishing of stream lines does not imply the diminishing of the flow velocities. In fact, in all our calculations, flow velocities generally increase in time. Additionally, in our models, mantle viscosity varies from about 10^{19} poise immediately beneath the lithosphere to about 10^{23} poise near the bottom. The viscosity variation takes a very steep gradient near the surface and a relatively shallow gradient away from the asthenosphere. However, the viscosity structure adjacent to the slab is very complicated. It can no longer be expressed in terms of depth only. Since viscosity is directly related to temperature, a complex viscosity structure around a slab is a direct reflection of the complex thermal structure around a descending lithosphere.

Figures 5—7 represent the enlarged view of the isotherm plots presented in the previous evolution sequence (Figs. 2—4). Only regions immediately surrounding the slab are included. Again, the shaded area indicates the lithosphere and the descending slab. These diagrams provide better resolution of the thermal structure in the vicinity of a subducting lithosphere. In general, these diagrams demonstrate the cooling of the mantle as the low temperature isotherms are being carried downward by the descending slab. Since the mantle on both sides of the slab is being cooled by the subducting cold lithosphere, the slab appears thicker thermally except in the region between 100 and 200 km depth. In this region, the cold slab is unable to cool the mantle because a convective current is being induced in such a way that hot mantle material is continuously being pumped towards the vicinity of the slab at this depth. As a result, a bottle neck formation appears in the thermal structure of a slab subduction environment. Because of the coarse grid system used for numerical computation, our models cannot resolve whether this induced flow is able to maintain a temperature sufficiently high in the area immediately adjacent to the slab surface. This is important for the generation of magmas in island-arc volcanoes. Nevertheless, our models demonstrate that the

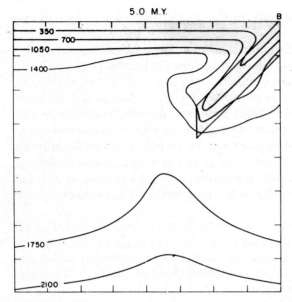

Fig. 5. Enlarged temperature plot of Fig. 2 in the vicinity of a descending slab.

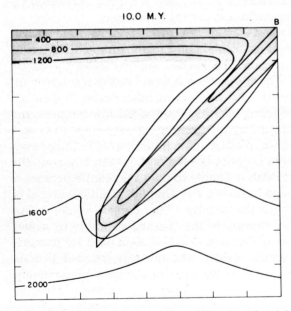

Fig. 6. Enlarged temperature plot of Fig. 3 in the vicinity of a descending slab.

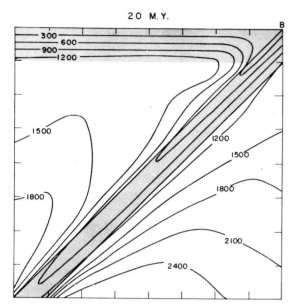

Fig. 7. Enlarged temperature plot of Fig. 4 in the vicinity of a descending slab.

asthenosphere above the slab will not be solidified by the descending cold lithosphere, simply because the induced current will continuously feed hot mantle material into the area.

The numerical results, showing that the temperature at the top of the wedge region remains relatively high, are consistent with the high attenuation of seismic waves and the high surface heat flow of the back arc region (see first section for references). It should be pointed out that all the models presented thus far are constructed with the shear heating effect incorporated not only between the slab and the mantle, but within the mantle flow as well. If the shear heating effect within the mantle flow is not included, the high temperature region at the top of the wedge area no longer exists as shown in Fig. 8. Therefore, viscous shear heating within the mantle flow appears to be important in maintaining a high temperature region beneath the back-arc basins. In fact, viscous shear heating can be as much as 2—3 times higher than the radiogenic heating. Therefore, the requirement of shear heating to maintain a high temperature wedge does not appear too surprising.

In order to see how much our slab temperature model differs from other models, a comparison is made (Fig. 9) between our model taken at 50 m.y. after the initiation of subduction and other published models (McKenzie, 1969; Toksöz et al., 1973; Turcotte and Schubert, 1973). The model by McKenzie (1969) represents the simplest slab model. It is analytic and considers only the effects of subduction and conduction. Therefore, it gives a lower

336

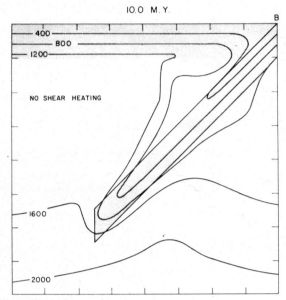

Fig. 8. Same as Fig. 6 except that viscous shear heating has not been incorporated in the model.

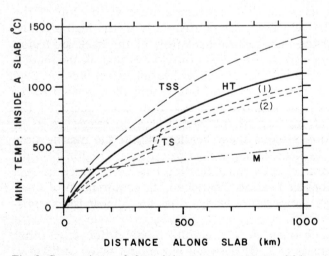

Fig. 9. Comparison of the minimum temperature within a slab as a function of distance along a slab among different slab models. *TSS* is from the model by Toksöz et al. (1973). *HT* is that of this paper. *TS* are those by Turcotte and Schubert (1973). Curve (*1*) is for constant shear stress while curve (*2*) is for constant frictional coefficient. *M* is that by Mc-Kenzie (1969).

bound of the thermal regime within a descending slab. Turcotte and Schubert (1973) have also determined the thermal structure within a slab using analytic methods. The effect of phase transition (olivine-spinel) has been incorporated. However, because of the absence of internal heat generation and utilizing a lower temperature at the base of the lithosphere, their models appear to be generally cooler than our models. Using finite-difference numerical techniques, Toksöz et al. (1973) have calculated the thermal structure within a slab using a convective mantle geotherm and including the effects of adiabatic compression, internal heat release, shear heating and phase changes. This model appears warmer than our present model. Since our model has not considered the effect of phase changes and because olivine-spinel phase change is an exothermic process, it can probably explain why our model is somewhat cooler. Meaningful comparison with the slab model of Toksöz et al. (1971) and Minear and Toksöz (1970a, b) cannot be made, however, because these models used conductive mantle geotherms that result in higher mantle temperatures. The present slab model is bracketed by other models as can be seen in Fig. 9, and it represents a good average temperature model for the subducting lithosphere.

OTHER MODELS

As pointed out previously, the condition of the mantle prior to slab subduction is subject to great uncertainty. Whether the thermal structure of the descending oceanic lithosphere and its environment are related to the initial mantle conditions is not immediately apparent. In order to investigate the relationship between the initial mantle conditions and the subsequent thermal evolution of a descending slab, we carried out models for which subduction has been initiated at different locations. Figures 10 and 11 show the thermal structure of a slab and its environment at two different times. In this case, subduction has been initiated at a point where downgoing mantle flow is taking place (point A in Fig. 1). Comparing Figs. 10 and 11 to Figs. 5 and 7, respectively, the thermal structures within the slab and within the wedge region are found to be very similar. When subduction is initiated at point C (in Fig. 1) where upwelling mantle flow is taking place, results (Figs. 12 and 13), again, are found to be similar to those where subduction is initiated either above a downgoing mantle flow or at the middle of a convection cell. From these results, it is evident that the thermal structure beneath the subduction area is controlled by the descending lithosphere and the resulting induced flow in the wedge area. The subduction process dominates whatever the mantle condition was prior to the descent of the cold oceanic lithosphere.

One of the most prominent features in plate convergent areas is the island arc volcanism. Unfortunately, due to the coarse grid system of the present models, a detailed study of island arc magma generation is not possible. Since both surface topography and gravity in these areas are dominated by

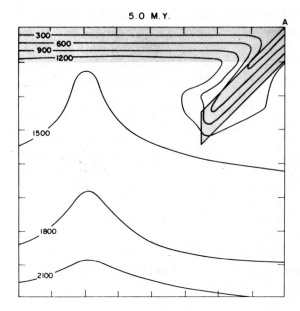

Fig. 10. Same as Fig. 5 except slab subduction is initiated at point A in Fig. 1.

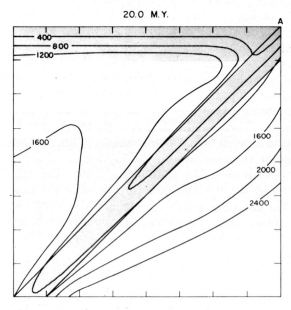

Fig. 11. Same as Fig. 7 except slab subduction is initiated at point A in Fig. 1.

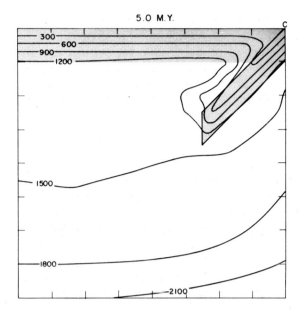

Fig. 12. Same as Fig. 5 except slab subduction is initiated at point C in Fig. 1.

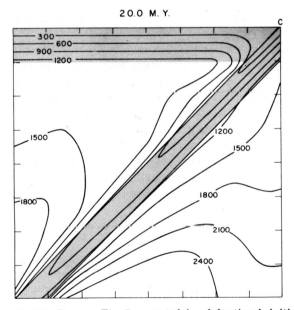

Fig. 13. Same as Fig. 7 except slab subduction is initiated at point C in Fig. 1.

island arcs, the failure to create island arc volcanoes prohibits a meaningful presentation of the surface topography and gravity anomalies predicted by the present models. However, the present model is able to confirm that the induced flow in the wedge area is indeed a strong convection flow, as discussed earlier. Even though by itself alone, the induced flow may not be sufficient to initiate back arc spreading (Toksöz and Hsui, 1978), it remains a plausible mechanism nevertheless.

CONCLUSIONS.

A thermal regime of a slab subducting into the mantle and its surrounding regions has been calculated taking into account the effect of convection with temperature, pressure dependent viscosity. Results indicate that the subduction process is primarily a cooling process. The cold descending lithosphere and the induced flow are the dominating factors in determining the thermal structure of the region at any given time. The temperature within the slab and in its immediate environment is insensitive to the mantle condition prior to slab subduction. In order to maintain a high temperature region beneath the inter-arc basins, as suggested by the high heat flow and the high attenuation of seismic waves, our model indicates that viscous shear heating within the mantle is probably an important factor. Without this, the wedge area is likely to be cooled rapidly by the descending cold lithosphere. Our model of the thermal structure of a downgoing slab is in good agreement with other slab models and it represents an average of these calculations. Finally, the induced flow within the wedge area is indeed a prominent feature of the subduction process. Its contribution to the mechanism of inter-arc spreading remains viable.

The primary purpose of the present model was to understand the thermal evolution of a slab and its surrounding mantle over a broad region. Because of the coarse grid system used for computational efficiency, detailed features such as magma generation and island arc volcanism cannot be studied. In order to have models which can explain detailed surface features, further work with improved computational techniques that yield finer resolutions must be carried out. In the mean time, the models presented here provide a better general understanding of the subduction process and they represent an advancement towards a fully dynamic and more realistic model of the subduction regions.

ACKNOWLEDGEMENTS

We would like to thank Prof. S. Uyeda and Dr. B. Hager for helpful discussions and suggestions. This research was supported by the National Science Foundation under Grant 76-12471EAR and by NASA under Grant NSG-7081 and Cooperative Agreement NCC5-8.

REFERENCES

Akimoto, S. and Fujisawa, H., 1968. Olivine-spinel solid solution equilibria in the system Mg_2SiO_4-Fe_2SiO_4. J. Geophys. Res., 73: 1467—1479.

Andrews, D.J. and Sleep, N.H., 1974. Numerical modelling of tectonic flow behind island arcs. Geophys. J. R. Astron. Soc., 38: 237—251.

Barazangi, M. and Isacks, B., 1971. Lateral variation of seismic wave attenuation in the upper mantle above the inclined earthquake zone of the Tonga Island Arc: Deep anomaly in the upper mantle. J. Geophys. Res., 76: 8493—8516.

Barazangi, M., Pennington, W. and Isacks, B., 1975. Global study of seismic wave attenuation in the upper mantle behind island arcs using P waves. J. Geophys. Res., 80: 1079—1092.

Cathles, L.M., III, 1975. The Viscosity of the Earth's Mantle. Princeton University Press, Princeton, N.J., pp. 155—196.

Griggs, D.T., 1972. The sinking lithosphere and the focal mechanism of deep earthquakes. In: Eugene C. Robertson (Editor), Nature of the Solid Earth. McGraw-Hill, New York, N.Y., 361—384.

Hasebe, K., Fujii, N. and Uyeda, S., 1970. Thermal processes under island arcs. Tectonophysics, 10: 335—355.

Hsui, A.T., 1978. Numerical simulation of finite amplitude thermal convection with large viscosity variation in axisymmetric spherical geometry: effect of mechanical boundary conditions. Tectonophysics, 50: 147—162.

Jones, G M., 1977. Numerical model of a subduction zone. EOS, Trans. Am. Geophys. Union, 58: 499.

Lachenbruch, A.H., 1968. Preliminary geothermal model of the Sierra Nevada. J. Geophys. Res., 72: 6977—6989.

McKenzie, D.P., 1969. Speculations on the consequences and causes of plate motions, Geophys. J. R. Astron. Soc., 18: 1—32.

Minear, J.W. and Toksöz, M.N., 1970a. Thermal regime of a downgoing slab and new global tectonics. J. Geophys. Res., 75: 1397—1419.

Minear, J.W. and Toksöz, M.N., 1970b. Thermal regime of a downgoing slab. Tectonophysics, 10: 367—390.

Molnar, P. and Oliver, J., 1969. Lateral variations in attenuation in the upper mantle and discontinuities in the lithosphere. J. Geophys. Res., 74: 2648—2682.

Oliver, J. and Isacks, B., 1967. Deep earthquake zones, anomalous structures in the upper mantle and the lithosphere. J. Geophys. Res., 72: 4259—4275.

Parmentier, E.M., Turcotte, D.L. and Torrance, K.E., 1976. Studies of finite-amplitude non-Newtonian thermal convection with application to convection in the earth's mantle, J. Geophys. Res., 81: 1839—1846.

Peltier, W.R., 1976. Glacial-isostatic adjustment, II. The inverse problem. Geophys. J. R. Astron. Soc., 46: 669—705.

Peltier, W.R. and Andrews, J.T., 1976. Glacial-isostatic adjustment, I, The forward problem. Geophys. J. R. Astron. Soc., 46: 625—668.

Richter, F.M., 1977. On the driving mechanism of plate tectonics. Tectonophysics, 38: 61—88.

Ringwood, A.E., 1970. Phase tranformations and the constitution of the mantle. Phys. Earth Planet. Int., 3: 109—155.

Ringwood, A.E., 1975. Composition and Petrology of the Earth's Mantle. McGraw-Hill, New York, N.Y., pp. 146—156.

Schubert, G. and Turcotte, D.L., 1971. Phase changes and mantle convection. J. Geophys. Res., 76: 1424—1432.

Schubert, G., Yuen, D.A. and Turcotte, D.L., 1975. Role of phase transitions in a dynamic mantle. Geophys. J.R. Astron. Soc., 42: 705—735.

Sclater, J.G., Jaupart, C. and Galson, D., 1979. The heat flow through oceans and continents. Rev. Geophys. Space Phys. in press.

Toksöz, M.N. and Bird, P., 1977. Formation and evolution of marginal basins and continental plateaus. In: M. Talwani and W.C. Pitman III (Editors), Island Arcs, Deep Sea Trenches, and Back-Arc Basins. Maurice Ewing Ser., Vol. 1, Am. Geophys. Union, Washington, D.C., pp. 379—393.

Toksöz, M.N. and Hsui, A.T., 1978. Numerical studies of back-arc convection and the formation of marginal basins. Tectonophysics, 50: 177—196.

Toksöz, M.N., Minear, J.W. and Julian, B.R., (1971). Temperature field and geophysical effects of a downgoing slab. J. Geophys. Res., 76: 1113—1138.

Toksöz, M.N., Sleep, N.H. and Smith, A.T., 1973. Evolution of the downgoing lithosphere and the mechanisms of deep focus earthquakes. Geophys. J.R. Astron. Soc., 35: 285—310.

Torrance, K., Davis, R., Eike, K., Gill, P., Gutman, D., Hsui, A., Lyons, S. and Zien, H., 1972. Cavity flows driven by buoyancy and shear. J. Fluid Mech., 51: 221—231.

Tovish, A., Schubert, G. and Luyendyk, B.P., 1978. Mantle flow pressure and the angle of subduction: non-Newtonian corner flows. J. Geophys. Res., 83: 5892—5898.

Turcotte, D.L. and E.R. Oxburgh (1969) Convection in a mantle with variable physical properties. J. Geophys. Res., 74: 1458—1474.

Turcotte, D.L. and Schubert, G., 1971. Structure of the olivine-spinel phase boundary in the descending lithosphere. J. Geophys. Res., 76: 7980—7987.

Turcotte, D.L. and Schubert, G., 1973. Frictional heating of the descending lithosphere, J. Geophys. Res., 78: 5876—5886.

Turcotte, D.L., Torrance, K.E. and Hsui, A.T., 1973. Convection in the Earth's mantle. In: B.A. Bolt (Editor), Methods of Computational Physics, Vol. 13. Academic Press, New York, N.Y., pp. 431—453.

Utsu, T., 1971. Seismological evidence for anomalous structure of island arcs with special reference to the Japanese region. Rev. Geophys. Space Phys., 9: 839—889.

Uyeda, S., 1977. Some basic problems in the trench-arc back-arc systems. In: M. Talwani and W.C. Pitman III (Editors), Island Arcs, Deep Sea Trenches and Back-Arc Basins. Maurice Ewing Ser., Vol. 1, Am. Geophys. Union, Washington, D.C., pp.1—14.

Verhoogen, J., 1965. Phase changes and convection in the earth's mantle. Trans R. Soc. London, Ser. A, 258: 276—283.

Watanabe, T., Langseth, M. and Anderson, R.N., 1977. Heat flow in back arc basins of the western Pacific. In: M. Talwani and W.C. Pitman III (Editors), Island Arcs, Deep Sea Trenches and Back-Arc Basins. Maurice Ewing Ser., Vol. 1, Am. Geophys. Union, Washington, D.C., pp. 137—161.

Chapter 16

A DETAILED CROSS-SECTION OF THE DEEP SEISMIC ZONE BENEATH NORTHEASTERN HONSHU, JAPAN*

TOSHIKATSU YOSHII

Earthquake Research Institute, The University of Tokyo, Tokyo (Japan)

SUMMARY

A cross-section of earthquakes located in northeastern Japan is presented by using pP-depths reported by the International Seismological Centre. Travel-time corrections for the water layer were used to recompute pP-depths of earthquakes located below the sea regions. Seven new focal-mechanism solutions, based on teleseismic and Japanese data, were determined for this region. The reconstructed cross-section shows a double seismic zone at intermediate depths of 80—150 km. Earthquakes located within the upper seismic plane are characterized by down-dip compression while those in the lower plane, located about 35 km below the other seismic plane, are characterized by down-dip extension. These observations suggest that, at these depths, stresses attributable to a simple "unbending" of a plate may contribute to the generation of earthquakes in addition to stresses generated by the gravitational sinking of the lithosphere. A detailed cross-section of shallow earthquakes in the same area between the trench and eastern coast of northeastern Honshu is presented along with focal-mechanism solutions. This cross-section delineates more clearly the seismic zones characterized by normal and low-angle thrust faulting.

INTRODUCTION

The spatial distribution of earthquakes beneath the trench—island-arc system is one of the more direct indications of the kinematic processes now operating there. Thus, a detailed cross-section of the seismicity combined with other geophysical data can contribute to our understanding of the behavior of the oceanic and continental lithosphere beneath the trench—arc system. The accurate determination of hypocenters, however, is difficult in the region between the trench and the island arc. Systematic errors in focal parameters cannot be excluded in this region, even if standard errors of least-squares determinations are very small, because seismic stations are usually limited to the land area and complex lateral heterogeneity exists in the crust and upper mantle.

---

*Originally published as: Yoshii, T., 1979. A detailed cross-section of the deep seismic zone beneath northeastern Honshu, Japan. Tectonophysics, 55: 349—360.

Hypocentral data reported by the Japan Meteorological Agency (JMA), the International Seismological Centre (ISC) and the National Earthquake Information Service (NEIS), are now available for earthquakes near Japan. Although ISC has determined focal parameters from arrival-time data at many world-wide stations, including Russian and Japanese stations, the focal parameters still have considerable systematic errors for earthquakes near the trench axis. Focal depths typically have larger errors than the other parameters, and the large errors in depth can seriously distort a cross-section of earthquakes beneath the trench—arc system. Fortunately, ISC has reported focal depths determined from pP—P times for the larger events. In this paper, an attempt is made to construct a cross-section of the earthquakes in northeastern Japan by using depths computed from pP phases, and seismo-tectonic features in this region are examined on the basis of the cross-section obtained and other geophysical data.

CROSS-SECTIONAL DISTRIBUTION OF EARTHQUAKES IN NORTHEASTERN JAPAN

In this study, a cross-section of earthquakes is constructed through northeastern Honshu between the western Pacific and Primorye. Figure 1 shows the area included by the cross-section (a 200 km wide band), focal-mechanism solutions of large, shallow earthquakes occurring in this region (Ichikawa, 1971; Kanamori, 1971a, b; Shimazaki, 1974; Hasegawa et al., 1975) and locations of the volcanic front (VF), the "aseismic front" (AF) and the trench axis (TA). The aseismic front corresponds to the eastern edge of the aseismic, low-Q—low-velocity wedge above the descending oceanic lithosphere (Yoshii, 1975). Almost all the large events between AF and TA have low-angle thrust solutions. Shallow events west of the AF are characterized by horizontal compression oriented in an east—west direction. The large event that occurred during 1933 near the axis of the Japan trench (Fig. 1), was investigated by Kanamori (1971b). He concluded that this event was due to large-scale normal faulting on a fault plane extending over the entire thickness of the oceanic lithosphere.

Figure 2A shows a cross-section of the earthquakes within the 200 km wide band shown in Fig. 1 based on the focal parameters reported by ISC for the period of 1964 through 1973. We can see that the inclined seismic zone at depths is characterized by a spatial concentration of events. It is not clear, however, whether the scattered distribution of the shallower events between northeastern Honshu and the trench axis is real. For example, ISC assumed the same depths determined by JMA for many small earthquakes; but, these depths are believed to be slightly overestimated in this region (Research Group for Explosion Seismology, 1977). Moreover, the cross-section in Fig. 2A contains events beneath the sea region whose depths were determined from pP—P times; the depths for these events are probably overestimated because of the low velocity of the water layer.

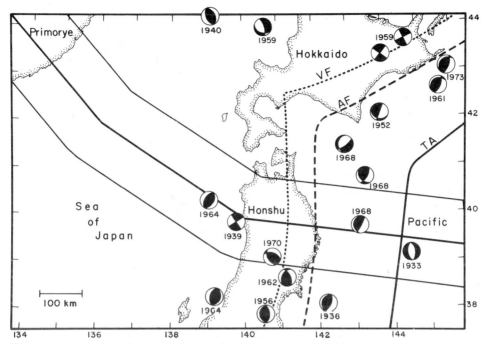

Fig. 1. A map of northeastern Japan showing the area included by the cross-section of earthquakes and focal-mechanism solutions and locations of large, shallow earthquakes (equal-area projection, lower hemisphere). VF, AF and TA indicate the volcanic front, aseismic front and the trench axis, respectively.

Figure 2B is a reconstructed cross-section. Epicenters reported by ISC were used without any change. Focal depths, however, were treated with the following simple criteria:

(1) Depths determined by ISC from pP—P times were used (closed circles in Fig. 2B). In the sea regions, corrections of the water depths were applied to the reported pP—depths because of the low velocity of the water layer. In the corrections, velocities of 7.0 km/sec and 1.5 km/sec were assumed for the crust—upper mantle and for the water, respectively.

(2) Beneath northeastern Honshu and the Sea of Japan, earthquakes located by the usual least-squares method were plotted when the standard error of their depth was reported to be less than 5 km (open circles in Fig. 2B). In these regions, systematic errors in the depth determination appear relatively small because of a better azimuthal distribution of stations.

By the use of the above criteria, many events with magnitude less than about 4 were omitted. The reconstructed cross-section of Fig. 2B shows a remarkable concentration of earthquakes as a plane, with a thickness of only 10—15 km, extending from the trench axis to a depth of about 300 km. Very thin seismic zones beneath trench—arc systems have been reported by several authors (e.g. Engdahl, 1977). As far as the present author knows,

346

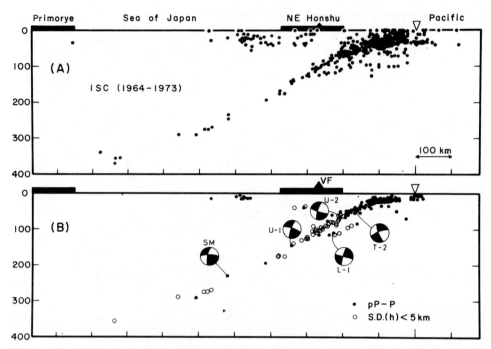

Fig. 2. Cross-sections of earthquakes along the 200 km wide band shown in Fig. 1. Triangles indicate the position of the Japan trench. A. A cross-section based on focal parameters reported by ISC. B. A reconstructed cross-section using pP-depths and well-determined depths (see the text for details). Focal-mechanism solutions of five events are presented in the form of an orthographic projection of the focal sphere on the side of the reader onto a vertical plane parallel to the profile. The solution denoted by SM is based on Stauder and Mualchin (1976).

however, the cross-section in Fig. 2B is a very rare example showing a thin, shallow seismic plane near the trench axis.

A very interesting feature of the cross-section is the existence of another seismic plane 30—35 km below the seismic plane discussed earlier. A double seismic zone beneath northeastern Honshu has been reported for local earthquakes recorded by an array on Honshu by Umino and Hasegawa (1975) and Hasegawa et al. (1976). Their cross-section is shown in Fig. 3; the location of their profile is nearly the same as that in the present study. In addition, they observed that the averaged P- and T-axes of focal-mechanism solutions of earthquakes in each of these two planes are completely reversed. To confirm this, an analysis of the focal mechanism of three events (denoted by U-1, U-2 and L-1 in Fig. 2B) was made.

Figure 4 shows focal-mechanism solutions of these events, and the solutions are also given in Fig. 2B in the form of an orthographic projection onto a vertical plane parallel to the profile. The solutions for U-1, U-2, and SM (Stauder and Mualchin, 1976) on the upper seismic plane indicate a dip of

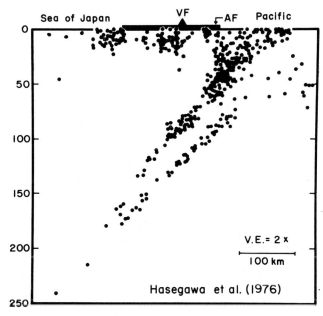

Fig. 3. A cross-section of earthquakes in northeastern Honshu showing the double seismic zone (Hasegawa et al., 1976).

the compressional axis of about 25° which is almost the same as the dip of the seismic plane. On the other hand, the solution for L-1 on the lower plane shows a completely reverse situation; the dip of the tensional axis is about 29°. This result is concordant with solutions of focal mechanisms for small earthquakes on the two planes obtained by Umino and Hasegawa (1975) and Hasegawa et al. (1976).

DETAILED CROSS-SECTION AND MECHANISMS OF EARTHQUAKES NEAR THE JAPAN TRENCH

Figure 5A presents a cross-section of earthquakes from the Pacific coast of northeastern Honshu to near the trench axis during 1968 and 1969. The position of the profile is the same as before (Fig. 1) and pP-depths were employed after correcting for the water layer. Almost all the events in Fig. 5A are aftershocks of the Off Tokachi Earthquake of May 18, 1968. In August 1969, a swarm of relatively small earthquakes took place near the trench axis.

The cross-section of Fig. 5A shows that most of the earthquakes occur within a very thin seismic plane with a thickness of only 10 or 15 km. We can see, however, an abrupt change in the depth of the earthquakes around the dotted line shown in the figure. Across this line, earthquakes on the trench side are slightly deeper on the average than those on the other side.

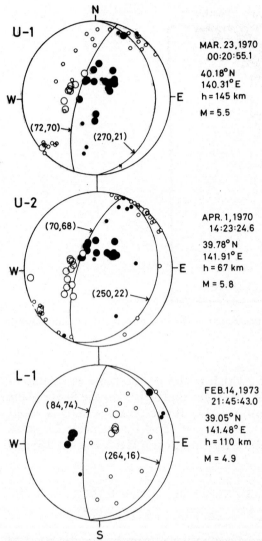

Fig. 4. Focal-mechanism solutions (equal-area projection, lower hemisphere) of events U-1, U-2 and L-1 shown in Fig. 2B. Open circles indicate that the ground motion was down at the station. Larger and smaller circles indicate the data obtained by the author from WWSSN records and those from JMA bulletins, respectively.

Considering the configuration of the seismic plane and the sea-bottom topography along with the mechanism shown later, the inferred upper surface of the oceanic lithosphere is given by the thin curve in Fig. 5A.

Also shown in Fig. 5A are the focal-mechanism solutions for several events with the same projection as in Fig. 2B. Solutions for T-1, T-2, N-1 and N-2 were obtained by the author (see Fig. 6), and those for the other

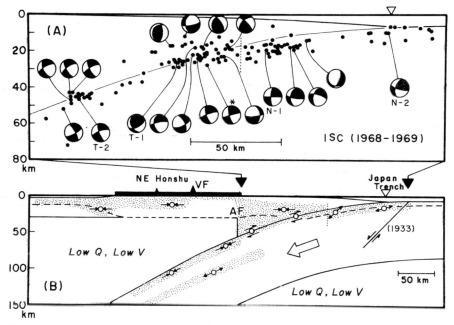

Fig. 5. A. A cross-section from the Pacific coast of northeastern Honshu to the Japan trench. pP-depths were used after correcting for water depth. Mechanism diagrams are orthographic projections onto a vertical plane. The solutions for T-1, T-2, N-1 and N-2 were determined by the author (see Fig. 6). The solution with an asterisk is after Kanamori (1971a), and the others after Ichikawa (1971). B. Schematic representation of the focal mechanism beneath northeastern Honshu. The fault plane for an event of 1933 is after Kanamori (1971b). Dashed curves indicate the M-discontinuity. The stippled areas schematically indicate the seismic zone beneath northeastern Honshu.

events were taken from studies by Ichikawa (1971) and Kanamori (1971a). It has been noticed that earthquakes characterized by thrusting and normal faulting take place very close to each other in this region (Sasatani, 1971; Ichikawa, 1971). Presumably the abrupt change, mentioned above, in the focal depth corresponds to the boundary of these two types of faulting.

In Fig. 5A we can see earthquakes characterized by low-angle thrusting along the inferred upper surface of the oceanic lithosphere, normal faulting below the surface and horizontal compression above the surface. It is well known that earthquakes that are due to normal faulting, often occur near or on the ocean side of the trench axis (e.g. Stauder, 1968a, b; Abe 1972; Kanamori, 1971b; Shimazaki, 1972). Northeastern Honshu may be a slightly special region where we observe normal faulting in a wide zone on the landward side of the trench axis.

Figure 5B is a summary of the focal mechanism around northeastern Honshu. The upper-mantle structure in the figure is the same as those described by Utsu (1971) and Yoshii (1972, 1975). The shallower portion of

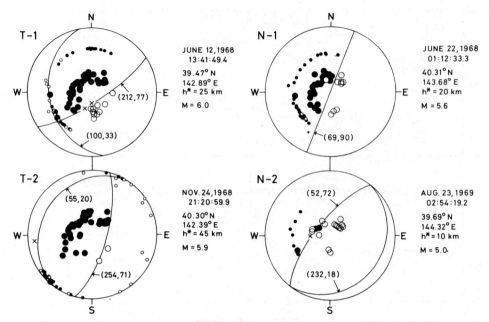

Fig. 6. Focal mechanism diagrams of T-1, T-2, N-1 and N-2 shown in Fig. 5A. Cross marks in the diagrams show nodal arrivals.

this summary is similar to the previous seismo-tectonic model of Isacks et al. (1968, 1969). The present work, however, shows a more detailed relationship between the cross-section and the focal mechanisms of earthquakes near the trench axis.

Figure 6 displays the mechanism solution for events T-1, T-2, N-1 and N-2 in Fig. 5A. Although the locations of T-2 and U-2 are very close to each other, their focal mechanisms are clearly different (see also Fig. 2B). This situation clearly suggests that T-2 took place on the boundary of the oceanic and continental lithosphere while U-2 occurred within the descending oceanic lithosphere. Thus, the aseismic front may mark the position for changes in the predominate type of faulting as shown in Fig. 5B.

DISCUSSION AND CONCLUSION

Figure 7 presents some geophysical profiles from the western Pacific to Primorye along a heavy line shown in Fig. 1 (Yoshii, 1977). Although these data are very important for understanding the present structure and evolution of the trench—arc system, the profile of earthquakes — which is the same as that of Fig. 2A — may be seriously distorted by undesirable errors in the focal depths as described. We have shown that employing the pP-depths is quite effective in removing such errors.

The crust and upper-mantle structure in Fig. 5B well explains all the

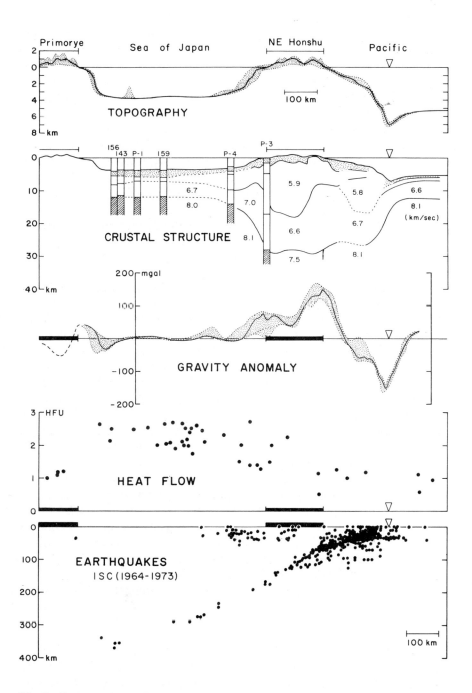

Fig. 7. Cross-sections of topography, crustal structure, gravity anomalies, heat flow and earthquakes along the heavy line shown in Fig. 1 (Yoshii, 1977).

geophysical data on Fig. 7 (Yoshii, 1972, 1977). Important features of this structure are the thin lithosphere of about 30 km thick beneath Honshu and the Sea of Japan, the descending oceanic lithosphere of about 100 km thick and an aseismic, low-Q—low-velocity wedge between them. The M discontinuity just beneath northeastern Honshu is believed to come directly into contact with the underlying low-Q—low-velocity zone or the "asthenosphere", because the P_n velocity in this region is only about 7.5 km/sec (see the crustal structure in Fig. 7).

Schematic representation of the forces operating on this crust and upper mantle, shown in Fig. 5B, strongly supports the previous models of faulting (e.g., Isacks et al., 1968), at least in the shallower portion. It should be noticed, however, that the tensional axes for events beneath the inner wall of the trench are not exactly parallel to the descending oceanic lithosphere, but are dipping slightly toward Honshu. A special explanation may be necessary for this situation although the normal faulting near the trench axis is believed to be a result of the bending of the oceanic lithosphere.

The focal mechanism of earthquakes located in the upper plane of the double seismic zone is of down-dip compression and that of earthquakes in the lower plane is of down-dip tension. A similar result was obtained for earthquakes between 70 and 200 km in depth in the Kurile region (Veith, 1974). These features at intermediate depths may be interpreted as the result of "unbending" (Isacks and Barazangi, 1977; Engdahl and Scholz, 1977) or "sagging" (Smith and Toksöz, 1972) of the oceanic lithosphere beneath the island arc, at least for the lower seismic plane. Moreover, the lower seismic plane may mark the lower boundary of the effectively elastic part of the oceanic lithosphere. The elastic thickness of the lithosphere inferred from the distance of the two seismic planes is 30—40 km and is concordant with the previous estimates from topography of the outer-trench high (e.g., Hanks, 1971; Caldwell et al., 1976).

It is remarkable that we do not observe large thrust earthquakes in the region of about 50 km wide just inside the trench axis. It seems plausible that this region corresponds to the "accretionary prism" (Karig and Sharman, 1975) and cannot cause large thrust earthquakes. An exception may be the Sanriku Earthquake of 1896 which seems to be caused by thrust faulting from tsunami observations (Kanamori, 1972; Hatori, 1974). This event caused a very strong tsunami while its seismological magnitude was rather moderate (JMA magnitude about 7.2). The special nature of this earthquake may be correlated with the mechanically weak "prism" mentioned above.

ACKNOWLEDGEMENT

The author wishes to thank Shuzo Asano and Hugh Rowlett for critically reading the manuscript. Discussions with L. Sykes, K. Nakamura and T. Watanabe were helpful. This research was partly supported by the Earth Science Section, National Science Foundation, NSF Grant EAR-75-03640.

REFERENCES

Abe, K., 1972. Lithospheric normal faulting beneath the Aleutian Trench. Phys. Earth Planet. Inter., 5: 190—198.

Caldwell, J.G., Haxby, W.F., Karig, D.E. and Turcotte, D.L. 1976. On the applicability of a universal elastic trench profile. Earth Planet. Sci. Lett., 31: 239—246.

Engdahl, E.R., 1977. Seismicity and plate subduction in the central Aleutians. In: M. Talwani and W.C. Pitman III (Editors), Island Arcs, Deep Sea Trenches and Back-Arc Basins. Am. Geophys. Union, Washington, D.C., pp. 259—271.

Engdahl, E.R. and Scholz, C.H., 1977. A double Benioff zone beneath the central Aleutians: an unbending of the lithosphere. Geophys. Res. Lett., 4: 473—476.

Hanks, T.C., 1971. The Kurile trench—Kokkaido rise system: large, shallow earthquakes and simple models of deformation. Geophys. J.R. Astron. Soc., 23: 173—189.

Hasegawa, A., Kasahara, K., Hasegawa, T. and Hori, S. 1975. On the focal mechanism of the southeastern Akita earthquake in 1970 (2). Zisin 2, 28: 141—151 (in Japanese with English abstract).

Hasegawa, A., Umino, N. and Takagi, T., 1976. Fine structure of a deep seismic plane in northeast Japan. Abstr. Spring Meet. Seismol. Soc. Jpn., 2: 18 (abstract, in Japanese).

Hatori, T., 1974. Tsunami sources on the Pacific side in northeast Japan. Zisin 2, 27: 321—337 (in Japanese with English abstract).

Ichikawa, M., 1971. Reanalyses of mechanism of earthquakes which occurred in and near Japan, and statistical studies on the nodal-plane solutions obtained, 1926—1968. Geophys. Mag., 35: 207—274.

Isacks, B. and Barazangi, M., 1977. Geometry of Benioff zones: lateral segmentation and downward bending of the subducted lithosphere. In: M. Talwani and W.C. Pitman III (Editors), Island Arcs, Deep Sea Trenches and Back-Arc Basins. Am. Geophys. Union, Washington, D.C., pp. 99—114.

Isacks, B., Oliver, J. and Sykes, L., 1968. Seismology and new global tectonics. J. Geophys. Res., 73: 5855—5899.

Isacks, B., Sykes, L. and Oliver, J., 1969. Focal mechanisms of deep and shallow earthquakes in the Tonga—Kermadic region and the tectonics of island arc. Geol. Soc. Am. Bull., 80: 1443—1470.

Kanamori, H., 1971a. Focal mechanism of the Tokachi-oki earthquake of May 16, 1968: contortion of the lithosphere at a junction of two trenches. Tectonophysics, 12: 1—13.

Kanamori, H., 1971b. Seismological evidence for a lithospheric normal faulting—the Sanriku earthquake of 1933. Phys. Earth Planet. Inter., 4: 289—300.

Kanamori, H., 1972. Mechanism of tsunami earthquakes. Phys. Earth Planet. Inter., 6: 346—359.

Karig, D.E. and Sharman III, G.F., 1975. Subduction and accretion in trenches. Geol. Soc. Am. Bull., 86: 377—389.

Research Group for Explosion Seismology, 1977. Regionality of the upper mantle around northeastern Japan as derived from explosion seismic observations and its seismological implications. Tectonophysics, 37: 117—130.

Sasatani, T., 1971. Distribution of the P-wave initial motions by earthquakes occurred on the Pacific side in the northeast of Japan. Geophys. Bull. Hokkaido Univ., 25: 243—257 (in Japanese with English abstract).

Shimazaki, K., 1972. Focal mechanism of a shock at the northwestern boundary of the Pacific plate: extentional feature of the oceanic lithosphere and compressional feature of the continental lithosphere. Phys. Earth Planet. Inter., 6: 397—404.

Shimazaki, K., 1974. Nemuro-oki earthquake of July 17, 1973: a lithospheric rebound at the upper half of the interface. Phys. Earth Planet. Inter., 9: 314—327.

Smith, A.T. and Toksöz, M.N., 1972. Stress distribution beneath island arcs. Geophys. J.R. Astron. Soc., 29: 289—318.

Stauder, W., 1968a. Mechanism of the Rat Island earthquake sequence of February 4, 1965, with relation to island arcs and sea-floor spreading. J. Geophys. Res., 73: 3847—3858.

Stauder, W., 1968b. Tensional character of earthquake foci beneath the Aleutian Trench with relation to sea-floor spreading. J. Geophys. Res., 73: 7693—7701.

Stauder, W. and Mualchin, L., 1976. Fault motion in the large earthquakes of the Kurile—Kamchatka arc and of the Kurile—Hokkaido corner. J. Geophys. Res., 81: 297—308.

Umino, N. and Hasegawa, A., 1975. On the two-layered structure of a deep seismic plane in the northeastern Japan arc. Zisin 2, 28: 125—139 (in Japanese with English abstract).

Utsu. T., 1971. Seismological evidence for the anomalous structure of island arcs with special reference to the Japanese region. Rev. Geophys. Space Phys., 9: 839—890.

Veith, K.F., 1974. The relationship of island-arc seismicity to plate tectonics. Eos, 55: 349 (abst.).

Yoshii, T., 1972. Feature of the upper mantle around Japan as inferred from gravity anomalies. J. Phys. Earth, 20: 23—34.

Yoshii, T., 1975. Proposal of the "aseismic front". Zisin 2, 28: 365—367 (in Japanese).

Yoshii, T., 1977. Crust and upper-mantle structure beneath northeastern Japan. Kagaku, 47: 170—176 (in Japanese).

Chapter 17

INTERRELATIONSHIPS BETWEEN VOLCANISM, SEISMICITY, AND ANELASTICITY IN WESTERN SOUTH AMERICA*

I. SELWYN SACKS

Carnegie Institution of Washington, Department of Terrestrial Magnetism, Washington, D.C. (U.S.A.)

SUMMARY

Anelasticity studies of the upper mantle beneath South America show that while the thickness of the continental lithosphere is generally in excess of 300 km, there are differences in the interaction between this lithosphere and the downgoing oceanic plate underthrusting the west coast. In the Chile—southern Peru region, there is asthenospheric material between the continental and oceanic lithospheres. The seismicity is mainly confined to the subducting slab, and volcanoes occur in the same relative position to this seismicity as in other subduction regions such as Japan. In central Peru, however, the asthenosphere seems to be absent, the seismicity is dispersed, and there are no volcanoes. This suggests that the contact of hot and weak material (asthenospheric) with the subducting plate is necessary for the release of magma from it.

INTRODUCTION

Subduction zones are regions where oceanic plates, generated at mid-ocean ridges, plunge down into the asthenosphere under an adjacent oceanic or continental plate. This interaction generates a characteristic pattern of seismic and volcanic activity as well as a clearly recognizable anelasticity structure. I study the distribution of these features along the extensive subduction zone of western South America using their expression in the Japan region as a reference for normal distribution. Regions of anomalous anelasticity structure are also found to have diffuse seismicity and an absence of volcanoes.

ANELASTICITY

The absorptive part of seismic energy propagation may be used to calculate an anelasticity structure of a region. The effect of an absorptive medium

---

*Originally published as: Sacks, I.S., 1977. Interrelationships between volcanism, seismicity, and anelasticity in western South America. In: S. Uyeda (Editor), Subduction Zones, Mid-Ocean Ridges, Ocean Trenches and Geodynamics. Tectonophysics, 37: 131—139.

on seismic waves is to attenuate the amplitudes at higher frequencies more than at lower. Seismographs with a wide frequency range (Sacks, 1966) are used and allow accurate absorption $(1/Q)$ determinations to be made. Records from local seismographs, though of smaller bandwidth, are used to expand the coverage. Spectral ratio techniques (e.g., Sacks, 1968) are used throughout to avoid dependence on the source spectrum of the earthquakes used for the study. Anelasticity is often used as a sensitive indicator of the lithosphere (high Q, or low absorption) and the asthenosphere (low Q, or high absorption).

Anelasticity studies of the upper mantle have been made for regions beneath Japan (Katsumata, 1960; Utsu, 1966; Sacks and Okada, 1974), Fiji (Isacks et al., 1968; Barazangi and Isacks, 1971), South America (Sacks, 1969; Sacks and Okada, 1974), and New Zealand (Mooney, 1970).

The anelasticity structure of a typical subduction zone consists of a high-Q $(Q_p \simeq 1000)$ lithosphere overlying a low-Q $(Q_p \simeq 200\text{---}500)$ asthenosphere which in turn overlies the dipping subducted oceanic lithosphere with Q_p of 1000 near the surface to 3000 or more at depth. The lithosphere beneath Japan is about 70 km thick, whereas that beneath South America is in excess of 300 km. The study leading to these conclusions is fully described in Sacks and Okada (1974), and the gross anelasticity structures are shown in Fig. 1.

In some regions such as northern Chile and southern Peru, the gross anelasticity structure is similar to that beneath Japan except for the much thicker lithosphere beneath South America. In these regions, a low-Q zone was found between the high-Q subducting lithosphere and the high-Q continental lithosphere. Intermediate depth (200---300 km) earthquakes occurring in the subducting slab were found to have low-Q paths to the broad-band seismograph at Toconce, which is 210 km from the coast in northern Chile. However, paths to the same seismograph originating at, or penetrating to, the same depth but staying wholly within the continental lithosphere have high Q. Figure 2 shows the seismic ray paths. Paths to a coastal seismograph at Antofagasta have high Q from the same intermediate depth subduction zone earthquakes which had low Q to Toconce. Therefore, there must be a low-Q region between the subducting lithosphere and the continental lithosphere. It is difficult to determine the thickness of this low-Q zone; it is less than 100 km but probably not less than 40 km. This zone has a Q_p of about 400, which is in the range of asthenosphere values found elsewhere, e.g., Japan. The low-Q zone can be traced from Puno, in southern Peru, to Peldehue in central Chile. Seismograms from these stations show the same absorption distribution as those from Toconce in northern Chile.

Beneath central Peru, which has similar seismic station coverage, we have not been able to find an equivalent low-Q slice. All paths from earthquakes at depths less than 300 km to the broad-band seismograph at Cusco (340 km from the coast) have high Q. This seems to be true also for the Huancayo seismograph further to the northwest. These paths are considered in greater detail in Sacks (1969).

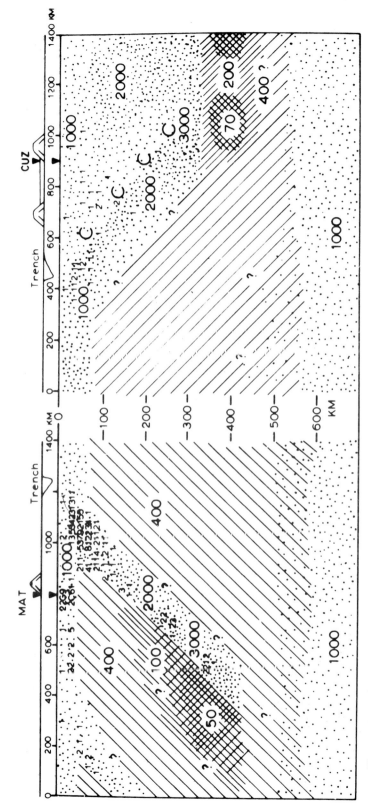

Fig. 1. The Q_p-structure beneath Japan (left) and South America (right). Dots indicate areas of high Q (1000–3000), slant lines intermediate Q-values (300–500), and cross-hatching low Q-values (50–100). The small numbers in the dotted region indicate earthquake hypocenters. The region above the dipping seismic plane is of intermediate Q in Japan, whereas in South America it has high Q. The high-Q slab in Japan is continuous down to the deepest earthquakes; in South America there is a substantial decrease in Q below 350 km even though there are earthquakes 600 km deep. In northern Chile and southern Peru, a distinct low-Q region, marked C, overlies the dipping plane.

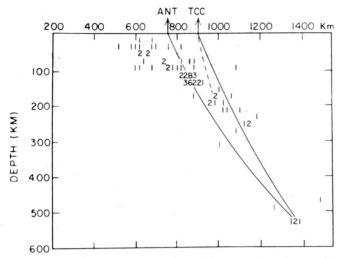

Fig. 2. A seismicity section, 50 km wide, normal to the trench in northern Chile. The numbers indicate earthquake locations (according to CGS) and the arrows show the position of two seismographs whose records were compared. TCC (Toconce) arrivals from 200—250 km events show asthenosphere-type Q-values.

The anelasticity structure described above is in agreement with the areal distribution of absorption given by Barazangi et al. (1975).

To summarize: South of Cusco, i.e., southern Peru and northern Chile, there is a low-Q zone between the subducting lithosphere and the continental lithosphere. Q-values in this zone are similar to those found in the asthenosphere. In the region north of Cusco, no low-Q zone between the high-Q subducting and continental lithospheres has been found, presumably because such a zone is either very thin or non-existent.

SEISMICITY

The seismicity of subduction regions is characterized by a dipping seismic plane some tens-of-kilometers thick, which is thought to delineate the upper surface of the subducting lithospheric plate. There are no earthquakes in the asthenosphere on either side of this subducting plate. Earthquakes do occur in the continental lithosphere above the subduction. The seismicity is not uniform along the subducting plate; regions of low or zero seismic release are found. In some regions, such as New Hebrides, the deeper part of the subducted slab appears to be detached (Barazangi et al., 1973). This is suggested to be the case beneath South America as well (Sacks and Okada, 1974; Snoke et al., 1974, and others). However, Isacks and Barazangi (1973) interpret the subducting plate to be continuous, though having zero seismicity, in the depth range 350—500 km.

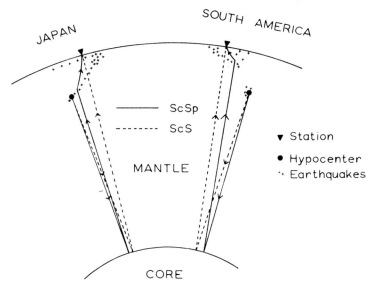

Fig. 3. A schematic model showing the paths of the ScSp phase (solid line) and the ScS phase (dotted line). The shear energy reflected from the earth's core, ScS, is efficiently converted to compressional energy at the dipping interface.

For regions in which the subduction is not clearly delineated by the seismicity, an alternate technique is desirable. Okada (1974) developed such a technique. It relies on the efficient conversion of near-vertical shear waves, ScS, into compressional waves at an inclined interface such as the upper surface of the subducting plate. Figure 3 shows the ray paths giving rise to the phase ScSp. The time difference between ScSp and ScS gives a locus of possible conversion points in the earth, indicated by the shaded line in Fig. 4, which also shows an example of the phase. The amplitude ratio ScSp/ScS is used to determine the depth, indicated by a circle, at which the conversion of shear-to-compressional energy takes place (Snoke and Sacks, 1975; Okada, in preparation). This method has been confirmed in areas where the subducting plate is clearly delineated by earthquakes, e.g., Japan and southern Peru (Okada, 1974).

At Peldehue, in central Chile, clear ScSp arrivals originate from the deeper part (150 km) of the dipping seismic plane. The interpretation of the Arequipa (southern Peru) records indicates an interface which coincides well with that delineated by seismicity. Further north, however, this is no longer the case. The Naña (central Peru) seismograms show clear ScSp arrivals (Fig. 4), which also indicates the probable conversion point. This conversion point lies on an extension of the shallow seismicity (near the trench) which seems to disappear below 100 km depth. This suggests that the subduction has a similar dip angle to the other regions discussed, southern Peru to central Chile, but is of low seismicity.

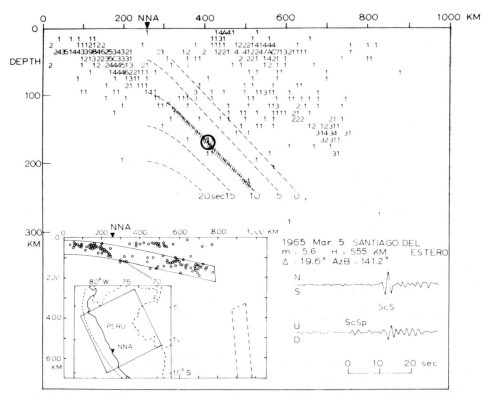

Fig. 4. Vertical seismicity cross-section and traced seismograms of ScS and ScSp arrivals at NNA (central Peru) with a circle representing the preferred conversion region. Isacks and Molnar's (1971) gently dipping slab model is shown in the inset at the lower left. Their seismicity cross-section is taken over the same region as the one above it, but they include only $M > 5$ (USCGS, 1961—1967).

There is considerable seismic activity in what we have interpreted to be the continental lithosphere. This is interpreted by some, e.g., Stauder (1975), Isacks and Molnar (1971), to be subducting oceanic lithosphere with a dip of about 15 degrees. The ScSp observations and the trend of shallow seismicity persuade us that this is not the case.

VOLCANISM

Volcanoes generally occur in subduction regions at some fairly well-defined position relative to the dipping seismic plane. In both South America and Japan (and other regions as well), the earthquakes in the dipping seismic plane are at a depth of about 150 km beneath the line of volcanoes. Figure 5 shows the recent volcanism in South America as well as seismic sections through Chile and Peru. In central Chile, there is a well-defined dipping seis-

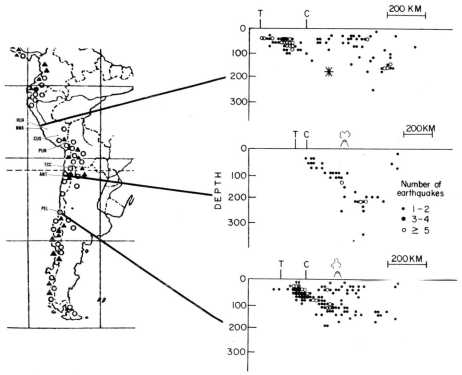

Fig. 5. Recent volcanism and seismicity of western South America. The circles on the map indicate extinct volcanoes, the triangles recent volcanism (reproduced from Fig. V.11, p. 105, in The Earth's Crust and Mantle by F.A. Vening Meinesz, 1964). The three seismicity sections of CGS hypocenters on the right-hand side of the figure are taken normal to the trench and are of equal width. T and C stand for trench and coast, respectively. The asterisk indicates the conversion point for the ScSp phase recorded by the NNA (Naña) seismograph. The seismograph stations mentioned are indicated.

mic plane down to about 150 km, above which the volcanoes occur. In northern Chile, there is a similar situation, i.e., volcanoes 150 km above the earthquake plane, except that the seismic plane extends down to about 300 km. In southern Peru, volcanoes occur about 150 km above the dipping seismicity, which now extends down to 250 km.

In central Peru, however, there are no recent volcanoes. The northern margin of volcanic activity coincides with the disappearance of a clear low-Q zone between the subducting and continental lithospheres as well as a poorer delineation of the subducting plate by the seismicity.

DISCUSSION

For the interaction between the subducting and continental lithospheres, three independent observations suggest that there is a difference between the

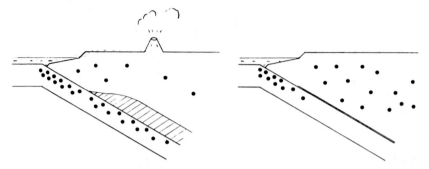

Fig. 6. Schematic drawing indicating the key features in the Chile—southern Peru region (left-hand diagram) and the central Peru region (right-hand diagram). The shaded zone indicates asthenosphere. Where this exists, the earthquakes (dots) are mainly confined to the subducting slab, and volcanoes occur above the 150—175 km deep earthquakes. Where asthenosphere is absent, such as in central Peru, volcanoes do not occur and the seismicity is more dispersed.

southern Peru—Chile region and that of central and northern Peru (see (Fig. 6).

In the southern regions, there is a well-developed low-Q zone, presumably indicating temperatures high enough to cause partial melt, between the subducting plate and the thick (approximately 300 km) continental keel. Most of the seismicity in these regions (see lower two seismicity sections in Fig. 5) is confined to the subducting plate, though there is also considerable activity in the continental block. Volcanoes occur, and the seismic plane is at a depth of about 150 km beneath them, as is the case in other regions such as Japan.

In the central Peru region, however, the low-Q zone (sliver of asthenosphere) is small or absent, and the subducting slab is not characterized by seismicity below about 100 km in depth. There is no recent volcanism in this region, see Fig. 5. Note that there are earthquakes at depths of 150 km, but these occur inland far from the subducting slab (as determined in the studies discussed above) and are considered to be in the continental lithosphere.

It appears that the contact of hot and, therefore, weak material with the subducting lithosphere provides the boundary conditions which give rise to a well-defined seismic plane and the generation of magma.

ACKNOWLEDGMENTS

The operation of the broad-band seismographs is partially supported by N.S.F. Grant No. DES 72-01295A02. Drs. A.T. Linde and J.A. Snoke critically reviewed the manuscript and made many helpful comments.

REFERENCES

Barazangi, M. and Isacks, B., 1971. Lateral variations of seismic-wave attenuation in the upper mantle above the inclined earthquake zone of the Tonga island arc: Deep anomaly in the upper mantle. J. Geophys. Res., 76: 8493—8516.

Barazangi, M., Isacks, B.L., Oliver, J., Dubois, J. and Pascal, G., 1973. Descent of lithosphere beneath New Hebrides, Tonga—Fiji and New Zealand: Evidence for detached slabs. Nature, 242: 98—101.

Barazangi, M., Pennington, W. and Isacks, B., 1975. Global study of seismic wave attenuation in the upper mantle behind island arcs using P-waves. J. Geophys. Res., 80: 1079—1092.

Isacks, B.L. and Barazangi, M., 1973. High-frequency shear waves guided by a continuous lithosphere descending beneath South America. Geophys. J. R. Astron. Soc., 33: 129—139.

Isacks, B.L. and Molnar, P., 1971. Distribution of stresses in the descending lithosphere from a global survey of focal-mechanism solutions of mantle earthquakes. Rev. Geophys. Space Phys., 9: 103—174.

Isacks, B., Oliver, J. and Sykes, L.R., 1968. Seismology and the new global tectonics. J. Geophys. Res., 73: 5855—5899.

Katsumata, M., 1960. The effect of a seismic zone upon the transmission of seismic waves. Kenshinjiho (Q. J. Seismol.), 25: 89—95 (in Japanese).

Mooney, H.M., 1970. Upper mantle inhomogeneity beneath New Zealand: Seismic evidence. J. Geophys. Res., 75: 285—309.

Okada, H., 1974. Geophysical implications of the phase ScSp on the dipping lithosphere underthrusting western South America. Carnegie Inst. Washington, Yearb., 73: 1032—1039.

Okada, H., in preparation. Ph. D. thesis, Hokkaido Univ., Sapporo, Japan.

Sacks, I.S., 1966. A broad-band large dynamic range seismograph. Am. Geophys. Union, Geophys. Monogr. 10: 543—553.

Sacks, I.S., 1968. Q for P-waves in the mantle. Carnegie Inst. Washington, Yearb., 66: 28—29.

Sacks, I.S., 1969. Distribution of absorption of shear waves in South America and its tectonic significance. Carnegie Inst. Washington, Yearb., 67: 339—344.

Sacks, I.S. and Okada, H., 1974. A comparison of the anelasticity structure beneath western South America and Japan. Phys. Earth Planet. Inter., 9: 211—219.

Snoke, J.A. and Sacks, I.S., 1975. Determination of the subducting lithosphere boundary by use of converted phases. Carnegie Inst. Washington, Yearb., 74: 266—273.

Snoke, J.A., Sacks, I.S. and Okada, H., 1974. A model not requiring continuous lithosphere for anomalous high-frequency arrivals from deep-focus South American earthquakes. Phys. Earth Planet. Inter., 9: 199—206.

Stauder, W., 1975. Subduction of the Nazca plate under Peru as evidenced by focal mechanisms and by seismicity. J. Geophys. Res., 80: 1053—1064.

Utsu, T., 1966. Regional differences in absorption of seismic waves in the upper mantle as inferred from abnormal distributions of seismic intensities. J. Fac. Sci. Hokkaido Univ., Ser. 7 (Geophys.), 2: 359—374.

Chapter 18

VARIATIONS IN ANDEAN ANDESITE COMPOSITIONS AND THEIR PETROGENETIC SIGNIFICANCE*

R.S. THORPE and P.W. FRANCIS

Department of Earth Sciences, The Open University, Walton Hall, Milton Keynes MK7 6AA (Great Britain)

SUMMARY

Three linear zones of active andesite volcanism are present in the Andes — a northern zone (5°N—2°S) in Colombia and Ecuador, a central zone (16°S—28°S) largely in south Peru and north Chile and a southern zone (33°S—52°S) largely in south Chile. The northern zone is characterized by basaltic andesites, the central zone by andesite—dacite lavas and ignimbrites and the southern zone by high-alumina basalts, basaltic andesites and andesites. Shoshonites and volcanic rocks of the alkali basalt—trachyte association occur at scattered localities east of the active volcanic chain.

The northern and central volcanic zones are 140 km above an eastward-dipping Benioff zone, while the southern zone lies only 90 km above a Benioff zone. Continental crust is ca. 70 km in thickness below the central zone, but is 30—45 km thick below northern and southern volcanic zones. The correlation between volcanic products and their structural setting is supported by trace element and isotope data. The central zone andesite lavas have higher Si, K, Rb, Sr and Ba, and higher initial Sr isotope ratios than the northern or southern zone lavas. The southern zone high-alumina basalts have lower Ce/Yb ratios than volcanics from the other zones. In addition, the central zone andesite lavas show a well-defined eastward increase in K, Rb and Ba and a decrease in Sr.

Andean andesite magmas are a result of a complex interplay of partial melting, fractional crystallization and "contamination" processes at mantle depths, and contamination and fractional crystallization in the crust. Variations in andesite composition *across* the central Andean chain reflect a diminishing degree of partial melting or an increase in fractional crystallization or an increase in "contamination" passing eastwards. Variations *along* the Andean chain indicate a significant crustal contribution for andesites in the central zone, and indicate that the high-alumina basalts and basaltic andesites of the southern zone are from a shallower mantle source region than other volcanic rocks. The dacite—rhyolite ignimbrites of the central zone share a common source with the andesites and might result from fractional crystallization of andesite magma during uprise through thick continental crust. The occurrence of shoshonites and alkali basalts east of the active volcanic chain is attributed to partial melting of mantle peridotite distant from the subduction zone.

*Originally published as: Thorpe, R.S. and Francis, P.W., 1979. Variations in Andean andesite compositions and their petrogenetic significance. In: S. Uyeda (Editor), Processes at Subduction Zones. Tectonophysics, 57: 53—70.

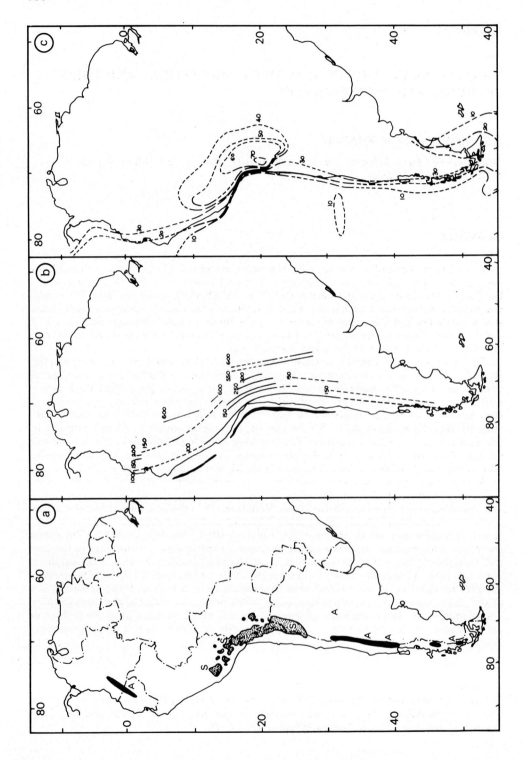

INTRODUCTION

The descent of the oceanic Nazca plate below western South America gives rise to the seismicity, tectonism and volcanism considered definitive of "Andean-type" destructive plate margins. The Andean volcanic rocks show important petrological and geochemical variations both along and across the Andean chain. Here we review the distribution and pattern of Andean volcanism and the variations of crustal thickness and mantle structure, the latter indicated by the depth to the Benioff zone below the volcanic chain. Although the magmas result from a complex interplay of processes in the mantle and crust, there are clear correlations between the compositions of the volcanic products with crustal thickness and thickness of the mantle wedge above the Benioff zone. In this paper we use these correlations to place constraints on the processes responsible for the origin of Andean andesites and their associated volcanic rocks.

DISTRIBUTION OF VOLCANIC ACTIVITY

There are several hundred andesite volcanoes along the length of the Andean Cordillera (Fig. 1). Only 47 of them are identified as "active" (Casertano, 1962; Macdonald, 1972), but many others have undoubtedly experienced recent activity for which no records exist. Three linear zones of recent activity are clearly defined; a northern zone between latitudes 5°N and 2°S (in Ecuador and Colombia), a central zone between 16°S and 28°S (in Peru, Chile, Bolivia and Argentina) and a southern zone between 33°S and 52°S (in Chile and Argentina).

The volcanoes of the northern zone are the least well known, despite the fact that some of them are among the highest and most active in the Andes, notably Sangay, Reventador and Cotopaxi in Ecuador (Hankte, 1966; Pichler et al., 1976). Ramirez (1968) has summarized the information available for southern Colombian volcanoes, and records many historic eruptions. Recent work in Ecuador suggests that the volcanoes of the northern zone are characterized by basaltic andesite eruptions, leading to the construction of simple composite cones (Pichler, 1976; Francis et al., 1977). Rocks of more silicic compositions are known both as pyroclastic fall and flow deposits, but in relatively minor amounts.

Although there is probably a greater density of volcanoes in the central

---

Fig. 1. Relationships between the distribution of volcanic rocks with crustal and mantle structure. a. Solid black = northern and southern active volcanic zones; stipple = central active volcanic zone and ignimbrite province; S = shoshonites; A = alkali-basalts. From data sources in text. b. Contours to top of Benioff zone (km). Dashed lines indicate that contours are based on fewer data than shown by solid lines (from Baranzangi and Isacks; 1976). c. Contours to the base of the crust based on recent seismic measurements (km) (from Cummings and Schiller, 1971).

zone than in any other continental area in the world, the number of major eruptions described is remarkably small. Three main suites of volcanic rocks may be identified. First, extrusive rocks ranging in composition from basaltic andesite to dacite occur throughout the zone, and volcanoes composed of rocks of broadly "andesitic" composition have been built up to heights of 6000 m in many places (Roobol et al., 1976). These major peaks are in general confined to the Western Cordillera of the Andes along the margin of the Bolivian *altiplano* (Fig. 1), but some individual cones occur on the *altiplano* itself, and a distinct chain of recently active cones extends eastwards at the north of the Salar de Uyuni in Bolivia.

Second, large-volume ignimbrite sheets also occur throughout the Western Cordillera, and are also known to outcrop extensively in the Eastern Cordillera of Bolivia (Frailes Formation). The compositions of the ignimbrites are generally dacitic or rhyodacitic (Pichler and Zeil, 1972). Francis and Rundle (1976) estimated that, in one representative part of this area, the ratio of the volumes of andesite to ignimbrite is about 1.3 : 1.

Third, locally there is a volumetrically insignificant but geochemically distinct suite of young alkali basalt extrusives. (Fernandez et al., 1973; Hörmann et al., 1973; Schwab and Lippolt, 1976).

The southern volcanic zone exhibits a markedly higher level of activity than the central zone. Casertano (1962) has summarized data on historic eruptions. Although the summit heights of volcanoes are much lower in this zone than in either the northern and central zones, the sizes of the largest volcanoes are generally similar. Typical volcanic products range from high-alumina basalts to basaltic andesites. More silicic rocks do occur, but in smaller volumes — large-volume ignimbrite sheets are unknown.

AGE AND MIGRATION OF VOLCANISM

All three zones have had long histories of volcanism. Extremely few radiometric dates are available for the northern zone, so it is impossible to delineate a date for the initiation of the youngest episode of volcanism. There is, however, some slight evidence of an eastward migration of volcanic activity with time, in that all the most active volcanoes today are located in the Eastern Cordillera of Ecuador; those in the Western Cordillera seem to be inactive.

The central zone is much better documented, perhaps because of the better preservation of volcanic features in the arid environment. Noble et al. (1974) have compiled radiometric data for volcanic rocks of Peru, and have suggested that there was a period of inactivity about 30 Ma ago followed by a major "pulse" which commenced about 25 Ma ago and "peaked" 12 Ma ago. They ascribed the onset of the pulse to a marked increase in the rotation rate of the Pacific plate 25 Ma ago. More detailed geochronological studies by Baker (1977) and Schwab and Lippolt (1976) demonstrates that construction of andesite volcanoes started at least 10 Ma ago, but that the initiation of widespread ignimbrite volcanism varied from 21 to 10 Ma in different parts of the Western Cordillera. A single date of 7 Ma exists for the major ignimbrite province of the Frailes Cordillera (Evernden, 1966; Baker, 1977). Ignimbrites less than 1 Ma old are known from the Western Cordillera, indicating that this episode of silicic volcanism may still be in progress. The alkali basalts of the Argentinian *puna* are all young.

Morphological data — especially that obtained from LANDSAT images — indicates that the oldest andesitic cones were located in the furthest west parts of the Cordillera, and subsequently migrated eastwards into the Bolivian *altiplano* and Argentinian *puna* (Baker, 1977). Although the eastward migration of activity in the central Andes seems quite clear, and parallels the eastward migration of intrusive centres described by Farrar et al. (1970), it also seems that a *reversal* has taken place in the recent geological past. Almost all the most recently active volcanoes are clustered along the western edge of the Western Cordillera, and even on individual volcanoes there is evidence that the centres of eruptive activity have shifted westwards (Francis et al., 1974).

Drake (1976) has made a detailed geochronological study of part of the southern zone. He demonstrated the existence of several episodes of volcanism from about 30 Ma onwards, separated by periods of folding, and that volcanism has been almost continuous over the last 2.4 Ma. He further showed that a marked westwards migration of volcanic centres has taken place over the last 1—2 Ma, but that no discernible migration of intrusive centres has taken place.

DEPTH TO THE BENIOFF ZONE AND CRUSTAL THICKNESS

The detailed work of Baranzangi and Isacks (1976) makes it possible to establish the relationships between active Andean volcanoes and the depth to the Benioff zone. They pointed out that the two volcanically "dead" areas, between northern and central and central and southern zones correspond exactly with two flat segments of the subducted Nazca plate. The volcanoes of the northern zone are located about 140 km above a well-defined Benioff zone dipping eastwards at about $25°$. A closely similar situation exists in the central zone. In the southern zone, the inclination of the Benioff zone is similar, but here the volcanoes are located only 90 km above it.

The variations in thickness of the continental crust are less accurately known. For the Central Andes, it is reported to be about 70 km (James, 1971b). Beneath the northern zone it is probably about 45 km (Meissnar et al., 1972), while it may be as little as 30 km beneath the southern zone (Cummings and Schiller, 1971). According to Baranzangi and Isacks (1976), a wedge of asthenospheric material between the descending Nazca plate and the overriding South American plate is necessary for the development of active volcanism. This suggestion is supported by the existence of material of extremely high attenuation beneath the central zone; the data for the other two zones is less clear. The occurrence of Pliocene—Recent volcanism (<4 Ma) between the active volcanic zones therefore indicates that the orientation of Benioff zones can change over a short time-scale (Noble and McKee, 1977).

GEOCHEMISTRY

The calc-alkaline lavas show systematic variations in chemical composition parallel to, and transverse to the main Andean volcanic chain. Some of these variations are summarized in Figs. 2 and 3 and chemical analyses of representative lavas are listed in Table I. Lavas in the central zone of the Andes (16—28°S) are dominantly andesite in composition (SiO_2 = 56—66%: Lefèvre, 1973; Francis et al., 1974; Roobol et al., 1976; Dostal et al., 1977a, 1977b). In contrast, lavas from Ecuador in the northern zone are of basaltic andesite and andesite composition (SiO_2 = 53—61%: Pichler, 1976; Pichler et al., 1976; Francis and Thorpe, unpublished). Similarly, lavas from S. Chile in the southern zone (31—42°S) are high-alumina basalts and basaltic andesites (SiO_2 = 50—60%) with less abundant andesites and dacites (Katsui, 1972; Vergara, 1972; Lopez-Escobar et al., 1976, 1977; Moreno, 1976). These variations are matched by trace-element and isotope variations. The lavas from the central zone have higher concentrations of K, Rb, Sr and Ba and have higher initial $^{87}Sr/^{86}Sr$ ratios (generally >0.705, see Fig. 3) than those of Ecuador and S. Chile (references above; Francis et al., 1977; Klerkx et al., 1977; initial $^{87}Sr/^{86}Sr$ of ca. 0.704; see Fig. 3); In addition there are important differences in REE patterns of volcanics from the three volcanic zones: andesites of the northern and central zones have steeper REE patterns with Ce/Yb > 20, while the high-alumina basalts of the southern zone have Ce/Yb < 20 (see Fig. 3).

In addition to these variations, there are important chemical changes in passing from west to east, across the volcanic chain (Fig. 2). These have been well described in the Central Andes: south Peru (Lefèvre, 1973) and north Chile—northwest Bolivia, northwest Argentina (Roobol et al., 1976; Dostal

Fig. 2. Relationships between crustal and mantle structure and chemical composition of lavas in a section across the Central Andes. Ornament is as follows: ticks = continental crust, dots = lithosphere, black = oceanic crust and fine stipple = asthenosphere. The schematic cross-section is at 21—22°S, based on data in James (1971a, b), Stauder (1973) and Baranzangi and Isacks (1976). Variations in SiO_2, K_2O, Sr and Rb for averages of groups of volcanic rocks at different distances from the oceanic trench are shown in the correct position relative to the cross-section. The circles are data from Roobol et al. (1976, table 2 analyses C and D), and Fernandez et al. (1973, tables 2 and 3). The squares are data from Dupuy and Lefèvre (1973, Table 2, Groups A_1, A_2 and B). In the lower graph the open symbols refer to Sr and the filled symbols refer to Rb. See text for further discussion.

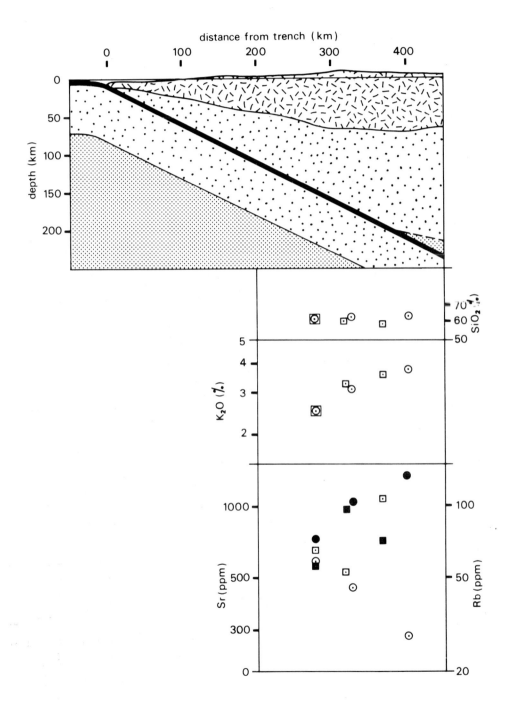

TABLE I

Chemical analyses of representative Andean lavas *

	1	2	3	4	5	6	7	8
SiO_2	62.50	61.9	65.21	53.4	55.72	52.70	52.88	59.20
TiO_2	0.63	0.8	0.98	1.45	0.89	0.89	0.68	1.06
Al_2O_3	15.90	16.6	16.01	15.5	16.89	20.81	18.96	17.08
Fe_2O_3	1.29	1.2	2.16	n.d.	n.d.	n.d.	2.92	2.41
FeO	3.31	4.0	2.03	7.4	7.85	6.11	2.88	3.34
MnO	0.07	0.07	0.07	0.16	0.10	0.17	0.13	0.15
MgO	3.53	2.9	1.56	5.91	5.12	4.80	2.22	1.48
CaO	4.57	5.2	3.65	8.02	7.51	10.50	6.40	2.44
Na_2O	4.19	3.7	2.33	3.8	3.86	3.32	5.09	6.00
K_2O	2.81	3.2	4.29	2.86	1.14	0.44	4.05	3.71
P_2O_5	0.17	0.2	0.28	n.d.	0.23	0.21	0.42	0.42
Others	1.79	n.d.	1.57	n.d.	n.d.	n.d.	2.61	2.40
Total	100.76	99.7	100.14	98.5	99.31	99.95	99.24	99.69
Rb	84	85	194	52	21	18	n.d.	100
Sr	555	510	375	2220	638	374	n.d.	470
$^{87}Sr/^{86}Sr$	0.7063	n.d.	0.7133	0.7042	0.7044	0.7039	n.d.	n.d.
Ce/Yb	30.45	n.d.	n.d.	79.7	27.0	9.9	n.d.	21.9

* Major elements in wt.%; trace elements in p.p.m.; n.d. = not determined. Column headings:
1. Andesite, San Pedro volcano, N. Chile. Westernmost chain of active volcanoes (No. 185, Francis et al., 1974, 1977; Thorpe et al., 1976).
2. Andesite, Ollague volcano, N. Chile (30 km east of analyses 1) (No. 414, unpubl. data).
3. "Rhyodacite", Uturuncu volcano, SW Bolivia (No. 1/5, Fernandez et al., 1973; Klerkx et al., 1977).
4. "Low-Si latite" (shoshonite), Ayacucho, Central Peru (No. AYA-1A, Noble et al., 1975).
5. Basaltic andesite, Cotopaxi volcano (No. EF7, Francis et al., 1977, and unpubl. data).
6. High-alumina basalt, Villarrica volcano, S. Chile (No. 802, Lopez-Escobar et al., 1977; Sr isotope ratio from "olivine andesite" No. V124 in Klerkx et al., 1977).
7. Hauyne-bearing "andesite tephrite", Sumaco volcano (No. 1, Colony and Sinclair, 1928).
8. "Trachyandesite", Pino Hachado, S. Chile (No. TH-34, Lopez-Escobar et al., 1976).

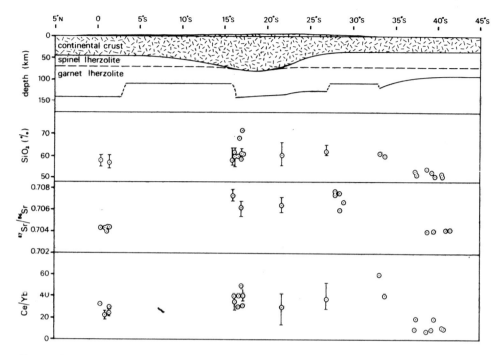

Fig. 3. Relationships between crustal and mantle structures and chemical composition of lavas along the Andean volcanic chain. The upper diagram shows variations in crustal thickness (ticked, from Cummings and Schiller, 1971) and the position of the Benioff zone (solid line, from Baranzangi and Isacks, 1976) between 5°N and 45°S along the approximate line of the active volcanic chain. The lower graphs show variations in SiO_2, $^{87}Sr/^{86}Sr$ and Ce/Yb for individual volcanoes and volcanic groups at different latitudes along the section shown above. Data are from McNutt et al. (1975), Thorpe et al. (1976), James et al. (1976), Lopez-Escobar et al. (1977), Dostal et al. (1977a, 1977b) and Klerkx et al. (1977). See text for further discussion.

et al., 1977b). These variations are a slight increase in SiO_2, accompanied by a significant increase in K and Rb, and a decrease in Sr in passing from west to east (Fig. 2). They are associated with the occurrence of shoshonitic lavas in south Peru (Lefèvre, 1973) and northwest Argentina (Hörmann et al., 1973). These lavas have SiO_2 = 50—60% with $K_2O/Na_2O \gtrsim 1$, but have higher Ti and Sr than other calc-alkaline lavas to the west (Hörmann et al., 1973; Dupuy and Lefèvre, 1974).

Alkaline lavas have been reported from scattered localities to the east of the main Andean volcanic chain, in Ecuador, Bolivia, Chile and Argentina (Fig. 1). In Ecuador, the Sumaco volcano is composed of hauyne tephrite (Colony and Sinclair, 1928). Further south, lavas belonging to the alkali basalt—trachyte association occur in western Bolivia (Francis and Thorpe, unpublished), Chile and Argentina (Vergara, 1972; Lopez-Escobar et al.,

1976, 1977). These data, together with those on distribution and tectonic setting, enable us to place some constraints on the origin of Andean andesites.

PETROGENESIS

As a starting point, we consider detailed studies of the andesite lavas of the San Pedro—San Pablo volcanoes of north Chile (Francis et al., 1974; Thorpe et al., 1976; Roobol et al., 1976; Francis et al., 1977).

We envisage an origin for these andesites using the model of Nicholls and Ringwood (1973) and Ringwood (1974). In this model, partial melting of subducted oceanic crust produces melts which rise into, and react with, the overlying mantle wedge. This produces pyroxene-rich garnet peridotite bodies which rise diapirically and partially melt to produce the parental andesite magmas. In this context, origin of andesite might involve some or all of the following processes:

(1) Partial melting of large-ion litholphile (LIL) element enriched garnetiferous mantle above the underlying zone.

(2) Fractional crystallization of the resultant melt at mantle depth.

(3) Interaction with the lower continental crust where it exceeds about 45-km thickness, involving "scavenging" of Sr (and probably some other elements).

(4) Fractional crystallization of olivine, pyroxene and (<35-km depth) plagioclase *within* the continental crust.

According to this model, the sharply defined western boundary of the volcanic chain is critically controlled by pressure/temperature conditions at the subduction zone. In view of the earlier discussion of migration of the volcanic chain, such migration might reflect changes in the depth of the subduction zone through time.

Transverse variations

To explore how these processes might account for the transverse variations described above, we use geochemical data from Roobol et al. (1976), Fernandez et al. (1973) and Hörmann et al. (1973), together with data on tectonic setting as summarized in Fig. 2. The increase in concentration of elements listed in Fig. 2 across the Andes can be accomplished by:

(1) a decrease in distribution coefficients (D);

(2) decrease of the degree of partial melting or;

(3) an enrichment of these elements in the mantle in passing from west to east.

Each of these is reviewed below.

(1) We consider first the possibility that a decrease in D occurs in the source regions of the lavas in passing from west to east in Fig. 2 (Jakěs and White, 1970; Marsh and Carmichael, 1974). The possible source regions of the lavas in Fig. 2 (subducted oceanic crust and mantle) are likely to be com-

posed largely of olivine, clinopyroxene, orthopyroxene and garnet with minor amphibole at depths of up to ca. 100 km (Green, 1973) and phlogopite at depth of up to ca. 175 km (Modreski and Boettcher, 1972). For these minerals, values of D for K, Rb and Ba are generally <0.01, except for amphibole and phlogopite where D is much greater and may even exceed 1.0 (Arth, 1975). Therefore, equilibration of the andesite magmas across a zone which has a diminishing proportion of amphibole and/or phlogopite might be responsible for an eastwards increase in K, Rb and Ba. For a small degree (ca. 5%) of partial melting, the presence of ca. 5% of amphibole and/or phlogopite in the source regions of the westernmost lavas, diminishing eastwards, might account for the eastward increase in K, Rb and Ba.

It is therefore possible that phlogopite in subducted oceanic crust at ca. 175 km could account for the eastward increase in K, Rb and Ba (Modreski and Boettcher, 1972, cf. Fig. 2). Similarly, equilibration of andesite magma in a zone traversing the breakdown of amphibole below 100 km and that of phlogopite at 175 km could also account for the chemical zonation. But we do not consider that andesites originate solely by partial melting of subducted oceanic crust, and argue that the petrogenesis of andesite takes place over a wide range of depths.

Possibilities (2) and (3) are more difficult to evaluate. Alternative (2) envisages that the andesite composition is controlled by the degree of partial melting (or fractional crystallization) at the source depth. If we assume that andesite is a "low-melting" fraction in equilibrium with hydrous peridotite (Mysen and Boettcher, 1975), and that K, Rb and Ba behave as constituents with $D = 0.1$, then to increase the concentration of these elements in the melt by a factor of 50% (cf. Fig. 2) the amount of melt produced must decrease by 40%. In view of decreasing temperatures away from the oceanic slab, this is the most likely possibility. But we cannot exclude other possibilities: these are discussed later.

Alternative (3), an increase of the concentration of K, Rb and Ba in the mantle away from the subduction zone, might occur in a number of ways. In view of the small degree of partial melting needed for production of andesite from peridotite (and the possible eastward decrease suggested above), it has been suggested that the subcontinental mantle might be "enriched" or possibly spatially zoned in K and associated elements (Lopez-Escobar et al., 1976, 1977; Thorpe et al., 1976; Dostal et al., 1977a, 1977b). The andesite (or its parental magma) might therefore derive its concentrations of K, Rb and Ba from the mantle by a zone-refining process (Harris, 1974). In this case, to effect the increase in K ($D = 0.01$) observed in Fig. 2 requires that the number of zone lengths traversed increases by a factor of ca. 1.6 in passing eastwards in Fig. 2. Since this is approximately the same as the eastward increase in mantle thickness from the Benioff zone to the base of the crust (see Fig. 2), then a zone-refining mechanism might contribute to the eastward increase in K, Rb and Ba.

The three possible processes so far discussed are responsible for determining the composition of andesite at *mantle* depths, but processes at *crustal* depths are also important. We have argued (Francis et al., 1977), that the contrast in $^{87}Sr/^{86}Sr$ ratios between north Chile (0.705—0.707) and Ecuador (ca. 0.704) indicate that andesite magmas become "contaminated" by up to ca. 20% "crustal" Sr. Moorbath has determined the $^{87}Sr/^{86}Sr$ ratios of seven Pliocene—Recent volcanics in our study area north of the Salar de Uyuni in Bolivia (19—20°S) (unpublished data). This area lies to the northeast of the San Pedro—San Pablo area and is 300 km from the Chile trench, lying close to the "crustal keel" of the Andes. The $^{87}Sr/^{86}Sr$ ratios fall into two groups — a single sample of alkali basalt from a Recent tuff ring has a ratio of 0.7041, while samples of andesite from the same area have ratios between 0.7055 and 0.7064, these being slightly lower than the ratios reported for north Chile (0.7058—0.7072; Francis et al., 1977). The Sr isotope ratio for the alkali basalt is within the range of ratios reported for andesites from Ecuador (Francis et al., 1977) and we interpret these ratios (ca. 0.704) as representative of the sub-Andean mantle in the Central Andes. Therefore, the Bolivian andesites, like those from north Chile, have been "contaminated" with crustal Sr, although the amount is somewhat less than for north Chilean andesites. In this respect it is interesting to note that, in contrast to K, Rb and Ba, Sr shows a *decrease* in andesites in passing from west to east (Dupuy and Lefèvre, 1974; Roobol et al., 1976). This might therefore be partly a reflection of diminishing crustal contamination. In view of our arguments regarding Sr, it is interesting to note that Dupuy et al. (1976) and Zentilli and Dostal (1977) have attributed lateral variations in U concentrations within volcanics in south Peru, Chile and Argentina to crustal contamination.

Finally, we consider shallow level crustal processes — these are fractional crystallization of olivine, clinopyroxene, orthopyroxene, amphibole at depths of less than 100 km, and plagioclase at depths of less than 35 km. Since the depths over which these processes operate are independent of overall crustal thickness, removal of these phases might contribute to the zonation of K, Rb and Ba ($D < 1$ for phases above) and decrease of Sr ($D_{plag} = 2$—3). The magnitude of overall fractionation at crustal depths can be estimated from the trace-element composition of the andesites: removal of ca. 10% olivine and pyroxene is required to reduce Cr and Ni contents from those expected for mantle-derived magmas to those characteristic of andesites (Dostal et al., 1977b), and removal of < ca. 20% plagioclase is necessitated by the REE data (Thorpe et al., 1976).

Andesitic magmas of the Central Andes cannot therefore be derived from peridotite mantle as the result of a single-stage process — such magmas are the result of a complex interplay of partial melting, fractional crystallization and "contamination" throughout depths between the subduction zone and the upper crust. But we emphasize that the bulk of andesite magma probably *originates* at mantle depths.

Longitudinal variations

The processes responsible for transverse chemical variations in Andean volcanics can also be used to explain variations *parallel* to the Andean chain (Fig. 3). Andesites from the northern zone are more basic than those in north Chile (Pichler, 1976; Francis et al., 1977). The REE abundances and patterns are similar to those in north Chile (Thorpe et al., 1976), but the $^{87}Sr/^{86}Sr$ ratios are lower (Francis et al., 1977). In view of the similar depth to the Benioff zone and the thinner continental crust (see earlier), we suggest that these andesites have a similar history to those from north Chile, but experienced less fractional crystallization and negligible crustal contamination during their ascent (cf. Noble et al., 1975). The high-alumina basalts and andesites of the southern zone (south Chile) also form a more basic association than that of north Chile (Katsui, 1972). In view of the shallower Benioff zone (ca. 90 km) and thinner crust (see earlier), the calc-alkaline magmas probably formed at a shallow depth and as the result of a higher degree of partial melting or a lower degree of fractional crystallization, and negligible crustal contamination during ascent (cf. Lopez-Escobar et al., 1977). The higher concentration of heavy REE and lower Ce/Yb (<20) in the south Chilean basalts suggest an origin involving equilibration with garnet-free peridotite (Fig. 3: cf. Thorpe et al., 1976; Dostal et al., 1977b, Lopez-Escobar et al., 1977).

The shoshonitic lavas identified in south Peru (Lefèvre, 1973; Dostal et al., 1977a), and northwest Argentina (Hörmann et al., 1973), and the alkaline basalt—trachyte lavas from Ecuador, Bolivia, Chile and Argentina (Colony and Sinclair, 1928; Vergara, 1972; Lopez-Escobar et al., 1976) all occur in volcanoes towards the eastern side of the Andean chain (see earlier). Some alkaline lavas form isolated volcanoes up to 400 km from the main Andean chain (Vergara, 1972). These alkalic lavas are formed by partial melting of mantle peridotite in tectonic settings which range from those connected with subduction to those in "mid-plate" environments.

The shoshonitic rocks of south Peru and northwest Argentina have significantly higher contents of Ti, Ba and Sr than the associated andesites (Hörmann et al., 1973; Dostal et al., 1977a), and are therefore probably produced by different processes. The chemical compositions of the shoshonites are consistent with an origin by a small degree of partial melting of garnetiferous mantle peridotite (Dostal et al., 1977a). We infer that the shoshonites and other alkaline lavas are probably produced by partial melting of mantle peridotite behind the Andean volcanic chain, where temperatures may be lower and the proportion of water in the volatile phase less. The distribution of such volcanic rocks might be related to fracturing or extension behind the Andes, possibly connected with decrease in the ratio of H_2O to CO_2 in the coexisting volatile phases (Mysen and Boettcher, 1975). The form of the irregular eastern margin of the Andean volcanic chain, which includes these alkaline volcanoes is therefore related to lithospheric inhomogeneities rather than to subduction.

IGNIMBRITES

Finally we consider the origin of the voluminous andesite—rhyolite ignimbrites which form a prominent part of the volcanic scenery in the central zone of the Andes (Fig. 1). Several authors (Zeil and Pichler, 1967; Pichler and Zeil, 1972; Fernandez et al., 1973; Klerkx et al., 1977) emphasize that crustal fusion is important in the formation of these magmas. Zeil and Pichler (1967), Pichler and Zeil (1972) and Fernandez et al. (1973) have advanced several arguments for interpreting "the magmas of the Rhyolite Formation . . . as the products of widespread crustal fusion" (Pichler and Zeil, 1972, pp. 434—435). These arguments include the occurrence of "partly melted quartz and plagioclase xenocrysts", the near-"eutectic" composition and the peraluminous normative composition of most rhyolitic rocks. Pichler and Zeil (1972) and Fernandez et al. (1973) also refer to the unusually high contents of Cu and Zn, and the occurrence of xenoliths of sedimentary and metamorphic rock in the ignimbrites. Klerkx et al. (1977) have also argued from Sr isotope data that the ignimbrites are derived in a two-stage melting process by remelting of older, mantle-derived igneous rocks which have been subducted.

Most of these observations are consistent with operation of normal crystal-melt equilibria or with minor assimilation. For example, corrosion of early precipitated quartz and plagioclase phenocrysts (or xenocrysts) can occur as a natural consequence of ascent of acid magma, and rhyolites may approach a "eutectic" composition as a result of formation by *either* partial melting *or* fractional crystallization processes. Our own observations suggest that xenoliths of sedimentary and metamorphic rocks in andesites and ignimbrites are very rare. We therefore suggest that the arguments for a crustal origin of ignimbrite are not persuasive, and suggest that the rhyolite and andesite magmas share a common origin (Pushkar et al., 1972).

The case for a common origin of andesites and ignimbrites is strengthened by both their compositional similarities (Pichler and Zeil, 1972; Fernandez et al., 1973) and by the fact that eruptions of andesite and ignimbrite overlap in space and time (Pichler and Zeil, 1972; Francis et al., 1974; Baker, 1977). In addition, the $^{87}Sr/^{86}Sr$ ratios of ignimbrite in north Chile (0.705—0.710; Klerkx et al., 1977) overlap those of the north Chilean andesites (0.705—0.707; Klerkx et al., 1977; Francis et al., 1977). The low Sr contents of the ignimbrites invoked by Klerkx et al. to support their two-stage melting model can be adequately explained by simple fractionation of plagioclase feldspar.

We therefore suggest that the rhyolite—dacite ignimbrite magmas originate by processes similar to those described for the origin of andesite magma, and that contamination by continental crust plays a minor role. Although the ignimbrite province is confined to the Central Andean zone where the continental crust is thickest (cf. Figs. 1 and 2), thick continental crust is not present beneath other major ignimbrite provinces (e.g. New Zealand, north

Mexico). We therefore speculate that in the case of the Andes, the thick continental crust favours the occurrence of more extensive fractional crystallization during the uprise of mantle-derived andesite magma, resulting in the formation and eruption of extensive dacite—rhyolite ignimbrites.

CONCLUSIONS

Andean andesites and associated volcanic rocks exhibit important petrological and geochemical variations both along and across the Andean volcanic chain. These variations provide important clues as to the variety of processes responsible for the origins of Andean andesites.

Recent models for petrogenesis of andesite magmas in continental margin settings have been reviewed by Nicholls and Ringwood (1973) and Ringwood (1974). These models emphasize that the processes responsible for andesite volcanism are initiated at mantle depths by subsolidus dehydration of descending oceanic crust initiated at depths of 60—100 km. The dehydration continues to greater depths, and at ca. 100 km, partial melting of the oceanic slab leads to formation of melts which rise into and contaminate the overlying mantle wedge. This process leads to production of garnet-bearing pyroxene-rich peridotites which rise diapirically through the subcontinental mantle. Melting of this ultramafic material and fractional crystallization of the derived melts, leads to the production of the andesitic magmas.

On such a model, the variation in composition of andesite and associated volcanic rocks depends on the depth to the underlying Benioff zone, and the thickness of the continental crust. Increases of K and associated elements with increasing depth to the Benioff zone across the Andean chain result from a decrease in partial melting and an increase in "contamination" as the melt traverses a greater thickness of mantle. The presence of a thick continental crust provides an opportunity for some crustal contamination and enhanced fractional crystallization during the rise of magma to the surface, the latter culminating in the formation of dacite—rhyolite magmas in the central zone of the Andes. Important variations along the Andes are linked with the shallow Benioff zone in south Chile, where the volcanics are formed from garnet-free peridotite at shallower depths. Finally, the occurrence of shoshonitic and alkaline volcanics in scattered areas to the east of the Andean chain reflects an origin by partial melting of mantle peridotite distant from the subduction zone.

ACKNOWLEDGEMENTS

We are grateful to Dr. S. Moorbath, University of Oxford, for determining the Sr isotope ratios on Bolivian andesite samples. Fieldwork in the Andes has been financed by grants from the Royal Society, Natural Environmental Research Council, the Open University and the University of London Central Research Fund. We are grateful to Dr. J.A. Pearce for reading and criticizing an earlier draft of the manuscript.

REFERENCES

Arth, J.G., 1976. Behaviour of trace elements during magmatic processes — a summary of theoretical models and their applications. J. Res. U.S. Geol. Surv., 4: 41—47.

Baker, M.C.W., 1977. Geochronology of Upper Tertiary volcanic activity in the Andes of North Chile. Geol. Rundsch., 66(2): 455—465.

Baranzangi, M. and Isacks, B.L., 1976. Spatial distribution of earthquakes and subduction of the Nazca plate below South America. Geology, 4: 686—692.

Casertano, L., 1962. General characteristics of active Andean Volcanoes and a summary of their activities during recent centuries. Bull. Seismol. Soc. Am., 53: 1415—1433.

Cummings, D. and Schiller, G.I., 1971. Isopach map of the earth's crust. Earth Sci. Rev., 7: 97—125.

Colony, R.J. and Sinclair, J.H., 1928. The lavas of the volcano Sumaco, eastern Ecuador, South America. Am. J. Sci., 16: 299—312.

Dostal, J., Dupuy, C. and Lefèvre, C., 1977a. Rare earth distribution in Plio—Quaternary volcanic rocks from southern Peru. Lithos, 10: 173—183.

Dostal, J., Zentilli, M., Caelles, J.C. and Clark, A.H., 1977b. Geochemistry and origin of volcanic rocks from the Andes (26°—28°S). Contrib. Mineral. Petrol., 63: 113—128.

Drake, R.E., 1976. The chronology of Cenozoic igneous and tectonic events in the central Chilean Andes. In: O. Gonzales-Ferran (Editor), Proc. Symp. on Andean and Antarctic Volcanology Problems, Santiago, Chile, 1974, pp. 670—697.

Dupuy, C. and Lefèvre, C., 1974. Fraccionnement des élements en trace Li, Rb, Ba et Sr dans les séries andesitiques et shoshonitiques du Perou-Comparison avec d'autres zones orogéniques. Contrib. Mineral. Petrol., 46: 147—157.

Dupuy, C., Dostal, J., Capredi, S. and Lefèvre, C., 1976. Petrogenetic implications of uranium abundances in volcanic rocks from southern Peru. Bull. Volcanol., 39: 363—370.

Evernden, J.F., 1966. Correlaciones de las formaciones tercerias de la cuenca altiplanica a base de etudes absolutos, determinados por el metodo potasio-argon. Serv. Geol. Boliv. hoja Inf., Vol. I. La Paz, Bolivia.

Farrar, E.H., Clark, S.J. Haynes, G.S., Quirt, H., Conn., and Zentilli, M., 1970. K—Ar evidence for the post-Paleozoic migration of granitic intrusion foci in the Andes of northern Chile. Earth. Planet. Sci. Lett., 10: 60—66.

Fernandez, A., Hörmann, P.K., Kussmaul, S., Meave, J., Pichler, H. and Subieta, T., 1973. First petrologic data on young volcanic rocks of S.W. Bolivia. Tschermaks Mineral. Petrogr. Mitt., 19: 149—172.

Francis, P.W. and Rundle, C., 1976. Rates of production of the main magma types in the central Andes. Geol. Soc. Am. Bull., 87: 474—480.

Francis, P.W., Roobol, M.J., Walker, G.P.L., Cobbold, P.R. and Coward, M.P., 1974. The San Pedro and San Pablo volcanoes of northern Chile and their hot avalanche deposits, Geol. Rundsch., 63: 357—388.

Francis, P.W., Moorbath, S. and Thorpe, R.S., 1977. Strontium isotope data for recent andesites in Ecuador and North Chile. Earth Planet. Sci. Lett., 37: 197—202.

Green, D.H., 1973. Contrasted melting relations in a pyrolite upper mantle under mid-oceanic ridge, stable crust and island arc environments. Tectonophysics, 17: 285—297.

Hantke, G., 1966. The volcanoes of Ecuador. In: Catalogue of Active Volcanoes of the World, Part 19. Int. Assoc. for Volcanol. and Chem. Earth's Interior, Naples, pp. 26—61.

Harris, P.G., 1974. Origin of alkaline magmas as a result of anatexis. A. Mantle Anatexis. In: H. Sorensen (Editor), The Alkaline Rocks. Wiley, London.

Hörmann, P.K., Pichler, H. and Zeil, Q., 1973. New data on the young volcanism in the Puna of N.W. Argentina. Geol. Rundsch., 62: 397—418.

Jakěs, P. and White, A.J.R., 1970. K/Rb ratios of rocks from island arcs. Geochim. Cosmochim. Acta, 34: 849—856.

James, D.E., 1971a. Plate tectonic model for the evolution of the central Andes. Geol. Soc. Am. Bull., 82: 3325—3346.

James, D.E., 1971b. Andean crustal and upper mantle structure. J. Geophys. Res., 76: 3246—3271.

James, D.E., Brooks, C. and Cuyubamba, 1976. Andean Cenozoic volcanism: magmagenesis in the light of strontium isotopic composition and trace-element geochemistry. Geol. Soc. Am. Bull., 87: 592—600.

Katsui, Y., 1972. Late Cenozoic petrographic provinces of the volcanic rocks from the Andes to Antarctica. In: R.J. Adie (Editor), Antarctic Geology and Solid Earth Geophysics. Oslo, pp. 181—185.

Klerkx, J., Deutsch, S., Pichler, H. and Zeil, W., 1977. Strontium isotope composition and trace element data bearing on the origin of Cenozoic volcanic rocks of the central and southern Andes. J. Volcanol. Geotherm. Res., 2: 48—71.

Lefèvre, C., 1973. Les caractères magmatiques du volcanisme plioquaternaire des Andes dans le Sud du Pérou. Contrib. Mineral. Petrol., 41: 259—272.

Lopez-Escobar, L., Frey, F.A. and Vergara, M., 1976. Andesites from central—south Chile: trace element abundances and petrogenesis. In: O. Gonzales-Ferran (Editor), Proc. Symp. on Andean and Antarctic Volcanology Problems. Santiago Chile, 1974, pp. 725—761.

Lopez-Escobar, L., Frey, F.A. and Vergara, M., 1977. Andesites and high-alumina basalts from the central—south Chile High Andes: Geochemical evidence bearing on their petrogenesis. Contrib. Mineral. Petrol., 63: 199—228.

MacDonald, G.A., 1972. Volcanoes. Prentice-Hall, N.J.

McNutt, R.H., Crocket, J.H., Clark, A.H., Caelles, J.C., Farrar, E., Haynes, S.J. and Zentilli, M., 1975. Initial $^{87}Sr/^{86}Sr$ ratios of plutonic and volcanic rocks of the central Andes between latitudes 26° and 29° south. Earth Planet. Sci. Lett., 27: 305—313.

Marsh, B.D. and Carmichael, I.S.E., 1974. Benioff Zone magmatism. J. Geophys. Res., 79: 1196—1206.

Meissnar, R.O., Flueh, E.R., Stibane, F.R. and Berg, F., 1972. Dynamics of the active plate boundary in Southwest Colombia. Tectonophysics, 35: 115—136.

Modreski, P.J. and Boettcher, A.L., 1972. The stability of phlogopite + enstatite at high pressures: a model for micas in the interior of the Earth. Am. J. Sci., 272: 852—869.

Moreno, R.H., 1976. The Upper Cenozoic volcanism in the Andes of southern Chile from 40°00′ to 41°30′ S). In: O. Gonzales-Ferran (Editor), Proc. Symp. on Andean and Antarctic Volcanology Problems, Santiago, Chile, 1974, pp. 143—171.

Mysen, B.O. and Boettcher, A.L., 1975. Melting of an hydrous mantle: II Geochemistry of crystals and liquids formed by anatexis of mantle peridotite at high pressures and as a function of controlled activities of water, hydrogen and carbon dioxide. J. Petrol., 16: 549—593.

Nicholls, I.A. and Ringwood, A.E., 1973. Effect of water on olivine stability in tholeiites and the production of silica saturated magmas in the island arc environment. J. Geol., 81: 285—300.

Noble, D.C., and McKee, E.H., 1977. Spatial distribution of earthquakes and subduction of the Nazca Plate beneath South America: Comment. Geology, 5: 576—578.

Noble, D.C., McKee, E.H., Farrar, E. and Peterman, U., 1974. Episodic Cenozoic volcanism and tectonism in the Andes of Peru. Earth Planet. Sci. Lett., 21: 213—220.

Noble, D.C., Bowman, H.R., Hebert, A.J., Silberman, M.L. Heropoulos, C.E., Fabbi, B.P. and Hedge, C.E., 1975. Chemical and isotopic constraints on the origin of low-silica latite and andesite from the Andes of central Peru. Geology, 3: 501—504.

Pichler, H., 1976. Cenozoic volcanic rocks of Ecuador, Abs. III Congr. Latinoam. Geol., p. 106.

Pichler, H. and Zeil, W., 1972. The Cenozoic rhyolite—andesite associations of the Chilean Andes. Bull. Volcanol., 35: 424—452.

Pichler, H. Hörmann, P.K. and Braun, A.F., 1976. First petrologic data on lavas of the volcano El Reventador (Eastern Ecuador): Münster. Forsch. Geol. Palaeontol., 38/39: 129—141.

Pushkar, P., McBirney, A.R. and Kudo, A.M., 1972. The isotopic composition of strontium in Central American ignimbrites. Bull. Volcanol., 35: 265—294.

Ramirez, J.E., 1968. Los Volcans de Colombia. Rev. Acad. Colomb. Ciencos Exactos, 13: 227—235.

Ringwood, A.E., 1974. Petrological evolution of island arc systems. J. Geol. Soc. Lond., 130: 183—204.

Roobol, M.J., Francis, P.W., Ridley, W.I., Rhodes, M. and Walker, G.P.L., 1976. Physicochemical characters of the Andean volcanic chain between 21° and 22° south. In: O. Gonzales-Ferran (Editor), Proc. Symp. on Andean and Antarctic Volcanology Problems, Santiago, Chile, 1974, pp. 450—464.

Schwab, K. and Lippolt, H., 1976. K—Ar mineral ages and late Cenozoic history of the Salar de Cauchari area (Argentina Puna). In: O. Gonzales-Ferran (Editor), Proc. Symp. on Andean and Antarctic Volcanology Problems, Santiago, Chile, 1974, pp. 698—714.

Stauder, W.M., 1973. Mechanisn and spatial distribution of Chilean earthquakes with relation to subduction of the oceanic plate. J. Geophys. Res., 78: 5033—5061.

Thorpe, R.S., Francis, P.W. and Potts, P.J., 1976. Rare earth data and petrogenesis of andesites from the N. Chilean Andes. Contrib. Mineral. Petrol., 54: 65—78.

Vergara, M., 1972. Note on the zonation of the Upper Cenozoic volcanism of the Andean area of central—south Chile and Argentina. Symp. on the Results of Upper Mantle Investigations with Emphasis on Latin America. International Upper Mantle Project, Buenos Aires, pp. 381—397.

Zeil, W. and Pichler, H., 1967. Die Känozoische Rhyolith-Formations im mittleren Abschnitt der Andes, Geol. Rundsch., 57: 48—81.

Zentilli, M. and Dostal, J., 1977. Uranium in volcanic rocks from the central Andes. J. Volcanol. Geotherm. Res., 2: 251—258.

Chapter 19

Sr ISOTOPIC STUDIES OF VOLCANIC ROCKS FROM ISLAND ARCS IN THE WESTERN PACIFIC*

JUN-ICHI MATSUDA, SHIGEO ZASHU and MINORU OZIMA

Geophysical Institute, University of Tokyo, Bunkyo-ku, Tokyo, 113 (Japan)

SUMMARY

Systematic measurements were made on initial ($^{87}Sr/^{86}Sr$) ratios and Rb, Sr and K contents of the rocks from Izu, Ogasawara (Bonin), Iwojima, Mariana, Yap and Palau Islands. The initial ($^{87}Sr/^{86}Sr$) ratios for volcanic rocks from Izu, Iwojima and Mariana Islands fall in a range from 0.7030 to 0.7050, which is similar to previously reported values for island-arc volcanics. However, a high initial ($^{87}Sr/^{86}Sr$) ratio (~ 0.7052) was found for rocks from Ogasawara Islands, suggesting that they are genetically different from other islands. This result may be related to the particular topographical feature of the trench discontinuity in the eastern side of the Ogasawara Islands. On the contrary, rocks from the Palau Islands show low initial ($^{87}Sr/^{86}Sr$) ratios (~ 0.7028), which are similar to those of oceanic ridge basalts, and Rb and Sr contents similar to those of normal island-arc volcanics. A low initial ($^{87}Sr/^{86}Sr$) ratio was found for the rock from the Yap Islands, supporting the oceanic origin of them as suggested by Shiraki. It is likely that Palau, Yap and Mariana Island arcs represent the three different stages of development of island arcs. These results suggest that except for the Ogasawara Islands, the island arcs in the western Pacific have not been formed by the splitting or drifting of continental margin, but have grown from oceanic structure.

INTRODUCTION

A series of island arcs are located in the western Pacific. They consist of northern Japan, Izu, Ogasawara (Bonin), Iwojima, Mariana, Yap and Paulu Islands from north to south (Fig. 1). Karig (1971a,b) proposed that the oceanic basins behind the island arcs of these areas are extensional in origin, new oceanic crust being formed behind the volcanic arc. The dredged samples from the inter-arc basin (Mariana Basin) showed similar characteristics to those of the mid-ocean spreading-ridge basalts (Hart, 1971; Hart et al., 1972), supporting Karig's idea.

Originally published as: Matsuda, J.-I., Zashu, S. and Ozima, M., 1977. Sr isotopic studies of volcanic rocks from island arcs in the western Pacific. In: S. Uyeda (Editor), Subduction Zones, Mid-Ocean Ridges, Oceanic Trenches and Geodynamics. Tectonophysics, 37: 141—151.

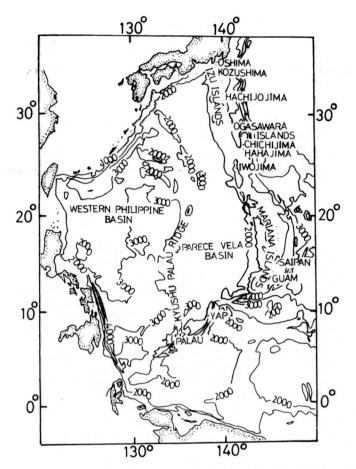

Fig. 1. Island arcs in the western Pacific (mainly after Pushkar, 1968).

The (^{87}Sr/^{86}Sr) ratios of oceanic-ridge basalts range mostly from 0.7010 to 0.7030, having a mean value of 0.7026, whereas (^{87}Sr/^{86}Sr) ratios of oceanic-island and island-arc volcanic rocks range from 0.7030 to 0.7060 with a mean value of 0.7038 (Peterman and Hedge, 1971). Oceanic-ridge basalts are more depleted in trace elements than are oceanic-island basalts. The depletion is larger for elements with larger ionic radius, resulting in high K/Rb ratios (Engel et al., 1965; Gast, 1968). Therefore, oceanic-ridge basalts may be distinguished from island-arc and oceanic-island volcanic rocks by their low (^{87}Sr/^{86}Sr) and high K/Rb ratios.

In this paper the origin and the development of island arcs in the western Pacific are discussed from the results of systematic measurements of Rb, Sr and K contents and (^{87}Sr/^{86}Sr) ratios for the rocks from Izu, Ogasawara, Iwojima, Mariana, Yap and Palau Islands.

TABLE I

($^{87}Sr/^{86}Sr$) ratios and trace-element contents of the rocks from the island arcs in the western Pacific

Locality	Sample No.	Rock type	K–Ar age * (m.y.)	K (ppm)	Rb (ppm)	Sr (ppm)	K/Rb	($^{87}Sr/^{86}Sr$)	($^{87}Sr/^{86}Sr)_0$ **
Oshima	56091919	andesite	0.42	2820				0.7039 ± 2	
	5609315D	diorite	2.14	5030	12.8	204	393	0.7033 ± 1	0.7033 ± 1
	NI6C03093b	basalt	2.41	2660				0.7034 ± 2	
Kozushima	NI6C071803	rhyolite	0.28	56700	85.6	51	662	0.7032 ± 2	0.7032 ± 2
Hachijojima	NI55071806	gabbro	2.12	5250	2.09	170	2512	0.7032 ± 5	0.7032 ± 5
	NI67091704	andesite	0.14	3900	3.80	250	443	0.7034 ± 3	0.7034 ± 3
Chichijima	R382	andesite	26.0	10800	24.3	115	443	0.7054 ± 4	0.7052 ± 4
Hahajima	NI69021002	andesite	40.4	16300	25.6	192	637	0.7040 ± 3	0.7038 ± 3
Iwojima	NI68082304	trachyandesite	0.03	36100	83.2	440	434	0.7039 ± 1	0.7039 ± 1
Saipan	131B5	dacite		12200	8.25	229	1479	0.7041 ± 3	0.7040 ± 3
	7141D	andesite		5900	8.04	215	734	0.7040 ± 3	0.7040 ± 3
Guam	1–7	basalt		4700	10.04	447	468	0.7037 ± 2	0.7037 ± 2
	2–4	basalt		6720	12.00	510	560	0.7038 ± 4	0.7038 ± 4
Yap		greenschist		2460	0.42	557	5857	0.7029 ± 1	0.7029 ± 1
Palau	1	basalt		2700	2.28	206	1184	0.7030 ± 1	0.7030 ± 1
	6-A	basalt		3820	8.29	194	461	0.7025 ± 2	0.7024 ± 2
	12	basalt		5630	5.41	177	669	0.7032 ± 1	0.7031 ± 1
	14	basalt		4850	5.60	239	866	0.7026 ± 3	0.7026 ± 3
Mariana Basin (mean)	***	basalt		3547	4.51	186	786	0.7028 ± 1	0.7028 ± 1

* Data from Kaneoka et al. (1970).

** ($^{87}Sr/^{86}Sr)_0$: the initial ($^{87}Sr/^{86}Sr$) ratio corrected by Rb/Sr ratio and age.

*** Data from Hart et al. (1972).

SAMPLES AND GEOLOGICAL SITUATION

The rock types of the samples are briefly described in Table I. The detailed petrographic descriptions for some samples from Izu, Ogasawara and Iwojima Islands are given by Kaneoka et al. (1970). Samples from Mariana, Yap and Palau Islands were collected for paleomagnetic studies. Among these samples we chose fresh ones for Sr isotopic analyses. Most of these samples are volcanic rocks, though the sample from the Yap Islands is a greenschist. The Yap Islands occupy a unique geological position, having exposures of pre-Tertiary metamorphic basement (Shiraki, 1971). The northern part of the Izu Islands from Oshima to Hachijojima consist primarily of basalt, basaltic andesite and andesite of tholeiitic series, though rhyolite of the calc-alkalic series is predominant in Kozushima (Kuno, 1960). The ages of these rocks are Quaternary (Kaneoka et al., 1970). The Iwojima Islands are composed largely of trachyandesites and volcanism is still active at present (Macdonald, 1948). The Ogasawara Islands consist of Eocene andesite lavas covered with a Miocene limestone layer (Yoshiwara, 1902). The Guam and Saipan Islands of the Mariana arc have been formed by volcanism of Eocene to Miocene age, and are mostly made up of calc-alkalic series (Stark, 1963; Tracey et al., 1964). The petrographic observations of andesites and dacites of calc-alkalic series from the Palau Islands are presented by Tsuboya (1932).

EXPERIMENTS

Sr isotopic ratios were measured with a 25 cm radius, single focussing mass spectrometer with an electron multiplier. A tantalum single filament was used. Peak height was measured with a digital voltmeter. Details for the data reduction were described by Ozima et al. (1971). The measurements of isotopic ratios were made twice for each sample and the error assigned for the ($^{87}Sr/^{86}Sr$) ratio in Table I is the difference from the mean value for the repeated analyses. The mean value was normalized to 0.7080 for the Eimer and Amend standard sample.

Rb and Sr contents were measured to calculate the initial ($^{87}Sr/^{86}Sr$) ratios. The determination of Rb and Sr contents was done by an isotope-dilution method with a 22 cm radius, single focussing mass spectrometer. The measurements were also made twice for each sample. K content was analyzed with a flame photometer with a Li internal standard.

RESULTS

The results are listed in Table I. The initial ($^{87}Sr/^{86}Sr$) ratios are calculated by Rb/Sr ratios and K—Ar ages. In case of the Saipan, Guam, Yap and Palau samples, fossil ages were assigned for aging correction. However, the difference between the measured ($^{87}Sr/^{86}Sr$) ratio and the initial ($^{87}Sr/^{86}Sr$) ratio was insignificant for all the samples because of their low Rb/Sr ratios and rel-

atively young ages. For comparison, we also list in Table I the average value
for the Mariana basin basalts obtained by other authors (Hart et al., 1972).

Sr isotopic ratios

Our results from Oshima, Hachijojima, Kozushima, Iwojima and Mariana
Islands are all within a range of 0.703—0.705, which agrees with the results
for the volcanic rocks from the island arcs by previous authors (Hedge, 1966;
Pushkar, 1968). However, a somewhat high ratio was obtained for rocks
from Chichijima of the Ogasawara Islands. Pushkar (1968) also reported high
ratios for two lavas from the Ogasawara Islands. Since he did not give Rb
contents, it is not possible to evaluate the aging correction for those two
samples. Our results show that Rb/Sr ratios from the Ogasawara Islands are
much larger than those from other islands. Rb contents are more than two
or three times as much as those from other islands. Since the change in $(^{87}Sr/$
$^{86}Sr)$ ratio due to an aging effect is almost negligible because of the young
age, we can conclude that the rock from Ogasawara Islands has a high initial
$(^{87}Sr/^{86}Sr)$ ratio of 0.7052. Though such high $(^{87}Sr/^{86}Sr)$ ratios are reported
from other island arcs (e.g. calc-alkalic rocks from Sunda arc (Whitford,
1975)), it is noteworthy that among the island arcs in the Western Pacific only the volcanic rock from the Ogasawara Islands has a high $(^{87}Sr/^{86}Sr)$ ratio.
The source material for volcanic rocks of the Ogasawara Islands must be different from that for other islands. This result does not agree with the conclusion, from the analyses of major-element contents, that the volcanic rocks
from the Ogasawara Islands are similar to some of the andesites of Saipan
(Larson et al., 1974).

The rocks from the Palau Islands show low $(^{87}Sr/^{86}Sr)$ ratios. Such low
$(^{87}Sr/^{86}Sr)$ ratios (<0.7030) were also reported for the dredged sample from
the Mariana basin (Hart et al., 1972). In Fig. 2, $(^{87}Sr/^{86}Sr)$ ratio is plotted
against K/Rb ratio. In this diagram samples from the Mariana basin seem to
occupy the similar region to that of ridge basalts (Hart, 1971). The results
for rocks from the Palau Islands show the same characteristics on the (K/
Rb) — $(^{87}Sr/^{86}Sr)$ diagram as the Mariana basin basalts (Fig. 2). The low
$(^{87}Sr/^{86}Sr)$ ratios seem to distinguish the Palau rocks from the volcanic rocks
from typical island arcs. The rock from Chichijima of the Ogasawara Islands
is situated in the upper part in this diagram. The rocks from Mariana, Izu and
Iwojima Islands have similar values of $(^{87}Sr/^{86}Sr)$ ratios, though the $(^{87}Sr/$
$^{86}Sr)$ ratio for a rhyolite from Kozushima is a little low.

Rb, Sr and K contents

K/Rb ratios of volcanic rocks from both island arcs (Jakes and White,
1970) and oceanic ridges (Kay et al., 1970) show a wide range, i.e., the
former range from 150 to 1000 and the latter from 400 to 2000. The K/Rb
ratios (460—1100) of rocks from the Palau Islands are in the high range for

388

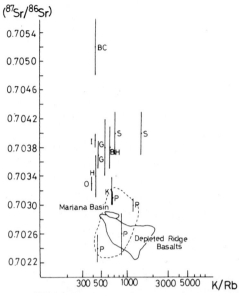

Fig. 2. K/Rb ratio vs. (^{87}Sr/^{86}Sr) ratio diagram for volcanic rock samples. The error bar of (^{87}Sr/^{86}Sr) ratio means the deviation from the mean (^{87}Sr/^{86}Sr) ratio for repeated analyses.
O = Oshima, K = Kozushima, BH = Hahajima, BC = Chichijima, H = Hachijojima, I = Iwojima, S = Saipan, G = Guam, P = Palau.
The dashed line encloses the rocks from the Palau Islands. The field of rocks from Mariana Basin and depleted ridge basalts are taken from Hart (1971).

the island-arc volcanic rocks, but are in the low range for the oceanic-ridge basalts. As the K/Rb ratios show such a wide range in both types of volcanic rocks, it is not possible to define unequivocally on the basis of K/Rb ratio the type of the mantle material which formed Palau Islands.

The change in Rb content with increasing K content (Fig. 3) for the volcanic rocks from the island arcs in the western Pacific seems to lie on the line defined for the various types of modern basalts (Hart et al., 1970). Except the data for Kozushima, Ogasawara and Iwojima Islands, present data are distributed around the average point for the tholeiitic basalts. Rocks from the Ogasawara Islands (BH and BC) occupy almost the same area as alkalic basalts in the diagram. The rocks from Iwojima and Kozushima Islands are plotted in the right-upper position of the trend line. Since the volcanic rocks from Iwojima and Kozushima are trachyandesite and ryholite, respectively, their high contents of alkalies are not anomalous. It is interesting that the trend that Rb content increases with increasing K content is consistent even in such high "alkalic" rocks.

Sr contents do not show a distinct correlation with K contents (Fig. 3). The Sr contents for the basalt and andesite from Izu and Mariana Islands, except for Guam Island, are about 200 ppm. The Sr contents of the volcanic

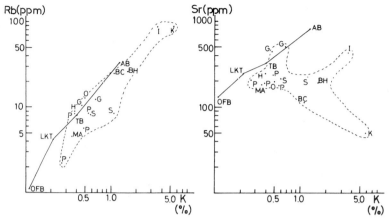

Fig. 3. Rb content vs. K content (left) and Sr content vs. K content (right) diagram for volcanic rock samples. *OFB, LKT, TB* and *AB* are taken from Hart et al. (1970). *OFB* = ocean-floor basalt, *LKT* = low potassium tholeiite, *TB* = tholeiitic basalt, *AB* = alkali basalt

These are mean values for various rock types of modern basalts, respectively. The field enclosed by the dashed line is for the result of island arcs in the western Pacific (this paper). The other symbols are the same as those given in Fig. 2.

rocks from Guam Island show about two times more enrichment in comparison with that of rocks from other islands. A low Sr content (~51 ppm) is obtained for rhyolite from Kozushima, which may be attributed to the rhyolite being a product of a late stage of differentiation (Baker, 1968). However, trachyandesite from Iwojima, which is also the late stage product in differentiation, has a high Sr content (~440 ppm). Hence, it is evident that there appears no simple relation between the Sr content and the degree of differentiation (Jakes and White, 1972).

With respect to Rb and Sr contents, volcanic rocks from the Palau Islands resemble the island-arc tholeiites. According to Jakes and White (1972), tholeiites from the island arcs have an average Rb content of 5 ppm, which is almost twice as high as that of abyssal tholeiites. The average Rb and Sr contents of the rocks from the Palau Islands are about 6 ppm and 204 ppm, respectively, which is also much higher than those of abyssal tholeiites. Summarizing, rocks from the Palau Islands are characterized by low ($^{87}Sr/^{86}Sr$) ratios similar to oceanic-ridge basalts, whereas their Rb and Sr contents are higher than those of the latter, falling in the range for island-arc tholeiites.

DISCUSSION

Ogasawara Islands

Hart (1969) showed that the chemical composition of oceanic basalt changes towards that of alkali basalt due to the weathering of sea water. If the Ogasawara Islands had been below sea level for a sufficiently long time, this weathering process might have resulted in the high ($^{87}Sr/^{86}Sr$) ratio and high contents of alkali elements. To test this possibility, the water contents and D/H ratios of the rocks from Ogasawara Islands were measured. The δD

$$\left(= \frac{(D/H)_{sample} - (D/H)_{SMOW}}{(DH)_{SMOV}} \times 1,000 \right)$$

obtained for Ogasawara Islands are very low, suggesting that the rocks from the Ogasawara Islands were not significantly contaminated by sea water (Table II).

The observed high ($^{87}Sr/^{86}Sr$) ratio may be explained by the fact that beneath the Ogasawara Islands, there exists sial material having high radiogenic Sr as debris of continental crust, which was assimilated with the ascending magma to give rise to the observed high ($^{87}Sr/^{86}Sr$) ratio in the volcanic rock. The Ogasawara island arc is anomalous as an island arc, because it has no deep trench and almost no deep earthquakes and it is characterized by alkalic volcanism (Miyashiro, 1974). Hence, it may not be entirely unreasonable to assume the continental remnant beneath the Ogasawara Islands. The above explanation may account for the topographical feature of trench discontinuity in the eastern side of the Ogasawara Islands. The large root of the continental remnant may fill up or now be floating on the trench. Unfortunately, there are few geophysical data on Ogasawara Islands, and more data on gravity, magnetization and seismology are needed to check the above interpretation.

Palau, Yap and Mariana Islands

A little higher ($^{87}Sr/^{86}Sr$) ratio for island-arc volcanic rock than that of oceanic-ridge basalt is interpreted as the contamination of sea-floor basalt with sialic material (Hart et al., 1970). Average oceanic sediment has a ($^{87}Sr/^{86}Sr$) ratio of approximately 0.7093, whereas that of sea-floor basalts is approximately 0.7026. Therefore, even a low degree of contamination by the

TABLE II

The H_2O and deutrium contents of the water for the rocks from the Ogasawara Islands

Sample		H_2O (wt. %)	D (‰) *
Chichijima	R382	4.94	−78.3
Hahajima	NI69021002	0.84	−77.3

* δD (‰) = $\dfrac{(D/H)_{sample} - (D/H)_{SMOW}}{(D/H)_{SMOW}} \times 1,000$

where SMOW means Standard Mean Oceanic Water.

oceanic sediment would result in high ($^{87}Sr/^{86}Sr$) ratio of the sea-floor basalt. We calculated the effect of sediment contamination on the K, Rb and Sr contents as well as the ($^{87}Sr/^{86}Sr$) ratio of sea-floor basalts. A simple numerical treatment shows that it is difficult to enrich K, Rb and Sr to the extent observed in the rocks from Palau Islands by adding a few percent of oceanic sediments to sea-floor basalts. About 4% contamination is the maximum permissible value, so that the ($^{87}Sr/^{86}Sr$) ratio of the rock does not exceed the observed value of 0.7030.

One of the possible interpretations may be that the Palau Islands represent the initial stage of the development of an island arc. If an arc formation begins at a deep ocean floor, abyssal tholeiites, which are the same as oceanic-ridge tholeiites, may be mixed with the newly erupted island-arc volcanic rocks. Therefore, in the initial stage of arc formation the erupted rock is likely to be similar to the abyssal tholeiite rather than the well-developed island-arc tholeiite (Miyashiro, 1974). Miyashiro suggested that such "initial-stage arc tholeiite" may be discovered in island arcs and that most such rocks are subjected to compositional changes, at least in alkalies, during the subsequent weathering and burial and regional metamorphism in the process of the growth of the arc. The rocks from the Palau Islands may be such remnant volcanic rocks.

After the metamorphism became more active in the subsequent arc development, the ophiolitic basement was formed. This ophiolitic basement may be representative of the greenschist from the Yap Islands. The major chemical compositions of rocks from Yap Islands also suggest that "the basic metamorphic rocks represent part of a lower layer of oceanic crust or have formed as the earlier manifestation of island-arc development" (Shiraki, 1971).

The initial formation of the island arc now considered may have started at the position of the Kyushu—Palau Ridge (Karig, 1971a,b). The volcanic rocks from the Palau Islands are transitional from abyssal tholeiite to island-arc tholeiite, where only the enrichment of alkalies has occurred without the change of ($^{87}Sr/^{86}Sr$) ratios. The Yap Islands developed eastwards at the stage of the active metamorphism, forming ophiolitic basement, then decreased its further development as an island arc. This stage of development must be a more advanced one than that which formed the Palau Islands. Hence, the Yap Islands occupy a more easterly position than the Palau Islands. Only the center portion of the Kyushu—Palau Ridge has continuously developed eastwards to have formed the Mariana island arc, and reached a more mature stage of island-arc formation. Therefore, the three island arcs, Palau, Yap and Mariana, may represent three stages of the development of the "Kyushu—Palau Island Arc", that is, the initial stage, the subsequent metamorphism and final maturity of the formation of island arc. The above interpretation seems to be consistent with the present topographical position for these island arcs in the western Pacific and the extensional model of marginal sea (Karig, 1971a,b; Uyeda and Ben-Avraham, 1972; Uyeda and Miyashiro, 1974).

ACKNOWLEDGEMENTS

We are grateful to Dr. N. Isshiki for providing samples from Izu, Iwojima and Ogasawara Islands. Thanks are also to Dr. T. Suzuoki and Dr. Y. Kuroda, who kindly measured the H_2O contents and D/H ratios of the rocks from Ogasawara Islands.

REFERENCES

Baker, P.E., 1968. Comparative volcanology and petrology of the Atlantic island-arcs. Bull. Volcanol., 32: 189—206.

Engel, A.E.J., Engel, C.G. and Havens, R.G., 1965. Chemical characteristics of oceanic basalts and the upper mantle. Geol. Soc. Am. Bull., 76: 719—734.

Gast, P.W., 1968. Trace element fractionation and the origin of tholeiitic and alkaline magma types. Geochim. Cosmochim. Acta, 32: 1057—1086.

Hart, S.R., 1969. K, Rb, Cs contents and K/Rb, K/Cs ratios of fresh and altered submarine basalts. Earth Planet. Sci. Lett., 6: 295—303.

Hart S.R., 1971. The geochemistry of basaltic rocks. Carnegie Inst. Year Book, 70: 353—355.

Hart, S.R., Brooks, C., Krogh, T.E., Davis, G.L. and Nava, D., 1970. Ancient and modern volcanic rocks: a trace element model. Earth Planet. Sci. Lett., 10: 17—28.

Hart, S.R., Glassley, W.E. and Karig, D.E., 1972. Basalts and sea-floor spreading behind the Mariana Island Arc. Earth Planet. Sci. Lett., 15: 12—18.

Hedge, C.E., 1966. Variation in radiogenic strontium found in volcanic rocks. J. Geophys. Res., 71: 6119—6126.

Jakes, P. and White, A.J.R., 1970. K/Rb ratios of rocks from island arcs. Geochim. Cosmochim. Acta, 34: 849—856.

Jakes, P. and White, A.J.R., 1972. Major and trace element abundances in volcanic rocks of orogenic areas. Geol. Soc. Am. Bull., 83: 29—40.

Kaneoka, I., Isshiki, N. and Zashu, S., 1970. K—Ar ages of the Izu—Bonin Islands. Geochem. J., 4: 53—60.

Karig, D.E., 1971a. Origin and development of marginal basins in the Western Pacific. J. Geophys. Res., 76: 2542—2561.

Karig, D.E., 1971b. Structural history of the Mariana Island Arc system. Geol. Soc. Am. Bull., 82: 323—344.

Kay, R., Hubbard, N.J. and Gast, P.W., 1970. Chemical characteristics and origin of oceanic ridge volcanic rocks. J. Geophys. Res., 75: 1585—1613.

Kuno, H., 1960. High-alumina basalt. J. Petrol., 1: 121—145.

Larson, E.E., Reynolds, R.L., Merrill, R., Levi, S., Ozima, M., Aoki, Y., Kinoshita, H., Zashu, S., Kawai, N., Nakajima, T. and Hirooka, K., 1974. Major-element petrochemistry of some extrusive rocks from the volcanically active Mariana Islands. Bull. Volcanol., 38: 361—377.

Macdonald, G.A., 1948. Petrography of Iwo Jima. Geol. Soc. Am. Bull., 59: 1009—1018.

Miyashiro, A., 1974. Volcanic rock series in island arcs and active continental margins. Am. J. Sci., 274: 321—355.

Ozima, M., Zashu, S. and Ueno, N., 1971. K/Rb and $(^{87}Sr/^{86}Sr)_0$ ratios of dredged submarine basalts. Earth Planet. Sci. Lett., 10: 239—244.

Peterman, Z.E. and Hedge, C.E., 1971. Related strontium isotopic and chemical variations in oceanic basalts. Geol. Soc. Am. Bull., 82: 493—499.

Pushkar, P., 1968. Strontium isotopic ratios in volcanic rocks of three island arc areas. J. Geophys. Res., 73: 2701—2714.

Shiraki, K., 1971. Metamorphic basement rocks of Yap islands, Western Pacific: possible oceanic crust beneath an island arc. Earth Planet. Sci. Lett., 13: 167—174.

Stark, J.T., 1963. Petrology of the volcanic rocks of Guam. U. S. Geol. Surv. Prof. Pap., 403-C.

Tracey, J.I., Schlanger, S.O., Stark, J.T., Doan, D.B. and May, H.G.,, 1964. General geology of Guam. U. S. Geol. Surv. Prof. Pap., 403-A.

Tsuboya, K., 1932. Petrographical investigation of some volcanic rocks from the South Sea Islands, Palau, Yap and Saipan. Jap. J. Geol. Geogr., 9: 201—213.

Uyeda, S. and Ben-Avraham, Z., 1972. Origin and development of the Philippine Sea. Nature Phys. Sci., 240: 176—178.

Uyeda, S. and Miyashiro, A., 1974. Plate tectonics and the Japanese Islands: a synthesis. Geol. Soc. Am. Bull., 85: 1159—1170.

Whitford, D.J., 1975. Strontium isotopic studies of the volcanic rocks of the Sunda arc, Indonesia, and their petrogenetic implications. Geochim. Cosmochim. Acta, 39: 1287—1302.

Yoshiwara, S., 1902. Geological age of the Ogasawara Group (Bonin Islands) as indicated by the occurrence of Nummulites. Geol. Mag., 9: 296—303.

Chapter 20

VOLCANIC EVOLUTION OF THE MARGINAL AND INTERARC BASINS*

M.B. LORDKIPANIDZE [1], G.S. ZAKARIADZE [1] and E.I. POPOLITOV [2]

[1] *Geological Institute, Academy of Sciences, Georgian S.S.R., Tbilisi (U.S.S.R.)*
[2] *Institute of Geochemistry, Siberian branch, Academy of Sciences, I.S.S.R., Irkutsk (U.S.S.R.)*

SUMMARY

Volcanic evolution of the interarc and marginal basins is analysed using the available data on volcanics from the presently existent and ancient back-arc basins of the western Pacific and Mediterranean. It is shown that in early (pre-spreading) stages of back-arc rifting, the character of volcanism is determined by "maturity" of the adjacent island arc. It is predominantly alkaline or mildly alkaline for back-arc basins related to the island-arcs with high-potash calc-alkaline and shoshonitic volcanism. The back-arc alkaline and mildly alkaline basalts strongly differ from the continental and oceanic rift volcanoes by constantly lower Ti, Nb and Zr contents. Because of these features these basalts are akin to the basaltic members of the island-arc volcanic series. As the latter, they are generally strongly enriched in K_2O and LIL elements, whereas Na_2O reveals comparatively small variability. With initiation of spreading a sharp depression of K_2O, LIL and light REE occurs in the axial basalts of back-arc basins, that progressively approach the MORB composition. But even tholeiites from the most evolved basins that underwent a considerable spreading reveal slight but detectable geochemical peculiarities, indicating their island-arc affinities. Origin of the low-Ti alkaline basaltic magmas of the active continental margins is discussed.

INTRODUCTION

Geodynamic models of marginal and interarc basins suggested in recent years (Karig, 1971; Packham and Falvey, 1971; Sleep and Toksöz, 1971 etc.) do not pay special attention to the petrologic aspects of their magmatism. It appears widely accepted that their magmatism is similar to the tholeiitic magmatism of mid-oceanic ridges. This view is based mainly on the chemistry of basalts from the central parts of "evolved" basins that underwent a considerable spreading (Woodlark, Lau, Parece Vela). But even these reveal greater variability in Ti, Nb, Zr and LIL element contents than the mid-

*Originally published as: Lordkipanidze, M.B., Zakariadze, G.S. and Popolitov, E.I., 1979. Volcanic evolution of the marginal and interarc basins. In: S. Uyeda (Editor), Processes at Subduction Zones. Tectonophysics, 57: 71—83.

oceanic ridge basalts (MORB) and have some geochemical features indicating their island-arc affinities (Gill, 1976). An analysis of the volcanic evolution of the marginal and interarc basins from the incipient to the "evolved" stages of their development is of interest for a better insight into the geodynamic and magmatic processes within the active continental margins.

Data on two groups of volcanic rocks are used in the present paper. The first group includes volcanics of the presently existent back-arc basins of the Mediterranean (Tyrrhenian Sea) (Dietrich et al., 1978) and western Pacific—New Georgian group of Solomon Islands (Stanton and Bell, 1969), New Hebridean interarc basalts (Colley and Warden, 1974), Lau volcanics (Gill, 1976; Hawkins, 1976), Parece Vela basalts (Ridley et al., 1974).

The second group comprises basaltic sequences of the possible ancient interarc and marginal basins of the Mediterranean and western Pacific belts. Situated on the rear side of the synchronous island-arc-type "andesitic" belts and associated with deep marine radiolarites, slates or turbidites, these basaltic sequences might reasonably be considered as back-arc volcanics of ancient active continental margins (Boccaletti et al., 1974. Adamia et al., 1977, Hsü et al., 1977; Khain, 1977). Among these are the volcanic series of the peri-Pacific basaltic belt comprising the Upper Cretaceous to Paleogene basaltic series of Kariakia, Central Kamchatka and the Kurile Island (Rotman and Markowski, 1976) and the Mesozoic—Cenozoic basaltic sequences of the Pontian—Caucasian active margin of the East European platform (Adamia et al., 1977; Hsü et al., 1977; Khain, 1977).

Of the above back-arc basaltic series, the Adjara-Trialetian Paleogene volcanic sequences are of the main interest because they allow us to characterize the possible form of volcanic evolution of a back-arc basin from the incipient rifting up to the early stages of spreading.

VERTICAL AND LATERAL PETROCHEMICAL VARIATIONS OF THE BACK-ARC BASALTIC SERIES

The Black Sea—Adjara-Trialetian rift was initiated in the central part of the Pontian—Transcaucasian Jurassic—Cretaceous island arc, splitting it into a southern active volcanic arc and a northern remnant nonvolcanic arc (Fig. 1). Volcanic activity started in the Lower Eocene and attained its climax in the Middle Eocene. In this period a thick (5—7-km) predominantly basaltic volcanic sequence was formed. Intensity of rifting and volcanic activity progressively increased from east to west, towards the Black Sea and presumably caused formation of the oceanic crust presently existent in the central part of the basin (Adamia et al., 1974a).

The first (Lower Eocene) volcanic manifestation of the Adjara-Trialetian interarc basin produced low TiO_2, low-Mg mildly alkaline basalts (Table I), accompanied by minor andesites and trachyandesites (Zakariadze et al., 1978). They differ from the synchronous volcanic series of the adjacent Pontian—Transcaucasian island arc only by an increased thickness (1500 m)

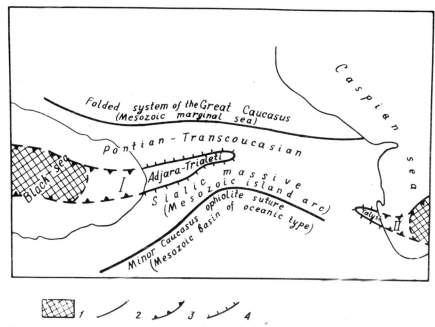

Fig. 1. Scheme of location of the Black Sea—Adjara-Trialetian (*I*) and Talysh-South Caspian (*II*) rift zones: *1* = areas of suboceanic crust in the Black Sea and South Caspian; *2* = inversed deep marine basins. Back-arc rift boundaries; *3* = marine areas, *4* = land.

and predominance of basalts. The latter are closely similar in composition to the basaltic members of the island-arc series, both being products of the H_2O and LIL element-enriched basaltic magmas and comprising hornblende-phyric varieties with abundant hornblende cumulates (Zakariadze et al., 1978). It is to be stressed that other Upper Cretaceous and Paleogene back-arc basins (Burgas, Talysh) situated behind the Mediterranean high-potash andesitic and shoshonitic volcanic arc are as well characterized by a wide development of alkaline and mildly alkaline potassic basalts (Asizbekov et al., 1969, 1972; Stanisheva-Vassileva, 1971).

All other known examples also indicate that on early (prespreading) stages of back-arc development volcanic activity reveals a strong resemblance to volcanism of the related island arc. For the "mature" arcs with high-potash calc-alkaline and shoshonitic series it is predominantly alkaline to mildly alkaline — Kariakian and Kamchatkan Upper Cretaceous—Paleogene back-arc basins (Rotman and Markovski, 1974, 1976), and Muriah lavas related to the incipient rifting behind the Sunda arc (Nicholls and Withford, 1976). For the island arcs with predominantly normal potassic andesitic series, back-arc basin volcanics are mildly alkaline to island-arc tholeiitic (New Georgian group of Solomon Islands — Stanton and Bell, 1969; New Hebridean interarc basalts — Colley and Warden, 1974). Only island-arc-type

TABLE I

Average composition of main types of basalts from the Mediterranean interarc and marginal basins (on water-free basis) *

	1	2	3	4	5	6	7	8
SiO_2	51.16	51.04	52.02	49.10	50.47	49.42	47.80	51.00
TiO_2	0.75	0.54	0.53	0.72	0.60	0.63	1.05	1.68
$Al_{22}O_3$	18.70	19.93	16.98	17.50	18.05	12.13	19.14	17.00
Fe_2O_3	6.35	6.71	5.92	5.30	4.90	6.47	8.65	10.20
FeO	3.20	2.12	3.41	4.39	4.42	3.92		
MnO	0.17	0.19	0.16	0.19	0.16	0.16	0.16	0.16
MgO	4.26	4.39	4.64	6.50	8.36	10.25	8.60	6.52
CaO	10.66	11.21	9.22	10.84	9.76	12.08	11.88	8.74
Na_2O	2.82	3.32	2.92	2.92	2.84	1.84	2.92	4.29
K_2O	1.67	0.40	2.94	2.00	0.35	2.68	0.29	0.34
P_2O_5	0.26	0.15	0.40	0.54	0.10	0.35	0.14	0.27
Rb	—	11	56	93	4.7	73	2.5	1
Ba	730	80	440	245	221	1328	58.5	58
Sr	420	230	585	800	258	827	178	181
Cr	35	26	25	176	258	432	206	154
Ni	39	22	30	65	44	159	6	—
Co	37	32	48	49	38	50	30	32
V	258	210	285	388	287	210	—	—
La	—	5.7	15	25	5.6	15	4.88	8.68
Ce	—	15	29	—	42	54	5.9	12.7
Nd	—	11.5	19	33	14	19	—	—
Y	—	21	25	25	20	17	21	35.5
Yb	—	1.8	1.8	2.7	3.9	2.3	1.72	3.2
Zr	52.10	7.0	31	82	57	135	54	122
Nb	3.15	0.8	3.7	5	4.0	8	3.1	6.5

* Column headings:
1—6. *Black Sea—Adjara-Trialeti Paleogene interarc basalts*: 1. Prespreading stage alkaline basalt (Lower Eocene) (10) **· Initial spreading stage basalts (Middle Eocene): 2.3. Axial tholeiite and periphreal alkaline basalt (22) of the eastern segment. 4—6. Southern flank mildly alkaline basalt (14), axial tholeiite (11) and potassic ancaramite (22) of the northern flank of the western segment.
8—9. Tyrrhenian Sea low-TiO_2 basalt (11) and high-TiO_2 basalt (9) (Dietrich et al., (1977).
** Here and in the notes to Table II figures in parentheses indicate number of analyses used for calculation of the average.

tholeiites or basalts closely similar to the MORB are as yet known from the back-arc basins related to the immature island arcs (Lau, Parece Vela and Mariana basins) (Gill, 1976; Hawkins, 1976; Ridley et al., 1974).

In the Black Sea—Adjara-Trialetian back-arc rift, intensity of tension reached its maximum in the Middle Eocene (Adamia et al., 1974a, b). For the thick (3—5-km) basaltic sequences of this period a distinct transversal lateral zoning is established (Adamia et al., 1974b), high-potash alkaline and

TABLE II

Average composition of main types of basalts from the western Pacific interarc and marginal basins (on water-free basis)

	1	2	3	4	5	6	7	8	9	10
SiO_2	50.3	47.4	49.3	48.94	48.84	49.64	49.64	52.83	49.29	49.31
TiO_2	0.6	0.7	1.3	1.03	0.77	0.90	1.78	1.45	1.10	1.04
Al_2O_3	14.3	10.5	16.2	16.65	14.49	19.28	16.16	18.30	17.87	17.07
Fe_2O_3	3.4	4.5	5.0	2.07	10.14	10.22	2.87	2.81	7.35	4.32
FeO	5.8	5.8	5.7	6.7			7.92	5.57	3.56	5.55
MnO	—	—	—	0.16	0.18	0.18	0.20	0.15	0.16	0.16
MgO	10.0	16.4	5.8	9.64	11.54	5.49	6.99	4.48	4.85	7.78
CaO	12.1	10.9	11.2	12.26	11.20	11.11	12.09	10.20	11.96	11.25
Na_2O	2.0	2.0	3.0	2.32	1.97	2.31	2.48	3.10	3.37	3.10
K_2O	0.5	1.0	1.4	0.15	0.64	0.70	0.34	0.91	0.38	0.26
P_2O_5	0.10	0.20	0.30	0.08	0.20	0.17	0.09	0.20	0.10	0.12
Rb	1	3	6	1	5	10.5	2.5	7.1	7.8	5.2
Ba	170	352	444	—	197	162	66.4	174	33.5	11.5
Sr	296	511	703	107	285	422	137	146	192	163
Cr	158	785	53	433	621	50	285	—	—	—
Ni	56	350	38	187	194	22	163	—	165	173
Co	—	—	—	85	66	34	67	—	—	—
V	256	254	333	244	250	234	285	—	—	—
La	—	—	—	—	—	—	4.6	16.8	—	—
Ce	—	—	—	—	—	—	12.4	3.2	8.0	7.8
Nd	—	—	—	—	—	—	8.2	16.7	6.7	6.6
Y	—	—	—	—	13	17	—	—	25	24
Yb	—	—	—	—	—	—	2.2	2.5	2.9	2.8
Zr	48	64	100	—	34	33	78.0	121	69	68
Nb	—	—	—	—	—	—	4.6	25	2	3

Column headings:

1–3. *New Hebridean interarc basalts*: 1. Banks tholeiite (4) from the central part. 2,3. Aoba mildly alkaline picrite (4) and basalt (4) from the southern margin (Colley and Warden 1974).

4–8. *Lau basin basalts*: 4. Central tholeiites (Sites 95, 64, 61, 74, 223, 225). 4,5. Western marginal melanocratic basalt (sample 382 of Korombasanga series — Lau ridge and 101–2 from Lau basin). 7,8. High-TiO_2 basalts from the north-central (Sites 106, 86, 89, 79) and northwestern (Site 103) periphery (after data by Hawkins, 1976 and Gill, 1976).

9–10. *Parece Vela*: N54-4 and 54-8 basalts (Ridley et al., 1974).

mildly alkaline basalts occurring on its southern and northern flanks and low-potash basalts in the axial parts (Table I). Beside transversal zoning regular compositional changes occur along the strike of the Adjara-Trialetian rift, where intensity of distension distinctly increased from east to west, towards the Black Sea (Adamia et al., 1974a, b). The volume of nondifferentiatted basalts increases in the same direction (Table I), whereas rocks of intermediate composition are strongly diminishing up to their complete disappearance.

Very pronounced and regular compositional variations occur in volcanic sequences of several other ancient and existent back-arc basins. In the Jurassic sequence of the Great Caucasian marginal sea, tholeiites are located in the central part, mildly alkaline to alkaline basaltic series with minor trachy-andesitic differentiates in the southern flank and high-potash picrites at the northern periphery (Zakariadze et al., 1978). In the New Hebridean inter-arc basin, Aoba Island basalts of the southern margin are mildly alkaline, and Banks Island basalts in the central part represent island-arc-type tholeiites (Table II).

Significant compositional differences among Lau basin basalts are evident from data by Hawkins (1976). The authors, in agreement with J.B. Gill (1976), are inclined to consider that these differences cannot be attributed to the alteration and differentiation processes, but are possibly indicative of variation, developed at magma genesis sites. Furthermore, notwithstanding alteration and differentiation, there seems to exist a definite relation between the sample chemistry and its location. Thus, all samples from the central part of the basin seem to bear a resemblance to the MORB, slightly differing from the latter only by lower TiO_2 and increasing Sr^{87}/Sr^{86} (Gill, 1976). Samples from the western marginal part of the basin (DSDP site 101) are low-TiO_2, LIL element-enriched basalts similar to the Korombasanga basalts of the Lau ridge, formed during the Lau rifting event (Gill, 1976). Both Korombasanga and Site 101 basalts bear a resemblance to the island-arc tholeiites (Table II). In the northern peripheral part of the basin basalts are characterized by enrichment in TiO_2, alkalines (primarily in Na_2O), LIL elements and light REE (Table II).

Thus, a distinct depression in K_2O and LIL elements in the central part of the basin seems to be a general rule for the back-arc basaltic series.

EVOLUTION OF LOW-K_2O BASALTIC MAGMAS WITH INCREASING INTENSITY OF SPREADING

In the Adjara-Trialetia, the axial low K_2O basalts of the eastern segment, where distension was the least intense, are represented by the quartz-normative high-Al_2O_3, low-MgO varieties. All rocks are strongly porphyric, and sometimes hornblende-bearing. By their geochemistry — a low TiO_2, Cr and Ni, high Co/Ni and elevated Ba, Sr and Rb (Table I, Figs. 2—4) — they bear resemblance to the basaltic members of the island-arc calc-alkaline andesitic series.

In the western Adjara-Trialetia, low-potash axial basalts are formed under conditions of initial spreading, when magma rose along the open fractures, carrying to the surface huge (50 × 800-m) blocks of Upper Cretaceous limestones, disrupted from the basement (Adamia et al., 1974b). In comparison to the axial basalts of the eastern segment, western tholeiites are enriched in MgO, Cr and Ni and depleted in K_2O, LIL elements and light REE (Table I, Figs. 2—4). However, these basalts still differ from the MORB in their geochemistry, manifesting lower TiO_2, Cr and Ni, and a higher LIL element, as well as an increased H_2O content, indicated by the presence of hornblende-phyric varieties and by no iron enrichment trends (Adamia et al., 1974b).

Black Sea floor basalts, formed in the western part of the Black Sea—Adjara-Trialetian rift, that underwent a strong spreading are as yet unavailable. But basalts from the abyssal plain of the Tyrrhenian Sea, formed in

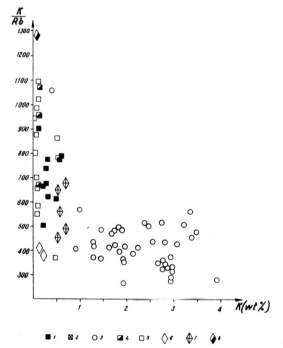

Fig. 2. K/Rb—K diagram for basaltic series of marginal and interarc basins *1—3* = Adjara-Trialeti basalts of initial spreading stage, axial low-potassic basalts of western (*1*) and eastern (*2*) segments, respectively, and marginal alkaline basalts (*3*); *4* = Tyrrhenian tholeiites (Dietrich et al., 1978); *5* = Mid-oceanic ridge tholeiites (Kay et al., 1970); *6* = Parece Vela tholeiites (Ridley et al., 1974); *7* = Korombasanga basalts — Lau ridge (Gill, 1976); *8* = Lau basin tholeiites (Gill, 1976).

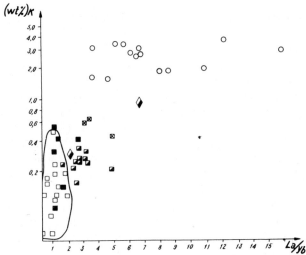

Fig. 3. K—La/Yb diagram for basaltic series of back-arc basins (for explanations of other symbols, see Fig. 2).

closely similar tectonic conditions as a result of the Pliocene back-arc spreading (Dietrich et al., 1978), may be considered as analogous in composition.

These basalts reveal mineralogy and structures typical of the MORB. Their TiO_2 is highly variable, but distinctly increased in comparison to the Adjara-Trialetian axial tholeiites, whereas K and Rb are reduced to the level of maximum concentrations of these elements in the MORB (Fig. 2). Nevertheless, they still differ from the latter by elevated Ba and Sr contents, higher La/Yb values (Table I, Figs. 3, 4) as well as by no iron enrichment (Dietrich et al., 1978).

The same tendency of evolution is characteristic of the axial low-potash

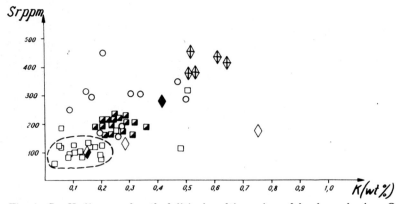

Fig. 4. Sr—K diagram for tholeiitic basaltic series of back-arc basins. Open circles indicate Adjara-Trialetian axial tholeiites; filled rhomb = tholeiite of New Hebriden interarc basin (Colley and Warden, 1974); for explanations of other symbols see Fig. 2.

basalts from the back-arc basins of the western Pacific. In the New Hebridean interarc basin, affected by a small-scale spreading, central tholeiites (from Banks Island) have very pronounced island-arc affinities (Table II, Figs. 2—4), whereas central tholeiites from the more "evolved" basins, formed as a result of an extensive sea-floor spreading and related to the oceanic island arcs, are closer to the MORB in their composition. But even these basalts slightly differ from the MORB, being characterized by a lower TiO_2 and higher Sr^{87}/Sr^{86} (Lau basin central tholeiites) or by elevated LIL and light REE contents (Parece Vela) (Table II, Figs. 2—4).

The above data may allow one to assume that along with increasing intensity of rifting and spreading, basalts from the central parts of the Mediterranean and western Pacific ancient and existent back-arc basins are gradually depleted in K_2O, LIL elements and light REE, and enriched in Cr and Ni, progressively approaching the MORB compositions. But even the tholeiites from the most evolved marginal seas (Parece Vela, Lau) maintain slight but detectable peculiarities, relating them to the island-arc magmatism.

LOW-TiO_2 ALKALINE BASALTS OF THE BACK-ARC BASINS AND THE PROBLEM OF THEIR GENESIS

Lateral petrochemical zoning of back-arc basaltic series with strong depression of K_2O, LIL and light REE in central parts of the basins is similar to the well-known zoning of volcanic series originated in the continental and oceanic rift structures (Kay et al., 1970; Mohr, 1971; Gass et al., 1973). But along with the general similarity, important geochemical differences also occur, expressed primarily in Ti, Nb and Zr contents.

In distinction from the continental and oceanic rift volcanics where increasing alkalinity is accompanied by a pronounced enrichment in Ti and geochemically closely related Nb and Zr, in the back-arc basalts Ti, Nb and Zr contents reveal only slight correlation to alkalinity and remain constantly low, even in the rocks with the highest K_2O content (Figs. 5, 6).

In these rocks, Na_2O reveals a relatively small variability, alkalinity rising with K_2O enrichment. The constantly low Ti, Nb and Zr, and the generally stronger enrichment in K_2O, LIL elements and H_2O may relate the alkaline basaltic magmatism of the back-arc basins to the shoshonitic volcanism of the island arcs. This correlation is suggestive of a common magma-generating mechanism for these two structural elements of the plate-consumption-type active margin, at least in the beginning.

Of the suggested models of island-arc magmatism, the hypothesis of a high degree of mantle melting, caused by the fluids or water-rich melts ascending from the subduction zone (Keller, 1974; Nicholls and Withford, 1976) seems most acceptable.

Absence of the deep mantle inclusions (garnet peridotites), as well as geochemical characteristics of the back-arc basalts such as low TiO_2, often com-

404

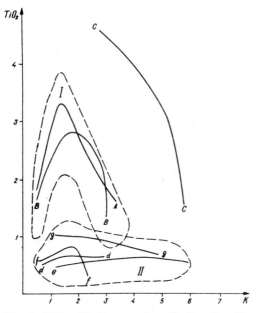

Fig. 5. TiO_2—K diagram for alkaline basalts of continental, oceanic and back-arc rift zones. Field 1 of oceanic and continental rift basalts is outlined after the following data: East Africa rift (Belousov et al., 1974); Afar and Red Sea rifts (Mohr, 1971; Coleman et al., 1975); Aden (Cox et al., 1970); Jan Mayen (data by authors); Saint Helena Island (Baker, 1969); Gough Island (Le Maitre, 1962); Hawaii (Macdonald and Katsura, 1964). Field II of back-arc basalts is outlined after data on the Adjara-Trialeti Paleogene back-arc basin, Great Caucasian Jurassic marginal sea (data by authors); Burgas Cretaceous back-arc basin (Stanisheva-Vassileva, 1971), Kamchatka Cretaceous—Paleogene basins (Rotman and Markovski, 1974, 1976); New Georgian group of Solomon Islands (Stanton and Bell, 1969); New Hebridean interarc basin (Colley and Warden, 1974); Talysh Paleogene basin (Asizbekov et al., 1969, 1972). Variation curves: A = Saint Helena; B = Jan Mayen; C = East African rift; d, e = southern and northern flanks of Adjara-Trialeti respectively; g = Talysh; f = New Hebridean interarc basin. Number of data points is seven for New Hebridean and 15—120 for other curves here and in Fig. 6.

bined with a high MgO content and wide development of picrites, require a high degree of melting at comparatively shallow levels of the upper mantle. At the same time, interaction with fluids or water-rich melts issued from the subducting lithospheric slab may account for enrichment of magmas in K_2O, LIL elements, REE and H_2O.

These fluids, presumably originating in the reducing environments (Marakushev and Perchuk, 1971), during their ascent would extract from the mantle primarily strong bases — K, Rb, Ba, Sr, etc. — whereas amphoteric Ti, Nb and Zr remain relatively immobile. With increasing depth of the subduction zone (i.e., away from the ocean) the fluids pass an increasing thickness of the mantle and therefore would be increasingly enriched in K and LIL elements. This effect accounts for the well-known lateral zoning of the

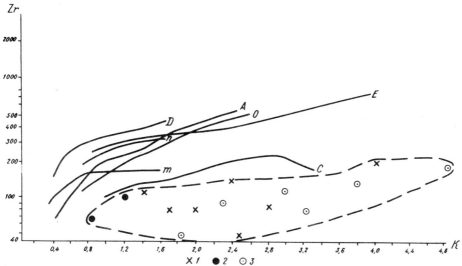

Fig. 6. Zr—K relations in alkaline basalts of oceanic, continental and back-arc rifts and island arcs: *1* = Adjara-Trialeti; *2* = New Hebridean interarc basin (Colley and Warden, 1974); *3* = Aeolian arc (Keller, 1974). Dotted line outlines field of back-arc and island-arc basalts. Variation curves: *A* = Saint Helena (Baker, 1969); *G* = Gough (Le Maitre, 1962); *D* = Red Sea islands (Gass et al., 1973); *E* = potassic series of East African rift (Belousov et al., 1974); *O* = Aden bay coastal basalts (Cox et al., 1970); *m* = As Sirat and *n* = Al Birc, Saudi Arabia (Coleman et al., 1975).

island-arc volcanic series, as well as for the elevated alkalinity of basalts, related to the prespreading stages of back-arc rifting.

With initiation of back-arc spreading, a sharp pressure decrease in axial parts of the back-arc basins might trigger upwelling and high-level melting of the mantle, more or less independent of fluid action. At this stage the fluid effect on magmatic melts seems to be strongly reduced.

REFERENCES

Adamia, Sh.A., Gamkrelidze, I.P., Zakariadze, G.S. and Lordkipanidze, M.B., 1974a. Adjara-Trialeti trough and problem of Black Sea origin. Geotektonika, 1: 74—98 (in Russian).

Adamia, Sh.A., Gamkrelidze, I.P., Zakariadze, G.S. and Lordkipanidze, M.B., 1974b. Lateral zoning of basaltoids in the Black Sea—Adjara-Trialetian paleorift. C.R. Acad. Sci. U.S.S.R., 216(4): 901—903. (in Russian).

Adamia, Sh.A., Lordkipanidze, M.B. and Zakariadze, G.S., 1977. Evolution of an active continental margin as examplified by the Alpine history of the Caucasus. Tectonophysics, 40: 183—199.

Asizbekov, Sh.A., Ismailzade, A.D. and Nijeradze, N.Sh., 1969. Leucite bearing effusives of Talysh. Izv. Akad. Nauk S.S.S.R., Ser. Geol., 1969. (4): 128—134 (in Russian).

Asizbekov, Sh.A., Valiev, M.M. and Ismailzade, A.D., 1972. Petrochemical characteristics of the Eocene mildly alkaline volcanic series of Talysh. Acad. Sci. U.S.S.R. Her. (Vestn.) (Ser. Earth. Sci.), 1: 312—319 (in Russian).

Baker, P., 1969. Petrology of volcanic rocks of the Saint Helena island, South Atlantic. Geol. Soc. Am. Bull., 80(7): 1283—1310.

Belousov, V.V., Gerasimovski, V.I., Goriachev, A.V., Dobrovolski, V.V., Kapitsa, A.P., Logachev, A.V., Milanovski, E.E., Poliakov, A.I., Rikunov, L.N. and Sedov, V.V., 1974. East African rift system, Vol. III. Nauka, Moscow, 328 pp. (in Russian).

Boccaletti, M., Manetti, P. and Peccerilo, A., 1974. The Balkanides as an instance of back-arc thrust belt: possible relation with the Hellenides. Geol. Soc. Am. Bull., 85: 1077—1084.

Coleman, R.G., Fleck, R.J., Hedge, C.E. and Ghent, E.D., 1975. The volcanic rocks of southwest Saudi Arabia and opening of the Red Sea. U.S. Geol. Surv. Saudi Arabia Proj. Rep. 194, 66 pp.

Colley, H. and Warden, A.J., 1974. Petrology of the New Hebrides. Geol. Soc. Am. Bull., 85: 1635—1646.

Cox, K.G., Gass, L.G. and Mallik, D.I.J., 1970. The peralkine volcanic suite of Aden and Little Aden, South Arabia. J. Petrol., 11(3): 433—462.

Dietrich, V., Emmerman, R., Keller, J. and Puchelt, H., 1978. Oceanic basalts from the Tyrrhenian basin. Earth. Planet. Sci. Lett., 39: 127—144.

Gass, L.G., Mallick, D.J.I. and Cox, G., 1973. Volcanic islands of the Red Sea. J. Geol. Soc., 129(3): 275—331.

Gill, G.B., 1976. Composition and age of Lau Basin and Ridge volcanics: implications for evolution of an interarc basin and remnant arc. Geol. Soc. Am. Bull., 87: 1384—1395.

Hawkins, J.W., 1976. Petrology and geochemistry of basaltic rocks from Lau Basin. Earth. Planet. Sci. Lett., 28: 283—297.

Hsü, K.J., Nachev, I.K. and Vuchev, V.T., 1977. Geological evolution of Bulgaria in light of plate tectonics. Tectonophysics, 40: 245—256.

Karig, D.E., 1971. Origin and development of marginal basins in the western Pacific. J. Geophys. Res., 74(11): 2543—2579.

Kay, R., Hubbard, N.J. and Gast, P.W., 1970. Chemical characteristics and origin of oceanic ridge volcanic rocks. J. Geophys. Res., 75(8): 1575—1613.

Keller, J., 1974. Petrology of some volcanic rocks series of the Aeolian arc, southern Tyrrhenian Sea: calc-alkaline and shoshonitic associations. Contrib. Mineral. Petrol., 46(1): 29—49.

Khain, V.E., 1977. Critical comparison of mobilistic models of tectonic development of the Caucasus. In: B. Biju-Duval and L. Montardet (Editors), Int. Symp. on the Structural History of the Mediterranean Basins, Split, 1976. Editions Technip, Paris, pp. 353—362.

Le Maitre, R.W., 1962. Petrology of volcanic rocks, Gough island, South Atlantic. Geol. Soc. Am. Bull., 73(11): 1309—1340.

Macdonald, G.A. and Katsura, T., 1964. Chemical composition of Hawaiian lavas. J. Petrol., 5: 82—133.

Marakushev, A.A. and Perchuk, L.L., 1971. Origin and evolution of the transmagmatic and metamorphic fluids. 1st Int. Geochem. Congr., Vol. II of theses. Academy of Sciences. U.S.S.R. Press, Moscow, pp. 513—515.

Mohr, P.A., 1971. Ethiopian rift and plateau: some volcanic petrochemical differences. J. Geophys. Res., 76(8): 1967—1984.

Nicholls, I.A. and Withford, D.J., 1976. Potassium-rich volcanic rocks of the Sunda islands arc. Java, Indonesia. 25th Int. Congr. Abstr., I: 58—59.

Packham, G.H. and Falvey, D.A., 1971. A hypothesis for the formation of marginal seas in the western Pacific. Tectonophysics, 11: 79—109.

Ridley, N.I., Rhodes, J.M., Reid, A.M., Jakes, P., Shih, C. and Bass, M.N., 1974. Basalts from Leg 6 of the Deep-Sea Drilling Project. J. Petrol., 15: 140—159.

Rotman, V.K. and Markovski, B.A., 1974. On the peri-Pacific geosynclinal basaltic rocks.

In: G.D. Afanasiev, O.A. Bogaticov and A.K. Simon (Editors), Actual Problems of Modern Petrology. Nauka, Moscow, pp. 184—195 (in Russian).

Rotman, V.K. and Markovski, B.A., 1976. Basaltic volcanic belts of the ocean-margins — a peculiar type of peri-Pacific geosynclines. C.R. Acad. Sci. U.S.S.R., 231(3): 702—706 (in Russian).

Sleep, N. and Toksös, H.N., 1971. Evolution of marginal basins. Nature, 233(5321): 548—550.

Stanisheva-Vassileva, G., 1971. Cretaceous magmatic formation of the Burgas synclinorium. C.R. Acad. Sci. Bulg., 24: 1509—1512.

Stanton, R.L. and Bell, J.D., 1969. Volcanic and associated rocks of the New Georgian group, British Solomon Islands protectorate. Overseas Geol. Miner. Resour., 10(2): 282—317.

Zakariadze, G.S., Lordkipanidze, M.B. and Popolitov, E.I., 1978. Some aspects of geochemical evolution of back-arc volcanism as examplified by the Black Sea—Adjara-Trialeti paleorift. Geochimia (Geochemistry), 6: 821—831 (in Russian).

Chapter 21

SEA-FLOOR SPREADING IN MARGINAL BASINS OF THE WESTERN PACIFIC *

A.B. WATTS, J.K. WEISSEL and R.L. LARSON

Lamont-Doherty Geological Observatory of Columbia University, Palisades, New York 10964 (U.S.A.)

SUMMARY

Observed magnetic lineations from the Shikoku, South Fiji, and West Philippine marginal basins are used to determine aspects of the tectonic evolution of these basins. In the western part of the Shikoku basin, anomalies 7 through 5E (~27—19 m.y. B.P.) are identified between the base of the Palau—Kyushu ridge and the center of the basin. In the eastern part of the South Fiji basin, anomalies 12 through 8 (~35—29 m.y. B.P.) are identified between the base of the Lau—Colville ridge and the center of the basin. In the West Philippine basin, magnetic lineations which trend ESE—WNW have been mapped north of the Central Basin Fault. These lineations and the previously mapped lineations south of the fault (Louden, 1976) may be one limb of a Mesozoic spreading system, but more probably represent a two-limb system of Early Tertiary age.

INTRODUCTION

Small, semi-enclosed seas which are termed "marginal basins" occur behind most of the major island arc—trench systems in the western Pacific Ocean (Fig. 1, see fold-out pages 417—418). Seismic refraction measurements (Murauchi et al., 1968; Shor et al., 1970) and petrology of dredged rocks (Hawkins, 1974) indicate the crust of marginal basins is oceanic in type. Karig (1970, 1971) and Packham and Falvey (1971) suggested that these basins formed by processes of crustal extension behind island arcs. On the basis of sediment thickness, depth and character of acoustic basement and mean heat flow Karig (1971) distinguished those basins in which active crustal extension is occurring (for example Lau—Havre basin and Mariana trough) from those in which extension has ceased (for example South Fiji,

—————

*Originally published as: Watts, A.B., Weissel, J.K. and Larson, R.L., 1977. Sea-floor spreading in marginal basins of the western Pacific. In: S. Uyeda (Editor), Subduction Zones, Mid-Ocean Ridges, Oceanic Trenches and Geodynamics. Tectonophysics, 37: 167—181.
Lamont-Doherty Geological Observatory Contribution No. 2249.

Shikoku, West Philippine, Coral Sea, Tasman Sea and Japan Sea).

As in the world's major ocean basins, the way in which oceanic crust is generated in marginal basins may be best reflected in magnetic lineation patterns. Magnetic lineations have been mapped in the Japan Sea (Isezaki and Uyeda, 1973), Coral Sea basin (Falvey, 1972), and the South China Basin (Ben-Avraham and Uyeda, 1973) but correlation of the lineations with the geomagnetic reversal time scale has proved difficult. Even though geological evidence from adjacent islands and JOIDES sites has indicated the general age of these basins, it has not been possible to determine details of crustal ages as it has in the world's major ocean basins.

However, in some marginal basins of the western Pacific magnetic lineations have been mapped and correlated with the geomagnetic time scale. Louden (1976) mapped ESE—WNW trending lineations south of the Central Basin Fault in the West Philippine basin which he identified as anomalies 18 (~45 m.y. B.P.) through 21 (~52 m.y. B.P.). The lineations trend obliquely to the trends of the Palau—Kyushu ridge and the active Izu-Bonin—Mariana island arc—trench system (Fig. 1). Watts and Weissel (1975) mapped NNW—SSE trending lineations in the western part of the Shikoku basin which they identified as anomalies 5E (~19 m.y. B.P.) through 7 (~27 m.y. B.P.). The lineations in this basin are nearly parallel to the trend of the adjacent active Izu—Bonin island arc and the remnant arc, the Palau—Kyushu ridge (Fig. 1). The basin is younger than the associated remnant and active arcs which were contiguous prior to crustal extension. Weissel and Watts (1975) mapped N—S trending lineations in the east-central part of the South Fiji basin which they identified as anomalies 8 (~29 m.y. B.P.) through 12 (~35 m.y. B.P.). These lineations are nearly parallel to the trend of the Lau—Colville ridge, a remnant arc (Fig. 1). Cooper et al. (1976) mapped generally N—S trending lineations in the Bering basin which they identified as anomalies M-0 (~108 m.y. B.P.) through M-13 (~130 m.y. B.P.). These lineations which trend obliquely to the Aleutian arc imply that the age of the basin is greater than the age of the arc. They have suggested that this basin was generated at a normal mid-oceanic ridge and was later trapped when the Aleutian island arc developed.

Magnetic lineation fabrics have not been mapped in all marginal basins of the western Pacific, for example the Parece Vela basin, Kuril basin and the Mariana trough. Several factors other than paucity of data may contribute to the absence of lineations in these basins. The Parece Vela basin probably formed in close proximity to the paleo-magnetic equator. Substantial thicknesses of sediments in the Kuril basin may have prevented the acquisition of a strong thermal remanent magnetization in the oceanic crust. The Mariana trough has shallow (about 3 km) rough basement and is located at low geomagnetic latitudes.

The purpose of this paper is to use observed sea-floor spreading-type magnetic lineations to determine aspects of the tectonic histories of three major inactive marginal basins in the western Pacific, the Shikoku, South Fiji and

West Philippine basins. The observations from marginal basins will be compared to those from mid-oceanic ridge systems of the world and we will attempt to determine whether the crust in these basins accreted asymmetrically (one-limb) or symmetrically (two limb) about a center of spreading.

SHIKOKU BASIN

Karig (1971) classified the Shikoku basin as an inactive marginal basin with high heat flow. JOIDES site 297 (Scientific Staff, 1973a) suggests the age of the basin is Early Miocene and site 296 (Scientific Staff, 1973a) indicates volcanism terminated on the Palau—Kyushu ridge by the Late Oligocene.

Tomoda et al. (1975), Watts and Weissel (1975) and Murauchi et al. (1976) have noted that magnetic lineations in the basin trend almost parallel to the Palau—Kyushu ridge. The main difference between these results is the identification of the anomalies which comprise the lineation pattern. Tomoda et al. (1975) proposed that a symmetric sequence comprising anomaly 6A (~22 m.y. B.P.) through 5 (~10 m.y. B.P.) occurs in the basin. Watts and Weissel (1975) using more data than were available to Tomoda et al. (1975) identified anomalies 7 (~27 m.y. B.P.) through 5E (~19 m.y. B.P.) in the western part of the basin. They claimed that anomalies could only be unequivocally identified in the western part of the basin even though the eastern part of the basin shows a fragmented lineation fabric. Murauchi et al. (1976) proposed two episodes of spreading in the basin, an early one from anomaly 6B (~24 m.y. B.P.) through 5E and a later episode from anomaly 5A (~12 m.y. B.P.) through 3' (~6 m.y. B.P.).

To illustrate the interpretation of Watts and Weissel (1975), a representative magnetic anomaly and seismic reflection profile from the western part of the Shikoku basin is shown in Fig. 2. There is good general agreement between the observed magnetic anomaly profile and profiles computed from magnetic block models (Fig. 3).

The shapes of lineated magnetic anomalies may be used to examine the horizontal distribution of magnetization within the oceanic crust. Instead of a standard block model with narrow and vertical transition zones between oppositely magnetized blocks, Blakely and Cox (1972), Schouten and McCamy (1972) and Atwater and Mudie (1973) used a model which assumed crustal material is emplaced at a spreading center with a normal distribution of standard deviation σ. The parameter σ is thus related to the width of the transition zone between blocks of opposite magnetization. Blakely and Cox (1972) determined $\sigma = 1-3$ km for moderate to fast spreading ridges in the Pacific. We estimated transition zone widths in the Shikoku basin using similar methods and found that in order to explain the shapes of observed anomalies values of $\sigma = 2$ to 3 km are required (Fig. 3). This close similarity between transition zone widths suggests that similar processes of crustal generation operate in marginal basins and in the main Pacific Ocean basin.

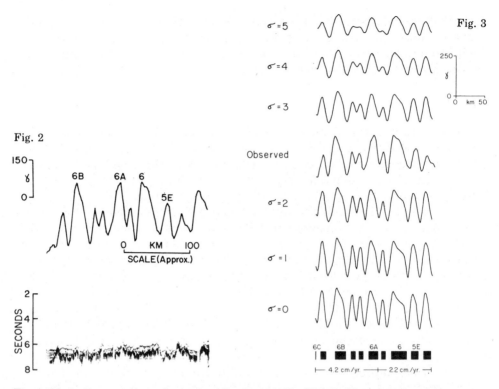

Fig. 2. Magnetic anomaly and seismic reflection profile V2817 (Fig. 1) across the western part of the Shikoku basin. Two-way travel time is indicated in seconds. The identification of the magnetic anomalies is from Watts and Weissel (1975).

Fig. 3. Comparison of magnetic anomaly profile V2817 (Figs. 1 and 2) from the western part of the Shikoku marginal basin to computed profiles based on the geomagnetic time scale of Heirtzler et al. (1968) and Blakely (1974). The computed profiles were generated for a model in which the binary function representing the standard magnetic models was convolved with a Gaussian distribution function with σ = 0, 1, 2, 3, 4 and 5 km. The effect on the computed profiles of increasing σ is to attenuate the amplitudes of higher frequencies (short events) in the profiles. The most satisfactory fit to the observed profile is for σ = 2 to σ = 3.

The presence of a fragmented lineation pattern in the eastern part of the Shikoku basin indicates that the basin evolved in two possible ways. If the basin developed by asymmetric (one-limb) accretion, the fragmented eastern lineations would represent part of a continuous anomaly sequence which gets younger towards the island arc. If the basin developed by symmetric (two-limb) accretion about a spreading center in the center of the basin, these lineations would be the counterparts of the well-identified western lineations (Tomoda et al., 1975; Watts and Weissel, 1975). Although we discussed both possibilities, on the basis of basement morphology and present-day seismicity and tectonics in Japan we concluded that the basin probably formed at a two-limb system and that the eastern limb may have been subsequently deformed (Watts and Weissel, 1975).

SOUTH FIJI BASIN

Karig (1970) described the South Fiji basin as an inactive marginal basin formed by extension between the Loyalty ridge—Three Kings Rise and the Tonga—Kermadec arc during the Early to Middle Tertiary. JOIDES sites 205 and 285 (Scientific Staff, 1972, 1973b) have generally confirmed the ages suggested by Karig and show that the basin is at least as old as Middle Miocene.

We recently identified a pattern of N—S trending magnetic anomalies in the east-central part of the South Fiji basin which we interpreted as anomalies 8 through 12 (Weissel and Watts, 1975). Anomaly 12 (~35 m.y. B.P.) is located near the base of the Lau—Colville ridge and anomaly 8 (~29 m.y. B.P.) near the center of the basin close to a morphologic feature identified

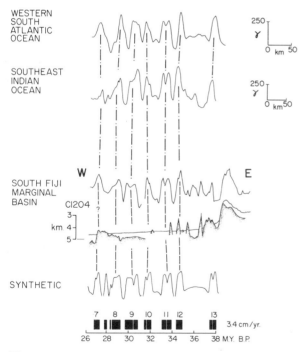

Fig. 4. Comparisons of magnetic anomaly profile C1204 (Fig. 1) from the South Fiji marginal basin to computed profiles based on the geomagnetic time scale (Heirtzler et al., 1968) and to two profiles of the world's main ocean basins (Weissel and Watts, 1975). The profiles from the main ocean basins have been adjusted to match the shapes and amplitudes of observed anomalies in the South Fiji basin.

on Scripps' bathymetry maps (Mammerickx et al., 1971) as the Bounty "channel" (Fig. 1).

A representative magnetic anomaly profile and a line drawing of a seismic reflection profile from the South Fiji basin are shown in Fig. 4. The observed magnetic anomaly profile is compared to a standard magnetic block model based on the Heirtzler et al. (1968) geomagnetic time scale and to two profiles from the major ocean basins. There is a close agreement between the observed profile from the South Fiji basin and the two profiles from major ocean basins.

In the east-central part of the basin the inferred age of the crust decreases toward the center of the basin. If the basin was generated at a simple two-limb system, magnetic lineation counterparts of the observed N—S lineations would be expected further west in the basin. However, in the west-central part of the basin E—W trending lineations have been tentatively mapped (Fig. 1) suggesting a more complicated tectonic evolution for the basin.

WEST PHILIPPINE BASIN

The West Philippine basin is bounded to the north by the Oki Daito ridge and to the east by the Palau—Kyushu ridge (Fig. 1). Along its western margin the basin is currently underthrusting the Asian plate at the Ryukyu and Philippine trenches (Katsumata and Sykes, 1969). The age of the basin as inferred from JOIDES sites 290—295 is Early Tertiary (Karig et al., 1973; Scientific Staff, 1973a).

From a study of magnetic anomaly profiles, Ben-Avraham et al. (1972) suggested that the age of the basin was Mesozoic and that the Central Basin Fault (Fig. 1) represented an extinct spreading center. Recently, Louden (in press) mapped ENE—WSW trending magnetic lineations in the basin south of the Central Basin Fault. He identified these lineations as anomalies 18 through 21 (~45—52 m.y. B.P.) in general agreement with the Early Tertiary age for the basin suggested by JOIDES sites 290—295.

We recently obtained three long profiles across the West Philippine basin from the Oki Daito ridge to the Philippine trench which complemented existing data discussed by Ben-Avraham et al. (1972) and Louden (in press). The profiles were obtained over deep and smooth basement morphology such as seen in a representative profile from north of the Central Basin Fault in Fig. 5. The new data have facilitated mapping of magnetic lineations north of the Central Basin Fault (Fig. 6, see fold-out pages 419—420) and have confirmed the previously mapped magnetic lineation trends south of the fault. Even though magnetic lineation trends are the same either side of the fault, anomalies north of the fault differ in shape from those south of the fault (Fig. 6). This can be established by phase-shifting observed anomaly profiles (Schouten and McCamy, 1972). Anomalies south of the fault are brought into symmetric shapes for $\theta = -170°$ to $-190°$ while anomalies north of the fault are brought into symmetry for $\theta = -220°$ to $-240°$

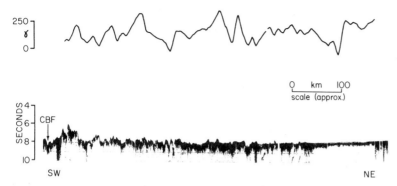

Fig. 5. Magnetic anomaly and seismic reflection profile C1711-4 (Fig. 1) from the West Philippine basin north of the Central Basin Fault. This part of the basin is about 6 km deep and is characterized by a relatively smooth acoustic basement. Sediment cover is thin (<0.7 km) except at the northern end of the profile near the base of the Oki Daito ridge.

(Fig. 7). The de-skewed profiles (Fig. 7) suggest that magnetic anomalies are repeated about the Central Basin Fault but the large difference in θ values required to produce symmetric shapes either side of the fault is difficult to explain.

In Fig. 7 we compare de-skewed magnetic anomaly profiles to standard magnetic block models for the part of the Cenozoic time scale (Heirtzler et al., 1968) used by Louden (in press). We also compare profiles from the north side of the Central Basin Fault with a block model based on the Late Mesozoic time scale (Larson and Hilde, 1975). The Early Tertiary model matches well the shapes of the de-skewed anomalies either side of the Central Basin Fault at least out to anomaly 20 (~50 m.y. B.P.). The longer wavelength de-skewed anomalies further from the Central Basin Fault (especially on the north) are poorly matched by the Early Tertiary model with a constant half-spreading rate of 4.4 cm/year. If the crust is of Early Tertiary age as indicated by JOIDES sites 290—295 (Fig. 6; Scientific Staff, 1973a) then spreading rates prior to anomaly 20 must have been substantially higher, probably in excess of 8 cm/year. The Mesozoic model matches the longer wavelength de-skewed anomalies north of the fault and the implied age is more consistent with the large observed depths (>6.1 km; Fig. 5) but the correlation between synthetic and observed anomalies south of the fault is poor (Fig. 7). Another problem with this model is in explaining the tectonic significance of the fault. A Mesozoic model would imply that the crust south of the fault is older than that to the north and that the Central Basin Fault is a younger strike-slip feature. On the basis of JOIDES results and the apparent repetition of de-skewed magnetic anomalies about the Central Basin Fault, we believe that the Early Tertiary two-limb model for the evolution of the basin is more probable.

416

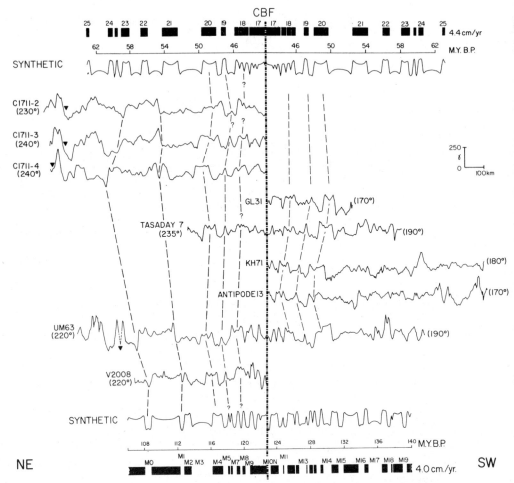

Fig. 7. Comparison of de-skewed magnetic anomaly profiles across the Central Basin Fault (Fig. 6) and to computed profiles based on the standard magnetic block model for the Early Tertiary (upper synthetic profile) and the Late Mesozoic (lower synthetic profile) parts of the geomagnetic reversal time scale. The computed models assume a depth to upper surface of the two-dimensional blocks of 6.0 km, a layer thickness of 0.5 km, uniform magnetization contrast of 0.005 emu/cm^3 and $\theta = 0$. The numbers in parenthesis indicate the values of θ (multiplied by -1) required to phase shift anomalies to symmetry.

CONCLUSIONS

This study of the Shikoku, South Fiji and West Philippine marginal basins allows the following conclusions to be made.

From observed magnetic lineation patterns in the Shikoku basin we conclude that the basin formed between anomaly 7 (~ 27 m.y. B.P.) and anomaly 5E (~ 19 m.y. B.P.) time probably by accretion at a two-limb spreading sys-

tem. The eastern limb of the inferred two-limb system was probably deformed by subsequent tectonic processes severely disrupting the magnetic fabric in the eastern part of the basin (Watts and Weissel, 1975). Analysis of the shapes of magnetic anomalies indicates that crustal accretion in this basin took place over horizontal distances which are comparable to distances inferred for accretion at normal mid-ocean ridges.

Magnetic lineations in the east-central part of the South Fiji basin which trend N—S indicate that at least part of the basin formed between anomaly 12 (~35 m.y. B.P.) and anomaly 9 (~29 m.y. B.P.) time. The presence of E—W trending lineations further west in the basin suggests that more than one simple two-limb spreading system generated the crust of this basin (Weissel and Watts, 1975).

Our study of magnetic lineations in the West Phillipine basin suggests two explanations for the area north and south of the Central Basin Fault. The entire area possibly represents one limb of a Mesozoic spreading system containing anomalies M-0 (~108 m.y. B.P.) through M-16 (~134 m.y. B.P.). In this model, the Central Basin Fault is a subsequent tectonic feature of a strike-slip nature. More probably the West Philippine basin formed from a two-limb spreading system during the Early Tertiary. Anomalies 18 through 20 are observed on both sides of the Central Basin Fault which represents a spreading center that became extinct about anomaly 17 (~42 m.y. B.P.) time. It is difficult to apply a simple model of spreading behind an island arc to the Early Tertiary evolution of the basin. Although the Oki Daito ridge trends parallel to the magnetic lineation fabric (Figs. 1 and 6) and may represent part of an old island-arc system, south of the Central Basin Fault there is no corresponding feature which may represent a rifted fragment of an old arc system.

ACKNOWLEDGEMENTS

The computer program which was used to estimate transition zone widths in the Shikoku basin was written by S. Cande and J. LaBrecque. We are grateful for their help in data analysis and to W.B.F. Ryan and E.M. Herron for critically reading the manuscript. This research was supported by National Science Foundation Grants DES-71-00214, OPP-74-02238 and IDO-75-16027 and Office of Naval Research Contract N00014-75-C-0210.

REFERENCES

Atwater, T.M. and Mudie, J.D., 1973. Detailed near-bottom geophysical survey of the Gorda Rise. J. Geophys. Res., 78: 8665.

Ben-Avraham, Z. and Uyeda, S., 1973. The Evolution of the China basin and the Mesozoic Paleogeography of Borneo. Earth Planet. Sci. Lett., 18: 365—367.

Ben-Avraham, Z., Bowin, C. and Segawa, J., 1972. An extinct spreading centre in the Philippine Sea. Nature Lond., 240: 453—455.

Blakely, R.J., 1974. Geomagnetic reversals and crustal spreading rates during the Miocene. J. Geophys. Res., 79: 2979—2985.

Blakely, R.J. and Cox, A., 1972. Evidence for short geomagnetic polarity intervals in the Early Cenozoic. J. Geophys. Res., 77: 7065—7072.

Bracey, D.R., 1975. Reconnaissance geophysical survey of the Caroline Basin. Geol. Soc. Am. Bull., 86: 775—784.

Cooper, A.K., Marlow, M.S. and Scholl, D.W., 1976. Mesozoic magnetic lineations in the Bering Sea marginal basin. J. Geophys. Res., 81: 1916—1934.

Falvey, D.A., 1972. On the Origin of Marginal Plateaux. Ph.D. thesis, Univ. of New South Wales, Australia.

Hawkins, J.W., 1974. Geology of the Lau basin, a marginal sea behind the Tonga arc. In: C.A. Burk and C.L. Drake (editors), The Geology of Continental Margins. p. 505—520.

Hayes, D.E. and Ringis, J., 1973. Seafloor spreading in the Tasman Sea. Nature Lond., 243: 454—458.

Heirtzler, J.R., Dickson, G.O., Herron, E.M., Pitman, W.C. III and Le Pichon, X., 1968. Marine magnetic anomalies, geomagnetic anomalies, geomagnetic field reversals and motions of the ocean floor and continents. J. Geophys. Res., 73: 2119—2136.

Hilde, T.W.C., 1973. Mesozoic Sea-floor Spreading in the North Pacific. Ph.D. thesis, Univ. of Tokyo, 84 p.

Isezaki, N. and Uyeda, S., 1973. Geomagnetic anomaly pattern of the Japan Sea. J. Mar. Geophys. Res., 2: 51—59.

Karig, D.E., 1970. Ridges and basins of the Tonga—Kermadec island arc system. J. Geophys. Res., 75: 239—254.

Karig, D.E., 1971. Origin and development of marginal basins in the western Pacific. J. Geophys. Res., 76: 2542—2559.

Karig, D.E., Ingle, J.C., Bouma, A.M., Ellis, H., Haile, N., Koizumi, I., MacGregor, I.D., Moore, C., Ujiie, H., Watanabe, T., White, S.M., Yasui, M. and Yi Ling, H., 1973. Origin of the West Philippine Basin. Nature, 246: 458—461.

Katsumata, M. and Sykes, L.R., 1969. Seismicity and tectonics of the western Pacific: Izu—Mariana—Caroline and Ryukyu—Taiwan regions. J. Geophys. Res., 74: 5923.

Larson, R.L., 1975. Late Jurassic sea-floor spreading in the eastern Indian Ocean. Geology, 3: 69—71.

Larson, R.L. and Chase, C.G., 1972. Late Mesozoic evolution of the western Pacific ocean. Geol. Soc. Am. Bull., 83: 3627—3644.

Larson, R.L. and Hilde, T.W.C., 1975. A revised time scale of magnetic reversals for the Early Cretaceous and Late Jurassic. J. Geophys. Res., 80: 2586—2594.

Louden, K., 1976. Magnetic anomalies in the West Philippine basin. AGU Monogr. 19, Int. Woollard Symp., Washington, D.C., pp. 253—267.

Mammerickx, J., Chase, T.E., Smith, S.M. and Taylor, I.L., 1971. Bathymetry of the South Pacific, Chart 12. Scripps Institution of Oceanography.

Markl, R.G., 1974. Evidence for the breakup of eastern Gondwanaland by the Early Cretaceous. Nature, 251: 196—200.

Murauchi, S., Den, N., Asano, S., Hotta, H., Yoshii, T., Asanuma, T., Hagiwara, K., Ichikawa, K., Sato, R., Ludwig, W.J., Ewing, J.I., Edgar, N.T. and Houtz, R.E., 1968. Crustal structure of the Philippine Sea. J. Geophys. Res., 73: 3143—3171.

Murauchi, S., Asanuma, T. and Saki, K., 1976. Seafloor spreading in the Shikoku basin, South of Japan. Abstr. with Progr., Int. Ewing Symp. Harriman, New York.

Packham, G.H. and Falvey, D.A., 1971. An hypothesis for the formation of marginal seas in the western Pacific. Tectonophysics, 11: 79—109.

Schouten, H. and McCamy, K., 1972. Filtering marine magnetic anomalies. J. Geophys. Res., 77: 7089—7099.

Scientific Staff, 1972. Deep Sea Drilling Project, Leg 21, Geotimes, 14: 14—16.

Scientific Staff, 1973a. Deep Sea Drilling Project, Leg 31, Geotimes, 18: 22—25.

Scientific Staff, 1973b. Deep Sea Drilling Project, Leg 30, Geotimes, 18: 19—21.

Shor, G.G., Kirk, H.K. and Menard, H.W., 1970. Crustal structure of the Melanesian area. J. Geophys. Res., 76: 2562—2586.

Tomoda, Y., Kobayashi, K., Segawa, J., Nomura, M., Kimora, K., Saki, T., 1975. Linear magnetic anomalies in the Shikoku basin, northeastern Philippine Sea. J. Geomagn. Geoelectr., 28: 47—56.

Watts, A.B. and Weissel, J.K., 1975. Tectonic history of the Shikoku marginal basin. Earth Planet. Sci. Lett., 25: 239—250.

Weissel, J.K. and Hayes, D.E., 1972. Magnetic anomalies in the southeast Indian Ocean. In: Antarctic Oceanology II: Australian—New Zealand Sector. Antarctic Res. Ser., v. 19, edited by D.E. Hayes. AGU, Washington, D.C., p. 165—196.

Weissel, J.K. and Watts, A.B., 1975. Tectonic complexities in the South Fiji marginal basin. Earth Planet. Sci. Lett., 28: 121—126.

Chapter 22

SEISMOTECTONICS AND TECTONIC HISTORY OF THE ANDAMAN SEA*

TAKAO EGUCHI, SEIYA UYEDA and TADASHI MAKI

Earthquake Research Institute, University of Tokyo, Tokyo (Japan)

SUMMARY

The Andaman Sea is considered as an actively spreading back-arc basin. Seismicity and newly determined focal-mechanism solutions in the Andaman Sea area support this view. The tectonic history of the region is inferred from magnetic lineations in the northeastern Indian Ocean and the northward motion of Greater India. The mid-oceanic ridge which migrated northward along the east side of the Ninetyeast Ridge collided with the western end of the "old Sunda Trench" in the Middle or Late Miocene (10—20 m.y. B.P.). This ridge—trench collision released much of the compressional stress in the back-arc area and the continued northward movement of India that collided with Eurasia exerted a drag on the back-arc region, causing the opening of the Andaman Sea. In appearance, the subducted ridge jumped to the back-arc area. Thus, the Andaman Sea is not an ordinary subduction-related back-arc basin, but probably a basin formed by oblique *extensional* rifting associated with both ridge subduction and deformation of the back-arc area caused by a nearby continental collision.

INTRODUCTION

A number of marginal basins, distributed along the western margin of the Pacific Ocean, are surrounded by a continent and an island arc (back-arc basins) or by two island arcs (inter-arc basins), and are frequently associated with a Benioff-Wadati zone of deep earthquakes. On the other hand, marginal seas such as the Gulf of California are associated with a mid-oceanic ridge system.

There are several hypotheses about the origin of back-arc basins (e.g., Terada, 1934; Carey, 1958; Murauchi, 1966; McKenzie, 1969; Hasebe et al., 1970; Sleep and Toksöz, 1971; Matsuda and Uyeda, 1971; Karig, 1971a, b; Hyndman, 1972; Moberly, 1972; Uyeda and Miyashiro, 1974; Uyeda and Kanamori, 1979). Karig (1971b) divided back-arc basins into two groups; those with active crustal extension (active marginal basins) and those in

---

*Originally published as: Eguchi, T., Uyeda, S. and Maki, T., 1979. Seismotectonics and tectonic history of the Andaman Sea. In: S. Uyeda (Editor), Processes at Subduction Zones. Tectonophysics, 57: 35—51.

which extension has ceased already (inactive marginal basins). The Andaman Sea (Rodolfo, 1969), the Lau-Havre Basin (Sclater et al., 1972; Gill, 1976; Lawver et al., 1976a), the Mariana Trough (Karig, 1971a; Le Pichon et al., 1975), Izu-Bonin zone (Karig and Moore, 1975), the Okinawa Trough (Herman et al., 1979) and the Bismarck Sea (Connelly, 1976) are the examples of active marginal basins. Behind the Andaman—Nicobar island arc in the Andaman Sea, geomorphic and seismic evidence shows that active extension is taking place along the central Andaman trough (Fitch, 1972). Lawver et al. (1976b) and Curray et al. (1979) also concluded that the Andaman Sea is an extensional basin from the results of Scripps geophysical surveys, including seismic profiling, and magnetic and heat flow measurements.

In this paper, we first examine the seismicity and the present tectonics in the Andaman Sea region. We then speculate on the tectonic history of this region as related to the spreading of the eastern Indian Ocean and the northward movement of Greater India.

SEISMICITY

All the earthquakes of B.I.S.C. magnitude (m_b) over 4.5 between latitudes $0°$N and $20°$N and longitudes $90°$E and $100°$E from January 1964 to December 1973 are examined. Major features of the seismicity of the Andaman Sea area can be observed from the epicenter plot shown in Fig. 1. First, there is a group of epicenters parallel to the Andaman—Nicobar island arc, but not to the convex side of the island arc. This may be related to the highly oblique subduction of the Indian plate beneath the Andaman Sea area. Second feature is the epicenters in a relatively narrow "back-arc seismic zone" trending NNE—SSW in the northern Andaman Sea. This seismic zone may be playing an important role in the spreading of the Andaman Sea. This activity seems to terminate at about $15.6°$N, $96.2°$E at least during the period of our study. These two seismic belts seem to intersect at about $9°$N, $94°$E. Only a few earthquakes of m_b over 4.5 occurred near the Central Andaman Sea Rift Valley (see Fig. 1). The earthquakes deeper than 150 km occurred only twice (1964 August 28, $7.1°$N, $95.1°$E, $d = 180$ km and 1973 April 12, $7.57°$N, $95.05°$E, $d = 229$ km) in the Andaman Sea during our study period. These two events, located in the southern part of the Andaman Sea, are probably related to the oblique subduction of the Indian plate. Most earthquakes in the Andaman Sea are shallower than 150 km, and the depth of events tend to become shallower to the north. We plotted the foci of earthquakes which occurred between $11°$N and $13°$N, on an east—west vertical section (Fig. 2). It is possible to identify the above-mentioned two main characteristic seismic belts on this section.

FOCAL MECHANISMS AND THEIR INTERPRETATION

The focal mechanisms of all the earthquakes of B.I.S.C. magnitude (m_b) over 5.0 that occurred in the study area during the period 1964—1973 are

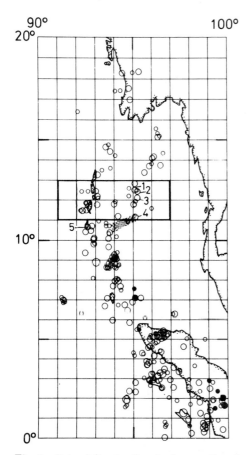

Fig. 1. Seismicity in the Andaman Sea from B.I.S.C. Open and closed circles indicate events with focal depth shallower and deeper than 150 km for the years from 1964 to 1973. Smaller circles show events with m_b over 4.5 and under 5.0. Larger circles show events with m_b over 5.0. The numbered circles are the events for which focal mechanism has been determined newly. Shaded area shows the region of the Central Andaman Sea Rift Valley.

examined. Filmchip copies of W.W.S.S.N. seismograms are also scanned with attention to the events occurred in the central Andaman Sea, including those studied by Fitch (1970, 1972). Only five new earthquakes were found to have sufficient number of teleseismic stations to permit mechanism determination. In Figs. 1 and 2, the locations of these events are shown by symbols 1—5. For these events (see Table I), the P-wave first motions were read on as many W.W.S.S.N. long-period seismograms as possible. Jeffreys-Bullen table was used for the calculation of travel times. Assuming the double-couple-point source model, the focal-mechanism solutions shown in Fig. 3 (on the equal-area projection of the lower hemisphere of the focal sphere) were obtained. The orientation of the nodal planes are given in Table II.

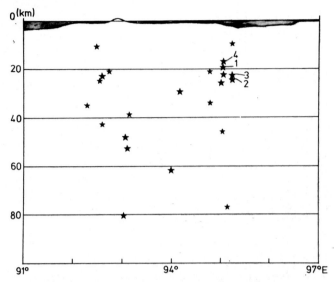

Fig. 2. Earthquake distribution on the vertical cross-section. Events occurred between 11°N and 13°N, see Fig. 1, are plotted. Smaller asterisks show events with m_b over 4.5 and under 5.0. Larger asterisks show events with m_b over 5.0.

All five mechanisms indicate strike-slip faulting. Mechanisms 1, 2, 3 and 4 are from earthquakes along the narrow and shallow (17—24 km) seismic zone trending nearly NNE—SSW in the central Andaman Sea, which we call "the Andaman Sea seismic zone". These four events have almost the same mechanism solutions (see Fig. 3), indicating the activity of shear motions in the Andaman Sea seismic zone. We consider the nodal plane striking nearly N—S trending as the actual shear fault plane in view of the N—S trending of the Andaman Sea seismic zone. Mechanism 5 seems to reflect the right lateral shear motion striking NW—SE. This solution resembles those of other earthquakes near the Nicobar island arc (Fitch, 1970, 1972), suggesting that

TABLE I

Earthquakes used for this study

Event *	Origin time				Lat. (°N)	Long. (°E)	d (km)	m_b
	Date	h	min	sec				
1	Aug. 12, 1971	04	17	03.00	12.50	95.08	19.7	5.3
2	Nov. 13, 1972	23	34	11.80	12.44	95.23	24.1	5.0
3	Mar. 28, 1971	08	23	21.50	12.12	95.22	22.2	5.2
4	Mar. 29, 1971	17	55	43.00	11.16	95.11	16.5	5.1
5	Jul. 09, 1973	16	19	46.70	10.66	92.59	43.8	5.6

* Epicenter determination from B.I.S.C. for earthquakes that yielded a mechanism solution.

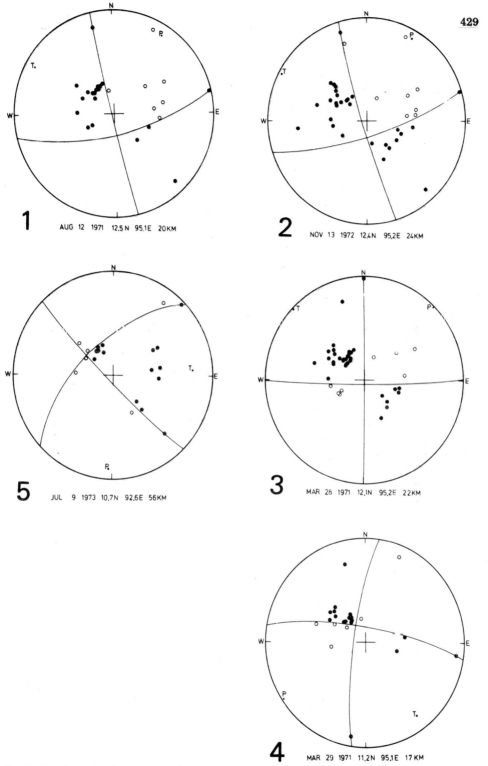

Fig. 3. Focal mechanisms represented by first-motion data on the lower focal hemisphere.

TABLE II

Orientations of nodal planes (in degrees)

Event	1st pole		2nd pole		P-axes		T-axes		B-axes	
	AZ	PL	AZ	PL	AZ	PL	AZ	PL	AZ	PL
1	78	2	347	17	34	13	300	12	174	17
2	74	4	297	4	30	12	297	4	182	13
3	0	3	90	0	45	2	315	2	180	3
4	98	9	190	11	236	1	144	15	327	15
5	45	4	137	25	182	12	86	22	305	27

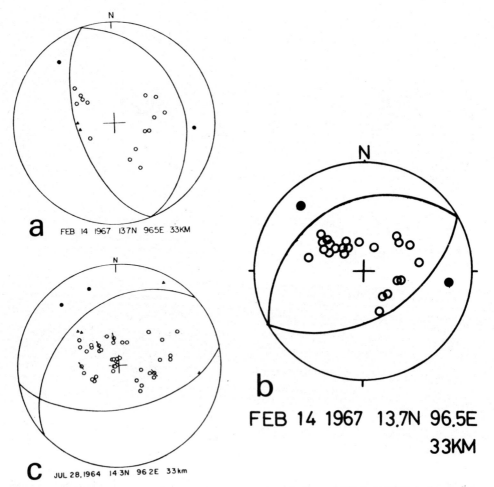

a FEB 14 1967 137N 965E 33KM

b FEB 14 1967 13.7N 96.5E 33KM

c JUL 28,1964 14 3N 96 2E 33km

Fig. 4. Focal-mechanism solutions a. and c. are the solution of Fitch (1970), b. is an alternative solution of a.

this event was a strike-slip earthquake in the decoupling zone associated with the oblique subduction of the Indian plate (Fitch, 1972).

Among the earthquakes studied by Fitch (1970), the mechanism solution of the event (1967, February 14, 13.7°N, 96.5°E, d = 33 km) shown in Fig. 4a may be interpreted differently as shown in Fig. 4b. Note only two compressional first motions of this event were read from the W.W.S.S.N. long-period seismograms to constrain the nodal planes. This alternative mechanism solution is quite similar to that of the event (1964, July 28, 14.3°N, 96.2°E, d = 33 km; Fitch, 1970), shown in Fig. 4c. These two events suggest that an extensional stress field trending NNW—SSE exists beneath the area around 14°N, 96°E in the northern Andaman Sea.

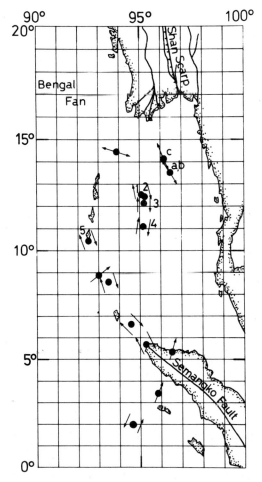

Fig. 5. Fault-plane solution (the present work and Fitch, 1970, 1972) represented by arrow(s) in the Andaman Sea region. Single arrow shows direction of underthrusting; pairs of arrows away from each other show orientation of T-axis for normal faulting. Numbered and alphabetized solutions correpond to those of Figs. 3 and 4, respectively.

INFERRED TECTONIC MOVEMENT

The directions of newly inferred fault motions and those of Fitch (1970, 1972) are shown in Fig. 5. One of the solutions determined by Fitch (1970) has been reinterpreted as explained above. From these mechanism solutions and other published information the tectonic movements occurring under the Andaman Sea may be deduced as shown in Fig. 6a.

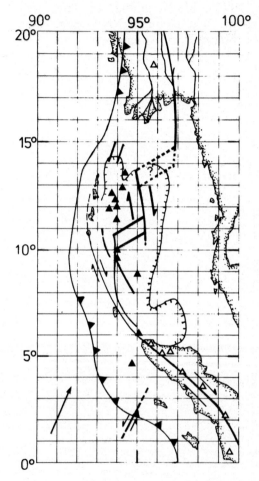

Fig. 6. a. Recent tectonics in the Andaman Sea region inferred from Fig. 5 and other geophysical data. Double lines represent spreading zones. One with broken lines is inferred from the present study, but not confirmed by Curray et al. (1979). Bold line represents fault(s) corroborated by focal-mechanism solutions, and bold arrows indicate sense of motion. Open and closed triangles represent volcanoes and seamounts, respectively. Barbed line indicates continental slope at about −1000 m.
b. Tectonic map of Burma and the Andaman Sea. (From Curray et al., 1979, with permission of the authors.)

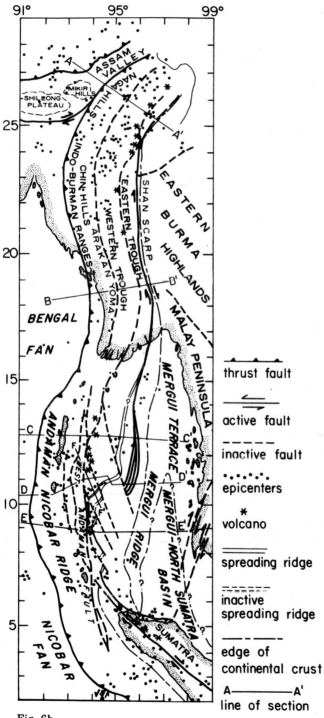

Fig. 6b

thrust fault

active fault

inactive fault

epicenters

* volcano

spreading ridge

inactive
spreading ridge

edge of
continental crust

A————A'
line of section

In Sumatra, earthquakes with right lateral shear motion are known (Katili and Hehuwat, 1967) to occur along the Semangko fault (see Fig. 5) trending NW—SE. This fault is one of the active seismic zones of transcurrent movements within active island arcs and is interpreted as representing the decoupled zone associated with the oblique subduction (Fitch, 1972). The submarine continuation of this fault may extend to the Nicobar island arc (Fitch, 1972).

The Central Andaman Sea Rift Valley, located in the region around $11°N$, $95°E$ and trending NE—SW, has been shown to be an active spreading center from seismic reflection profile and magnetic lineations (Curray and Moore, 1974; Curray et al., 1979). The N—S striking seismicity pattern in the region $95.2°E$, $11—13°N$ and the right lateral strike-slip-type source mechanisms indicate the existence of N—S trending ridge—ridge type transform fault or faults. This fault (or faults) is considered to link the Central Andaman Sea Rift Valley spreading center with another spreading center which exists in the area around $14—15°N$, $94—96°E$. The latter segment of spreading ridge is inferred from the normal-fault-type source mechanisms as cited above. But this proposed ridge is not apparent in the submarine topography. It may probably be because of the high sedimentation rate from the nearby big estuaries in Burma. Unfortunately, this spreading center has not been discovered even by the detailed seismic reflection survey by Curray et al. (1979), as shown in Fig. 6b, forcing our proposal remain quite speculative. A transform fault seems to run northward from this hidden spreading center through the Shan Scarp in Burma (see Fig. 5), defining the eastern border of the Burma plate. We thus infer that the main tectonic feature of the Andaman Sea may be represented by segments of active spreading centers that are connected by right lateral transform fault (or faults).

From the vertical section (Fig. 2), it may be observed that the seismic activity of the transform fault appears to be occurring along a plane that is inclined about $15°$ downward from the western horizon. This activity seems to join with the west-side activity at the depth of about 50 km, forming a V-shaped vertical section. Verma et al. (1976a, b) pointed out that the seismic zone underlying Burma is also V-shaped in the vertical section. If these lines of seismic activity represent the actual fault planes, the V-shape may be defining the cross-section of the Burma plate. The tectonic significance of this peculiar cross-section of the Burma plate is not clear.

TECTONIC HISTORY IN THE ANDAMAN SEA

In the following, we infer the tectonic history of the Andaman Sea based on two lines of information. One is the spreading history in the Indian Ocean and Greater India's northward movement away from Australia and Antarctica (Sclater and Fisher, 1974; Curray and Moore, 1974; Veevers and McElhinny, 1976; Johnson et al., 1976). The other is the recent tectonic movement described in the previous section.

From the magnetic anomaly lineations, it is clear that the age of the ocean bottom to the east of the Ninetyeast Ridge increases southward, while that to the west of the Ninetyeast Ridge increases northward (Sclater and Fisher, 1974). Magnetic lineations to the east of the Ninetyeast Ridge are summarized in Fig. 7. In this figure, we tentatively added the lineations number 13 (~38 m.y. B.P.) and 12 (~35 m.y. B.P.) that we consider identifiable on LSD7 (END) in fig. 2 of Sclater and Fisher (1974). If the Indian plate had not migrated northward, the lineations in Fig. 7 would indicate the past position of the E—W trending ridge. However, the result of the D.S.D.P. indicated that the drilled segment of the Ninetyeast Ridge has moved northward during its sinking in the Late Maastrichtian—Middle Paleocene time (Pimm et al., 1974). The Ninetyeast Ridge was an active right lateral transform fault during the period between 53 m.y. B.P. and 32 m.y. B.P., and at about 32 m.y. B.P. relative motion on either side of

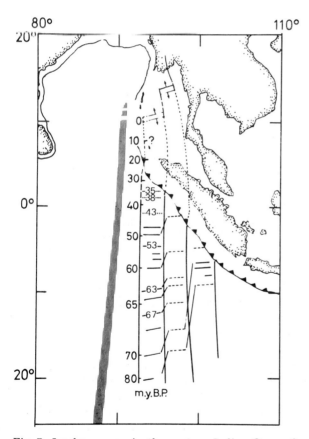

Fig. 7. Isochron map in the eastern Indian Ocean from Sclater and Fisher (1974), our identification and the assumption of rigid plate. Spreading centers in the Andaman Sea are from Fig. 6a. The Ninetyeast Ridge is represented by the shaded band.

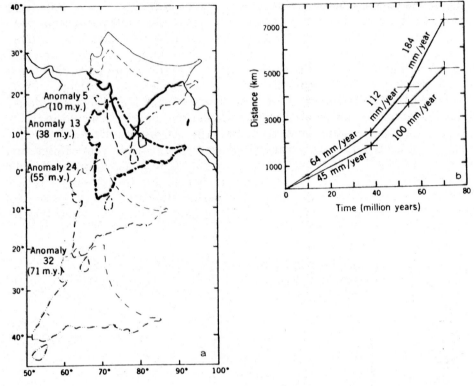

Fig. 8. a. Positions of India with respect to Eurasia. b. Distance from present position of northeast tips of India as a function of time. (From Molnar and Tapponnier, 1975.)

this ridge terminated (Sclater and Fisher, 1974), although it shows small left lateral motion at present (Stein and Okal, 1978). On the other hand, the motion of the Indian plate relative to Eurasia has been inferred by Molnar and Tapponnier (1975), as shown in Fig. 8a, b. Supposing that there was little relative movement between either side of the Ninetyeast Ridge since about 32 m.y. B.P., it is possible to reconstruct the position of the E—W trending ridge to 32 m.y. B.P., and such a reconstruction will enable us to evaluate the time of the collision of the spreading ridge with the "old Sunda Trench". For example, the positions of the Greater India and the E—W trending ridge relative to Eurasia at 20 m.y. B.P. are estimated as shown in Fig. 9. At 20 m.y. B.P. the ridge was still located to the south of the Sunda arc. Assuming that Sumatra was located at the same position as at present, we can thus infer that the spreading ridge might have arrived at the point B in Fig. 9 and collided with the "old Sunda Trench" at about 8 m.y. B.P. But if the 20-degree clockwise rotation of Sumatra around the center of the Sunda Strait (point C in Fig. 9) as proposed by Ninkovich (1976) had occurred, our estimate would demand that the ridge—trench collision and

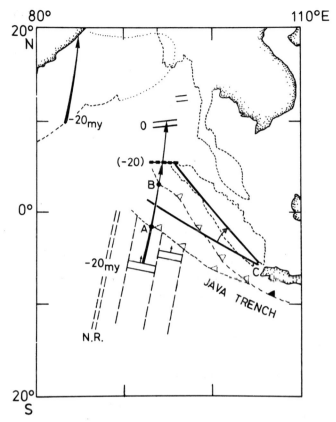

Fig. 9. Inferred locations of spreading ridge, Greater India and Sunda arc at 20 m.y. B.P. Long arrows show the movement of the Indian plate relative to Eurasia for 20 m.y., short arrow shows the amount of spreading from ridge for 20 m.y. and curved arrow denotes the rotation of Sumatra, proposed by Ninkovich (1976). Point $A(B)$ is the collision position of the spreading ridge and "old Sunda Trench" at about 15(8) m.y. B.P. Point C is the Sunda Strait.

rotation of Sumatra began at about 15 m.y. B.P. (point A in Fig. 9) and not 10 m.y. B.P. as proposed by Ninkovich (1976). The duration of the collision process, therefore, might have been extended from about 15 m.y. B.P. to several million years later, because this process seems highly complicated and may not have been completed in a short period. During this period, the rotation of Sumatra and the landward retreat of the "old Sunda Trench" might have occurred. Probably, the collision of the ridge was completed at about 10 m.y. B.P. and this was apparently the time when the spreading center in the Andaman Sea started its activity. The process may be viewed as a jumping of a subducted ridge.

The possibility of back-arc extension due to the subduction of an active spreading center has been suggested already (Scholz et al., 1971; Uyeda and

Miyashiro, 1974). Certainly, collision of a ridge with a trench changes the relative motion of the plates across the boundary. Generally, it reduces the normal convergent component, so that the compressive force originated from the plate convergence will be reduced. However, the plate motion after the collision does not always become divergent to make the stress scheme in the back-arc region tensile (Uyeda and Kanamori, 1978). In the case of the western North America after the collision of the Pacific—Farallon Ridge with the trench, the relative motion became almost purely translational (Atwater, 1970), so that Scholz et al. (1971) had to assume that the subduction previous to the ridge—trench collision had provided some extensional stress. Whether Scholz et al. were right or not, there is no doubt that the active spreading of the Gulf of California is a result of ridge—trench collision, since the ridge must have been to the west of Baja California until just several million years ago and it is now in the Gulf. Here again, the actual process might have caused the birth of a ridge in the back-arc region shortly after the death of ridge upon collision and could be viewed as a ridge jump. In the case of the Andaman Sea, the collision was between the E—W trending ridge and the old Sunda arc. The relative motion after the collision, as can be seen from the present motion at the Java Trench, happens to be still highly convergent and apparently no back-arc opening has taken place behind the Java Trench. Back-arc opening has taken place only at the western end of the Sunda Arc, i.e., the Andaman Sea area.

Perhaps, there were some additional factors that were operative only in the western end of the Sunda Arc. One significant feature to be noted is the change in the strike of the arc. Because of the bend of the arc to the N—S direction in the Sumatra region, the relative motion there is much less convergent and almost translational. The northward motion of Greater India is probably related to the bend of the arc. Namely, since it is inferred

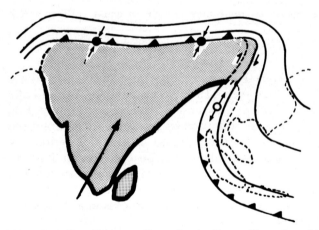

Fig. 10. Inferred stress pattern due to the northward pushing of Greater India on Eurasia. Open and closed circles with arrows represent normal and thrust faulting, respectively.

that Greater India, although its exact northern extent before collision is uncertain, collided with Eurasia also at about 30—10 m.y. B.P. (Fig. 8a) and has continued to push itself into the Asian continent up to the present, the Burma part of Asia has been dragged to the north to make the Andaman Sea area stretch extensionally (Fig. 10). This might have caused both the opening of the Andaman Sea and the clockwise rotation of Sumatra.

Thus, we conclude at this stage that the Andaman Sea has appeared in the area between Burma and western Sumatra as a combined result of the ridge subduction and the collision of India with Asia, the former and the latter probably starting at about 20—10 and 30—10 m.y. B.P., respectively. We would like to point out here that a general view similar to the one proposed here on continental collisions was mentioned by Alfred Wegener in his classic book on continental drift (Wegener, 1924).

ACKNOWLEDGEMENT

We thank J. Curray of Scripps Institution of Oceanography, University of California, San Diego, for critically reading our manuscript and giving us permission of using Curray et al.'s (1979) results of the survey in the Andaman Sea (Fig. 6b) before publication. Drs. R. Geller and S. Stein of Stanford University have kindly given useful suggestions to the authors to improve the text of the paper.

REFERENCES

Atwater, T., 1970. Implications of plate tectonics for the Cenozoic tectonic evolution of western North America. Geol. Soc. Am. Bull., 81: 3513—3535.

Carey, S.W., 1958. A tectonic approach to continental drift. In: S.W. Carey (Editor), Continental Drift. A Symposium. University of Tasmania, Hobart.

Connelly, J.B., 1976. Tectonic development of the Bismarck Sea based on gravity and magnetic modelling. Geophys. J.R. Astron. Soc., 46: 23—40.

Curray, J.R. and Moore, D.G. 1974. Sedimentary and tectonic processes in the Bengal Deep-Sea Fan and Geosyncline. In: C.A. Burk and C.L. Drake (Editors), The Geology of Continental Margins. Springer, New York, N.Y., pp. 617—627.

Curray, J.R., Moore, D.C., Lawver, L., Emmel, F., Raitt, R.W. and Henry, M., 1979. Tectonics of the Andaman Sea and Burma. In: J. Watkins and L. Montadert (Editors), Memoir on Continental Margins. Am. Assoc. Pet. Geol., in press.

Fitch, T.J., 1970. Earthquake mechanisms in the Himalayan, Burmese and Andaman regions and continental tectonics in Central Asia. J. Geophys. Res., 75: 2699 2709.

Fitch, T.J., 1972. Plate convergence, transcurrent faults and internal deformation adjacent to southeast Asia and western Pacific. J. Geophys. Res., 77: 4432—4460.

Gill, J.B., 1976. Composition and age of Lau Basin and Ridge volcanic rocks: implications for evolution of an interarc basin and remanent arc. Geol. Soc. Am. Bull., 87: 1384—1395.

Hasebe, K., Fujii, N. and Uyeda, S., 1970. Thermal process under island arcs. Tectonophysics, 10: 335—355.

Herman, B.M., Anderson, R.N. and Truchan, M., 1979. Extensional tectonics in Okinawa Trough. In: J. Watkins and L. Montadert (Editors), Memoir on Continental Margins. Am. Assoc. Pet. Geol., in press.

Hyndman, R., 1972. Plate motions relative to the deep mantle and the development of subduction zones. Nature, 238: 263—265.

Johnson, B.D., Powell, C.McA. and Veevers, J.J., 1976. Spreading history of the eastern Indian Ocean and Greater India's northward flight from Antarctica and Australia. Geol. Soc. Am. Bull., 87: 1560—1566.

Karig, D.E., 1971a. Structural history of the Mariana island-arc system. Geol. Soc. Am. Bull., 82: 323—344.

Karig, D.E., 1971b. Origin and development of marginal basins in the western Pacific. J. Geophys. Res., 76: 2542—2559.

Karig, D.E. and Moore, G.F., 1975. Tectonic complexities in the Bonin arc system. Tectonophysics, 27: 97—118.

Katili, J.A. and Hehuwat, F., 1967. On the occurrence of large transcurrent faults in Sumatra, Indonesia. J. Geosci., Osaka City Univ., 10: 5—15.

Lawver, L.A., Hawkins, J.W. and Sclater, J.G., 1976a. Magnetic anomalies and crustal dilation in the Lau Basin. Earth. Planet. Sci. Lett., 33: 27—35.

Lawver, L.A., Curray, J.R. and Moore, D.G., 1976b. Tectonic evolution of the Andaman Sea. EOS, Trans. Am. Geophys. Union, 57: 333.

Le Pichon, X., Francheteau, J. and Sharman III., G.F., 1975. Rigid plate accretion in an inter-arc basin: Mariana Trough. J. Phys. Earth, 23: 251—256.

Matsuda, T. and Uyeda, S., 1971. On Pacific-type orogeny and its model-extension of the paired belts concept and possible origin of marginal seas. Tectonophysics, 11: 5—27.

McKenzie, D.P., 1969. Speculations on the consequences and causes of plate motions. Geophys. J. R. Astron. Soc., 18: 1—32.

Moberly, R., 1972. Origin of lithosphere behind island arcs, with reference to the western Pacific. Geol. Soc. Am. Bull., Mem., 132: 35—55.

Molnar, P. and Tapponnier, P., 1975. Cenozoic tectonics of Asia: effects of a continental collision. Science, 189: 419—426.

Murauchi, S., 1966. On the origin of the Japan Sea. (Read at Monthly Meeting of Earthquake Research Institute, Univ. of Tokyo.)

Ninkovich, D., 1976. Late Cenozoic clockwise rotation of Sumatra. Earth Planet. Sci. Lett., 29: 260—275.

Pimm, A.C., McGowran, B. and Gartner, S., 1974. Early sinking history of Ninetyeast Ridge, northeastern Indian Ocean. Geol. Soc. Am. Bull., 85: 1219—1224.

Rodolfo, K.S., 1969. Bathymetry and marine geology of the Andaman basin, and tectonic implications for southeast Asia. Geol. Soc. Am. Bull., 80: 1203—1230.

Scholz, C.H., Barazangi, M. and Sbar, M.L., 1971. Late Cenozoic evolution of the Great Basin, western United States, as an ensialic interarc basin. Geol. Soc. Am. Bull., 82: 2979—2990.

Sclater, J.G. and Fisher, R.L., 1974. Evolution of the east central Indian Ocean, with emphasis on the tectonic setting of the Ninetyeast Ridge. Geol. Soc. Am. Bull., 85: 683—702.

Sclater, J.G., Hawkins, J.W., Mammerickx, J. and Chase, C.G., 1972. Crustal extension between the Tonga and Lau Ridge: petrologic and geophysical evidence. Geol. Soc. Am. Bull., 83: 505—518.

Sleep, N. and Toksöz, M.N., 1971. Evolution of marginal basins. Nature, 233: 548—550.

Stein, S. and Okal, E.A., 1978. Seismicity and tectonics of the Ninetyeast Ridge area: evidence for internal deformation of the Indian plate. J. Geophys. Res., 83: 2233—2246.

Terada, T., 1934. On bathymetrical features of the Japan Sea. Bull. Earthquake Res. Inst., 12: 650—656.

Uyeda, S. and Kanamori, H., 1979. Back-arc opening and the mode of subduction. J. Geophys. Res., No. 3, in press.

Uyeda, S. and Miyashiro, A., 1974. Plate tectonics and the Japanese islands: a synthesis. Geol. Soc. Am. Bull., 85: 1159—1170.

Veevers, J.J. and McElhinny, M.W., 1976. The separation of Australia from other continents. Earth-Sci. Rev., 12: 139—159.

Verma, R.K., Mukhopadhyay, M. and Ahluwalia, N.S., 1976a. Earthquake mechanisms and tectonic features of northern Burma. Tectonophysics, 32: 387—399.

Verma, R.K., Mukhopadhyay, M. and Ahluwalia, N.S., 1976b. Seismicity, gravity and tectonics of northeast India and northern Burma. Bull. Seismol. Soc. Am., 66: 1683—1694.

Wegener, A., 1924. The Origin of Continents and Oceans. Methuen, London, 246 pp.

Chapter 23

THE STRUCTURE AND ORIGIN OF THE OKHOTSK AND JAPAN SEA ABYSSAL DEPRESSIONS ACCORDING TO NEW GEOPHYSICAL AND GEOLOGICAL DATA*

S.L. SOLOVIEV, I.K. TOUEZOV and B.I. VASILIEV

Sakhalin Complex Scientific Research Institute, Far East Science Center, Academy of Sciences of the U.S.S.R., Novoalexandrovsk, Sakhalin 694050 (U.S.S.R.)

SUMMARY

During the three years 1972–1974 the Sakhalin Complex Scientific Research Institute, Far East Science Center, Academy of Sciences of the U.S.S.R., conducted a large number of combined geological and geophysical investigations in the Kurile abyssal depression of the Okhotsk Sea and on the adjacent shelves. The investigations included seismic profiling, gravimetry, magnetometry, echo sounding and dredging. About 20 thousand combined geophysical measurements and 376 dredge stations were obtained. The results of these investigations, along with the previously available information on the crust and upper-mantle structure, heat flow and geological structure of the island and continental regions adjacent to the Okhotsk and Japan seas, allow us to approach the problem of the structure and origin of the above abyssal depressions. The problem has a great importance for correct understanding of the essence of the tectonic processes taking place within the transition zone from the continent to the Pacific.

THEORY

Within the Okhotsk and Japan seas, three abyssal depressions are distinguished: the Kurile depression in the Okhotsk Sea (Fig. 1) and the Central and the Honshu abyssal depressions in the Japan Sea. The Kurile depression occupies only a small southern part of the Okhotsk Sea, while the Central and Honshu depressions are distributed nearly all over the Japan Sea area, being separated from each other by the abyssal rise, the so-called Yamato Bank. The depressions have a flat, practically horizontal floor with maximal depths of about 3400–3600 m. Only the Central Japan Sea depression is complicated by the Bogorova underwater uplift. According to the dredge data, this uplift is composed of volcanogenic rocks, which are intermediate to basic in composition and with an absolute age of 18 m.y. (Ueno et al., 1972; Vasiliev et al., 1975); they belong to the calc-alkalic series and are

---

*Originally published as: Soloviev, S.L., Touezov, I.K. and Vasiliev, B.I., 1977. The structure and origin of the Okhotsk and Japan Sea abyssal depressions according to new geophysical and geological data. In: S. Uyeda (Editor), Subduction Zones, Mid-Ocean Ridges, Oceanic Trenches and Geodynamics. Tectonophysics, 37: 153–166.

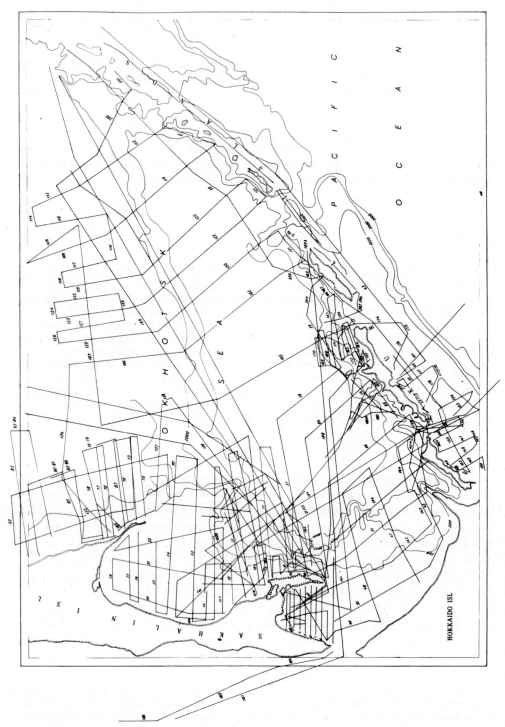

Fig. 1. The scheme of profile locations.

similar to the andesitic basalts of the Kurile—Kamchatka arc (Sakhno and Vasiliev, 1974). Separate parts of the depressions are sharply discordant with the margin structures.

The crust here belongs to the sub-oceanic type (Touezov, 1969). It consists of two layers: the sedimentary one, with thicknesses from 1 to 2 km in the Central Japan Sea depression and up to 5.5 km in the Kurile depression, and a basaltic layer having thicknesses of 8—12 km. The Honshu depression with the reduced granitic layer of small thickness is distinguished as being an exception. The seismic wave velocities in the mantle tops, according to deep seismic sounding, vary from normal (8.2 km/sec) in the Central depression to the decreased values in the Honshu and Kurile depressions (Touezov, 1972). The densities in the tops of the depression mantle are increased relative to the normal, although the values of the excess density as a rule are low and do not exceed 0.05—0.1 g/cm^3 (Gainanov et al., 1974). Within the crust of the abyssal depression, numerous magnetoactive bodies are noted, which not only penetrate the consolidated crust, but also complicate the upper mantle and sedimentary layer (Touezov et al., 1967; Touezov, 1975).

The abyssal depressions are practically a-seismic, but are characterized by two to three times higher (in comparison with the global mean value) heat flow of 2.0—3.0 μcal. cm^{-2} sec^{-1} (Veselov and Touezov, 1972). The values of the Bouguer anomalies reach 260 mGal, but free-air anomalies do not exceed 40 mGal in average (Stroyev, 1975). The magnetic field varies rather sharply over the Japan Sea abyssal depressions, but it is relatively uniform in the Kurile depression (Krasniy, 1972). The magnetic anomalies are predominantly directed northeastward, conformably to the anomaly trends in the Pacific regions near Japan and the Kuriles. The crust of the depressions is characterized by isostatic equilibrium.

The sedimentary layer of the abyssal depressions is divided into two strata (Figs. 2, 3). * The upper stratum, with a thickness of 0.7—1.0 km, contains a great number of sub-horizontal and, as a rule, rather long reflecting boundaries. In the lower stratum their number is significantly less, their extent being also less. The characteristics of the lower stratum are similar to those of the acoustically transparent oceanic layer.

As to the composition and age of the sediments composing the upper stratum of the Japan Sea abyssal depressions, these may be revealed from the data obtained by the 1973 *Glomar Challenger* drilling program (Summary of Deep Sea Drilling Project..., 1973). Pleistocene—Late Miocene turbidite sands, silts, clays, diatomic and mudstone-like clays, having general thickness of about 500 m were found in one of the wells. Within a 275—529 m section interval of another well supposedly Miocene—Oligocene zeolitic clays were found.

The acoustically transparent stratum of the abyssal depressions is underlain by the sharply reflecting horizon, coinciding with the roof of the

---

* See for Fig. 2 fold-out pages 449—452.

446

acoustic basement (Snegovskoy, 1974). The reflecting horizon and its corresponding reflection have a complex, but characteristic record form allowing us to distinguish easily this horizon through correlation discontinuities (Fig. 3). The relief of the acoustic basement of the depression is rather rough. In some parts there are arch-like structures, in others there are smooth structures similar to platform syneclises or gentle depressions. The calm behavior of the acoustic basement is distorted by the underwater mountains, outcropping from beneath the sea bottom (Fig. 3, prof. 38), or buried by the sediment strata (fig. 3, prof. X-a, prof. I-M). The cone-like form of these moun-

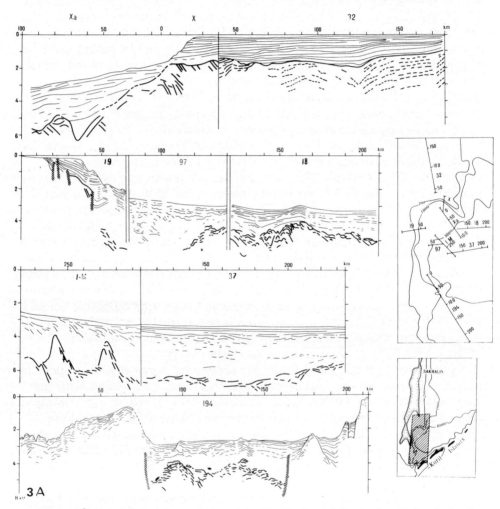

Fig. 3. Montage of seismic cross-sections of the Kurile abyssal depression. Legend: *1* = the reflecting boundaries in the sedimentary cover, *2* = the reflecting boundaries in the acoustic basement, *3* = fractures, *4* = isobaths(m), *5* = seismic profiles.

447

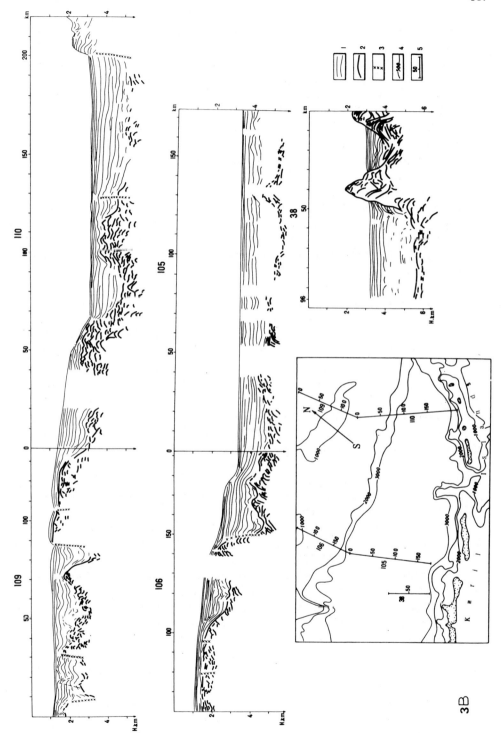

3B

tains and their complex internal structure evidently show them to be both buried and submerged volcanoes.

In the Japan Sea the roof of the acoustic basement coincides with the roof of the basaltic layer, in the Kurile depression it is 1.5—3.0 km higher than the latter, clearly corresponding to the roof of the acoustic basement of the Academia Nauk SSSR Rise (Fig. 3, prof. I10-I09, I05-I06), which is within the epi-Paleozoic platform of the Okhotsk Sea. The age of granodiorites which intruded through the uplifted acoustic basement, was determined by the dredge data to be Late-Early Cretaceous (absolute age of 87—113 m.y.), and the enclosing formations, represented by phyllites and shales with a seismic wave velocity of 4.8—6.0 km/sec are hypothetically dated as Early Mesozoic—Paleozoic. The identity of the acoustic basement in the Kurile depression to the folded basement of the Okhotsk Sea platform is shown by the presence of granite at a depth of 880—1200 m on the edge of the depression close to the Kurile Islands. According to the content of potassium and predominance of K_2O over N_2O, these granites belong to the continental but not the Kurile type.

Moreover, the Kurile depression basement is evidently heterogeneous, which is shown by its correlation with the Cenozoic (pre-Miocene) acoustic basement of Terpeniya Bay on the Sakhalin Shelf (Fig.3, prof. X-a,X,32) and the Upper Cenozoic rocks of the Kurile Islands. The arch-like uplifts strecth from the surrounding shelf to the depression. This fact is very interesting for the determination of the geological nature of the abyssal depression. The East-Sakhalin arch-like uplift of the Hokkaido—Sakhalin folded zone is an example of such a structure (Figs. 4, 5).

The Central and Honshu abyssal depressions of the Japan Sea are separated by the Yamato Bank. Unlike adjacent depressions, the Yamato Bank is characterized by a subcontinental crust which is composed of three layers: sedimentary, granitic and basaltic, their total thickness being 20—22 km. The thickness of the sedimentary cover of this bank varies sharply, decreasing to zero at some of its tops. The acoustic basement outcropping in such places is composed, as dredging showed, by metamorphic and igneous Mesozoic—Upper Paleozoic rocks having absolute ages of up to 220 m.y. (Ueno et al., 1972; Vasiliev et al., 1975).

The distribution of rock debris within islands and continental land adjacent to the abyssal depressions, the simultaneous appearance of living organisms within it, and the absence of marine deposits on the shore of the Japan Sea in some periods, for example, in Paleogene, all show that continental and subcontinental conditions existed in the second part of the Paleozoic, in the Mesozoic and in the first half of the Cenozoic in the area of the Japan Sea (Minato et al., 1968) and in some periods of the Tertiary in the area of the Kurile abyssal depression. We can state that the Okhotsk and Japan Sea abyssal depressions are uncompensated, newly formed, superimposed depressions, formed in the place of a folded massif in the Japan Sea and in the southern part of the Okhotsk Sea platform in the Okhotsk Sea in the geolog-

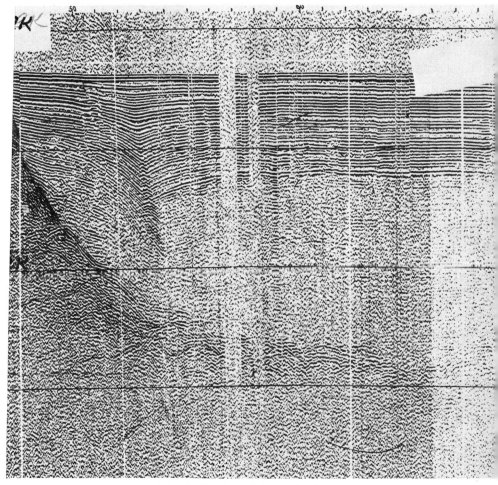

Time section along profile 38 (the Kurile abyssal depression).

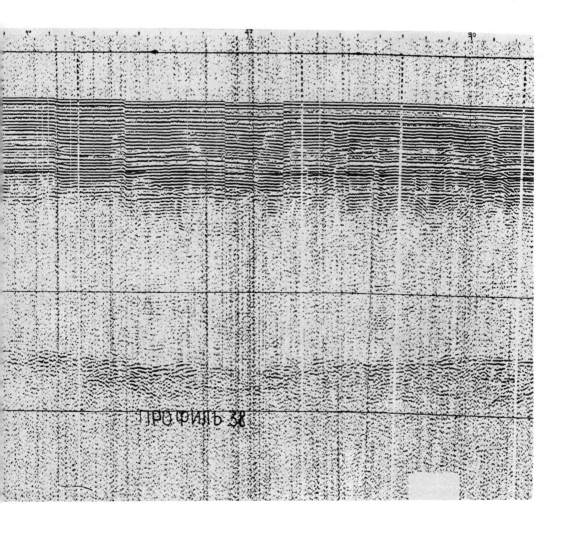

ПРОФИЛЬ 38

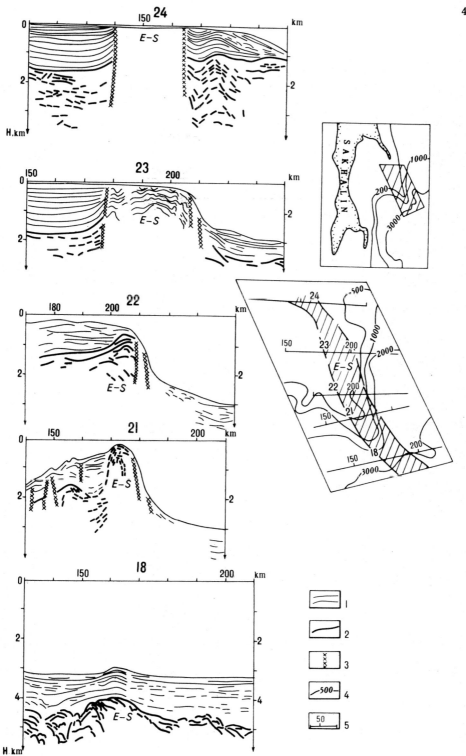

Fig. 4. Seismic cross-sections across the East-Sakhalin uplift (E—S). Legend: *1* = reflecting boundaries in the sedimentary cover, *2* = reflecting boundaries in the acoustic basement, *3* = fractures, *4* = isobaths(m), *5* = seismic profiles.

454

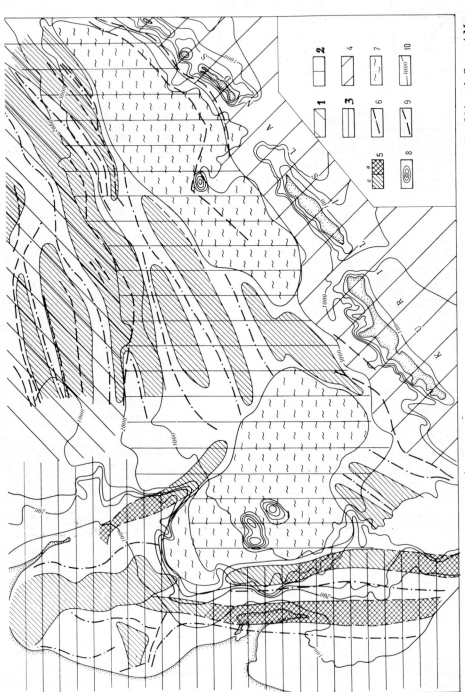

Fig. 5. The structural tectonic scheme of the Kurile abyssal depression and adjacent areas. Legend: *1* = the Okhotsk Sea epi-Mesozoic platform, *2* = the Kurile abyssal (uncompensated) depression, *3* = Sakhalin–Hokkaido Cenozoic folded system, *4* = the Kurile Islands system, *5* = arch-like and horst-like uplifts of the roof of the acoustic basement and sedimentary cover (a—the acoustic basement is overlaid by sedimentary cover, b—the acoustic basement roof outcrops on sea floor or at earth surface), *6* = axes of troughs, *7* = depressions (syneclesis), *8* = cone-like mountains (buried volcanoes ?), *9* = ruptures, *10* = isobaths(m).

ical past. This is supported by the following facts: the abyssal depressions are discordant relative to the surrounding structures; the acoustic basement of the Kurile depression is similar to the basement of the epi-Mesozoic platform of the Okhotsk Sea; the continental granites are present on the Kurile flank of this depression; the structures of the shelf continue into the abyssal depressions and are cut by the sea floor; in the geological past continental conditions existed in the region of the depressions; *Glomar Challenger* found shallow-type sediments in the depressions; on the Yamato Bank the metamorphic and igneous Lower Mesozoic—Upper Paleozoic rocks are present, etc. Correspondingly, the Yamato Bank is a remnant of a folded structure, which was in the place of the Japan Sea in the geological past. The geological structure of the surrounding shelf zones, in which both Tertiary deposits and older rock-complexes take part, also testifies to the existence of old structures in the place of the abyssal depressions.

The analysis of the areas of distribution of fresh-water fishes (Lindberg, 1955), of abyssal and pseudo-abyssal fauna, of the character and species composition of foraminifera (Saidova, 1961), the presence of beach-type pebbles on the Yamato Bank, the deeply submerged strato-volcanoes with truncated summits, and other facts, show the repeated variations of sea level in the water areas occupied now by the abyssal depressions. The juvenile formation of the depressions is supported by shallow Miocene organisms, present on the north slope of the Japan Sea and west slope of the Okhotsk Sea depressions (Vasiliev and Vasilkovsky, 1971), both submerged at present to a depth of 1000—1500 m, and also by the calculated rates of sedimentation within these depressions (Kaseno, 1972). The above-mentioned high heat flow evidently testifies to the activity of tectonic processes, causing the formation of abyssal depressions.

All the above-mentioned facts and the data on repeated sea-level variations show that not horizontal, but vertical movements predominated during the formation of the abyssal depressions. To explain these data, we do not need to use the ideas of eastward drift of the Japanese and the Kurile Islands.

Joint action of surface and inter-crust magmatic melts, crust-penetrating solutions, compaction and dehydration of the rocks under the action of high temperatures and pressures (Tikhomirov, 1960; Beloussov, 1968; Dortman and Mahyd, 1968) may be adopted as a possible mechanism of transformation of the former continental crust within the borders of the modern abyssal depressions into the present sub-oceanic crust.

In conclusion it must be noted that the formation of the abyssal depressions of the East Asiatic seas and the related phenomenon of transformation of the crust from one type to another, evidently occur against a more general process connected with the thrust of a continental block on the oceanic crust along the Benioff zone or underthrust of the latter under the former. It is possible to suppose that degradation of the crust in the depression is one of the components of the geosynclinal process occurring, evidently, at present within the Pacific mobile belt.

REFERENCES

Beloussov, V.V., 1968. The Earth's Crust and Upper Mantle of the Oceans. 253 pp.

Dortman, N.B. and Mahyd, M.Sh., 1968. New data on the elastic wave velocities in the crystalline rocks and their dependence on humidity. Sov. Geol., 5: 123—129.

Gainanov, A.C., Pavlov, Yu.A., Stroyev, P.A., Sychev, P.M. and Touezov, I.K., 1974. The anomalous gravity fields of the Far East seas and adjacent part of the Pacific. Izd. "Nauka", SO, Novosibirsk, 108 pp.

Kaseno, Y., 1972. Geological features of Japan sea floor: a review of recent study. Pac. Geol., 4: 91—111.

Krasniy, M.L., 1972. The anomalous magnetic field of the northwest sector of the Pacific mobile belt. Sb. "Glubinnaya struktura dalnevostochnikh morey i ostrovnikh dug". Tr. SakhKNII, N. 30: 272—279

Lindberg, G.U., 1955. Quaternary period in the light of biogeographical data. 34 pp.

Minato, M., Gorai, M. and Fanatsahi, M., 1968. Geological development of the Japan Islands. Izd. "Mir", Moscow, 719 pp.

Saidova, Kh.I., 1961. Ecology of foraminifera and paleogeography of the Far East seas of the USSR and northwest part of the Pacific. 231 pp.

Sakhno, V.G. and Vasiliev, B.I., 1974. Basaltoids from the Japan sea bottom. In: sb. "Voprosi geologii i geofiziki okrainnihk morey severo-zapadnoy tchasti Tikhogo okeana". p. 52—55.

Snegovskoy, S.S., 1974. The seismic reflection measurements and tectonics of the Okhotsk sea south part and the adjacent Pacific part. Izd. "Nauka", SO, Novosibirsk. 87 pp.

Stroyev, P.A., 1975. On the character of gravity anomalies in free air in the Japan Sea transition zone. Sb. "Morskie gravimetricheskie issledovaniya", vip. 8: 136—144.

Summary of Deep Sea Drilling Project, LEG-31, Univ. of California, 1973.

Tikhomirov, V.V., 1960. The question of earth's crust development and the role of meta-somatosis in the processes. 21st Sess. Int. Geol. Congr. The reports of Soviet geologists, p. 75—83.

Touezov, I.K., 1969. Geophysical investigations of the Far East sector of the circum-Pacific. Tr. SakhKNII, vip. 20: 5—26.

Touezov, I.K., 1972. Structure of the earth's crust of Okhotsk and Japan sectors accord-ing to the data of regional seismic measurements. Tr. SahkKNII, N. 33: 129—145.

Touezov, I.K., 1975. The lithosphere of the Asiatic—Pacific zone of transition, SO izd. "Nauka". 153 pp.

Touezov, I.K., Krasniy, M.L., Pavlov, Yu.A. and Soloviev, O.M., 1967. Distribution of magnetic active bodies within the earth's crust and upper mantle of the Far East sector of the Asiatic—Pacific transition zone. Geotektonika, 4: 95—101.

Ueno, N., Kaneoki, I., Ozima, M., Dzashu, S., Sato, T. and Iwabuchi, E., 1972. K—Ar age, strontium isotope relation and relation K—Rb in volcanic rocks from the Japan sea bottom. Tr. SakhKNII, vip. 33: 312—316.

Vasiliev, B.I. and Vasilkovsky, N.P., 1971. The presence of marine Miocene deposits on the continental slope of the Piotr Velikiy bay (Japan Sea). Dokl. Akad. Nauk SSSR, vol. 198, No. 5.

Vasiliev, B.I., Karp, B.Ya., Shevaldin, Yu.V. and Stroyev, P.A., 1975. The structure of the Yamato underwater uplift (Japan Sea) according to geophysical data. 97 pp.

Veselov, O.V. and Touezov, I.K., 1972. The heat flow in the northwest sector of the Pacific mobile belt. Tr. SahkKNII, vip. 26: 171—180.

Chapter 24

NUMERICAL STUDIES OF BACK-ARC CONVECTION AND THE FORMATION OF MARGINAL BASINS*

M. NAFI TOKSÖZ and ALBERT T. HSUI

Department of Earth and Planetary Sciences, Massachusetts Institute of Technology, Cambridge, MA. 02139 (U.S.A.)

SUMMARY

Marginal basins that occur behind island arcs display geological and geophysical features that imply an extensional origin and a spreading mechanism, possibly driven by the heating of the basin floor from below. The origin and evolution of these basins appear to be associated with the subduction process of the oceanic lithosphere. The mechanism we propose for the formation and spreading of marginal basins is the convective flow induced in the mantle by the subducting slab. This convective current brings hot mantle material to the base of the lithosphere behind island arcs. The combination of material upwelling, heating of the lithosphere and flow induced tension initiates spreading and the formation of marginal basins.

Quantitative investigation of this mechanism has been carried out utilizing numerical calculations using both constant and variable viscosity models. Numerical results indicate that the induced flow can generate a 1 km topographic rise and a tensile stress of about 100 bars. Together with the thinning of the lithosphere due to heating, this induced current is probably sufficient to generate inter-arc spreading at about 5—10 million years after the initiation of subduction. The horizontal scale of the spreading center obtained from the numerical models is also consistent with the observations over many marginal basins.

INTRODUCTION

Marginal basins behind island arcs are prominent features that are related to the subduction of the oceanic lithosphere. A majority of these basins possess properties which imply that they might have been formed by spreading. One mechanism which has been proposed to generate the inter-arc spreading is the convection flow induced in the asthenosphere by the subducting lithosphere (Sleep and Toksöz, 1971; Andrews and Sleep, 1974; Toksöz and Bird, 1977; Hsui et al., 1977). Figure 1 schematically demonstrates this mechanism and the relationship between the slab, the induced

---

*Originally published as: Toksöz, M.N. and Hsui, A.T., 1978. Numerical studies of back-arc convection and the formation of marginal basins. In: M.N. Toksöz (Editor), Numerical Modeling in Geodynamics. Tectonophysics, 50: 177—196.

458

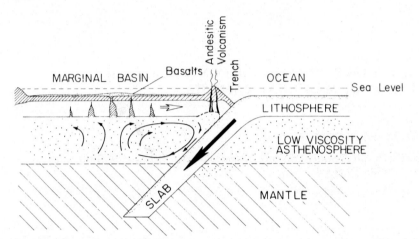

Fig. 1. Schematic diagram of convection induced by a downgoing slab and its role in the formation of marginal basins.

flow and marginal basin. In this paper, we explore this mechanism further by calculating the induced convection using both constant and variable viscosity models.

A large set of geologic and geophysical data on marginal basins of the Western Pacific exists. Sea floor topography, heat flow, sediment distribution, magnetic anomalies, properties of basalts, seismic wave velocities and attenuation, and earthquake mechanisms suggest either presently active or past spreading of the basin floor. Some of these data have been reviewed by Toksöz and Bird (1977). More detailed information can be found in papers by Yasui et al. (1967), Molnar and Oliver (1969), Watanabe et al. (1970, 1977), Karig (1970, 1971a, b, 1974), Barazangi and Isacks (1971), Barker (1972), Sclater (1972), Sclater et al. (1972a, b), Uyeda and Ben-Avraham (1972), Tomoda et al. (1975), Watts and Weissel (1975), Cooper et al. (1976, 1977), Hawkins (1977), Uyeda (1977).

Making a somewhat simplified generalization, the primary characteristics of the marginal basins can be summarized as follows:

(1) Marginal basins occur behind island arcs and are associated with the subduction process.

(2) Most basins have high heat flow as demonstrated in Fig. 2 (Watanabe et al., 1977). A few exceptions are the Aleutians (Bering Sea), South Fiji and West Philippines.

(3) Seismic velocities in the upper mantle under the basins are low and attenuation of both compressional and shear waves is high. These indicate high temperatures and possibly some partial melting.

(4) Some basins (Lau, Mariana, East Scotia Sea) show definite evidence of spreading, while others (Sea of Japan, Okhotsk Sea, Fiji Plateau, Parece Vela) possess some features of past spreading.

A satisfactory mechanism for the formation and evolution of marginal

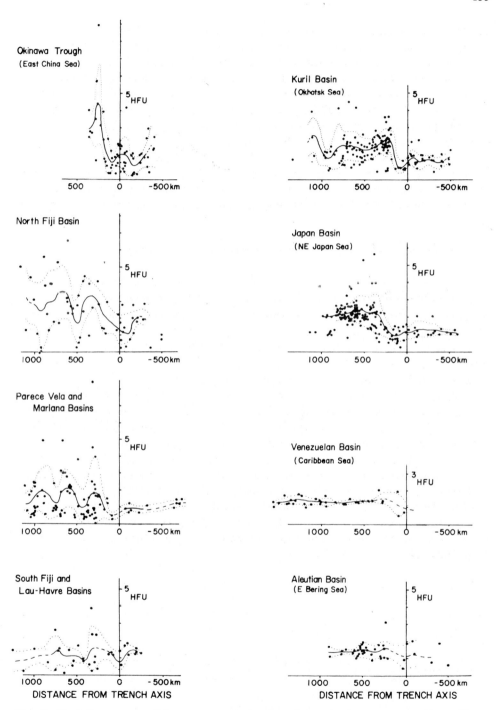

Fig. 2. Heat flow versus distance across major trench-arc—back-arc systems (from Watanabe et al., 1977).

basins must explain these properties. It is clear that the heating of the lithosphere and the basin floor from underneath are required to explain the observations. What is not clear are the sources of such heating.

Some likely mechanisms for the formation of marginal basins have been reviewed by Uyeda (1977). These include: (1) subduction of an active ridge (Uyeda and Miyashiro, 1974), (2) diapiric rise of heat above the subducting slab (Matsuda and Uyeda, 1971), (3) entrapment of old sea floor (Uyeda and Ben—Avraham, 1972), (4) very leaky transform fault (Lawver et al., 1976), and (5) back-arc spreading (Sleep and Toksöz, 1971; Karig, 1974).

Although the first four mechanisms are applicable to one or more specific cases, the most general mechanism is probably the back-arc spreading. In this paper we investigate the role of induced convection as the cause of back-arc spreading.

CONVECTION INDUCED BY THE SUBDUCTING LITHOSPHERE

Induced flow behind subducting slabs has been modeled by several investigators (McKenzie, 1969; Sleep and Toksöz, 1971; Andrews and Sleep, 1974; Toksöz and Bird, 1977). The present work represents a new model constructed by solving simultaneously the equations of motion and the energy equation in a closed system and incorporating both the shearing by the subducting slab and the buoyancy effects.

The models we calculate are two-dimensional. They consist of two parts — a high-viscosity conducting layer (lithosphere) over a convecting layer (mantle). The governing equation for the lithosphere is the two-dimensional time dependent conduction equation (Carslaw and Jaeger, 1959) with appropriate initial and boundary conditions. The equations for the convecting layers are similar to those described by Turcotte et al. (1973) for the flow within the earth's mantle. However, since the flow is driven by both buoyancy and shear, the equations are non-dimensionalized following Torrance et al. (1972). Utilizing an x—z coordinate system, with the origin defined at the tip of the wedge, and the positive x-axis pointing towards the back arc basins and the positive z-axis pointing downward, the final non-dimensional equations are:

$$u = -\partial \psi/\partial z \tag{1}$$

$$w = \partial \psi/\partial x \tag{2}$$

$$\eta = -\nabla^2 \psi \tag{3}$$

$$0 = \nabla^2(\pi\eta) + \frac{Gr}{Re}(\partial T/\partial x) - 2\left[\frac{\partial^2\pi}{\partial x^2}\frac{\partial u}{\partial z} - \frac{\partial^2\pi}{\partial z^2}\frac{\partial w}{\partial x} - \frac{\partial^2\pi}{\partial x\partial z}\left(\frac{\partial u}{\partial x} - \frac{\partial w}{\partial z}\right)\right] \tag{4}$$

$$\frac{DT}{Dt} - DiwT = \frac{1}{PrRe}\nabla^2 T + \frac{1}{PrRe} + \frac{ReDi}{Gr}\pi\Phi \tag{5}$$

where u is the horizontal velocity, w is the vertical velocity, ψ is the stream function, η is the vorticity, T is the temperature, Φ is the viscous dissipation function defined as $(\dot{e}_{ij})^2$, and $\pi = \nu/\nu_0$ is the nondimensionalized viscosity. The parameters appearing in the equations are the Reynold's number (Re), the Prandtl number (Pr), the compression—dissipation number (Di), and the Grashof number (Gr). These are defined as: $Re = Ud/\nu$, $Pr = \nu/K$, $Di = \alpha gd/c_p$ and $Gr = \alpha gH_0 d^5/(k\nu^2)$, where U = velocity of the subducting slab, d = depth of the convection region, ν = kinematic viscosity, K = thermal diffusivity, α = coefficient of thermal expansion, g = gravitational acceleration, c_p = specific heat at constant pressure, H_0 = reference heat source strength, and k = thermal conductivity. The inertia terms in the momentum equation have been neglected because infinite Prandtl number is assumed. However, it should be noted that the product of Prandtl number and Reynold's number is of order one since the Reynold's number is very small. Because of the assumption of two dimensional flow, vorticity can be treated as a scalar quantity.

The problem is cast in time dependent form so that the evolution of the flow patterns can be investigated. The initial condition chosen is a no-flow condition while the temperature field follows the adiabatic geotherm. Mechanical boundary conditions for the horizontal boundaries and the slab boundary are the impermeable, no-slip conditions (i.e. $\underline{u} = 0$, except at the moving slab, where $\underline{u} = 1$). On the vertical boundary, a shear stress free condition is imposed. Besides mechanical boundary conditions, thermal boundary conditions are also required. Temperature is specified to be "zero" at the top of the conducting layer (surface of the earth). No horizontal heat flux is allowed at either side of the conducting layer and at the vertical boundary of the convection region. The bottom of the convecting region is calculated with a prescribed heat flux input from the lower mantle. Finally, on the top of the slab, temperature is taken from that determined by Toksöz et al. (1973) with a convecting geotherm.

In the case of variable viscosity models, viscosity is expressed by:

$$\nu = a \, \exp(bT_m/T) \tag{6}$$

where T = local absolute temperature, T_m = absolute melting temperature of pyrolite with 0.1% water content taken from Ringwood (1975). Constants a and b are chosen ($a = 1.125 \cdot 10^{12}$ and $b = 16$) such that the viscosity in the asthenosphere is approximately 10^{19} poise while that of the mantle below 300 km is approximately 10^{22} poise. It should be noted that constant viscosity models can be recovered by merely setting $b = 0$ and $a = \nu_0$.

The equations (1)—(5) represent a non-linear system of five scalar equations for five unknowns u, w, ψ, η and T. The non-linearity is embedded in the convective term, and the viscous heating term in the energy equation (5), and the viscosity derivative terms in the vorticity equation (4). Finite difference numerical techniques described in the appendix, are employed to solve these equations. The explicit time marching approach (Torrance et al., 1972) was utilized with some modifications. Forward time, central space and a special three point non-central (also known as upwind) difference technique were used to approximate the time derivative, the space derivative and the convective terms respectively. Following Parmentier and Torrance (1975), vorticities, especially at the boundary, were calculated according to the circulation theorem and velocities were determined at the boundaries of each grid element. This approach ensures the conservation of vorticity, especially at the slab boundary where the grid elements are non-rectangular.

The equations are solved in a x—z grid system. In order to have the grid points coincide with the inclined surface of the slab, Δx and Δz must be related by $\Delta z = \Delta x \tan \theta$, where θ is the dip angle of the subducting slab measured from the horizontal $+x$ axis. Numerical results discussed are obtained using a 36×21 x—z grid for the convecting region and a 36×6 x—z grid for the surface conducting layer. For most calculations $\theta = 45°$ and $\Delta x = \Delta z$ were used.

In the computation, the energy equation (5) is first advanced by a time step, Δt. Δt is determined to conserve numerical stability (Torrance and Rockett, 1969). Vorticity equation (4) is then iterated using a factor of Jacobi scheme (Torrance and Turcotte, 1971). Between each vorticity iteration, the stream function equation (3) is solved by the method of Jacobi iteration. The new stream function is then used to update the velocity fields (eqs. 1 and 2) and the boundary vorticities. After the iteration of the vorticity equation (4) is completed, the calculation is advanced to the next time step. Our numerical experiments indicate that the conservative Jacobi scheme is the only stable scheme for both vorticity and stream function iterations in the case of variable viscosity models. However, in the case of constant viscosity models, successive over relaxation techniques can be used to iterate the stream function equation while the Jacobi scheme is required for vorticity iteration.

NUMERICAL MODELS OF INDUCED CONVECTION

To investigate in detail the convection induced by the subducting lithosphere and its role in the formation of marginal basin, numerical calculations were carried out for two different models. Each of these models had an 80 km thick lithosphere. The difference between models was in the convecting layer. In order of complexity the models were: (1) a single-layer constant viscosity model, and (2) a variable viscosity model. These two models, although different, give very similar results as described below.

Single layer constant viscosity model

In order to investigate the effects of buoyancy and viscous dissipation of the induced flow below marginal basins, computation of finite amplitude thermal convection of a constant viscosity Boussinesq fluid within a wedge area has been carried out. The scale of the flow and the flow velocities are normalized with respect to the depth of the convection region and the subduction velocity, respectively. Calculated results presented in this section are obtained using a convection depth of 300 km and a subduction velocity of 8 cm/y. Figure 3 represents the evolution of the induced flow as a function of time. The figures a, b, c show both the stream lines and isotherms. At 5 m.y. after the initiation of subduction, an induced flow in the wedge area is established. Hotter material is being carried towards the surface by the flow. The upwelling is at about 300 km from the wedge corner. As a result, the lithosphere above starts to heat up and the topography begins to rise. At 10 m.y. after the subduction, the subducting slab has penetrated through the asthenosphere, and horizontal temperature gradients are established by the induced flow. Consequently, the thermal buoyancy effect begins to dominate. An active flow adjustment starts to take place. After this period of adjustment, convection starts to settle into a more steady state like situation as shown in Figs. 3b and 3c. A single cell structure with 700 km horizontal extent is formed. This flow structure is distinctly different from that predicted by pure kinematic models of Sleep and Toksöz (1971), Andrews and Sleep (1974) and Toksöz and Bird (1977). Asthenospheric material adjoining the subducting slab is cold. When this material moves downward with the slab, it remains at the bottom of the convection layer. The negative buoyancy prohibits its return to the surface immediately as would have been the case in kinematic flows. Instead, it remains at the bottom until the internal heat sources and the heat flux from the bottom heats it up and upward buoyancy combined with kinematic forces moves it towards the surface. This demonstrates the relative effects of thermal and kinematic forces on the evolution of convection.

In order to better demonstrate the feasibility of this mechanism for inter-arc spreading, the topographic deformation resulting from this flow and the stresses exerted on the lithosphere are calculated. Topographic changes are computed by considering both the thermal expansion effect and the balance of the normal stresses exerted at the lithosphere by the flow field (McKenzie et al., 1974). The stresses in the lithosphere are calculated by combining the forces due to topography and the shear stresses exerted at the bottom of the lithosphere by the flow. Results shown in Fig. 4 are for 5 m.y. after subduction. The topography curve indicates that the flow is able to generate an elevation of about 1 km at about 250—300 km from the volcanic arc. The stress curve indicates that about 100 bar of stresses can be generated at the ridge crest. Together with the thinning of the lithosphere due to thermal effect, this stress may be sufficient to break the lithosphere and initiate spreading. The diagram at the bottom of this figure illustrates schematically this plausible physical process of inter-arc spreading.

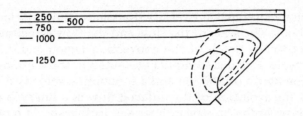

5 m.y.

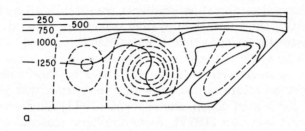

10 m.y.

a

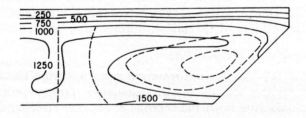

20 m.y.

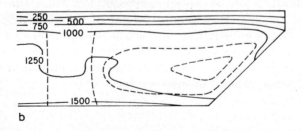

30 m.y.

b

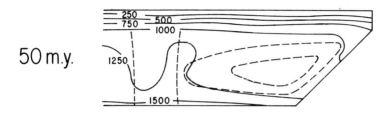

50 m.y.

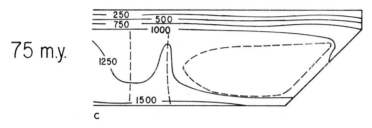

75 m.y.

c

Fig. 3. Numerical results showing the evolution of the induced flow. The vertical dimension of the convection layer is 300 km. A constant viscosity of 10^{20} poise is used for this model. Solid lines are isotherms. Number associated with each isotherm is temperature in °C. Dashed lines are the stream lines. Spacing between stream lines is inversely proportional to the flow velocity between them. Slab velocity is 8 cm/y. Time in millions of years after the initiation of subduction is given at the left of each diagram. (a) 5 and 10 m.y., (b) 20 and 30 m.y. and (c) 50 and 75 m.y.

Variable viscosity model

An alternate and preferred approach to model the induced convection is to use variable viscosity for the mantle. Viscosity is low in the asthenosphere and increases below. One would expect that major convection will be confined to use the low viscosity region with the bottom of the asthenosphere behaving as a barrier to the induced flow.

In order to investigate this model, calculations of induced thermal convection within a varianle viscosity fluid layer have been carried out. The viscosity variation and its relationship to melting temperature were the same as that given by eq. 6. Calculated results are shown in Fig. 5. The depth of the convection layer is taken to be 600 km with a subduction velocity of 8 cm/y. The assumption of the depth of this convection layer is arbitrary. However, numerical results do not seem to be very sensitive to this assumption as long as it is about 600 km or greater and viscosity below the asthenosphere is greater than 10^{22} poise.

Numerical results indicate that the induced flow is relatively narrow in the

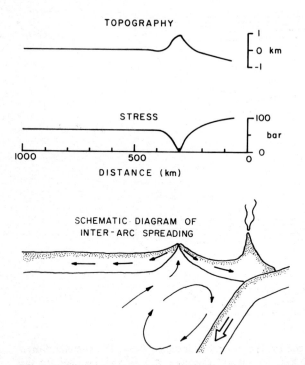

Fig. 4. Calculated topography and vertically averaged horizontal deviatoric compressive stresses within the lithosphere as a function of the distance from the wedge corner. Bottom diagram is a vertically exaggerated schematic sketch of the inter-arc spreading center and its relationship to the induced flow beneath. It also shows the direction of compression forces within the lithosphere.

horizontal extent, and the most active (highest velocity) region is indeed confined to the top 200—250 km. Again, the highest vertical velocity at the top of the convection layer is located about 250—300 km away from the wedge corner. These results are not much different from the constant viscosity models. The variable viscosity models show that the viscosity increase below 300 km depth acts as a barrier to flow.

Comparison with other models

Other models of convection induced by subducting lithosphere have been carried out by McKenzie (1969), Sleep and Toksöz (1971), Andrews and Sleep (1974), and Toksöz and Bird (1977). Utilizing an analytic wedge solution, McKenzie (1969) was able to calculate the thermal structure of the downgoing slab and in the wedge region. Andrews and Sleep (1974) have carried out a fully dynamical calculation of the region using analytic wedge solutions as boundary conditions. Comparison between the models presented in this paper and those presented by Andrews and Sleep (1974) indicates

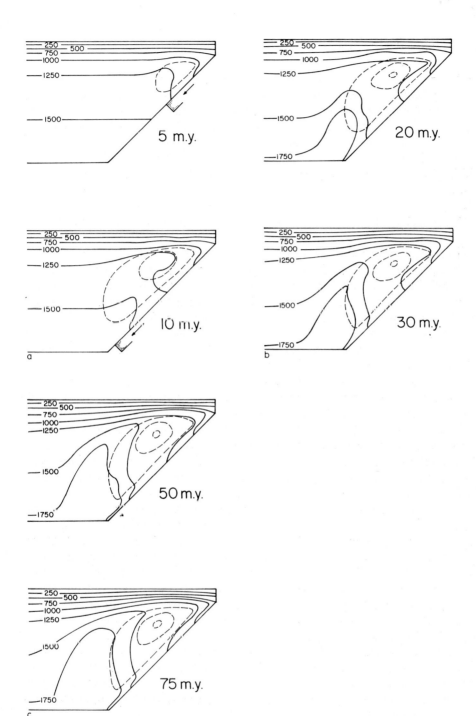

Fig. 5. Same as Fig. 3 except that variable viscosity is used (see text for details of the viscosity function used). The vertical scale of the convection region is 600 km.

that the induced flow is strongly influenced by the kinematic boundary conditions.

Models proposed by Sleep and Toksöz (1971) and Toksöz and Bird (1977) represent models with a closed system. However, in these models only the kinematic flows due to the subducting slabs were considered. In other words, the equations of motion and the energy equation are decoupled in these models. The effects of thermal buoyancy are important as shown in this paper. Therefore these latest results are more complete than those given in our earlier paper (Toksöz and Bird, 1977), although the main features are similar.

IMPLICATIONS FOR FORMATION AND EVOLUTION OF MARGINAL BASINS

The models presented in this paper demonstrate the effects of convection induced by the subducting lithosphere. With both the kinematic and thermal effects included, the calculations represent reasonably well the conditions inside the earth. The results obtained with both constant and variable viscosity models are similar in their general characteristics as shown in Figs. 3 and 5. Furthermore, they support the previous models. Low viscosity of the asthenosphere and a sharp viscosity increase at the bottom play the most important role on shape and scale of the induced convection.

The consequences of subduction and induced flow in the asthenosphere and their role in the formation of marginal basins are schematically illustrated in Fig. 6. After the subducting slab penetrates into the asthenosphere, island-arc volcanism begins. Mechanical effects of the subducting slab may cause features in the overlying lithosphere. Heating and erosion from below may further weaken the lithosphere to allow upwelling of magma from the asthenosphere. Partial melting and release of water and volatiles from the slab will contribute to the melting and magma generation.

The induced convection becomes effective in the formation of marginal basins when the subducting slab passes through the asthenosphere (sequence C, in Fig. 6). At first, kinematic effects or the "drag" by the moving slab drives the convection. However, once lateral temperature gradients are established thermal buoyancy effects become important. As convection continues, the lithosphere overlying the upwelling flow is heated, weakened and elevated. Stresses are generated by the elevation and the shear forces exerted at the bottom of the lithosphere by the lateral flow. According to calculations, the elevation of lithosphere at the top of upwelling flow is about 1 km 5—10 m.y. after the subduction. The deviatoric stresses within the lithosphere at this time can then be estimated. For a lithosphere of thickness h and at equilibrium, the deviatoric stress in the plate is produced by the stresses due to topographic elevation (σ_t) and the shear stresses exerted at the bottom of the plate by the mantle flow from beneath (τ_{xz}):

$$\sigma_{xx} = \sigma_t + \frac{1}{h} \int \tau_{xz} \mathrm{d}x$$

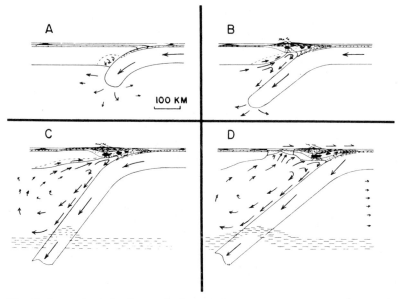

Fig. 6. Four stages in the evolution of the marginal basins. A. Initiation of a subduction zone. B. Generation of island-arc volcanism possibly due to mechanical erosion and the release of volatiles from the slab. C. Establishment of an induced convection as slab penetrates through the asthenosphere. D. Failing of lithosphere behind the island arc because of heating and flow induced tension. Inter-arc spreading is initiated at this stage and marginal basin is formed.

Our results show that the tensile stresses acting on the back-arc spreading center are about 50—100 bars.

These stresses may eventually break the lithosphere and initiate the spreading as shown in sequence D of Fig. 6. Once lithosphere breaks, the convection would drive the spreading. However, because of the smaller scale of convection, as well as resistance from sides (the adjacent continental lithosphere and the arc) the spreading should be weaker than that at mid-ocean ridges. An example of topography and high heat flow associated with such spreading in the Mariana Basin is shown in Fig. 7. The symmetry and longevity of spreading depends on the strength of the forces that cause such spreading. Under marginal basins such forces are weak. As a result, one would not expect the active spreading to continue indefinitely. As the resistance to the spreading lithosphere increases, spreading may cease. Generally this will occur in about 30 m.y. Although marginal basins will continue to have high heat flow after this time, they may not show signs of active spreading.

The calculated numerical models are especially helpful to quantize several aspects of this model. Actual times of sequence of events and values of heat flow, topographic elevation and stresses come out of the calculations and are important for the interpretation of observations and the test of the models.

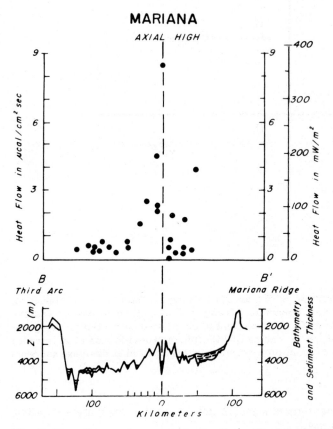

Fig. 7. Heat flow and topography in the active Mariana Basin. Data reproduced from Anderson (1975).

The model predictions are in agreement with the majority of the observations associated with marginal basins.

The applicability of theoretical models after the initiation of spreading requires some discussion, since conditions generated by the spreading were not incorporated in the modeling. For example, at the spreading center, there is very high heat flow and more heat loss than other parts of the basin. However, the "ridge" area is small and the net effect it has is not dominant. The motion of the basin floor and the effect it may have on the convection is another aspect that theoretical models did not incorporate. The lithospheric motion is in the same direction as the convective flow in the asthenosphere. Thus it would not impede the flow. Since spreading velocities are most likely small, it would not help it either. It is possible that as the basin flow widens, the location of the upwelling current relative to the spreading center may change. This may result in more complicated topography of basin floor, and a broad zone of high heat flows as observed in many basins (Fig. 2).

CONCLUSIONS

The key results of the numerical calculations of the induced convection and their implications for the marginal basins are:

(1) For a slab subducting at 8 cm/y and at a 45° angle, back arc spreading can take place about 5—10 m.y. after the initiation of subduction.

(2) The vertical scale of the induced flow is determined by the sharp viscosity increase at the bottom of the asthenosphere.

(3) According to our calculation and asthenosphere models, back arc spreading centers will be located about 200—300 km away from the volcanic arcs. This distance depends on the thickness of the low-viscosity zone.

(4) Assuming an elastic lithosphere, the induced flow generates an approximately 1 km topographic rise in the basin floor, at a distance of 200—300 km from the volcanic arc.

(5) The induced flow exerts about 35 bars of shear stresses at the bottom of the lithosphere. The stress due to topographic elevation at the center is about 50 bars. Resultant forces together with the thinning of the lithosphere due to thermal effects may break the lithosphere and initiate back-arc spreading.

(6) The induced convection raises the average surface heat flux in the marginal basins by about 20 ergs/cm^2 s without allowance for active spreading. The effect of active spreading is to increase the surface heat flux further, especially at the spreading center where magma appears. The magnitude of the surface heat flux will depend on the rate of intrusion and the rate of spreading.

(7) Induced convection is a necessary consequence of slab subduction. However, convection alone is not a sufficient condition to initiate back-arc spreading. Spreading will occur only if stresses due to convection can overcome the lithospheric strength and other regional tectonic stresses.

ACKNOWLEDGEMENTS

This research was supported by the National Science Foundation under Grant 76-12471 EAR.

APPENDIX: FINITE DIFFERENCE SCHEME

In this appendix we describe briefly the finite difference equations and the computational procedure used for the numerical calculations. The grid system is a rectangular cartesian system (Fig. A1) except at the shearing slab boundary where triangular grid elements are used. i and j are location indices of a grid point in the x and z directions, respectively. Velocities u and w are calculated at the boundaries of each grid element, while stream function (ψ), vorticity (η) and temperature (T) are calculated for the nodes of the grid system. The equivalent finite difference equations for eqs. 1—5 in the text,

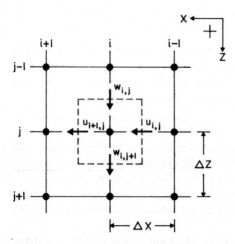

Fig. A1. A representative grid node used for the numerical computations.

are:

$$u_{i,j} = (\psi_{i,j-1} + \psi_{i-1,j-1} - \psi_{i,j+1} - \psi_{i-1,j+1})/4\Delta z \tag{A.1}$$

$$w_{i,j} = (\psi_{i+1,j} + \psi_{i+1,j-1} - \psi_{i-1,j-1} - \psi_{i-1,j})/4\Delta x \tag{A.2}$$

$$\psi_{i,j} = [\eta_{i,j} + (\psi_{i+1,j} + \psi_{i-1,j})/\Delta x^2 + (\psi_{i,j+1} + \psi_{i,j-1})/\Delta z^2]/S \tag{A.3}$$

where:

$$S = 2(1/\Delta x^2 + 1/\Delta z^2)$$

$$\phi_{i,j}^{n+1} = \{(\phi_{i+1,j}^n + \phi_{i-1,j}^n)/\Delta x^2 + (\phi_{i,j+1}^n + \phi_{i,j-1}^n)/\Delta z^2$$

$$+ Gr(T_{i+1,j}^{n+1} - T_{i-1,j}^{n+1})/(2\Delta x \cdot Pr) - 2[D_{i,j}^1(2\psi_{i,j} - \psi_{i,j+1} - \psi_{i,j-1})/\Delta z^2$$

$$+ D_{i,j}^2(\psi_{i+1,j} - 2\psi_{i,j} + \psi_{i-1,j})/\Delta x^2 - D_{i,j}^3(u_{i+1,j} - u_{i,j})/\Delta x$$

$$- D_{i,j}^3(w_{i,j+1} - w_{i,j})/\Delta z]\}/S \tag{A.4}$$

where $\phi_{i,j}^n = \pi_{i,j} \cdot \eta_{i,j}^n$, S is as given in (A.3), and $D_{i,j}^1$, $D_{i,j}^2$, and $D_{i,j}^3$ are the finite difference expression of $\partial^2\pi/\partial x^2$, $\partial^2\pi/\partial z^2$ and $\partial^2\pi/\partial x\partial z$, respectively. The finite difference approximation of these three parameters will be discussed later. The finite difference approximation of the energy equation (5) is:

$$T_{i,j}^{n+1} = \Delta t\{(1/\Delta t - [S/PrRe + C_1 + C_2 + C_3 + C_4 - \tfrac{1}{2} Di(W_{i,j+1} + W_{i,j})]) \cdot$$

$$T_{i,j}^n + [C_5 + 1/(PrRe \cdot \Delta x^2)] \cdot T_{i+1,j}^n + [C_6 + 1/(PrRe \cdot \Delta x^2)] \cdot T_{i-1,j}^n$$

$$+ [C_7 + 1/(PrRe \cdot \Delta z^2)] \cdot T_{i,j+1}^n + [C_8 + 1/(PrRe \cdot \Delta z^2)] \cdot T_{i,j-1}^n$$

$$+ 1/PrRe + Re \cdot Di \cdot \pi_{i,j} \cdot \Phi_{i,j}/Gr\} \tag{A.5}$$

As defined in the text, Pr, Re, Di and Gr are the Prandtl number, the Reynolds number, the compression—dissipation number, and the Grashof number, respectively. S is as given in (A.3). $\Phi_{i,j}$ is the dissipation function which can be calculated according to:

$$\Phi_{i,j} = 4\{(\psi_{i+1,j+1} - \psi_{i-1,j+1} - \psi_{i+1,j-1} + \psi_{i-1,j-1})/(4 \cdot \Delta x \cdot \Delta z)\}^2$$
$$+ \{(\psi_{i,j+1} - 2\psi_{i,j} + \psi_{i,j-1})/\Delta z^2 - (\psi_{i+1,j} - 2\psi_{i,j} + \psi_{i-1,j})/\Delta x^2\}^2$$

The coefficients C_1 to C_8 are related to the convective terms, and they are defined as follows:

if $u_{i+1,j} > 0$ then $C_1 = u_{i+1,j}$, $C_5 = 0$; else $C_1 = 0$ and $C_5 = -u_{i+1,j}$;

if $u_{i,j} > 0$ then $C_2 = 0$, $C_6 = u_{i,j}$; else $C_2 = -u_{i,j}$ and $C_6 = 0$;

if $w_{i,j+1} > 0$ then $C_3 = w_{i,j+1}$, $C_7 = 0$; else $C_3 = 0$ and $C_7 = -W_{i,j+1}$;

if $w_{i,j} > 0$ then $C_4 = 0$, $C_8 = w_{i,j}$; else $C_4 = -w_{i,j}$ and $C_8 = 0$.

Before coding eqs. A.1—A.5 for computation, it is necessary to derive the finite difference expression of $D^1_{i,j}$ $D^2_{i,j}$ and $D^3_{i,j}$ in eq. A.3. Since viscosity difference between grid nodes can be large (i.e., exceeds one order of magnitude), it is in general more accurate to transform viscosity derivatives into temperature derivative, because of the smaller difference in temperature between nodes. The transformation is straight forward once the viscosity function is defined. Following eq. 6 in the text:

$$\pi = \frac{a}{\nu_0} \exp(bT^m/T_c \cdot T)$$

where:

$$T_c = H_0 d^2/k \text{ and } T^m = T_m \tag{A.6}$$

and the second derivative of π becomes:

$$\frac{\partial^2 \pi}{\partial x^2} = -bT^m\pi[\partial^2 T/\partial x^2 - (2/T + bT^m/T_c \cdot T^2)(\partial T/\partial x)^2]/(T_c \cdot T^2) \tag{A.7}$$

In terms of finite difference formulation, eq. A.7 can be expressed as:

$$D^1_{i,j} = -b \cdot T^m_{i,j} \cdot \pi_{i,j}\{(T^{n+1}_{i+1,j} - 2T^{n+1}_{i,j} + T^{n+1}_{i-1,j})/\Delta x^2 - [2/T^{n+1}_{i,j}$$
$$+ b \cdot T^m_{i,j}/(T_c \cdot T^{n+1}_{i,j})] [(T^{n+1}_{i+1,j} - T^{n+1}_{i-1,j})/(2\Delta x)]^2\}/(T_c \cdot T^{n+1}_{i,j}) \tag{A.8}$$

similarly, $D^2_{i,j}$ and $D^3_{i,j}$ can be expressed as below:

$$D^2_{i,j} = -b \cdot T^m_{i,j} \cdot \pi_{i,j} \{(T^{n+1}_{i,j+1} - 2T^{n+1}_{i,j} + T^{n+1}_{i,j-1})/\Delta z^2 - [2/T^{n+1}_{i,j}$$
$$+ b \cdot T^m_{i,j}/(T_c \cdot T^{n+1}_{i,j})] [(T^{n+1}_{i,j+1} - T^{n+1}_{i,j-1})/(2\Delta z)]^2\}/(T_c \cdot T^{n+1}_{i,j}) \tag{A.9}$$

$$D^3_{i,j} = -b \cdot T^m_{i,j} \cdot \pi_{i,j}\{(T^{n+1}_{i+1,j+1} + T^{n+1}_{i-1,j-1} - T^{n+1}_{i+1,j-1} - T_{i-1,j+1})/(4 \cdot \Delta x \cdot \Delta z)$$
$$- [2/T^{n+1}_{i,j} + b \cdot T^m_{i,j}/(T_c \cdot T^{n+1}_{i,j})] [(T^{n+1}_{i,j+1} - T^{n+1}_{i,j-1})/(2\Delta z)]$$
$$\times [(T^{n+1}_{i+1,j} - T^{n+1}_{i-1,j})/(2\Delta x)]\}/(T_c \cdot T^{n+1}_{i,j}) \tag{A.10}$$

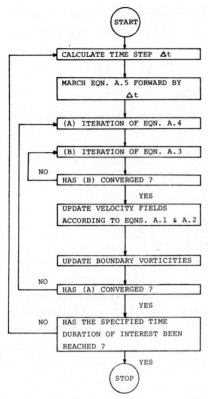

Fig. A2. A flow diagram of the numerical ,computation procedure.

Substituting (A.8)—(A.10) into (A.3), the finite difference equations A.1—A.5 represent the governing system to be solved numerically. Computational sequence has been described in the text. For the purpose of a graphic illustration, a flow diagram of the computational procedure is given in Fig. A2. Finally, numerical stability as discussed by Torrance and Rockett (1969), implies that the coefficient associated with $T_{i,j}^n$ in eq. A.5 must be positive. As a result, the time step can be calculated according to the following relations:

$$\Delta t \leqslant \min 1/[S/PrRe + C_1 + C_2 + C_3 + C_4 - \tfrac{1}{2} Di(w_{i,j+1} + w_{i,j})]$$

All the coefficients and parameters are the same as in eq. A.5. i and j indicate here all the interior grid nodes of the domain of interest.

REFERENCES

Anderson, R.N., 1975. Heat flow in the Mariana marginal basin. J. Geophys. Res., 80: 4043—4048.

Andrews, D.J. and Sleep, N.H., 1974. Numerical modelling of tectonic flow behind island arcs. Geophys. J.R. Astron. Soc., 38: 237—251.

Barazangi, M. and Isacks, B., 1971. Lateral variations of seismic wave attenuation in the upper mantle above the inclined earthquake zone of the Tonga Island Arc: Deep anomaly in the upper mantle. J. Geophys. Res., 76: 8493—8516.

Barker, P.F., 1972. A spreading center in the east Scotia Sea. Earth Planet. Sci. Lett., 15: 123—132.

Carslaw, H.S. and Jaeger, J.C., 1959. Conduction of Heat in Solids. Oxford University Press, London, p. 10.

Cooper, A.K., Scholl, D.W. and Marlow, M.S., 1976. A plate tectonic model for the evolution of the eastern Bering Sea basin. Geol. Soc. Am. Bull., 87: 1119—1126.

Cooper, A.K., Marlow, M.S. and Scholl, D.W., 1977. The Bering Sea — a multifarious marginal basin. In: M. Talwani and W.C. Pitman III (Editors), Island Arcs, Deep Sea Trenches and Back-Arc Basins. Maurice Ewing Ser., 1, Am. Geophys. Union, Washington, D.C., pp. 437—450.

Hawkins Jr., J.W., 1977. Petrologic and geochemical characteristics of marginal basin basalts. In: M. Talwani and W.C. Pitman III (Editors), Island Arcs, Deep Sea Trenches and Back-Arc Basins. Maurice Ewing Ser. 1, Am. Geophys. Union, Washington, D.C., pp. 355—365.

Hsui, A.T., Roecker, S.W. and Toksöz, M.N., 1977. Induced flow behind subducting slabs — a new kinematic model. EOS, Trans., Am. Geophys. Union, (abstract), 59: 499.

Karig, D.E., 1970. Ridges and basins of the Tonga—Kermadec island arc system. J. Geophys. Res., 75: 239—254.

Karig, D.E., 1971a. Origin and development of marginal basins in the western Pacific. J. Geophys. Res., 76: 2542—2561.

Karig, D.E., 1971b. Structural history of the Mariana island arc system. Geol. Soc. Am. Bull., 82: 323—344.

Karig, D.E., 1974. Evolution of arc systems in the western Pacific. Ann. Rev. Earth Planet. Sci., 2: 51—75.

Lawver, L.A., Curray, J.R. and Moore, D.G., 1976. Tectonic evolution of the Andaman Sea. EOS, Trans., Am. Geophys. Union, (abstract), 57: 333.

Matsuda, T. and Uyeda, S., 1971. On the Pacific-type orogeny and its model-extension of the paired belts concept and possible origin of marginal seas. Tectonophysics, 11: 5—27.

McKenzie, D.P., 1969. Speculations on the consequences and causes of plate motions. Geophys. J.R. Astron. Soc., 18: 1—32.

McKenzie, D.P., Roberts, J.M. and Weiss, N.O., 1974. Convection in the Earth's mantle: towards a numerical simulation. J. Fluid. Mech., 62: 465—538.

Molnar, P. and Oliver, J., 1969. Lateral variations of attenuation in the upper mantle and discontinuities in the lithosphere. J. Geophys. Res., 74: 2648—2682.

Parmentier, E.M. and Torrance, K.E., 1975. Kinematically consistent velocity fields for hydrodynamic calculations in curvilinear coordinates. J. Comput. Phys., 19: 404—417.

Ringwood, A.E., 1975. Composition and Petrology of the Earth's Mantle. McGraw-Hill, New York, N.Y., p. 152.

Sclater, J.G., 1972. Heat flow and elevation of the marginal basins of the Western Pacific. J. Geophys. Res., 77: 5705—5719.

Sclater, J.F., Hawkins, J.W., Mammerickx, J. and Chase, C.G., 1972a. Crustal extension between the Tonga and Lau Ridges: Petrologic and geophysical evidence. Geol. Soc. Am. Bull., 83: 505—518.

Sclater, J.G., Ritter, U.G. and Dixon, F.S., 1972b. Heat flow in the southwestern Pacific. J. Geophys. Res., 77: 5697—5704.

Sleep, N.H. and Toksöz, M.N., 1971. Evolution of marginal basins. Nature, 33: 548—550.

Toksöz, M.N. and Bird, P., 1977. Formation and evolution of marginal basins and continental plateaus. In: M. Talwani and W.C. Pitman III (Editors), Island Arcs, Deep Sea

Trenches and Back-Arc Basins. Maurice Ewing Ser., 1, Am. Geophys. Union, Washington, D.C., pp. 379—393.

Toksöz, M.N., Sleep, N.H. and Smith, A.T., 1973. Evolution of the downgoing lithosphere and the mechanisms of deep focus earthquakes. Geophys. J.R. Astron. Soc., 35: 285—310.

Tomoda, Y., Kobayashi, K., Segawa, J., Nomura, M., Kumura, K., and Saki, T., 1975. Linear magnetic anomalies in the Shikoku Basins. J. Geomagn. Geoelectr., 28: 47—56.

Torrance, K.E. and Rockett, J.A., 1969. Numerical study of natural convection in an enclosure with localized heating from below — creeping flow to the onset of laminar instability. J. Fluid Mech., 36: 33—54.

Torrance, K.E. and Turcotte, D.L., 1971. Thermal convection with large viscosity variations. J. Fluid Mech., 47: 113—125.

Torrance, K.E., Davis, R., Eike, K., Gill, P., Gutman, D., Hsui, A., Lyon, S. and Zien, H., 1972. Cavity flows driven by buoyancy and shear. J. Fluid Mech., 51: 221—231.

Turcotte, D.L., Torrance, K.E. and Hsui, A.T., 1973. Convection in the Earth's mantle. In: B.A. Bolt (Editor), Methods of Computational Physics. Academic Press, New York, N.Y., 13: 431—453.

Uyeda, S., 1977. Some basic problems in the trench-arc—back-arc systems. In: M. Talwani and W.C. Pitman III (Editors), Island Arcs, Deep Sea Trenches and Back-Arc Basins. Maurice Ewing Ser., 1, Am. Geophys. Union, Washington, D.C., pp. 1—14.

Uyeda, S. and Ben—Avraham, Z., 1972. Origin and development of the Philippine Sea. Nature, Phys. Sci., 240: 176—178.

Uyeda, S. and Miyashiro, A., 1974. Plate tectonics and Japanese islands: a synthesis. Geol. Soc. Am. Bull., 85: 1159—1170.

Watanabe, T., Epp, D., Uyeda, S., Langseth, M. and Yasui, M., 1970. Heat flow in the Philippine Sea. Tectonophysics, 10: 205—224.

Watanabe, T., Langseth, M. and Anderson, R.N., 1977. Heat flow in back arc basins of the western Pacific. In: M. Talwani and W.C. Pitman III (Editors), Island Arcs, Deep Sea Trenches and Back-Arc Basins. Maurice Ewing Ser., 1, Am. Geophys. Union, Washington, D.C., pp. 137—161.

Watts, A.B. and Weissel, J.K., 1975. Tectonic history of the Shikoku marginal basin. Earth Planet. Sci. Lett., 25: 239—250.

Yasui, M., Kishii, T. and Sudo, K., 1967. Terrestrial heat flow in the Sea of Okhotsk, 1. Oceanogr. Mag., 19: 87—94.

Chapter 25

SUTURE ZONE COMPLEXITIES: A REVIEW*

JOHN F. DEWEY

Department of Geological Sciences, State University of New York at Albany, Albany, N.Y. (U.S.A.)

SUMMARY

Suture zones, marking the sites of obliteration of oceanic lithosphere by subduction and the consequent intracontinental welding of continental masses, are rarely simple, single, easily recognizable lines. Although the obliteration of a major oceanic tract may be marked by an ophiolite-bearing suture or a cryptic suture, many mini-sutures with a wide variety of origins may be present within the convergent zone in addition to many kinds of intracontinental transform, graben, and fold/thrust zones in zones up to several thousand kilometers from the main sutures. This great variety of high-strain zones associated with wide zones of basement reactivation makes it difficult, particulary at deep structural levels in eroded older orogenic systems, not only to recognize a line along which continental masses may be snipped and dismembered to allow continental fragments to be reassembled to former relative positions, but also even to recognize the plate boundary facies indicators that enable one to make a qualitative plate-tectonic analysis because most of these facies indicators are developed at shallow structural levels.

INTRODUCTION

A major consequence of finite relative plate motion in complex plate mosaics that carry continents, continental fragments, island arcs, seamount chains, and seamounts, is that these masses commonly run into subduction zones where they cannot be subducted smoothly on account of their buoyancy (McKenzie, 1969) or merely that they form a topographic irregularity that inhibits subduction. The result is that collisional strains occur that lead to the welding or suturing of these masses against the subduction zone with which they collide. Sutures (Gansser, 1964) therefore mark the zones along which oceanic lithosphere has been totally subducted. Sutures and suture zones, however, are rarely simple well-defined easily recognizable lineaments. Continental collision, the 'terminal' form of suturing, is generally preceded by a long history of suturing on various scales and itself results in the generation of a great array of intracontinental high-strain zones that may resemble sutures sensu stricto, but many of which do not mark the sites of

---

*Originally published as: Dewey, J.F., 1977. Suture zone complexities: a review. In: M.W. McElhinny (Editor), The Past Distribution of Continents. Tectonophysics, 40: 53–67.

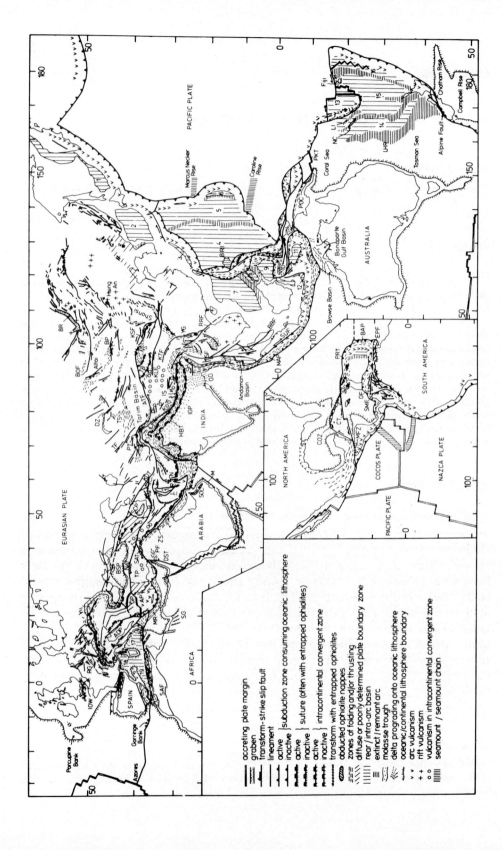

the obliteration of oceanic lithosphere and some of which may not penetrate the lithosphere.

There is little general disagreement that plate tectonics and, hence, suturing has occurred since the early Palaeozoic, but greater controversy has centered on the question of Precambrian plate tectonics (Dewey and Spall, 1975). Discussions of the possible distribution of continents and oceans during the Precambrian and Palaeozoic depend, critically, upon the recognition of sutures sensu stricto, along which continental masses may be dismembered and, hence, redistributed. Burke, Dewey and Kidd (1977) presented in their study a distribution of Precambrian and Palaeozoic sutures and suture zones; it is the purpose of this paper to outline some of the criteria by which sutures may be recognized and distinguished from other high-strain zones associated with island arcs and intracontinental convergent zones and to outline some of the complicated processes that lead to, and that occur during, terminal continental suturing. Examples of these structures and processes are chosen mainly from the Alpine System of Europe and Asia where their complexity is readily discernible.

The geometric complexity of the Alpine System (Fig. 1) bears witness to the fact that there is no simple, single 'template' for collisional suturing and related strain. The Alpine System has a protracted history, since the late Triassic, of generation and subduction of oceanic lithosphere that involved complex changes in relative motion between the major plates between which the System evolved and continual rearrangement of smaller plates within the System (Dewey et al., 1973). The present Alpine System shows major con-

---

Fig. 1. Schematic map of the Alpine System of Europe and Asia, and (inset) the Caribbean region, showing Palaeogene to present tectonics of the orogenic belt and its forelands. Key to abbreviations: ADP = Adriatic Plate, AP = Aegean Plate, ATF = Altyn Tagh Fault, RAP = Barbados accretionary prism, BBF = Bok Bak Fault, BOF = Bolnai Fault, BP = Bogpo Fault, BR = Baikal Rift, BRI = Benham Rise, BSP = Black Sea Plate, COZ = Cuban Ophiolite Zone, CT = Cayman Trough, DST = Dead Sea Transform, DF = Dzhungarian Fault, EPF = El Pilar Fault, GD = Ganges Delta, GP = Greek Plate, HBT = Himalayan Boundary Thrust, IGP = Indogangetic Plain, IOW = Isle of Wight Steep Belt, IS = Indus Suture, K = Karakorum Fault, KDT = Kopet Dagh Fault, KLF = Kun Lun Fault, KSF = Kansu Fault, KTF = Kang Ting Fault, L = Ladakh Massif, LP = Lut Plate Complex, LHR = Lord Howe Rise, LI - Loyalty Island Ridge, M = Masitah Island Zone, MAP = Mentawei accretionary prism, MFZ = Motagua Fault Zone, MR = Mediterranean Ridge, N = Neh Faults, NAT = North Anatolian Transform, NC = New Caledonia, NS = Nan Shan, OF = Oca Fault, PF - Palmyrian Fold Zone, PKT = Pocklington Trough, POC = Papuan Ophiolite Complex, PT = Pamir Thrust, PTR = Puerto Rico Trench, RG = Rhine Graben, RRF = Red River Fault, SAF = South Atlas Fault, SAT = South Anatolian Transform, SCP = South Caspian Plate, SF = Shan Fault, SG = Sirte Graben, SH = Shillong Plateau, SMF = Santa Marta Fault, SOC = Semail Ophiolite Complex, SUF = Sumatra Fault, TB = Tsaidam Basin, TF = Talas Fergana Fault, TP = Turkish Plate, TS = Tien Shan, YG = Yunman Graben, Z = Zagros Fold Belt, ZS = Zagros Suture. Key to rear and intra-arc basins: 1 = Okhotsk, 2 = Japan, 3 = Okinawa, 4 = West Philippine, 5 = Parece Vela, 6 = Mariana, 7 = South China, 8 = Sulu, 9 = Celebes, 10 = Corontalo, 11 = South Banda, 12 = Flores, 13 = New Hebrides, 14 = New Caledonia, 15 = South Fiji, 16 = Lau Havre, 17 = Tyrrhenian, 18 = Balearic, 19 = Alboran.

tinents in various stages of convergence; Australia beginning to tuck its northern margin under the Timor accretionary prism, oceanic remnants, such as the Eastern Mediterranean, passing laterally into the Zagros Crush Zone, and India boring its way into Asia since the early Oligocene. Thus, continental collision is a capricious irregular phenomenon along a major mountain system, depending upon the time of impingement and irregular shapes of colliding margins, which give rise to complex diachronous strains (Dewey and Burke, 1974; Dewey and Kidd, 1974). Further, it is evident from the geometry and history of the Alpine System in Europe that numerous sutures of various kinds contribute to the evolution of the zone; it is rare for a single suture to characterize a terminal collision zone. Terminal suturing is the last stage of, usually, a long history of subduction at various sites, with different and changing polarities and geometries (e.g., arc systems of Indonesia). These pre-terminal-collision relationships possess not only their own intrinsic historical complexities, but may become appallingly mangled during terminal suturing. Lastly, terminal suturing results in a great variety of foreland strains from simple foreland fold thrust zones (e.g., Zagros) to zones several thousand kilometers wide (e.g., Tibet and Western China) (Molnar and Tapponnier, 1975).

SUTURES AND RELATED STRAIN

The simplest kind of orogenic suture is a high-strain zone, containing mangled ophiolite remnants and, occasionally, blueschist melanges, that separates two continental terrains with dissimilar pre-collisional strain histories (Indus Suture, Zagros Crush Zone). The irregularity of colliding continental margins is shown in the Alpine System by sutures passing laterally into remnant oceanic tracts (Zagros Crush Zone—Eastern Mediterranean) and into subduction zones consuming large oceanic tracts (Indus Suture—Andaman Subduction Zone). The Black and South Caspian Seas appear to be oceanic tracts trapped within the Alpine System; such small remnant oceans may become filled with thick little-deformed sedimentary sequences that contrast with the intensely deformed suture zone assemblages, into which they pass laterally, and may develop into intra-montaine successor basins.

Continental collision may be regarded as only a terminal, though spectacular, form of suturing that is preceded by an array of sutures dating from the initiation of subduction in the oceanic tract, whose demise occurs during terminal suturing. The seismic work of Seely et al. (1974) has shown that the outer parts of arc—trench gaps consist of listric thrust assemblages that accrete oceanward (perhaps by diachronous recumbent folding) with complex internal strains (Moore, 1973) to progressively increase the width of arc trench gaps (Dickinson, 1973). Such subduction accretionary prisms (Fig. 2A) characterize fore-arc bulges, such and the Mentawei Islands, and involve a diachronous assemblage of facies from hemipelagic layer 1 sediments transported on the subduction conveyor belts, trench turbidites, perched sedi-

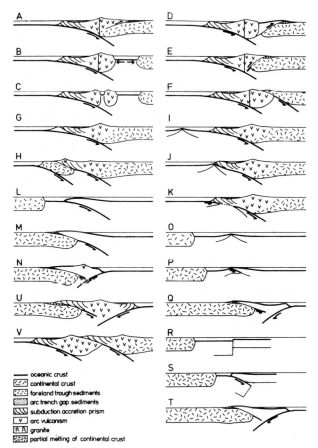

oceanic crust
continental crust
foreland trough sediments
arc trench gap sediments
subduction accretion prism
arc vulcanism
granite
partial melting of continental crust

Fig. 2. Schematic sections illustrating a variety of non-terminal sutures discussed in text.

ments, ophiolite fragments stripped from the downgoing slab and, perhaps, ophiolite basement rocks of arc—trench gaps.

Small-scale collisions and, hence, sutures will be generated at subduction zones by the attempted subduction of small, topographic irregularities on the ocean floor, such as seamounts (Fig. 2G) and microcontinents (Fig. 2H). The Benham Rise on the eastern edge of the Philippine plate appears to have recently collided with the northern Philippines where it jammed into a west-dipping subduction zone and may have been responsible for cessation of subduction at that site. The impingement of ridges and subduction zones, although unlikely to cause suturing, may have important effects on the leading edge of the subduction zone (Fig. 2I, J, K). These effects may include emergence and submergence of the frontal parts of the arc and short cessations in arc magmatism (DeLong and Fox, 1977).

Magmatic arcs may lie either at continental margins or be separated from continental margins by oceanic rear-arc basins. Some rear-arc basins (Bering

Sea) may be portions of ocean trapped behind an arc by the nucleation of a subduction zone/arc complex away from the continental margins, although most appear to be the result of back-arc spreading (Fig. 2B) (Karig, 1971). The process of back-arc spreading may be a repetitive phenomenon involving the continual splitting of arcs, perhaps near the magmatic axis where the arc lithosphere is thinnest and weakest, so that a trail of remnant arcs is left behind an oceanward migrating subduction zone (Fig. 2C) (e.g., remnant arcs of the Philippine Sea and south of Fiji, Fig. 1). Intra-arc basins vary in scale from the small, shallow summit basins of the Aleutians to the larger, deeper en-echelon troughs of the New Hebrides and may, accordingly, vary in the character of their floors from slightly distended arc rocks to oceanic lithosphere. It has been suggested from geological data (Dewey and Bird, 1971; Tarney et al. 1976) and from theoretical considerations (Armstrong and Dick, 1974; Dewey, 1976) that rear-arc basins generated by behind-the-arc spreading may close shortly thereafter. Closure and suturing may be accomplished by a variety of mechanisms, depending on the age and, therefore, thickness of the rear-arc lithosphere and the polarity and siting of the rear-arc subduction zone. If a young lithosphere breaks near the continental edge, a thin hot ophiolite sheet may be obducted across the continental margin (Fig. 2D). If the lithosphere breaks near the arc, a thin hot ophiolite sheet may be transported onto the back of the arc. In either case, obduction occurs shortly after subduction begins, so that there is insufficient time for arc magmatism and subduction accretion prisms to develop on the obducted slab. If, on the other hand, subduction polarity is such that the rear-arc basin is totally subducted, an arc/continental margin suture will develop with little or no ophiolite emplacement with either arc rocks thrust over continental margin (Fig. 2E) or vice-versa (Fig. 2F). Arc obduction and suturing following subduction of a rear-arc basin (Fig. 2E) may be very hard to distinguish from back-arc thrusts that do not mark the site of rear-arc basin obliteration (Fig. 2A). The latter appears to characterize several continental arcs and may be related to lateral and uphill spreading of hot weak arc rocks under compression or to a zone of thrust detachment between an upper thin, hot arc lithosphere and a lower thick, cold foreland lithosphere. Another form of closure has been suggested by Tarney et al. (1976) and Packham and Falvey (1971) that involves no subduction but a general compressional collapse; this may be important for very narrow rear-arc basins and particulary important for very narrow intra-arc basins. In such zones, complete transitions may be seen between an ophiolite complex sensu stricto through arc assemblages cut by mafic dikes, to 'pristine' arc rocks. Sutures generated by the subduction and obduction of rear-arc basins generally will be characterized by a short interval between ophiolite generation and emplacement and between the origin and destruction of the rifted margins and, perhaps, may be distinguished by an association with deformed thick volcanogenic stratigraphic sequences and thin high-temperature obduction aureoles.

Pre-terminal ophiolite obduction commonly takes the form of emplacement of giant high-level ophiolite thrust sheets (e.g., Papua, Davies, 1971); the sutures generated by this form of obduction are between continental and oceanic lithosphere. There has been considerable speculation about the mechanism for this form of obduction (e.g., Dewey, 1976). The problem is related to how intra-oceanic lithosphere detachment occurs close to a continental margin so that the oceanic lithosphere between continental margin and subduction zone is consumed with the ensuing attempt to subduct the continental margin and consequent obduction of a wedge of oceanic crust and mantle. Some possibilities are as follows: detachment in a seamount chain or an old fracture zone to produce a Macquarie Ridge-like subduction zone (Fig. 2L, M, N); detachment on a young fracture zone or ridge/ridge transform, the high young hotter side being obducted over the lower older colder side (Fig. 2R, S, T); or the conversion of a ridge to a subduction zone (Fig. 2, O, P, Q) (Dewey, 1976). Pre-terminal sutures will be generated also by arc collisions of the following types that will be characterized by distinctive structural polarity timings and sequences: collision of an arc with a rifted continental margin followed by a subduction polarity 'flip' (Fig. 2U); collision of an arc with an Andean-type continental margin, both subduction zones having the same polarity (Fig. 2V); and arc—arc collision, the associated subduction zones having either the same or opposing polarities.

The suture types discussed above predate terminal continental collision, at which time they are likely to be extensively and progressively modified by the superposition of complex intense strains associated with terminal suturing. These strains appear to vary greatly, both in the nature of deformation close to the site of suturing and in the extent and distance to which cratonic forelands are affected by collisional strains. These variations may result from the rate of post-suturing convergence, the length of time during which post-suturing convergence takes place and, critically, the shapes of colliding continental margins.

Continental collision (Fig. 3) may be regarded as the final stage of frontal subduction accretion (Fig. 3A). The northern Australian shelf is, presently, sliding beneath the Timor accretion prism with the development of broad-wavelength folds on the shelf (Fig. 1). Although obduction of ophiolite sheets does not appear to be accompanying this obduction of the accretion prism onto the shelf, it has been suggested (Dewey and Bird, 1970; Dewey, 1976) that the obduction of ophiolite nappes may occasionally characterize terminal suturing. The inception of such nappes may be related to the development of tectonic welts, such as the Mediterranean Ridge (Fig. 1) just prior to collision (Fig. 3A) or to the intervention of a spreading ridge. Where ophiolite nappes are developed, they occur at or near the top of a stacked nappe sequence whose stacking order is the reverse of the assembly order (Sales, 1971; Williams, 1971; Dewey, 1976). The growth of these nappe sequences is probably related to the progressive arrival of more and more proximal continental-margin assemblages at the subduction site, which become progressively understacked in the sequence subduction—accretion

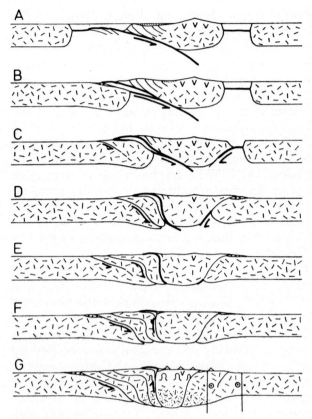

Fig. 3. Schematic sections illustrating the evolution of a terminal suture. Discussion in text. Key to ornament in Fig. 2.

prisms, ophiolite wedges, continental-rise allochthons, continental-shelf allochthons. The progressive migration of a growing nappe pile onto the continental margin is accompanied by migrating wildflysch, flysch and molasse sequences, which the nappe pile progressively overrides (Fig. 3B, C).

The extent to which basement is involved in foreland nappe sequences varies greatly. In the Himalayas (Fig. 1) basement is extensively involved in the nappe pile (Gansser, 1964), while in the Zagros (Fig. 1) the foreland zone is characterized by a fairly simple fold style, few thrusts, and little, if any, basement involvement. Perhaps salt and/or thick shale sequences low in the cover stratigraphy may induce decollement. However, the Zagros terminal collision is of Pliocene age and the Himalayan terminal collision of late Eocene or early Oligocene age; perhaps, cover folding is a precursor to basement nappe development. The development of basement nappes and frontal boundary thrusts in the Himalayas progressed from north to south, from the Indus Suture to the Indo-Gangetic Plain (LeFort, 1975), a sequence of nappes breaking loose within the foreland (Figs. 3C—G). This breaking sequence is a mechanism whereby intra-continental convergence can continue beyond suturing and suggests (LeFort, 1975) that successive boundary thrusts may penetrate the full thickness of continental crust to allow crust/mantle detachment, thereby isolating successive wedges of non-subductable continental crust that become rotated and shortened into a sigmoidal form (Fig. 3). The initial suture zone, along which subduction ceases shortly after collision may become the site of retrocharriage (Fig. 3F, G) (rotation zone of Roeder, 1973) following sigmoidal rotation of the thrust zone (Fig. 3D, E). This produces a narrow zone of high-level reversal of structural polarity.

A major effect of intense suturing and retrocharriage is that the earliest obducted subduction accretion, ophiolite, and continental rise, allochthons become isolated at a high structural level above a downward tapering suture zone at deep levels across which intense convergence may reduce the zone to a very narrow high-strain zone (cryptic suture — Dewey and Burke, 1973) (Fig. 4) that may be hard to recognize in deeply eroded older orogenic terrains unless ophiolite shreds are preserved along it. Another feature of terminal suturing may be the upwedging of sub-continental mantle and lower continental crust assemblages (Fig. 4), perhaps represented by the lherzolites of Sesio-Lanzo and the kinzigites and pyroxene granulites of the Ivrea Zone in the French/Italian Alps.

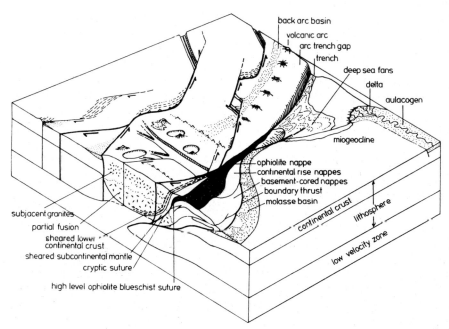

Fig. 4. Block diagram illustrating schematic tectonic outlines of a continental collision zone between irregular continental margins.

Incorporated into the collision sequence in Fig. 3 is the closure of a back-arc basin during the early stages of collision, which will produce synchronous collisional strains parallel with and close to the main suture. Of particular possible importance is the compressional smashing of narrow intra-arc basins during terminal collision to give narrow irregular lensing zones of deformation associated with dismembered ophiolite sequence that may be hard to distinguish from the main suture if their local fragmentation and convergence history is obscure. Some of the irregular ophiolite zones of Iran, such as those around the Lut Block and in the Grand Kavir, may be sited on narrow irregular rear-arc basins that collapsed during Pliocene collision across the Zagros Crush Zone.

Post-suturing intra-continental convergence may be taken up in a variety of ways apart from frontal and retrocharriage thrusting. Large-scale mechanisms for convergence must involve the downward (subduction), upward, or sideways, escape of continental lithosphere. Extensive subduction of continental lithosphere, at least with its continental crust, is prohibited by buoyancy considerations (McKenzie, 1969) so that major continental under-thrusting solutions (e.g., Powell and Conaghan, 1973) are unlikely to generate a double-thickness continental crust such as is supposed to characterize the Tibetan Plateau. If the Tibetan continental crust is very thick, it is more likely to have been thickened by north—south homogeneous shortening (Fig. 3G; Fig. 4), a process that would account for the widespread shallow folding and thrusting that characterize the plateau (Burke, Dewey and Kidd, in preparation). It may also account for the highly potassic silicic vulcanism and shallow granitic plutonism that may result from partial melting of the thickened lower continental crust to leave a depleted anorthosite granulite residue that may be represented in deeply eroded Precambrian reactivated basement terrains (Dewey and Burke, 1973) (Fig. 4).

Non-homogeneous behavior of continental forelands during intra-continental convergence, however, may dominate over the more homogeneous thickening process, whereby a foreland (e.g., Tibet and China, Fig. 1) may segment into a number of plates whose margins are, mainly, giant intra-continental transforms that appear to be the major mechanism whereby continental lithosphere escapes sideways from the collisional vise (Molnar and Tapponnier, 1975) (Fig. 4; Fig. 5A). These giant transform boundaries may have divergent (graben) and convergent (thrust/fold) segments. Depending upon the amount of lateral motion on the transforms, the divergent segments will vary from intra-continental graben floored by slightly extended continental crust, through heavily extended continental crust injected by mafic dike swarms, to narrow oceanic tracts, such as Lake Baikal (Fig. 1). A particularly important relationship involving transform-bounded smaller plates is illustrated by the Turkish and Aegean Plates (Fig. 1) shown schematically in Fig. 5B. The westward motion of these two small plates relative to Europe appears to be caused by the intra-continental convergence of Arabia and Eurasia resulting in westward escape of continental lithosphere

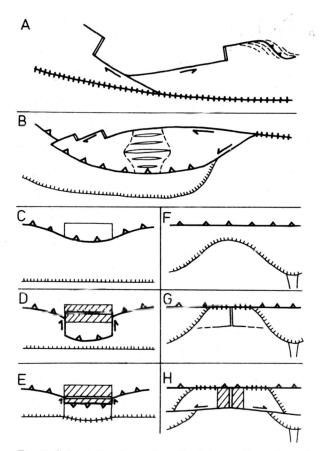

Fig. 5. Schematic plans of small plate configurations developed during terminal suturing. C—E from McKenzie and Weiss (1975). Discussion in text.

into the remnant oceanic tract of the Mediterranean (McKenzie, 1972). Short, narrow, divergent, oceanic segments occur in the northern Aegean and a system of lensing east—west graben forms a wide diffuse boundary between the Aegean and Turkish Plates. These various graben and, particularly, short narrow oceanic tracts are likely, during continued intra-continental convergence, to be rotated and compressed forming a host of, occasionally, ophiolite-bearing mini-sutures. Collisional cracking and fragmentation of forelands into complex small-plate systems followed by local rotation, convergence and suturing may be an alternate explanation for the Lut block style tectonics of Iran.

The transforms associated with suture zones may have entrapped ophiolite slivers, resulting either from an original suture zone developing into a transform zone (e.g., Motagua Fault Zone, Fig. 1, inset) or from transforms terminating at suture zones and dragging ophiolite slivers along them from the suture zone (e.g. Karakorum Fault, Fig. 1).

Two further ways of developing narrow oceanic tracts just prior to and during collisional process are illustrated in Figs. 5 C—H. McKenzie and Weiss (1975) have suggested the importance of rear-arc basin plate accretion behind convex-oceanward subduction zone segments just prior to collision. Portions of the continental margin arc move out between transforms (Fig. 5C, D) to collide with the opposing continental margin followed by the progressive subduction of the rear-arc basin lithosphere (Fig. 5E). Another possible geometry involves the fragmentation of a 'headland' in one of the colliding continental margins by rifting perpendicular to the suture zone so that slivers of continental lithosphere move sideways away from an oceanic tract that terminates at trench/trench ridge and transform/transform/ridge junctions (Fig. 5F, G). As convergence continues, this oceanic tract itself becomes a site of suturing (Fig. 5H). These varied forms of narrow oceanic tract, formed and destroyed during the collisional process, will be marked by a complex array of small-scale sutures that do not persist laterally and are characterized by ophiolite suites whose emplacement follows shortly after their origin by plate accretion. The rifted margins of these narrow basins and the sedimentary sequences that fill them do not predate collision and have a short life, unlike the rifted margins and sedimentary sequences of the larger oceans whose progressive subduction leads to terminal collision.

Other forms of collisional foreland deformation involve the development of graben at high angles to sutures (e.g., Rhine Graben) conjugate fault systems (e.g., Tertiary strike-slip faults of the European Alpine foreland) (Fig. 1) and extensional cross-fractures normal to sutures. The graben and cross-fractures may be characterized by silicic and basaltic vulcanism (e.g., Baikal and Hsing An Provinces, Fig. 1, and the north—south volcanic ridges of Tibet; Burke Dewey and Kidd, in preparation). Localized zones of steep faulting, thrusting, and monocline development (such as the Isle of Wight Zone (Fig. 1) in foreland cover sequences may signify the importance of the reactivation of older foreland basement structures.

Striking features of the Alpine system in Europe and Asia are the very variable extent to which forelands become involved in collisional strains and the importance of strike-slip and extensional zones during collision. In particular, the eastern Asian foreland of the Himalayan system is affected by these strains to distances of several thousand kilometers (Molnar and Tapponier, 1975). This variation is most likely the result of the irregular form of colliding margins. Peninsula India appears to be acting as a plunger, progressively indenting eastern Asia and causing giant wedges of lithosphere to escape eastward toward the Pacific (Molnar and Tapponier, 1975). Similarly, the Turkish plate is sliding westwards from the Arabia/Eurasia vise, into an oceanic tract. 'Headlands' along colliding margins, along which early suturing occurs, are likely to be the preferential sites of widespread foreland deformation while embayment forelands may remain largely unaffected by extensive strains.

Irregular continental collision produces orogenic zones that not only vary greatly laterally in style and width, but also in which strain is strongly dia-chronous from early and intense, where headlands impinge, to late and weak, in embayments (Dewey and Burke, 1973, 1974) (Fig.4). The embayments may have aulacogens whose structural/stratigraphic history dates from the time of rifting that produced the continental margin (Burke and Dewey, 1973; Hoffman et al., 1974), while the headlands may have collisional graben dating from the inception of collision.

The various types of sutures, cryptic sutures and transforms discussed above produce intra-continental high-strain zones. However, other forms of high-strain zone may be preserved within collisional orogens that may or may not mark the sites of destruction of oceanic lithosphere. Intra-arc transforms, such as the Sumatra Fault (Fig. 1), may be important where the slip vector is at a low angle to the subduction zone (Fitch, 1972). Such transforms may have compressional segments and ridge segments, the latter probably being developed at the northern termination of the Sumatra Fault in the Andaman Basin (Fig. 1). Thus, intra-arc basins may be, occasionally, nucleated on short divergent segments of intra-arc transforms, which them-selves are likely to form along the volcanic axis where the lithosphere is hottest, thinnest, and weakest. Continental borderland transforms, such as the San Andreas Fault with ridge segments (Gulf of California) and con-vergent segments (Transverse Ranges), are also likely to end up, eventually, near intra-continental suture zones. Lastly, transform segments of conti-nental margins, formed during initial continental rifting, are commonly zones of complex, folding and thrusting (Lowell, 1972; Dewey, in prepara-tion) which are preserved, eventually, in suture zones.

The enormous geometric complexity and irregularity along strike in the Alpine System (Fig. 1), an orogenic system that is witnessed in various stages of collisional evolution, results from inherited pre-collisional arc/microconti-nent geometries and from the growth of micro-plate systems and foreland deformation strains that reflect the irregularity of colliding margins and that intensify and spread as post-suturing intra-continental convergence proceeds. It is interesting to note that several Paleozoic orogens (e.g., Urals and Appalachian/Caledonian System), although possessing a protracted and com-plex structural history do not show the great geometric complexity and strike variation of the Alpine System. It is possible that the progressive con-vergence of irregular continental margins, although producing complex structures over wide zones, gradually cleans up the convergence zone partly by the lateral motion of continental plates, slivers, and flakes from headlands into embayments and partly by progressive rotation of structures during foreland to foreland shortening.

CONCLUSIONS

Although sutures (sensu stricto) mark the lines along which continents collide, the great irregularity of colliding continental margins and progressive intra-continental convergence combine to produce associated deformation zones of extraordinary complexity and width, involving a great array of mini-sutures and transform-related high-strain zones. Wide zones of Precambrian basement reactivation bounded by high-strain zones with shredded ultramafic and mafic assemblages may be deeply eroded remnants of suture bounded foreland deformation zones.

REFERENCES

Armstrong, R.L. and Dick, H.J.B., 1974. A model for the development of thin overthrust sheets of crystalline rock. Geology, 2: 35—40.

Burke, K. and Dewey, J.F., 1973. Plume-generated triple junctions: key indicators in applying plate tectonics to old rocks. J. Geol., 81: 406—433.

Davies, H.L., 1971. Peridotite-gabbro-basalt complex in eastern Papua: an overthrust plate of oceanic crust and mantle. Bur. Min. Res. Geol. Geophys. Aust. Bull., 128: 1—48.

DeLong, S.E. and Fox, P.J., 1977. Geological consequences of ridge subduction (in press).

Dewey, J.F., 1976. Ophiolite obduction. Tectonophysics, 31: 93—120.

Dewey, J.F. and Bird, J.M., 1970. Mountain belts and the new global tectonics. J. Geophys. Res., 75: 2625—2647.

Dewey, J.F. and Bird, J.M., 1971. Origin and emplacement of the ophiolite suite: Appalachian ophiolites in Newfoundland. J. Geophys. Res., 76: 3179—3206.

Dewey, J.F. and Burke, K., 1973. Tibetan, Variscan and Precambrian basement reactivation: products of continental collision. J. Geol., 81: 83—692.

Dewey, J.F. and Burke, K., 1974. Hot spots and continental breakup: implications for collisional orogeny. Geology, 2: 57—60.

Dewey, J.F. and Kidd, W.S.F., 1974. Continental collisions in the Appalachian—Caledonian orogenic belt; variations in style related to complete and incomplete suturing. Geology, 2: 543—546.

Dewey, J.F. and Spall, H., 1975. Pre-Mesozoic plate tectonics. Geology, 3: 422—424.

Dewey, J.F., Pitman, W.C. III, Ryan, W.B.F. and Bonnin, J., 1973. Plate tectonics and the evolution of the Alpine System. Geol. Soc. Am. Bull., 84: 3137—3180.

Dickinson, W.R., 1973. Widths of modern arc—trench gaps proportional to past duration of igneous activity in associated magmatic arcs. J. Geophys. Res., 78: 3376—3389.

Fitch, T.J., 1972. Plate convergence, transcurrent faults, and internal deformation adjacent to Southeast Asia and the Western Pacific. J. Geophys. Res., 77: 4432—4460.

Gansser, A., 1964. The Geology of the Himalayas. Interscience, New York, 289 pp.

Hoffman, P., Dewey, J.F. and Burke, K., 1974. Aulacogens and their genetic relation to geosynclines, with a Proterozoic example from Great Slave Lake, Canada. In: R.H. Dott and R.H. Shaver (editors), Modern and Ancient Geosynclinal Sedimentation. Soc. Econ. Paleontol. Mineral. Spec. Publ., 19: 38—55.

Karig, D.E., 1971. Origin and development of marginal basins in the western Pacific. J. Geophys. Res., 76: 2542—2560.

LeFort, P., 1975. Himalayas: the collided range. Present knowledge of the continental arc. Am. J. Sci., 275-A: 1—44.

Lowell, J.D., 1972. Spitzbergen Tertiary orogenic belt and the Spitzbergen Fracture Zone. Geol. Soc. Am. Bull., 83: 3091—3102.

McKenzie, D.P., 1969. Speculations on the consequences and causes of plate motion. Geophys. J.R. Astron. Soc., 18: 1—32.

McKenzie, D.P., 1972. Active tectonics of the Mediterranean region. Geophys. J.R. Astron. Soc., 30: 109—185.

McKenzie, D.P. and Weiss, N., 1975. Speculations on the thermal and tectonic history of the earth. Geophys. J.R. Astron. Soc., 42: 131—174.

Molnar, P. and Tapponnier, P., 1975. Cenozoic tectonics of Asia: effects of a continental collision. Science, 189: 419—426.

Moore, J.C., 1973. Complex deformation of Cretaceous trench deposits, southwestern Alaska. Geol. Soc. Am. Bull., 84: 2005—2020.

Packham, G.H. and Falvey, D.A., 1971. An hypothesis for the formation of marginal seas in the western Pacific. Tectonophysics, 11: 79—109.

Powell, C.Mc.A. and Conaghan, P.J., 1973. Plate tectonics and the Himalayas. Earth Planet. Sci. Lett., 20: 1—12.

Roeder, D.H., 1973. Subduction and orogeny. J. Geophys. Res., 78: 5005—5024.

Sales, J.K., 1971. The Taconic allochthon — not a detached gravity slide. Geol. Soc. Am. Abstr. Progr. Ann. Mtg., Washington, p. 693.

Seely, D.R., Vail, P.R. and Walton, G.G., 1974. Trench slope model. In: C.A. Burk and C.L. Drake (editors), The Geology of Continental Margins. Springer, New York, pp. 249—260.

Tarney, J., Dalziel, I.W.D. and De Wit, M.J., 1975. Marginal basin 'Rocas Verdes' Complex from S. Chile: a model for Archaean Greenstone Belt formation. In: B.F. Windley (editor), The Early History of the Earth. Wiley, New York, pp. 131—146.

Williams, H., 1971. Mafic—ultramafic complexes in western Newfoundland Appalachians and the evidence for their transportation: a review and interim report. Geol. Assoc. Can. Proc., 24: 9—25.

Reference added in proof:

Burke, K., Dewey, J.F. and Kidd, W.S.F., 1977. World distribution of sutures — the sites of former oceans. In: M.W. McElhinny (Editor), The Past Distribution of Continents. Tectonophysics, 40: 69—99.

Chapter 26

THE EASTERN MEDITERRANEAN AND THE LEVANT: TECTONICS OF CONTINENTAL COLLISION*

AMOS NUR and ZVI BEN-AVRAHAM

Department of Geophysics, Stanford University, Stanford, Calif. 94305 (U.S.A.)
Geophysical Laboratory, Weizmann Institute of Science, Rehovot, and
Israel Oceanographic and Limnological Res. Ltd., Haifa (Israel)

SUMMARY

By subdividing the Alpine belt into the Europe, Persia and Tibet segments, we recognize a linear relationship between slip rate and width of the Persia and Tibet segments, and their plateau-like morphology. The proposed explanation is purely geometrical — depending only on mass inflow of the nearby plates. The study of the Levant fracture zone, at the boundary between two of these plates, reveals that as we approach the collision zone, more and more horizontal deformation is taken up by branching faults. The mode of plate consumption is very sensitive to the nature of colliding crusts. As the thickness of the Africa plate increases eastward, the consumption becomes more and more collision-like, with the disappearance of deep earthquakes. In contrast, consumption is not strongly influenced by the rate of convergence. This suggests again that geometry and mass are the important quantities controlling consumption. We discern four zones in the Eastern Mediterranean with diffuse boundaries. These zones are not microplates, but rather the product of the breakup of the colliding Africa plate.

INTRODUCTION

Although general agreement exists that continental collisions do take place at present and did so in the past, the geometrical and mechanical details of collisions are still obscure and controversial. One controversial aspect recently summarized by Powell and Conaghan (1975) is the nature of lithospheric and crustal deformation within major mountain belts: is there substantial underthrusting of one continental plate beneath the other, or are mountain belts the result of multiple thrust-fault sheets? The confusion here is well reflected in various attempts to treat the deformation within mountain belts as 'micro plate tectonics' (e.g. McKenzie, 1972) as superposition of successive underthrust sheets (Chang and Cheng, 1973) or as a pile-up of island arcs (Freund et al., 1975).

---

*Originally published as: Nur, A. and Ben-Avraham, Z., 1978. The Eastern Mediterranean and the Levant: tectonics of continental collision. In: O.H. Oren (Editor), Structure and Tectonics of the Eastern Mediterranean. Tectonophysics, 46: 297—311.

Another unclear aspect of continental collision is the process by which coherent plates break up, deform, and transform into the mountain belt mass.

The region encompassing the Eastern Mediterranean and the Middle East may provide important insight into the processes of continental collisions. Here the Levant fracture zone is a well-exposed and accessible fault zone, transforming the spreading motion of the Red Sea into continental collision in eastern Turkey. The faster moving Arabia plate, as well as India further east, provide information about continent—continent collision. In the Eastern Mediterranean on the other hand, there is a complicated mixture of collisions between oceanic and continental crusts. This has led to numerous models and interpretations of the region. Freund et al. (1975) suggest that the Eastern Mediterranean is a marginal sea, whereas Neev (1975) believes that the tectonics here are controlled by a fracture zone extending through Africa into eastern Turkey. In contrast, McKenzie et al. (1970), McKenzie (1972), Le Pichon et al. (1973) and Dewey et al. (1973), for example, suggest that simple Afro-European plate collision models may be adequate. These models, as well as other ones, are not necessarily mutually exclusive.

In any case we believe that this complexity is the consequence at least in part of the relatively slow motion between Africa and Europe, which has not matured as yet into full-fledged continental collision. Several oceanic crustal chunks still remain here. This complication should yield some evidence for the role of crustal nature in the consumption process and plate breakup.

In this paper we explore some of the problems of collision tectonics by combining previously obtained results from the Alpine system, Near East tectonics and observations in the Eastern Mediterranean Sea. First, we consider the deformation in the Alpine belt itself, in which plates are being crushed and severely deformed. Next, we analyze the deformation in plates as they approach the continental collision zone. Here we attempt to understand the processes by which coherent and rigid plates break up as they collide. Finally, we attempt to explain the tectonics of the Eastern Mediterranean in relation to the crustal structure of the colliding continental segments.

CONTINENTAL COLLISION: THE TIBET AND PERSIA PLATEAUS

The complex Alpine—Himalaya mountain chain, which extends from North Africa through Southern Europe, Turkey, Iran and the Himalayas, is presumably the zone of collision between the continental blocks of Eurasia to the north and Africa, Arabia and India to the south.

In an attempt to clarify the mechanism of continental collision, Ben-Avraham and Nur (1976) explored the relations between the overall geologic features of the deformed Alpine—Himalaya region — height, width and style of deformation, and the gross elements of the neighboring plates — slip rates, slip direction, and zone of seismic activity. They found that active deformation, as indicated by seismicity, takes place throughout the entire regions of morphological highs and is not confined to zones of deformed young sediments. By subdividing the collision zone into the Himalaya, Persia and Europe segments which are defined by the nearby India, Arabia and Africa plates, they found that the width of the deformed segments increases with their collision rate. The simplest reason for the width-rate correlation is that both increase with distance from the poles of relative rotations of the colliding plates (Fig. 1). Ben-Avraham and Nur (1976) showed that the morphology and the height difference between the Tibet and Iran plateaus, which are 5 and 3 km high, respectively, cannot be explained by a viscous or an elastic-plastic lithosphere. Instead, the difference in plateaus height can be explained by proportionality to the total mass flow into the collision belt. This implies that the consumed collision mass remains within the deformed segments.

We suggest therefore that the simplest model for the morphology of the Alpine belt is a purely geometrical one, in which the details of rheology and stress may play only a secondary role. Furthermore, we may conclude from these results that the gross features width, height and styles of deformation

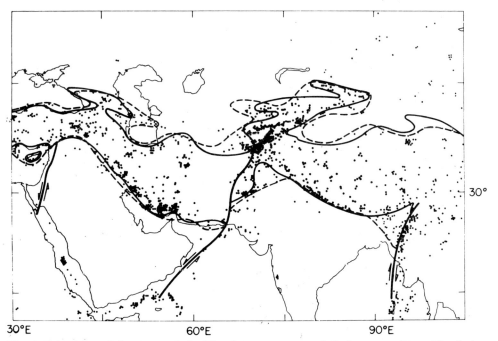

Fig. 1. Seismicity of the Iran and the Himalaya segments and their surroundings. The dash lines outline the seismic zones. Mountain regions, shown by shaded zones outlined by solid lines, include deformed rocks of varying ages. The Assam, Owen and the Levant faults are shown by heavy lines (after Ben-Avraham and Nur, 1976).

of the mountain belts in continental collision are simply related and therefore controlled by a few gross elements of the motion of nearby plates — slip directions, slip rates and total displacements.

The internal deformation in the belt, which accommodates the mass influx, is wide-spread and complex. It involves large-scale thrust, transcurrent and some normal faulting, as well as extensive folding of sedimentary rocks. It is therefore clear that the rock mass, which was adequately described as rigid while part of a continental plate, has totally lost its rigidity, and is here highly deformed as well as deformable. We may say therefore that the twin concepts of plates and rigidity may be completely irrelevant inside the mountain belt.

PLATE BREAKUP: HORIZONTAL DEFORMATION

The total crushing in the collision belt must be the final product of a transformation which converts a rigid plate into a highly deformed rock mass. It is reasonable to expect this transformation to occur mainly in the neighborhood of the collision suture, and the evidence should be looked for in the details of the geology in these areas. Of particular interest are the transform regions — the Assam seismic belt, the Quetta—Owen zone, and the Levant fracture zone.

Because of a long historical record of seismic activity, good ground exposure, and detailed geological mapping, the Red Sea—Levant region was particularly amenable to a regional study (Ben-Menahem et al., 1976).

The results of the seismic analysis can be divided into several areas:

The Levant rift zone

The Levant rift zone, including the Dead Sea, is one of the most prominent morphological features in the entire Middle East. Recent geological investigations revealed that past horizontal motion with large offsets took place along this fracture zone. The most obviously relevant feature is the offset, at the entrance to the Gulf of Eilat, of the Arabian coast relative to Sinai by about 100 km. Dubertret (1932), Quennell (1959) and Freund (1965) have suggested on the basis of stratigraphic and structural evidence that the same shift can be found further north on land. Furthermore, Freund et al. (1970) have traced this shift for stratigraphic units of various ages, showing the progressive increase of total displacement with time. Zak and Freund (1966) discovered very young stream offsets in the southern Arava near Eilat, which indicate horizontal strike-slip motion along the local segment of the Levant fracture zone. The estimates of the average rate of slip varies from 0.65 cm/year (Freund et al., 1970) to 1.0 cm/year for the past 3—4 million years (Girdler, 1958).

Ben-Menahem et al. (1976) have studied the seismograms of six earthquakes whose epicenters are located on the rift zone. The strongest of these events ($M \approx 6.5$) occurred on July 11, 1927, with epicenter at 32.0°N, 35.5°E (ISS 1927, p. 247). The average amount of slip was estimated as 40 ± 10 cm. Both macroscopic evidence of Brawer (1928) and directivity length yield a fault length of approximately 45 km. The shock was followed by a series of large aftershocks. Since 1927 the first earthquake that was properly studied occurred on Oct. 8, 1970, preceeded by a smaller foreshock, 10 hours earlier. Three more earthquakes, of Sep. 7 1971, Nov. 8, 1971, and Sep. 2, 1973 which were well-recorded at Eilat, show a radiation pattern of Love and Rayleigh waves and indicate strike-slip motion along a N—S trending fault.

All these shocks delineate a major fault which runs from the Damia bridge down to the middle of the Dead Sea, opposite En-Gedi, with an overall length of some 75 km. This fault, when slipping along its entire length, may produce an earthquake of magnitude $7-7\frac{1}{4}$. Fault-plane solutions indicate left-lateral strike-slip motion with strike of N8°—10°E.

Palmyra Chain and the Amanus—Taurus Mountains

The study of the Palmyra zone was done via two earthquakes recorded at Eilat during 1970—1974. The focal depth of 32—36 km was obtained from the inversion of the Love and Rayleigh spectra in agreement with the determinations of USCGS (33 km) and ISC (34 ± 6 km). Taking the direction of the strike to coincide with major fault traces (N50°—70°E), observed in ERTS photos, one finds strong left-lateral strike motion, accompanied with thrust faulting. The relatively shallow dip also indicates thrust-type motion for both events.

The tectonics of the Amanus—Taurus junction has been summarized by Nowroozi (1971, 1972). It is based mainly on the alignment of epicenters along the margins of the Zagros—Taurus thrust and a few fault-plane solutions (Canitez and Ücer, 1967; Shirokova, 1967), and long-period results at Eilat of the July 11, 1971 earthquake (Ben-Menahem et al., 1976). A sensible choice is the E—W strike solution parallel to the tectonic thrust line.

Off-shore faulting

Historical earthquakes off the Eastern Mediterranean coast are suggested from Sieberg's (1932) isoseismals of some of the major events in the Near East. More recently the Ksara, Lebanon, map of epicenters for 1961—1970 shows marine epicenters which are perhaps related to faults responsible for numerous tsunamis that wrecked the Levant coastal cities repeatedly.

Two recent events (M = 4.7—4.8 of March 26, 1968 and April 6, 1971), were studied by Ben-Menahem et al. (1976). The direction of the strikes of these are uncertain. Freund (1965) suggested from geological evidence that a series of right-lateral strike-slip faults, trending NE on land in Lebanon, are participating actively in the deformation of the crust. This pattern is consistent with one set of solutions (λ = 199°, δ = 76°, N144°E). The alternate set of solutions with N—NW strike directions yield normal faulting as well as left-lateral strike slip, indicating subsidence of the sea side — as implied by historical evidence. Both solutions have a dip-component of about 50%, which is highly tsunamigenic for magnitudes above 6.

The Sinai triple junction and its NW extension

While the motion of the African plate relative to the Arabian plate has been established, the role of the Sinai Peninsula in the regional tectonics is still unclear. Although historical earthquakes in the Suez region are known, only very few local events were adequately recorded in modern times outside Egypt. Fault-plane solutions are available for four earthquakes only: Two events with magnitudes above $4\frac{3}{4}$, were recorded at Eilat with sufficient power over a wide spectral band: One on March 31, 1969, with $M = 6\frac{1}{2}$—$6\frac{3}{4}$ at the junction of the Gulfs of Suez and Eilat (Ben-Menahem and Aboodi, 1971). The second earthquake of April 29, 1974, near Zagazig, Egypt was felt strongly in the Suez, Egypt and southern Israel and was recorded by LASA (Δ = 94.27°, Az = 332.5°). The other two were at: Sep. 12, 1955 and Jan. 30, 1951.

The choice of the correct fault-plane solution in this case is difficult. Nevertheless the fault-plane solutions obtained by Ben-Menahem et al. (1976) for the Zagazig 1974 and the Alexandria 1955 earthquakes, in combination with the established solution for the March 31, 1969 Red Sea earthquake, provide tight constraints not only on the motion across the Gulf of Suez, but also upon the relative motion between the Arabian, Sinai, and African plates. These earthquakes have one common set of fault-plane solutions with strikes N30°—40°W, steep dips to the NE, and a pronounced left-lateral strike-slip component of motion. Furthermore, the three events lie on a line which coincides with the general direction of their strikes, as shown in Fig. 2.

A possible interpretation of the results is that part of the motion appears as relative left-lateral strike-slip movement between Arabia and Sinai, and the rest is taken up by a combination of left-lateral slip combined with opening in the Gulf of Suez in qualitative agreement with the suggestion by Freund (1965) and Abdel Gawad (1969).

Consequently, it appears that an opening of the Gulf of Suez of about 1/10 of the Red Sea since Cretaceous time, combined with left-lateral strike slip of about $\frac{1}{4}$ the slip of the Arava fault, provides an explanation for the tectonic movements associated with the triple junction of the Arabian, African and the so-called Sinai plates. In particular, the model predicts a left-lateral slip rate of 1—2 mm/year along the Gulf of Suez, given the 6 mm/year along the Arava—Dead Sea fault system (Freund et al., 1970).

OVERALL HORIZONTAL DEFORMATION

From the calculated crustal structure and fault-plane solutions for the Afro-Eurasian junction, we obtain some estimates of the pattern of the tectonic motions which are currently taking place.

The overall pattern of seismic deformation associated with the Levant fracture zone, is shown in Fig. 2. With the selections described previously, we obtain a pattern in which the horizontal strike-slip component of motion is left-lateral everywhere — like the left-lateral motion along the Dead Sea fault itself. Aside from the events on the main transform fault which are purely strike slip, all other events have a dip-slip component. West of the Jordan—Dead Sea transform the motion invariably has a normal faulting component — observed for six events. On the east into the Arabia plate, the two Palmyra events have clear thrust components. This pattern, antisymmetric

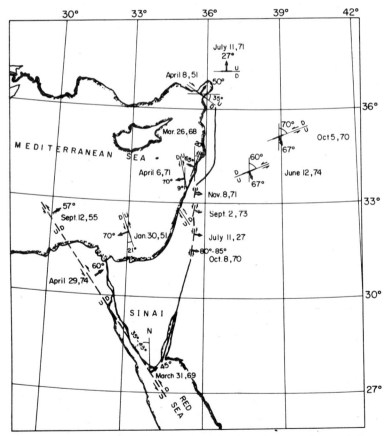

Fig. 2. Active fracture zones in the Afro-Eurasian junction and their corresponding slip vectors (after Ben-Menahem et al., 1976).

about the main fault, suggests a simple tectonic picture in which left-lateral motion with compression occurs on the faster moving Arabia plate along NE trending faults and left-lateral motion with tension takes place on NW trending branch faults on the slower moving Africa plate — including the Gulf of Suez and its extension. This picture is supported by available geologic evidence indicating normal faulting in the Gulf of Suez region. Furthermore, Freund et al. (1970) have shown left-lateral strike-slip motion on several faults in northern Israel as well as overall north—south elongation.

A rough model gives a simple interpretation of the observed motions. As the Dead Sea transform leaves the Red Sea, the relative slip along it diffuses sideways into the array of branching faults. These faults are thus responsible for the breakage of the plate regions adjacent to the transform fault. As we approach the collision zone — the Amanus—Taurus mountains — less and less slip between Arabia to the east and Africa to the west is accommodated by the main transform fault, and more slip takes place on the branch faults. Figure 3 shows, in a schematic way, the nature of the motion associated with breakup of the plate boundaries. Point P, which was situated at the intersection of a NE branch fault with the main transform, has moved with time to Q with a component of motion parallel to the main fault, plus a component along the branch fault, which has itself moved northward as part of the eastern Arabia plate. A point P' on the western plate has moved only to Q', as the Africa plate moves slower than the Arabia plate.

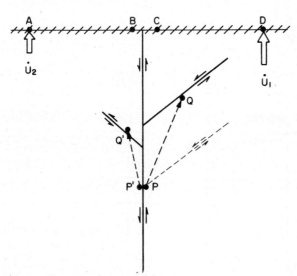

Fig. 3. A schematic model for the diffusion of slip from the main transform fault to branching faults. A particle at point P, at the intersection of the branch fault (broken line) with the main fault moves with time to point Q, while the fault itself, being attached to the plate also moves toward the consuming boundary. Points P' and Q' represent the analogous motion on the slower moving side of the transform fault.

This interpretation predicts that the slip rate, or total slip along the main transform, decreases from south to north. As shown earlier, the Gulf of Suez motion approximately explains the discrepancy between the slip rate (at the Red Sea) of $\Delta \dot{u}_0 \cong 1.1$ cm/year between Arabia and Africa, and the estimated 0.65 cm/year along the Arava fault. Assuming that the faulting in northern Israel, Lebanon and Palmyra has accommodated a total of approximately 40 km (Freund, 1965), we find that slip rate on the Bikaa fault should be only about 0.45 cm/year. This is qualitatively consistent with observed geologic offsets which also show smaller total displacement in the northern part of the Levant fracture zone.

Another major effect of this kind of breakup is to smear out the jump in collision rate between the faster Arabia plate and the slower Africa plate. Thus the collision rate at the Amanus—Taurus Mountains is probably 3.3— 3.7 cm/year, significantly less than the 4 cm/year for the Arabia—Zagros collision, and more than the 2.9 cm/year of the Africa—Europe collision (Le Pichon et al., 1973).

In summary, the seismic and geologic evidence associated with the junction zone shows that this is the region in which the edges of the coherent Arabia and Africa plates break up in the neighborhood of their boundary as they approach the region of continental collision with the Eurasian plate. This breakup consists of gradual loss of coherency of the plate, and deterioration of its rigidity — as more and more deformation is taken up by branching faults. At the collision zone itself, the plates loose their rigidity entirely. It is noteworthy that the region of breakup extends away from the collision zone by a distance comparable to the length of the irregularities in the shape of the collision-induced mountain belt of a few hundred kilometers.

TECTONICS OF THE EASTERN MEDITERRANEAN

The scheme suggested in the previous section is clearly much simpler than the actual tectonics of the Eastern Mediterranean. In particular, we notice a great variety of features at the Africa—Europe boundary region in the Eastern Mediterranean, with a mix of oceanic and continental crusts, deep and shallow seismicity, and variable sea-floor depth. Thus the large-scale segmentation of the entire Alpine belt must be further divided here. We now concentrate our attention on the Eastern Mediterranean and exclude the western part of the sea, which presumably consists mostly of young oceanic crust.

In the complex region extending from Italy to the Dead Sea transform we can recognize subtle but clear systematic changes in several geophysical phenomena. We summarize these changes from west to east:

Bathymetry

The bathymetry of the Ionian Sea, at the western end is the deepest in the entire region with typical depth of 3—4 km. The Ionian Sea is bordered to the west by Italy, Sicily, and the shallow Sicilian—African platform. To the east, the sea south of Crete is measurably shallower with depth of 2—3 km. Further east the Levantine basin south of Cyprus is even shallower, 2—2.5 km deep. The easternmost portion of the region consists of the shallow floor of the Levantine continental margin and the Levant itself which is mostly above sea level, reaching to several hundred meters.

Crustal structure

The crustal structure of the Eastern Mediterranean changes gradually from oceanic in the west to continental in the east: The Ionian Sea is underlain by a normal oceanic crust. Refraction data from the Ionian Sea indicate only 17 km to the mantle (Weigel and Hinz, 1971). Taking into account the large amount of sediments there — almost 10 km (Finetti and Morelli, 1973) and the water depth, the crust is indeed truly oceanic. In the sea south of Crete, in comparison, the crust is of a more continental character, with depth to the mantle of about 21 km (Papazachos, 1969), whereas in the Levantine Sea the depth to the mantle is even greater, some 25—27 km (Lort et al., 1974; Finetti and Morelli, 1973). The easternmost portion, the Levantine continental margin and the Levant itself is characterized by the thickest crust. Here the depth to the mantle is 35 km along the Dead Sea transform (Ben-Menahem et al., 1976).

Seismicity

The seismic activity also shows characteristic changes from west to east. In the west, the oceanic portion of the plate is clearly underthrusting below Italy, with steep dip (Papazachos, 1973). This area is characterized by large and deep earthquakes. Further east around Crete there are no deep events and only shallow and intermediate earthquakes occur. The seismicity indicates nevertheless a well-defined zone of underthrusting. In the Levantine Sea the seismic activity is mostly shallow with a few intermediate earthquakes. There is no clear underthrusting zone and some of the deformation along the Cyprian Arc consists of strike-slip faulting. In the Levantine continental margin and the Levant the entire seismic activity is shallow.

Magnetics

The magnetic field in the Ionian Sea is composed of anomalies with large amplitudes resembling oceanic anomalies with northeast—southwest trend (U.S. Naval Oceanographic Office, 1967). To the east in the sea south of Crete and in the Levantine Sea the magnetic field is remarkably smooth.

On the basis of these systematic variations in the region surrounding the Eastern Mediterranean, we can divide the region roughly into four parts: The Ionian Sea, sea south of Crete, the Levantine Sea and the Levant itself.

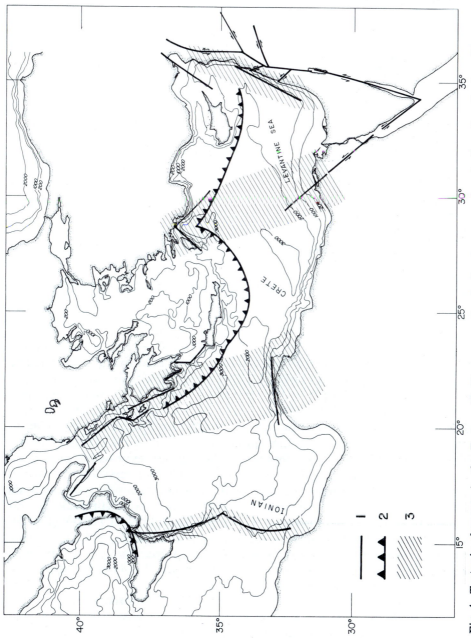

Fig. 4. Tectonic elements in the Eastern Mediterranean: *1* = faults; *2* = zone of underthrusting; *3* = transition zone.

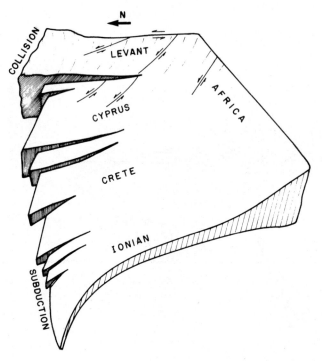

Fig. 5. Block diagram showing the disruption of the African plate.

The Ionian part is more or less a typical region of oceanic subduction. On the other end, the Levant part is a typical zone of continental collision. The part around Crete and the Levantine Sea are intermediate cases. As shown in Fig. 4 the parts are separated by transition zones of diffuse deformation, with many intra-plate faults. These faults accommodate the relative motion between the parts as they approach their respective consuming boundaries. One clear example is the Levantine continental slope (Ben-Avraham, 1978). It is likely that this feature is the surface expression of a fault whose vertical offset intensifies to the north. This fault may be the expression of the two types of interaction at the northern boundary of the Sinai plate, with the Cyprian arc underthrusting with some strike-slip faulting, and the purely continental collision of the Taurus Mountains in Turkey (Fig. 5). As a result of these interactions the lithosphere is tearing and the oceanic side sinks relatively to the continental side as it approaches Cyprus. Although inconclusive, observed gravity anomalies (Woodside, 1974) are roughly consistent with the proposed deformational model.

For lack of data it is more difficult to detail the deformation in the other two transition zones. It is nevertheless possible to delineate broad zones of transition in water depth and sea-floor morphology as shown in Fig. 4.

CONCLUSIONS

We have discussed some patterns and trends within and next to the Alpine mountain belt which may shed light on the processes of continental colli- sion. By subdividing the Alpine belt into the Europe, Persia and Tibet seg- ments, we recognize some repeated features: the linear relation between slip rate and width of the Persia and Tibet segments, and their plateau-like mor- phology. The proposed explanation for these features is purely geometrical — depending only on mass inflow. We emphasize the relation to nearby plate boundaries. One of these boundaries in particular, the Levant fracture zone, reveals some details of the breakup process of rigid plates: As continental plates approach the collision zone they tend to develop a branching system of faults, which permit the transfer of uniform shearing motion into a dis- tributed deformation. As we approach the collision zone more and more horizontal deformation is taken up by the branching faults, and the main transform becomes less important.

We find, however, that the mode of plate consumption is very sensitive, even on local scale, to the nature of colliding crusts. Where the crust is oceanic, as in the Ionian Sea, consumption is locally by subduction. As the crust of the African plate thickens eastward, the consumption becomes more and more collision-like, with the corresponding disappearance of deep earth- quakes, and the appearance of elevated regions such as in Turkey and Iran.

In contrast, the nature of the crust does not seem to influence the local velocity of consumption. Roughly speaking at least, these velocities are uniform from the Ionian Sea to the Levant fracture zone, although as pointed out already, the crust thickens significantly. This suggests again that velocities of consumption do not play a major role in the mechanical pro- cesses involved, whereas density and geometry do.

The geometry of the consumption region in the Eastern Mediterranean is

TABLE I

Summary of geophysical properties of parts of the Eastern Mediterranean

	Ionian Sea	South of Crete	Levantine Sea	Levant
Water depth	3—4 km	2—3 km	2—2.5 km	—
Seismicity	deep and large events	shallow and intermediate events	shallow and a few inter- mediate events	shallow and low-magnitude events
Crustal thickness: depth to mantle	17 km	21 km	25—27 km	≥35 km
Magnetic field	short wave- length anom- alies	smooth	smooth	—

very complex, but we discern four zones, based on its morphology and geophysical properties (see Table I): The Ionian Sea, sea south of Crete, the Levantine region and the Levant. The zones have diffuse boundaries which consist probably of numerous faults with sizeable vertical offsets. The deformation in these boundary regions is too diffuse to permit the definition of a 'micro plate'. We suggest therefore that in these regions plates break up, and the twin concepts of plates and rigidity cease to apply, or serve a useful purpose.

The differentiation into regions and transition zones near the consumption boundary leads probably to complex geology within the collision zone itself. Here we anticipate superposition of oceanic and continental crusts, faulted blocks and fault systems inherited from the nearby breaking plates (Fig. 5). It is difficult to imagine that plates exist here, particularly when the colliding boundaries have irregular shapes.

In summary, it appears to us from simple evidence in the relatively well-known Eastern Mediterranean and Near East, that continental collision can be characterized as a process in which plates break up and deform extensively, some time before and some distance away from the collision region itself.

ACKNOWLEDGEMENT

This work was supported by grants from NSF's Earth Science Division and from the Arthur L. Day fund, the U.S. National Academy of Science.

REFERENCES

Abdel-Gawad, M., 1969. New evidence of transcurrent movement in Red Sea area and petroleum implications. Am. Assoc. Pet. Geol., Bull., 55: 1466—1479.

Ben-Avraham, Z., 1978. The structure and tectonic setting of the Levant continental margin — Eastern Mediterranean. Tectonophysics, 44: 313—331.

Ben-Avraham, Z. and Nur, A., 1976. Slip rates and morphology of continental collision belts. Geology, 4: 661—664.

Ben-Menahem, A. and Aboodi, E., 1971. Tectonic patterns in the northern Red Sea region. J. Geophys. Res., 76: 2674—2689.

Ben-Menahem, A., Nur, A. and Vered, M., 1976. Tectonics, seismicity and structure of the Afro-Eurasian junction — the breaking of an incoherent plate. Phys. Earth Planet. Inter., 12: 1—50.

Brawer, A.J., 1928. Earthquakes in Palestine from July 1927, to August 1928. Jew. Palest. Explor. Soc., 316—325 (in Hebrew).

Canitez, N. and Ücer, S.B., 1967. Computer determinations for the fault-plane solutions in and near Anatolia. Tectonophysics, 4: 235—244.

Chang, Cheng-fa and Cheng, Hsi-Ian, 1973. Some tectonic features of the Mt. Jolmo Lungma area, southern Tibet, China. Sci. Sinica, 16: 257—265.

Dewey, J.F. and Burke, K.C.A., 1973. Tibetan, Variscan, and Precambrian basement reactivation: Products of continental collision. J. Geology, 81: 683—692.

Dewey, J.F., Pitman, W.C., III, Ryan, W.B.F. and Bonnin, J., 1973. Plate tectonics and the evolution of the Alpine system. Geol. Soc. Am. Bull., 84: 3137—3180.

Dubertret, L., 1932. Les formes structurales de la Syrie et de la Palestine, leur origine. C.R. Acad. Sci. Paris, 195: 66—68.

Finetti, I. and Morelli, C., 1973. Geophysical exploration of the Mediterranean Sea. Boll. Geofis. Teor. Appl., 15: 263—341.

Freund, R., 1965. A model of the structural development of Israel and adjacent areas since upper Cretaceous times. Geol. Mag., 102: 189—205.

Freund, R., Garfunkel, Z., Zak, I., Goldberg, M., Weissbrod, T. and Derin, B., 1970. The shear along the Dead Sea rift. Philos. Trans. R. Soc. London, Ser. A, 267: 107—130.

Freund, R., Goldberg, M., Weissbrod, T., Druckman, Y. and Derin, B., 1975. The Triassic—Jurassic structure of Israel and its relation to the origin of the eastern Mediterranean. Geol. Surv. Isr. Bull., 65.

Girdler, R.W., 1958. The relationship of the Red Sea to the East African rift system. Q.J. Geol. Soc. London, 114: 79—115.

ISS, 1927. International Seismological Summary for 1927. Kew Observatory, Richmond, Surrey.

Le Pichon, X., Francheteau, J. and Bonnin, J., 1973. Plate Tectonics. Elsevier, Amsterdam, 300 pp.

Lort, J.M., Limond, W.Q. and Gray, F., 1974. Preliminary seismic studies in the eastern Mediterranean. Earth Planet. Sci. Lett., 21: 355—366.

McKenzie, D.P., 1972. Active tectonics of the Mediterranean region. Geophys. J.R. Astron. Soc., 30: 109—185.

McKenzie, D.P., Davies, D. and Molnar, P., 1970. Plate tectonics of the Read Sea and east Africa. Nature, 226: 243—248.

Neev, D., 1975. Tectonic evolution of the Middle East and the Levantine basin. Geology, 3: 683—687.

Nowroozi, A.A., 1971. Seismo-tectonics of the Persian Plateau, eastern Turkey, Caucasus, and Hindu-Kush regions. Bull. Seismol. Soc. Am., 61: 317—341.

Nowroozi, A.A., 1972. Focal mechanism of earthquakes in Persia, Turkey, West Pakistan, and Afghanistan and plate tectonics of the Middle East. Bull. Seismol. Soc. Am., 62: 823—850.

Papazachos, B.C., 1969. Phase velocities of Rayleigh waves in southeastern Europe and eastern Mediterranean Sea. Rev. Pure Appl. Geophys., 75: 47—55.

Papazachos, B.C., 1973. Distribution of seismic foci in the Mediterranean and surrounding area and its tectonic implications. Geophys. J.R. Astron. Soc., 33: 421—430.

Powell, C., McA., and Conaghan, P.J., 1975. Tectonic models of the Tibetan plateau. Geology, 3: 727—731.

Quennell, A.M., 1959. Tectonics of the Dead Sea rift. Congr. Geol. Int. Mexico, 1956. Session 20, Assoc. Serv. Geol. Afr., pp. 385—405.

Shirokova, F.I., 1967. General features in the orientation of principal stresses in earthquakes foci in the Mediterranean—Asian seismic belt. U.S.S.R. Earth Phys. Bull. (Phys. Bull. Acad. Ser.), 1: 22—36.

Sieberg, A., 1932. Untersuchungen über Erdbeben und Bruchschollenbau im östlichen Mittelmeergebiet. Gustav Fischer, Jena, 113 pp.

U.S. Naval Oceanographic Office, 1967. Total and residual magnetic contour of the Mediterranean Sea 1967-8, Washington, D.C.

Weigel, W. and Hinz, K., 1971. Seismic measurements in the Ionian Sea and on the Malta shelf. Comm. Seismol. Europe, 15: 5.

Woodside, J.M., 1974. Gravity anomalies in the eastern Mediterranean and their implications. XXIV Congress-Assemblée Plénière de la C.I.E.S.M. (abstract).

Zak, I. and Freund, R., 1966. Recent strike-slip movements along the Dead Sea rift. Isr. J. Earth Sci., 15: 13—37.

Chapter 27

FINITE ELEMENT MODELING OF LITHOSPHERE DEFORMATION: THE ZAGROS COLLISION OROGENY*

PETER BIRD

Department of Earth and Space Sciences, University of California, Los Angeles, Calif. (U.S.A.)

SUMMARY

In the Zagros Mountains a formerly stable continental margin is being suddenly deformed. Finite element models of this tectonic flow are used to relate known surface deformations, earthquake locations, and fault plane solutions to unknown rock flow parameters and driving forces, in order to determine the latter. These two-dimensional plane-strain flow models incorporate the effects of nonlinear dislocation creep, frictional faulting, geological inhomogeneity, density anomalies, and varying temperature. All models driven by a subducting slab are unsuccessful because they require subduction of continental crust, which does not match present seismicity. Therefore the former oceanic slab must be detached, and the orogeny must be driven by horizontal compression in the lithosphere. Models also show that the subcrustal lithosphere is not shortened but acts as a stabilizing foundation. These results imply a simple geometry of crustal deformation which can be analytically modeled. The creep strength of the lower crust (75—100 bars) determines the topographic slope of the Zagros. The fact that subduction is not occurring on the old plate boundary places a limit on the shear stress deforming the cold upper crust. This limit is 300 bars if there is limestone at depth in the Crush Zone; otherwise 800 bars. These results are confirmed by a final finite element model. The total driving force of the orogeny associated with these limits is $2.8—5.5 \cdot 10^{15}$ dyne/cm, and the smaller amount could be provided by the gravitational spreading of the Red Sea rift. Shear—strain heating caused by the orogeny to date is less than $20°C$. These results imply that even unheated continental crust is considerably weaker than laboratory friction measurements imply, and that it is mechanically decoupled at the Moho from the stronger mantle lithosphere.

INTRODUCTION

The Zagros Mountains are forming today on what was until Plio-Pleistocene times the continental shelf margin of the Arabian shield. Although this shield has been converging on the Eurasian plate since Jurassic time with only a Paleocene interlude (Dewey et al., 1973), the recency of this conti-

---

*Originally published as: Bird, P., 1978. Finite element modeling of lithosphere deformation: the Zagros collision orogeny. In: M.N. Toksöz (Editor), Numerical Modeling in Geodynamics. Tectonophysics, 50: 307—336.

nental collision is shown by the youth of the molasse, the young marine sediments deformed in the Crush Zone (Wells, 1969), the line of extinct volcanoes northeast of and parallel to the ophiolite belt, and the lack of alternative sites for Tertiary plate convergence between Africa and Eurasia. Between the Late pre-Cambrian and the Pliocene this Atlantic-type continental margin facing the former Tethys was affected by only one known tectonic event: a Campanian collision with an offshore arc that forced the subduction of the Tethys to jump northeast into Iran. This event emplaced ophiolites onto the outer edge of the shelf in Oman and in the Neyriz and Kermanshah regions of the Zagros, leading some authors (e.g., Haynes and McQuillan, 1974) to conclude that the continental collision came in the Cretaceous.

This geologically simple history is apparently accompanied by near-normal, steady-state temperatures in the crust. Coster (1947) reported a heat flow of 37 mW/m^2 from the center of the Zagros, and relative heat flows estimated by Bird et al. (1975) range only from 37 to 48 mW/m^2. Toksöz and Bird (1977) have shown that these fluxes can be explained by a simple model of a shield with 50 mW/m^2 initial heat flow that is tectonically thickened, reducing the gradient. Furthermore, there are no known Tertiary igneous rocks southwest of the Crush Zone, and seismic velocities and attenuation (discussed below) have normal values for a stable platform area.

All of this means that in the Zagros we have an unparalleled opportunity to observe rapid deformation of a simple, cool piece of continental lithosphere. By constructing finite element models of this deformation process we can establish a link between the unknown flow laws and driving forces of the lithosphere, and such observables as surface strains, earthquake locations, and fault plane solutions. Comparing model predictions with actual seismicity, we distinguish a few acceptable models from among many that fail. Also, we can develop insights into the governing mechanisms of the orogeny which enable the major conclusions to be justified without the use of models.

FINITE ELEMENT METHOD FOR NONLINEAR CREEP OF ROCKS

The objective is to solve for the present velocity vectors of the rocks in the lithosphere, given their geometry, densities, flow laws and boundary conditions. Differentiation of these velocities yields strain rates and stresses. To make this solution feasible, three approximations are required:

(1) The flow vectors lie in a vertical plane perpendicular to the strike of the range. This plane-strain approximation, suggested by the simple folding in the Zagros, makes possible a two-dimensional treatment, with coordinates x (horizontal and perpendicular to strike) and z (downwards).

(2) Elastic strain rates are negligible in comparison to anelastic deformations. This assumption allows the use of the Navier—Stokes equation instead of a more difficult third-order nonlinear differential equation. Its corollary

(since the anelastic strain rates from creep and faulting are non-dilational) is the incompressibility condition:

$$\frac{dV_x}{dx} + \frac{dV_z}{dz} = 0 \tag{1}$$

where V_x and V_z are the two components of the velocity vector V.

(3) The rock is isotropic and has a flow law with shear stress given as a (non-decreasing) function of strain rate:

$$\tau^* = [\tfrac{1}{4}(\sigma_{xx} - \sigma_{zz})^2 + \sigma_{xz}^2]^{1/2} = f(\dot{e}^*) = f([\tfrac{1}{4}(\dot{e}_{xx} - \dot{e}_{zz})^2 + \dot{e}_{xz}^2]^{1/2}) \tag{2}$$

where σ_{ij} are the components of the stress tensor and $\dot{e}_{ij}$ are components of the rate-of-strain tensor. This assumption implies coincidence of principal stress and strain rate axes, and that the flow law at each point can be represented by a single parameter, the effective viscosity η:

$$\eta(\dot{e}^*) = \frac{\tau^*}{2\dot{e}^*} = \frac{f(\dot{e}^*)}{2\dot{e}^*} \tag{3}$$

Of course, the correct value of $\dot{e}^*$, and thus of η, is not known at the beginning of the calculation. This means that an iterative approach is required.

Within each step of the iteration, we solve for the velocities from the Navier—Stokes equation, neglecting accelerations:

$$2\frac{\partial}{\partial x}\left(\eta \frac{\partial V_x}{\partial x}\right) + \frac{\partial}{\partial z}\left[\eta\left(\frac{\partial V_x}{\partial z} + \frac{\partial V_z}{\partial x}\right)\right] - \frac{\partial P}{\partial x} = 0 \tag{4a}$$

$$2\frac{\partial}{\partial z}\left(\eta \frac{\partial V_z}{\partial z}\right) + \frac{\partial}{\partial x}\left[\eta\left(\frac{\partial V_x}{\partial z} + \frac{\partial V_z}{\partial x}\right)\right] - \frac{\partial P}{\partial z} + g\rho = 0 \tag{4b}$$

where P is pressure, g is gravitational acceleration, and ρ is density. The incompressibility condition is incorporated by expressing the velocity components as space derivatives of a scalar stream function Φ:

$$V_x = \frac{\partial \Phi}{\partial z} \tag{5a}$$

$$V_z = -\frac{\partial \Phi}{\partial x} \tag{5b}$$

Then eq. 4 is transformed into an equivalent weak form by multiplying (4a) by an arbitrary velocity perturbation δV_x, and (4b) by δV_z, adding them together, and integrating over the domain. Using integration by parts and bringing in the requirement of stress continuity with applied force densities on the boundaries (h_i), the Navier—Stokes equation is transformed to a single integral which does not involve pressure:

$$
\iint_A \left[\rho g \frac{\partial}{\partial x} (\delta\Phi) + 4\eta \frac{\partial^2\Phi}{\partial x \partial z} \frac{\partial^2}{\partial x \partial z} (\delta\Phi) + \eta \left(\frac{\partial^2\Phi}{\partial z^2} - \frac{\partial^2\Phi}{\partial x^2} \right) \left(\frac{\partial^2}{\partial z^2} (\delta\Phi) \right. \right.
$$

$$
\left. \left. - \frac{\partial^2}{\partial x^2} (\delta\Phi) \right) \right] dx\,dz - \oint_s \left[h_x \frac{\partial}{\partial z} (\delta\Phi) - h_z \frac{\partial}{\partial x} (\delta\Phi) \right] ds = 0 \tag{6}
$$

This expression is equivalent to eqs. 1–4 if it is required to hold for all possible doubly-differentiable perturbations of the stream function $(\delta\Phi)$. The derivation may be found in Bird (1960).

The essential approximation of the finite element method is to replace the infinite-dimensional space of possible stream functions by an N-dimensional subspace spanned by orthogonal component stream functions Φ_i:

$$
\Phi(x, z) = \sum_{i=1}^{N} C_i \Phi_i(x, z) \tag{7}
$$

Then the velocity solution is completely described by the constants C_i, and the space of possible $(\delta\Phi)$ is spanned by the N functions Φ_i. In order to satisfy eq. 6 for the whole model subspace, we require it to hold for each basis function separately. This gives a solvable matrix equation:

$$
[K_{ij}] \vec{C}_i = \vec{F}_j \tag{8}
$$

if we define:

$$
K_{ij} = \iint_A \left[4\eta \left(\frac{\partial^2\Phi_i}{\partial x \partial z} \right) \left(\frac{\partial^2\Phi_j}{\partial x \partial z} \right) + \eta \left(\frac{\partial^2\Phi_i}{\partial z^2} - \frac{\partial^2\Phi_i}{\partial x^2} \right) \left(\frac{\partial^2\Phi_j}{\partial z^2} - \frac{\partial^2\Phi_j}{\partial x^2} \right) \right] dx\,dz \tag{9}
$$

and:

$$
F_i = \iint_A \left[\rho g \frac{\partial\Phi_i}{\partial x} \right] dx\,dz + \oint_s \left[h_x \frac{\partial\Phi_i}{\partial z} - h_z \frac{\partial\Phi_i}{\partial x} \right] ds \tag{10}
$$

In this work the Φ_i employed are cubic functions of the space variables within a polygon surrounding a node and defined by other nodes, and zero outside. This type of element is described by Bazeley et al. (1965) in a different application. It allows for a quadratic variation of velocity and a linear variation of strains within each triangular element. Since there are three unknowns (C_i) and three associated Φ_i for each node, rigid-body translations and rotations and uniform strains are exactly represented by the grid. For more complex strains, normal-slope discontinuities in Φ appear on element boundaries between nodes, but test problems have shown that velocities and stresses are still accurate to $\pm 5\%$ (Bird, 1976).

The area integrals in eqs. 9 and 10 are performed numerically using seven points per element. With the proper locations and weights, this numerical integration is exact (Zienkiewicz, 1971) for quintic integrands. This means η and ρ can vary as cubic polynomials across each element as a function of temperature in order to more accurately represent the actual geologic structure. While it has been conventional in earlier work to use constant parameters within each element, nothing in eqs. 6 or 9 requires constant or continuous properties.

The iteration by which one settles on the correct effective viscosity η from an arbitrary flow law has been attempted in two ways, both described by Zienkiewicz (1971). The "initial stress" method was used on some early models, but did not live up to its reputation for efficiency. It is also less accurate since it involves summing solutions. Therefore, most of the models presented were calculated with the "variable viscosity" method. In this technique, each iteration inherits only an improved effective viscosity from the last solution:

$$\eta_{i+1}(x, z) = \frac{\tau_i^*(x, z)}{2\dot{c}_i^*(x, z)} \tag{11}$$

Seven viscosities converge separately at the integration points of each element.

It must be emphasized that this process converges on the flow law, not the Navier—Stokes equation. Thus, any of the iterations gives a possible velocity field consistent with stress equilibrium. Because the flow laws are not very accurately known, it is a waste of computing time to force complete convergence. The RMS departure of the model stress from the assumed flow laws (given in Table IV) is usually less than 6% of ambient stress and does not affect the character of the solution, or the element-averaged stresses shown.

The purely numerical error in the solution of eq. 8 can be tested by back-substitution. It is controlled by the stiffness of the equations, which depends on the viscosity of the hardest rock in the model. In order to keep these errors below 5 bars (0.5%), an upper vscosity limit of $1.3 \cdot 10^{24}$ poise was imposed on eq. 11. The deformation rate of rocks this stiff is of the same order as that due to elastic effects which have been neglected.

GEOMETRY OF THE ZAGROS MODELS

A cross-section of the gross geology of the Zagros range is shown in Fig. 1, along with the grid of elements used to represent it. The section extends from the edges of the Persian Gulf in the southwest to just before the extinct volcanic line to the northeast. Obviously the scale is such that individual mountains, anticlines, etc. cannot be included. This section details the various geophysical and geologic evidence which constrain the geometry of the model.

The Main Zagros Thrust forms the boundary between the Arabian and Eurasian plates and the northeast limit of the Zagros. The dip of this fault at

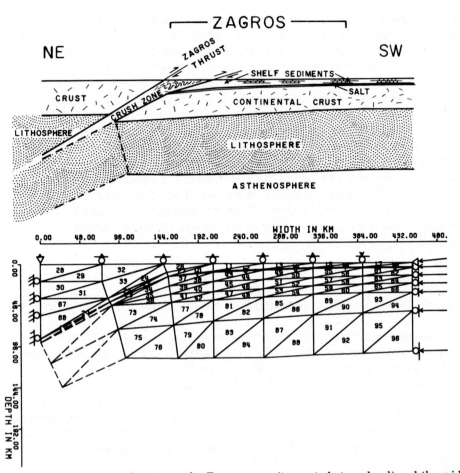

Fig. 1. A transverse section across the Zagros range (truncated at sea level) and the grid of finite elements used to represent it. The portion of the Arabian lithosphere outlined with dashes was deleted from later models. Engineering symbols show boundary conditions for the successful models.

depth is not known, so the general principle that volcanoes lie above the 100 km contour of subduction zones (Isacks and Barazangi, 1977) has been invoked, implying a dip of about 30°. The volcanic line has been left just outside the model domain, since its structure and temperature are unknown.

Intact shield-type lithosphere has been drawn below the crust of the Zagros and (possibly) connecting with subducted Tethyan oceanic lithosphere to the northeast. The lithosphere thickness is based on an analysis of surface-wave dispersion in the Arabian shield by Knopoff and Fouda (1975). The integrity and normal temperature of this lithosphere are established by the author's observation of clear S_n phases on the short-period seismograms at Shiraz, from earthquakes in southeastern Turkey which send waves along the strike of the range. A high velocity (4.65 km/sec) and high frequency, like that evident on the July 1, 1971 record from an event at 36.4°N and 43.4°E, have been found by Molnar and Oliver (1969) to correlate with low temperatures. There is no evidence for the persistence of an oceanic slab attached to the Arabian lithosphere, so it has been included in some models and deleted in others.

The thickness of sedimentary rocks in the Zagros is not well known; estimates range from 7 km (Lees, 1952) to 12 km (Falcon, 1969). A crucial question is whether the shortening and thickening of these sediments is sufficient to account for the topography on the top of the crust and the roots at its base. Three structure sections by Lees reported in Abdalian (1963) have been used to estimate a shortening of 21% for the southwestern half of the range, which appears to be too little by a factor of four. An important feature of the sedimentary sequence is the layer of Cambrian evaporites near the base, which may cause decoupling of sediments and basement. This evaporite is included in the model with a very thin 100 meter layer of elements, not visible in Fig. 1.

Because of this decoupling, it is impossible to determine from surface geology whether the basement is also shortened (Falcon, 1969). Because this question is basic to understanding the tectonics and selecting successful finite-element models, the available geophysical evidence is reviewed below. It suggests that there has been tectonic shortening of Precambrian basement by about the same amount as in the sediments.

Bouguer gravity anomalies are the strongest tool that we have. Since the predominantly limestone sediments are likely to have densities of 2.5–2.8 g/cm^3 (Berman et al., 1942), comparable to the basement, and since seismic evidence favors a normal mantle, this anomaly probably results from downwarping of the Moho. A two-dimensional gravity model which fits the available data is shown in Fig. 2. It has an increasing amount of crustal thickening, up to 17 km, going across the Zagros towards the suture. This amount of thickening would not be possible if the basement were not involved.

Second, there is the evidence of heat flow, which is depressed within the range below normal values. Toksöz and Bird (1977) showed this decrease to be consistent with shortening of the whole crust by up to 34% near the suture; but if the thickening was entirely in the sediments much greater strains would be necessary to explain the gravity, and a much greater depression of heat flow should result.

Third, the inferred crustal thickening is supported by teleseismic and local P-wave residuals at Shiraz. Akasheh (1975) found an average teleseismic delay of +0.5 sec, which would imply a thickening of about 14 km if crust and mantle velocities were normal. A thickness of 44 ± 5 km based on local travel times was reported by Eslami (1974). If the Zagros were underlain by thin or oceanic crust, as Haynes and McQuillan (1974) suggest, the travel-time residuals should be negative and the gravity anomaly positive.

This model, of an otherwise normal lithosphere with the crust tectonically shortened, was tested using group velocities of crustal Rayleigh waves.

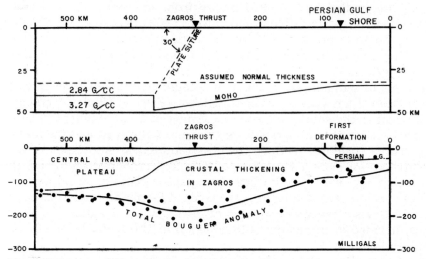

Fig. 2. Bouguer gravity anomaly data from the Zagros and an interpretation in terms of crustal thickness. Dots in lower half are 1° by 1° gravity means projected along strike onto the cross-section. Model at top assumes density contrast of 0.43 g/cm³ from Woolard (1959). Division of model curve shows the effects of different areas of the model.

Events were selected at short distances (Table I) giving propagation paths entirely within the Zagros to Shiraz. As Fig. 3 shows, there is excellent agreement between the observed and predicted group velocities. Of course the velocity model used (Table II) is non-unique, but it would be difficult to construct another with a thinner crust that still employed reasonable velocities. If the average crustal thickness was reduced from 46 km, a shear velocity lower than 3.67 km/sec would be needed in the lower crust. Likewise, if the sediments were made thicker, their shear velocity would have to exceed 3.0 km/sec, which is already high. The shear velocity in the upper mantle

TABLE I

Events used to determine Zagros Rayleigh wave group velocities

Date	Time (GMT)	Lat. (N)	Long. (E)	Magn.	Δ to Shiraz
9—04—64	3:39:36.7	39.8	40.3	5.0	14.4°
8—31—65	7:29:45.8	39.3	40.8	5.1	13.7°
9—13—66	20:23:50.4	39.1	40.7	4.5	13.7°
10—20—67	6:47:38.0	37.9	37.7	4.8	14.9°
10—30—68	16:51:33.5	37.9	38.6	4.9	14.3°
7—02—70	2:24:35.7	38.8	36.7	4.8	16.1°
9—03—70	5:32:09.7	39.6	38.7	5.1	15.2°

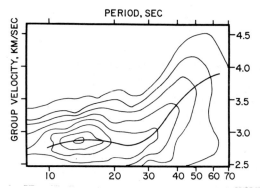

Fig. 3. Composite group velocities of seven Rayleigh waves propagating in the Zagros (Table I). Contours at 5 db intervals map the summed spectral amplitudes obtained by the method of Dziewonski et al. (1969). Model group velocity curve is for the structure in Table II.

was fixed at the velocity determined for S_n waves.

A final indication of basement deformation is the location of thrusting earthquakes in the basement and/or mantle. Even if we allow the usual error of ±50 km in the earthquake locations of Nowroozi (1971), many of the events in Fig. 4 are too deep to be in the sedimentary layer. The distribution of these events (apparently random, but confined to the Zagros folded belt) and their focal mechanisms (Fig. 4) are both consistent with distributed shortening of the basement. To confirm that at least some of these earthquakes are deep, the thrust event of June 23, 1968 was studied by the Rayleigh-wave-spectrum technique of Tsai and Aki (1970). Using the source mechanism of Nowroozi (1972) and the velocity structure of Table II, vertical-component spectra were predicted for various source depths at the stations AAE and JER. A comparison in Fig. 5 of the actual spectrum (corrected for instrument response) shows that the event is clearly deep (30—70 km) and not in the shallow sedimentary layer.

TABLE II

Seismic velocity model for Rayleigh wave group dispersion in the Zagros Mountains

Depth (km)	V_p (km/sec)	V_s	Density (g/cm^3)
0—9.0	5.20	3.00	2.60
9.0—27.3	6.20	3.58	2.70
27.3—45.6	6.40	3.67	2.87
45.6—125.0	8.17	4.65	3.40
125.0—325.0	8.17	4.45	3.40
325.0—∞	8.80	4.60	3.65

518

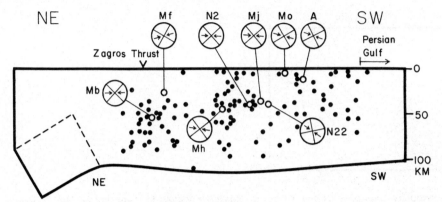

Fig. 4. Earthquake hypocenters in the Zagros from Nowroozi (1971) projected onto a single cross-section and superimposed on the outline of the finite element grid from Fig. 1. Focal mechanisms are from Akasheh (1973) {A}, McKenzie (1972) {M}, and Nowroozi (1972) {N}. Dashed portion of grid was dropped in later models.

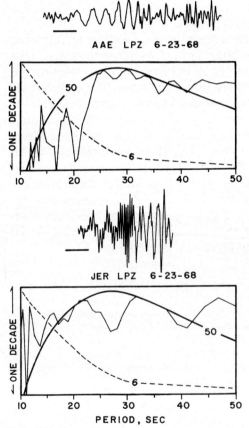

Fig. 5. Rayleigh waves from a Zagros thrusting earthquake at two stations and their amplitude spectra (corrected for instrument response). Solid bars indicate length of one minute. Model curves for this source at depths of 6 km and 50 km are compared to actual spectra. Note that there is probably multipath interference in top record and higher-mode interference in the lower.

INITIAL FLOW LAW ESTIMATES

The flow laws of the various rock types are considered variables in the modeling process, as well as the boundary conditions. However, it is not possible to try all possible combinations; so it is important to identify laboratory flow laws which may be relevant for initial estimates. These laws give upper limits on the strength of the rocks, since the addition of unknown deformation mechanism can only increase strain-rate and lower effective viscosity in eq. 3.

The mechanisms which are likely to be important are frictional sliding on pre-existing faults and dislocation creep (with recovery) in grains. Elastic compliance and cold-working (creep without recovery) may have contributed small initial strains early in the orogeny, but would not produce a continuing strain-rate relevant to a flow model. Initiation of new faults probably occurred in the recent past, but is too complex and transient a process to include in these models. Therefore we assume that in the Precambrian basement some recent or ancient fault of correct orientation will probably exist, and can be reactivated.

Frictional sliding is described by a rigid-plastic flow law in which the strain-rate does not enter:

$$\tau^* \leqslant \mu(\sigma_n - P_p) \tag{12}$$

where σ_n is normal stress on the fault, μ is the coefficient of friction, and P_p is the pore pressure. In this work a constant μ of 0.60 is assumed (Stesky et al., 1974). Normal stress is computed from a Mohr's-circle diagram, assuming that vertical stress is equal to the overburden pressure. Pore pressure is calculated from the weight of a column of water extending up to the surface, which is correct if all the porous rocks are also permeable. This frictional plastic flow law dominates the stress levels in the cold upper crust.

In each of the geologic formations included in the model, one well-studied mineral is identified and assumed to contribute most of the total creep strain of the rock. Then a flow law for the whole rock can be obtained in the form:

$$\tau^* \leqslant A(\dot{e}^*)^{1/n} \exp(B/T) = A(\dot{e}^*)^{1/n} \exp(Q_a/nRT) \tag{13}$$

by using the crude approximation:

$$[\dot{e}_{ij}]_{rock} = c[\dot{e}_{ij}]_{mineral} \tag{14}$$

where c is the volume concentration of the mineral. The assumed value of c is not critical because of theoretical and experimental evidence than $n \simeq 3$ (Weertman and Weertman, 1975).

In the mantle, the creeping mineral is probably olivine; we assume $c = 0.85$ and use the flow law of Kohlstedt and Goetze (1974). In the crust this mineral is probably quartz; we assume $c = 0.30$ (upper crust) to 0.15 (lower crust) and use a flow law derived from data in Tullis (1971) and Parrish et al. (1976). In the Crush Zone and below a large amount of limestone has probably been subducted; this is given the calcite flow law of Schmid (1976). In

TABLE III

Mechanical parameters of the initial Zagros model

Rock	Density (g/cm^3)	Elements	A (dyne/cm^2)	B (°K)	n	τ_{max}
Peridotite	3.27	67—96	$6.84 \cdot 10^4$	21710—22668	3	1.5kb
Lower crust	2.84	30,31,33, 36,41,42, 47,48,53, 54,59,60, 65,66	$4.44 \cdot 10^8$	7445	3	none
Middle crust	2.75	35,39,40, 45,46,51, 52,57,58, 63,64	$4.00 \cdot 10^8$	7445	3	none
Upper crust	2.67	28,29,32, 34,37,38, 43,44,49 50,55,56, 61,62	$3.48 \cdot 10^8$	7445	3	none
Crush Zone Limestone	2.67	19—27	867	12391	3	none
Evaporites	2.16	1—9	$2.27 \cdot 10^{11}$	0	3	none
Continental Shelf Sediments	2.67	10—18	$1.0 \cdot 10^{13}$	0	3	none

the evaporites the dominant mineral is halite, and the flow law of Heard (1972) was used. Deformation in the surface sediments is not understood at all well; it could be any combination of faulting, dilational cracking, stress solution, and dislocation creep. Because the style of deformation seen at the surface is predominantly ductile, the sediments are assumed to be weak, and given a flow law allowing rapid deformation at stresses of 200—300 bars. The parameters of all these flow laws are given in Table III. The temperatures needed in eq. 13 are taken from the Zagros model of Toksöz and Bird (1977; Fig. 2).

In anticipation of results below, it should be stated that mere reduction of the constants A and B in eq. 13 was not sufficient to produce a successful model. In order to explain deep earthquakes, it was necessary to modify eq. 12, and allow plastic deformation at lower stresses. This could be done by assuming a very low μ, high pore pressure, or an independent earthquake mechanism which is not observed in the laboratory because of scaling problems. The last alternative was chosen, and this unknown mechanism was described by the simplest possible flow law that is insensitive to strain-rate:

$$\tau^* \leqslant \tau_{max} \tag{15}$$

Equations 12, 13 and 15 were all considered in the evaluation of $f(\dot{e}^*)$ defined by eq. 2, with the mechanism producing the lowest stress limit assumed dominant.

TESTS FOR SUCCESSFUL MODELS

A "successful" model of the Zagros is distinguished from others by its ability to meet three tests:

(1) Horizontal shortening of the surface sediments must be occurring in the southwestern half of the range (i.e., in the range $275 \leqslant x \leqslant 400$ km on Fig. 1). Evidence for this includes upwarping of Quaternary molasse by rising folds, and vertical uplift by 20 m of an irrigation canal constructed ca. 300 A.D. (Lees and Falcon, 1952). Shortening may also be occurring in the northeast half, but this is less certain.

(2) The areal distribution of predicted seismicity must coincide with the actual distribution in Fig. 4. There must be evenly-scattered activity coinciding with the folded belt ($120 \leqslant x \leqslant 400$ km) and none in the old plate suture or Persian Gulf areas. In order to determine which areas of a model are "potentially seismic", each integration point of each element is checked to see if its stress intensity is governed by the plastic equations 12 and 15, or the creep equation 13. The former are possible earthquake sites when elastic strain-accumulation-and-release cycles are considered. At these places, a symbol with size proportional to the volumetric deformational power is plotted as a qualitative indicator of "predicted" earthquake frequency. The depths of the events are not expected to match because of the large depth errors inherent in the location of shallow earthquakes.

(3) Earthquake fault planes predicted from model stresses and flow laws must match the orientation and sense of slip inferred from fault-plane solutions (Fig. 4). At points governed by eq. 12 the two possible fault planes form an acute angle around the direction of greatest compression, with an angular separation whose tangent is the inverse of the coefficient of friction. Equation 15 does not involve normal stress, so the possible fault planes at points governed by eq. 15 are the planes of maximum shear stress, and lie 45° from the principal compression direction. Of course, a plane-strain model can only predict fault planes perpendicular to its plane; the actual source mechanisms of Fig. 4 are selected because both planes fit that criterion within 15°. Fortunately this type of earthquake is dominant in the Zagros.

MODELING EXPERIMENTS

Boundary conditions

All of the models computed shared the same lower boundary condition. The asthenosphere below the plates was treated as a perfect fluid, providing a normal force proportional to depth on the bottom boundary. This is a good approximation because the flow law of Kohlstedt and Goetze (1974), which was used to represent the mantle, implies negligible strength at above 1000°C. If there is water and/or partial melt in the asthenosphere, it would be even weaker.

It is necessary to impose the vertical component of velocity at the upper surface. Otherwise, the smallest errors in model densities cause departures from isostasy which result in huge vertical velocities. In nature, vertical displacement results in stabilizing resistive forces from the weight of topography. However, this force is proportional to displacement and not to velocity; hence it could not be incorporated in the stiffness matrix K_{ij}. Instead, zero vertical velocity was imposed on the top of the early models. Later, when an understanding of the governing principles was acquired, vertical uplift of 1 mm/yr was allowed in the Zagros, and 6 mm/yr of subsidence in the Persian Gulf. However, the horizontal gradient of uplift (with dimensions of strain-rate) is small enough in comparison with other strain-rates to cause very little rotation of principal axes.

The two classes of models computed may be called "slab-pull" and "ridge-push" models. In the former, the descending end of the Arabian lithosphere (at lower left in Fig. 1) is required to descend at 30° dip with a velocity of 4.7 cm/yr (Bird et al., 1975). This is intended to simulate the pull of a possible descending lithospheric slab not included in the model. In slab-pull models the southwestern end of the lithosphere is subject only to normal lithostatic pressure. However, horizontal velocity is required to be uniform with depth to simulate the constraint imposed by the adjacent Arabian shield which is not included in the model.

In ridge-push models, on the other hand, the slab end is not constrained. This is appropriate if the former Tethyan slab has detached and fallen away, as the lack of a seismic Benioff zone suggests. These models are driven by an imposed horizontal velocity of 4.7 cm/yr applied on the southwestern boundary. Forces of this type could come from many sources, acting on any of the other boundaries of the Arabian plate (further discussion below).

In all models of both types, the horizontal velocity component of the Eurasian plate was set to zero at the left or northeastern end, to provide a reference point. The choice of a reference point has no effect on stresses.

Parameters and results of all the models are summarized in Table IV. Because of space limitations, only a few can be illustrated. However, the great majority failed in very similar ways.

Slab-pull models

The first model to consider is *Z15*, computed with the initial flow-law estimates described above and listed in Table III. It resulted in pure subduction of the Arabian plate (Fig. 6). All of its predicted seismicity falls in the Crush Zone, which is incorrect. No crustal thickening occurs, and the surface

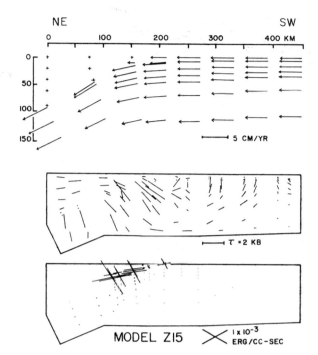

Fig. 6. Results of the unsuccessful model *Z15*. At top, velocity vectors of nodes are plotted. A cross indicates negligable motion. In center, directions of maximum compression are shown by lines, with the length of each line proportional to the local maximum shear stress. At bottom, a symbol is plotted at each integration point that is potentially seismic. Size of symbol is proportional to deformational power, and the two lines in each "X" show the possible fault planes. This model shows simple oceanic-type subduction into the Crush Zone.

sediments are under horizontal tension in the southern half of the Zagros. Deviatoric tension of up to 3 kb is required in the subducting slab. This model is not acceptable under any of the three criteria.

Compared to the large faulting stresses required in the cold upper crust of this model, the forces produced by crustal buoyancy are small. As a result, the natural solution is hardly different from the simple subduction of oceanic lithosphere. To investigate the possibility that a smaller faulting stress is the cause of crustal deformation, an upper limit (τ_{max}) was placed on the plastic faulting stress for all crustal rocks, including the sediments and Crush Zone material. This limit was successively reduced to 1000 bars, 500 bars, and 250 bars in models *Z16*, *Z17* and *Z18*, respectively. (In each case, the new rheology of the Crush Zone required a recomputation of pre-collision temperatures in that region, using the steady-state method of Bird, 1978.)

The results in each case were very similar. As detailed in Table IV, the crust simply subducted in each model, and the seismicity remained unaccept-

524

TABLE IV

Parameters and results of Zagros finite element models

Name	Fig.	Boundary conditions	Solution technique	Lower crustal A (dyne/cm^2)	Crust τ_{max}, (bars)	Mantle Q_a (kcal/mole)	RMS unconverged stress (bars)	Velocity solution	Maximum seismicity region, X = (km)—(km)	Fault plane sense and percentage
Z04	—	ridge-push	initial stress	$4.4 \cdot 10^8$	none	125	18% *	pure subduction	120—210	90% thrust 10% normal
Z05	—	ridge-push	initial stress	$1.9 \cdot 10^7$	none	125	7% *	pure subduction	140—210	70% indeterminate ** 30% thrust
Z06	—	ridge-push	initial stress	$1.9 \cdot 10^7$	200	125	31% *	pure subduction	150—210	70% indet. 30% thrust
Z11	—	R—P new grid	initial stress	$1.9 \cdot 10^7$	200	125	35% *	shortens in Persian Gulf	0—240	80% thrust 20% indet.
Z15	6	slab-pull	variable viscosity	$4.4 \cdot 10^8$	none	125	33	pure subduction	110—210	100% thrust (shallow dips)
Z16	—	slab-pull	variable viscosity	$4.4 \cdot 10^8$	1000	125	18	pure subduction	110—210	100% thrust 30° NE dips
Z17	—	slab-pull	variable viscosity	$4.4 \cdot 10^8$	500	125	29	pure subduction	80—210	90% thrust 10% normal (in SE)
Z18	—	slab-pull	variable	$4.4 \cdot 10^8$	250	125	48	pure subduction	80—210	100% thrust 30% NE dips
Z19	—	slab-pull	variable viscosity	$4.4 \cdot 10^8$	250	94	17	pure subduction	90—200	80% thrust 10% indet. 10% normal
Z21	7	ridge-push	variable viscosity	$4.4 \cdot 10^8$	250	94	16	shortening in Persian G. minor subduction	350—450 (80—200)	100% thrust 45° dip in SW 30° dip in NE

Z22	—	ridge-push	variable viscosity	$4.4 \cdot 10^8$	250	110	11.	pure sub-duction	80—200	85% thrust 10% indet. 5% normal
Z23	—	ridge-push	variable viscosity	$4.4 \cdot 10^8$	250	100	19	pure sub-duction	80—200 (380—450)	85% thrust 10% indet. 5% normal
Z24	—	R—P new grid	variable viscosity	$4.4 \cdot 10^8$	250	110	13	50% short-ening in NE Iran; 50% sub-duction	0—260	95% thrust 5% indet.
Z25	10, 11	R—P new grid ***	variable viscosity	$4.0 \cdot 10^7$	250	110	18	80% short-ening in Zagros; 20% sub-duction	100—370	100% thrust 45° NE dips
Z26	—	R—P new grid ***	variable viscosity	$8.0 \cdot 10^7$	250	110	27.	70% subduc-tion, 30% shortening in Zagros	100—250 (200—380 in sedi-ments)	100% thrust 35° NE dips
Z27	—	R—P new grid ***	variable viscosity	$1.5 \cdot 10^8$	250	110	25	70% subduc-tion, 35% shortening in Zagros	100—260 (200—380 in sedi-ments)	100% thrust 40° NE dips
Z28	—	R—P new grid ***	variable viscosity	$4.0 \cdot 10^7$	750	110	58.	90% short-ening in Zagros, 10% in Persian Gulf	150—450	100% thrust 45° NE dips
Z29	—	R—P new grid ***	variable viscosity	$8.0 \cdot 10^7$	750	110	38	80% short-ening in Zagros, 20% subduction	130—370	100% thrust

* For initial-stress models, percentage of points with errors over 10% of stress is given.
** One possible fault plane is normal, the other a thrust.
*** 0.11 cm/year uplift and 0.67 cm/yr subsidence added to boundary conditions.

ably far northeast. The only change was a slight rotation of the fault planes to steeper dips, but this was a trivial result of the change in flow law. When the faulting-stress equation 12 is replaced by eq. 15 it is no longer consistent to use the same equation for the angle between the fault planes. Because the lowered strength assumes no dependence on normal stress, this angle becomes 90° in all cases.

In model $Z19$ the creep strength of the mantle was also reduced, to that of water-saturated olivine, by reducing the activation energy to 94 kcal/mole (Post, 1973). The low crustal faulting stress of 250 bars was retained. Still, the stress field was dominated by deviatoric tension radiating from the downgoing slab, the crust simply subducted, and the seismicity pattern was unchanged. The lower crust was in a state of vertical compression/horizontal extension, neither promoting nor hindering the detachment of the upper crust from the lithosphere. Therefore no models with reduced lower crustal strength were computed. The defect of these models seems to be in the boundary conditions, since the attempts to scale down the rock strengths and so balance them against buoyancy had no effect.

Ridge-push models

Within this class of models the same procedure was tried: the strength of the various rock types was reduced singly or in combination in an attempt to move the deformation from the plate suture area into the Zagros. Models $Z04$, $Z05$ and $Z06$ were computed with initial parameters, with lower crustal creep strength reduced by twenty times, and with both faulting and creep strengths reduced (respectively). None of these combinations produced any change from the pattern of simple subduction with seismicity in the Crush Zone, too far northeast. The only change from "slab-pull" models was that the deviatoric stress in the crust was now uniformly horizontal compression instead of vertical. However, this produced no shortening (except by less than 1 cm/year in the sedimentary layer). These models still fail test (2) for seismicity and cannot explain the thrust faulting that was shown to occur in the basement.

Keeping the crustal faulting stress at a low 250 bars, the mantle creep activation energy was also reduced to 110 kcal/mole (model $Z22$) without any effect. When it is further reduced to 100 kcal/mole ($Z23$) a change in seismicity occurs. Receiving less support from the weak mantle, the crust shortens slightly at the southwest end, under the Persian Gulf. By reducing Q_a further to the water-saturated value of 94 kcal/mole, a dramatic change can be induced ($Z21$, Fig. 7). No longer able to support the stress, the crust and mantle both are shortened together in the Persian Gulf area, with considerable seismicity. This seismicity falls in an area which is supposed to be inactive, while little or none is predicted in the Zagros range. The shortening at the southwestern end occurs where it does because the crust there has no buoyant roots to resist further thickening by compression. This shows the

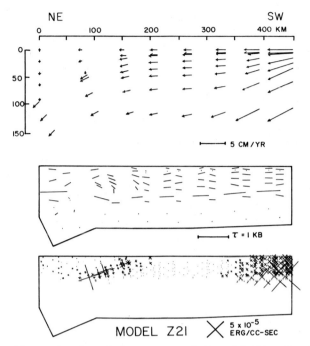

Fig. 7. Results of the unsuccessful model *Z21*. Symbols explained in caption 6 and parameters in Table IV. In this model the Arabian shield lithosphere is too weak, and collapses below the Persian Gulf.

futility of attempting to find a good model by adjusting the mantle strength parameters. If the mantle were very weak, shortening would occur in Arabia, where the crust is thinnest. If the mantle is made strong enough to stabilize the crust, it also forces it to subduct (regardless of crustal parameters) by a kind of "suction". This "suction" is caused by the Arabian lithosphere being in contact with Iranian lithosphere across the Crush Zone. Here, a part of the imposed horizontal velocity is translated into downward velocity which is transmitted to the crust. This appears to be the basic defect of all models described so far.

A modification of the model geometry

Intuition suggests that the Arabian continental shelf (the Zagros) is shortening and not subducting because continental crust is less dense than the mantle. Thus some kind of buoyancy force holds it up, opposing the stresses exerted upon it by the rest of the lithosphere. However, the buoyancy concept only applies to bodies in contact with a fluid supporting medium (i.e., the asthenosphere). In the models above this was not the case. The crust was preceded down the subduction zone by rigid mantle lithosphere and simple conservation of volume (eq. 1) required it to subduct also.

To correct this problem, the Arabian and Eurasian mantle lithospheres were decoupled by deleting those elements which are dashed in Fig. 1. If the Tethyan oceanic slab to which they belong had broken off, these rocks might well have been carried away as a part of it. There are two justifications (besides the model results) for this deletion. First, the seismicity cross-section (Fig. 4) shows no events in this area, implying a lack of cold lithosphere. Second, Wells (1969) reports that "partly kaolinized but very young acidic intrusives are found in four separate outcrops (probably connected at depth) in the Crush Zone between Bandar Abbas and Sirjan, some 190 km southeast of Neyriz. They are true granites, which intrude and thermally metamorphose both earlier basic intrusives and crushed sediments, and they are totally unfoliated even though they outcrop within the margin of the Crush Zone." Considering the very low temperatures found in subduction shear zones (Ernst, 1975; Bird, 1978), these granites could only be produced by bringing subducted sediments or crust into contact with hot asthenosphere. For this to happen below the Crush Zone requires disruption and separation in the mantle lithosphere.

STRESS AND DEFORMATION IN THE CRUSTAL WEDGE

A simple analytical model and the strength of lower continental crust

At this point, many potentially complicating factors have been disposed of. If the Tethyan slab has dropped away, then only "ridge-push" models are possible. It has been shown that the subcrustal lithosphere must be too strong to participate in the deformation, or else the orogeny would spread across the whole Arabian plate. Finally, there is nothing in the surface geology or fault-plane solutions to suggest other than simple and homogeneous horizontal compression of the crust. This leaves us with a very simple geometry (Fig. 8) which has been analytically modeled (on a smaller scale) before.

Recently both Chapple (1975) and Elliott (1976) have produced analyses of thin-skinned overthrust belts deforming above a passive basement. While they were discussing sedimentary sections, the same geometry applies to the crust of the Zagros. It is thin-skinned (33—49 km thick vs. 300 km wide) and (as model *Z21* showed) must be deforming independent of the more rigid sub-crustal lithosphere. Assuming plane strain and homogeneous material with a rigid-plastic constitution, both authors found that the shear stress (τ_b) on the base of the overthrust belt is independent of the internal plastic limit. Instead, in the case of simultaneous plasticity (faulting) of the whole "thin skin", the derivation of τ_b proceeds like this:

$$\tau_b = \int_0^l \frac{\partial \sigma_{xz}}{\partial z}\, dz = \int_0^l \frac{-\partial \sigma_{xx}}{\partial x}\, dz$$

$$\approx \int_0^l \frac{-\partial \sigma_{zz}}{\partial x}\, dz = -l\,\frac{\partial \sigma_{zz}}{\partial x} = l\rho g\,\frac{\partial h(x)}{\partial x}$$

(16)

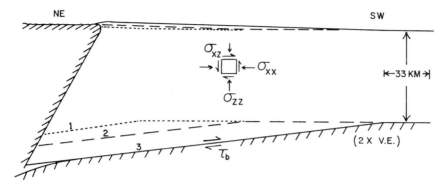

Fig. 8. A schematic cross-section of the Zagros crustal wedge, which is deforming by shortening and sliding over its supporting lithosphere. Dotted, dashed, and solid lines (*1—3*) show its successive outlines as the wedge grows without change of surface or Moho dip angles.

where $h(x)$ is the topography above sea level and l is the layer thickness. In the case of the Zagros this implies $\tau_b \approx 75–100$ bars, increasing from the Persian Gulf toward the Crush Zone. This result implies a lower crustal creep strength lower by a factor of five than that of the initial strength model that assumed the only creep was in a fractional content of dry quartz. This could be a result of water-weakening (Balderman, 1974) or creep in some other mineral.

This model successfully explains the constant dip of the Moho inferred from gravity (Fig. 2). Because of isostasy, it mirrors in exaggerated form the average slope of the surface topography (about 0.5°). This slope in turn is determined by the uniform low strength of the lower crust. The historical implication is that during the formation of the Zagros, both slopes have remained constant. As more crustal material is brought into the Zagros by plate convergence, the wedge gets thicker (Fig. 8) and wider without any fundamental change in the deviatoric stresses or style of deformation.

If this analysis is correct, then two predictions ought to be fulfilled. First, the age of the onset of deformation should decrease consistently to the southwest, because those areas have been within the spreading wedge for the least time. Second, the total shortening strain should decrease in the same way and for the same reason. Unfortunately, both predictions are weakened by the possibility of a second detachment of sediments from the basement, just as the crust was detached from the lithosphere.

Strength of the upper continental crust

As an inspection of eq. 16 will show, nothing in the analysis above requires the strength of crust to be the same at all depths. Conversely, nothing in the topography or crustal thickness can be used to determine the level of anomalous horizontal compressive stress ($\Delta\sigma_{xx}$) responsible for the orogeny. Yet this is a number we would like very much to know, both to determine whether the frictional law (eq. 12) applies to upper crust, and to determine the "driving force" required by this orogeny.

However, limits can be placed on $\Delta\sigma_{xx}$ by following the following logic:

(1) $\Delta\sigma_{xx}$ in the upper 40 km of the Crush Zone must be about equal to $\Delta\sigma_{xx}$ in the adjacent crust of the Zagros, since in the revised geometry there is no other mechanical connection between the two plates.

(2) If a creep law for Crush Zone material is specified, the maximum anomalous horizontal force that could have been transmitted across the Crush Zone before the continental collision can be calculated. This is because the force is self-limiting; high friction produces heat which raises the temperature and activates creep, which then reduces the stress exponentially with depth.

(3) The present value of $\Delta\sigma_{xx}$ in the Crush Zone today must be less than it was in pre-collision times, because the Crush Zone is no longer active (Fig. 4), and there has not been time for major temperature changes. Together, these imply that the creep law of the material in the Crush Zone can be used to place an upper limit on the stress that now causes earthquakes and deformation in the upper crust of the Zagros.

The basic tool in this analysis is a numerical technique for finding temperature (and thus stress) in a moving subduction zone of known rheology, by equating heat produced by friction to heat conducted away or used to warm the shear zone. The method is described by Bird (1976, 1978) and will not be repeated here. Besides the dip, velocity, and thickness already given, parameters include a thermal conductivity of $2.5 \cdot 10^5$ ergs/cm sec $^\circ$C, and a heat capacity of $8 \cdot 10^6$ erg/g $^\circ$C, and the flow law.

The trick is to find a rheology (weaker than or equal to laboratory values) which makes the top 40 km of the Crush Zone stronger under steady-state subduction conditions than the adjacent crust is now with a near-normal geotherm. The friction equation 12 taken in conjunction with the granite flow law of Table III does not fit this criterion, because the initiation of creep would be more shallow (12 km) in the Crush Zone than in the adjacent crust (24 km).

One can delay the onset of creep in the subduction zone by invoking an unknown mechanism to limit stress (eq. 15). Using a granite creep law for the Crush Zone, the maximum average stress in the upper 40 km is attained when τ_{max} is 765 bars (Fig. 9). This means that the present deformation and seismicity of the Zagros crust are caused by shear stresses of no more than 800 bars. If the crust were stiffer, it would be stronger than the Crush Zone, and it would still be subducting.

This is an upper limit, but a more likely situation is that creep in the Crush Zone is controlled by deformation of subducted limestones, in which case the Schmid (1976) calcite flow law is appropriate. Calcite creeps at lower temperatures than quartz, and so τ_{max} must be reduced to 250 bars to delay activating this stress-reducing mechanism (Fig. 9). Unless all of the sediment has been scraped off the basement at shallow depths in the Crush Zone, we must conclude that the strength of the adjacent crust is less than 300 bars in shear.

Successful finite element models

The predictions made using these analytical and comparative arguments are borne out by the success of the final group of finite element models. This group consisted of ridge-push models with the revised geometry, and with a subcrustal lithosphere sufficiently strong to resist major shortening. Model Z24 (Table IV) showed that reduction of τ_{max} in the crust is not alone sufficient; it resulted in subduction as before. The creep constant A in eq. 12 must also be reduced, so that the shear stress at the base of the crust may drop to the 75—100 bars which can be resisted by the topographic gradient.

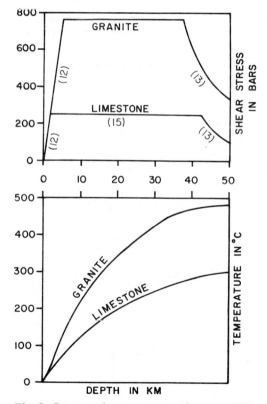

Fig. 9. Stress and temperature (for two different flow laws) prior to continental collision in the Crush Zone. At top, the number of the equation determining the stress is shown in parentheses. Flow laws are from Table III, with τ_{max} adjusted to cause creep initiation around 40 km deep.

532

When the creep stress of the crust is reduced by a factor of five in model *Z26*, the results begin to improve. Some crustal shortening results, but it is not evenly distributed. Seismicity falls in the prescribed range in x but is strongly concentrated to the northeast. (Since the proper parameters for the sediments are unknown, the seismicity in that shallow layer may be artificial. Neglecting it, the activity is too strongly concentrated near the plate suture.) This indicates that the creep strength is still slightly too high.

When the creep parameter A was reduced by a full factor of ten, to $4 \cdot 10^7$ dynes/cm^2, the first truly successful model was obtained. This is model *Z25*, which is shown in Figs. 10 and 11. Its major features are: relatively uniform shortening of the crust between the Crush Zone and the Persian Gulf, independent shortening of the sedimentary layer at higher rates in the SW Zagros, and uniformly distributed crustal seismicity on predominantly 45°-dipping thrust planes. This model satisfies all three tests. It disagrees with the data only in requiring all seismicity to occur at 60 km depth or less (Fig. 11). Some events have been located deeper by Nowroozi (1971), but these could actually be considerably shallower considering the limitations of body-wave location methods. No attempt was made to reproduce them by weakening the lithosphere, because a strong lithosphere is required to stabilize and localize the deformation.

It has been stated that the upper crustal τ_{max} might be 800 bars if the

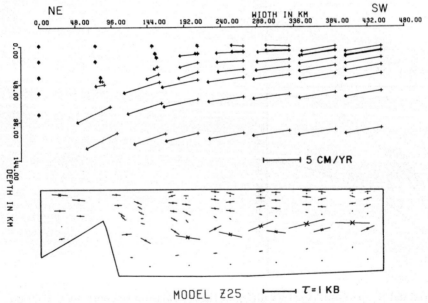

Fig. 10. Velocity vectors and principal compression directions for the successful Zagros model *Z25*. Note detachment of sediments from basement and of crust from lithosphere, shown by differences in vector length across boundaries.

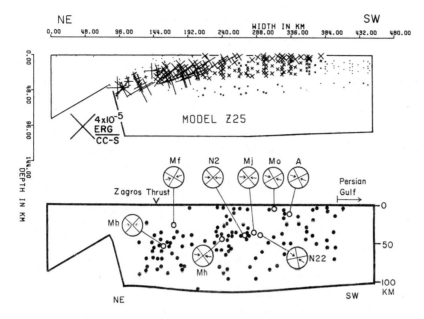

Fig. 11. Predicted seismicity of model *Z25* compared with actual Zagros earthquakes. See captions 4 and 6 for explanation of symbols. Uniform thrust faulting, distributed through the folded belt, occurs in the model and in nature. It is not known whether the discrepancy of some hypocentral depths is due to model errors or earthquake location errors.

Crush Zone were full of granite. To test this, τ_{max} was raised to 750 bars and models *Z28* and *Z29* were computed. In both, the Crush Zone elements were given the parameters of granite, and the creep strength in the lower crust was reduced ten and five times, respectively, from Table III values. *Z28* gave a more even distribution of seismicity in the appropriate band, but predicted a small amount in the Persian Gulf which should be inactive. *Z29* did not have this flaw, and was therefore more acceptable, although it predicts more of a concentration of seismicity in the NE Zagros than is observed. This model is an alternative to model *Z25*. The crustal shortening, sediment detachment, fault plane solutions, and stress patterns are almost identical. The most significant difference is that in model *Z29* with the higher faulting stress, twice as much driving force is required to cause the convergence of the plates.

DRIVING FORCE OF THE OROGENY

The driving force (D) can be defined as the vertical integral through the lithosphere of the deviatoric horizontal compression:

$$D(x) \equiv \int_0^{100 \text{ km}} [\sigma_{xx}(x, z) - (\int_0^z g\rho(x, z)\mathrm{d}z)]\mathrm{d}z \qquad (17)$$

Note that the driving force may vary from place to place because of density (ρ) changes, even if there is conservation of total horizontal force:

$$\frac{\partial}{\partial x}\left[\int_{0}^{100\ km}\sigma_{xx}(x,z)dz\right] = 0 \tag{18}$$

Taking a point in the Arabian shield ($x = 450$ km), we calculate D values for the successful finite element models $Z25$ and $Z29$ to be $2.8 \cdot 10^{15}$ and $5.5 \cdot 10^{15}$ dynes/cm, respectively. These are equivalent to compression of 280 and 550 bars distributed through a 100 km lithosphere. Lower values are possible only if the shear strength of continental crust is less than $250b$, which is unlikely.

It is interesting to compare this with the driving force that the Red Sea rift could exert on the Arabian Shield by simple gravitational spreading. Following Frank (1972), we assume that

(1) a spreading center is a place where fluid asthenosphere with no strength extends up to the seafloor;

(2) the ridge and other topography are isostatically compensated no deeper than some universal compensation depth Z_c;

(3) all deviatoric stresses vanish at $z = Z_c$; and

(4) the structure is two-dimensional ($\sigma_{xy} = 0$).

Assumption (1) implies that, at the Red Sea rift, $D = 0$. Assumptions (3) and (4) imply eq. 18. Putting them together, we obtain an alternative estimate of D in the Arabian Shield that considers only the effect of the Red Sea divergent plate boundary:

$$D(x = 450\ km) = \int_{0}^{Z_c}\left[\int_{0}^{Z}g(\rho_R(z) - \rho_A(z))dz\right]dZ \tag{19}$$

In this equation $\rho_R(z)$ and $\rho_A(z)$ are the density profiles in the Red Sea and Arabian shield. Assumption (2) is a useful constraint on these profiles.

In Fig. 12 the lithostatic pressure curves have been plotted, so that the value of D appears as the area between the curves. This type of plot was invented by Frank (1972). It has been assumed that the Red Sea is 1.5 km deep, that the Arabian crust is 33 km thick with a density of $2.8\ g/cm^3$, and that the density differences between the lithosphere and asthenosphere are caused solely by volumetric thermal contraction with a coefficient of $5 \cdot 10^{-5}\ °C^{-1}$ (constrained by the elevation of ridges worldwide). With these parameters the driving force D is $2.95 \cdot 10^{15}$ dyne/cm, or enough to drive model $Z25$ but not $Z29$.

Although either of these numbers which we are comparing could reasonably be out by a factor of two, it is encouraging that the model derived here for the Zagros orogeny does not require the invocation of untestable and hypothetical forces applied to the bottoms of the plates. Such asthenospheric tractions would have to be larger than 15 bars if the Red Sea rift did

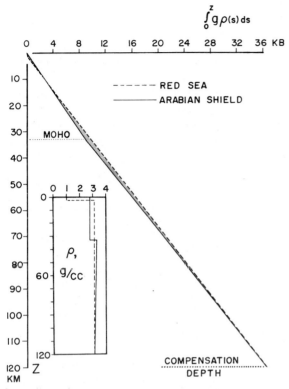

$$\int_0^z g\rho_{(s)}\,ds$$

RED SEA
ARABIAN SHIELD

MOHO

$\rho,$
g/cc

COMPENSATION
DEPTH

Fig. 12. Plot of overburden pressure versus depth for the center of the Red Sea rift and for the Arabian shield. The net shaded area represents the tectonic compression exerted by the former on the latter; the driving force available to compress the Zagros range. This is about $3 \cdot 10^{15}$ dynes/cm, enough to provide the stresses in Fig. 10. Inset shows the density profiles assumed.

not contribute any driving forces. Using the dry olivine flow law of Table III (which works well in the models) and estimating the asthenosphere temperature as 1350°C, such a stress would imply a sub-lithospheric strain rate of $7 \cdot 10^{-12}$ sec^{-1}. This is quite high, requiring convection currents that move with velocities two orders of magnitude higher than that of the plate above. It is, however, a possible alternative explanation.

CONCLUSIONS

When the last remnant of the Tethys Ocean lithosphere was subducted beneath Central Iran a few million years ago, it apparently detached itself from the rest of the Arabian plate at the edge of that continent, and sank independently into the mantle. This left a gap at depth between the lithospheres of Arabia and Eurasia approximately under the Zagros Crush Zone, and concentrated all the stress of the collision into the shallow continental crust.

This crust, formerly the continental shelf of Arabia, did not subduct in response to compression because of its low density. Rather, it was shortened horizontally and thickened vertically as the plates continued to converge. The subcrustal lithosphere, many times stronger than the crust, has stabilized this deformation and prevented it from spreading across the Arabian Shield. In order to shorten further, the crust must slide over the lithosphere, probably by intense shearing in the hottest part of the crust just above the Moho. The shear stress required to continue this crustal decollement is provided by gravity acting on the southwestward slope created by the thickening of the edge of the crust.

The single new thrust fault proposed by Bird et al. (1975) is probably an oversimplification of the present tectonics. The finite element models suggest instead that the whole crustal wedge of the Zagros is simultaneously deforming. Individual basement thrust faults may underlie the folds of the Asmari limestone, although a second decollement on the basal evaporite layers may have masked any connection between surface and basement deformation. As more and more crust is brought into the range by plate convergence, the deforming wedge grows wider and thicker. The height of the mountains continues to increase, as isostasy converts a fraction of the crustal thickening into uplift.

The driving force that powers this process is no more (perhaps less) than the force which formerly sheared one plate under the other at the Crush Zone. The most likely source of that force at present is the gravitationally-powered spreading of the Red Sea rift, which is in most general terms a form of thermal convection.

Finite element modeling combined with approximate analytical treatment has placed limits on the flow laws of the continental lithosphere involved in this orogeny. The subcrustal lithosphere supports shear stresses of a kilobar without substantial shortening, implying that its rheology is closer to "dry" olivine than to "wet". On the other hand, the detachment of the crust from the lithosphere at shear stresses of 75—100 bars implies that the flow law based on dry quartz is about five to ten times too strong. Water-weakening of quartz may be responsible, or some other mineral may creep more readily. In any case, the lower crust is surprisingly weak, and this means that in continental tectonic areas detachment and independent mobility of the crust may be a general phenomenon.

Another surprising result is that earthquakes and upper crustal deformation are caused here by shear stresses of no more than 800 bars; more probably 250 bars. This means that in applying the results of laboratory friction measurements to the crust, we may overestimate its strength by a factor of ten. The mechanism which facilitates such deformation in nature but not in the laboratory may well be dependent on the high velocities and large slips which occur in earthquakes. Whatever it is, the discovery of its physics and flow law must obviously precede any quantitative progress in continental tectonophysics or earthquake prediction.

ACKNOWLEDGEMENTS

I thank M. Nafi Toksöz, my former advisor, for his support and guidance throughout the gradual development of these computational methods and tectonic concepts. Christopher Goetze also helped greatly with suggestions and data on the relevant deformation mechanisms of rocks. Jerome Connor and Gilbert Strang suggested improvements in the finite element method. Elmer J. Hauer of the Defense Mapping Agency kindly sent the gravity data that was used, and Howard Patton provided a program for Rayleigh wave spectrum calculations. This work was supported by the Advanced Research Projects Agency and administered by the Air Force Office of Scientific Research under contract F44620-75-C-0064 with the Massachusetts Institute of Technology.

REFERENCES

Abdalian, S., 1963. Seismo-tectonique de l'Iran. Inst. Geophys. Univ. Tehran Publ., 16: 105 pp.

Akasheh, B., 1973. Mechanism of the earthquake of April 10, 1972 in Qir (Iran). Z. Geophys., 39: 1055—1061.

Akasheh, B., 1975. Travel time residuals in the Iranian Plateau. J. Geophys., 41: 281—288.

Balderman, M.A., 1974. The effect of strain-rate and temperature on the yield point of hydrolitically weakened quartz. J. Geophys. Res., 79: 1647—1652.

Bazeley, G.P., Cheung, Y.K., Irons, B.M. and Zienkiewicz, O.C., 1965. Triangular elements in bending- conforming and non-conforming solutions. In: Proceedings of the Conference on Matrix Methods in Structural Mechanics. Air Force Institute of Technology, Wright Patterson AFB, Ohio.

Berman, H., Daly, R.A. and Spicer, H.C., 1942. Density at room temperature and one atmosphere. In: F. Birch (Editor), Handbook of Physical Constants. Geol. Soc. Am., Spec. Pap., 36, pp. 7—26.

Bird, P., 1976. Thermal and Mechanical Evolution of Continental Convergence Zones: Zagros and Himalayas. Thesis. Mass. Inst. Technol., Cambridge, Mass., 423 pp.

Bird, P., 1978. Stress and temperature in subduction shear zones: Tonga and Mariana. Submitted to Geophys. J. R. Astron. Soc., in press.

Bird, P., Toksöz, M.N. and Sleep, N.H., 1975. Thermal and mechanical models of continent—continent convergence zones. J. Geophys. Res., 80: 4405—4416.

Bird, R.B., 1960. New variational principle for incompressible non-Newtonian flow. Phys. Fluids, 3: 539—541.

Chapple, W.M., 1975. Mechanics of thin-skinned fold and thrust belts. EOS, Trans. Am. Geophys. Union, 56: 457.

Coster, H.P., 1947. Terrestrial heat flow in Persia. Mon. Not. R. Astron. Soc., Geophys. Suppl. 5: 131—145.

Dewey, J.F., Pitman III, W.C., Ryan, W.B.F. and Bonnin, J., 1973. Plate tectonics and the evolution of the Alpine System. Geol. Soc. Am. Bull., 84: 3137—3180.

Dziewonski, A., Bloch, S. and Landisman, M., 1969. A technique for the analysis of transient seismic signals. Bull. Seismol. Soc. Am., 59: 427—444.

Elliott, D., 1976. The motion of thrust sheets. J. Geophys. Res., 81: 949—963.

Ernst, W.G., 1975. Summary. In: W.G. Ernst (Editor), Metamorphism and Plate Tectonic Regimes. Halsted Press, New York, N.Y., pp. 423—426.

Eslami, A.A., 1974. Detecting the thickness of the crust in Shiraz area using hypocenter situated below the crust. J. Earth Space Phys., 3: 1—12.

Falcon, N.L., 1969. Problems of the relationship between surface structure and deep displacements illustrated by the Zagros Range. In: Time and Place in Orogeny. Geol. Soc. London, London, pp. 9—22.

Frank, F.C., 1972. Plate tectonics, the analogy with glacier flow, and isostasy. Geophys. Monogr., Am. Geophys. Union, 16: 285—292.

Haynes, S.J. and McQuillan, H., 1974. Evolution of the Zagros suture zone, southern Iran. Geol. Soc. Am. Bull., 83: 739—744.

Heard, H.C., 1972. Steady-state flow in polycrystalline halite at pressure of 2 kilobars. Geophys. Monogr., Am. Geophys. Union, 16: 191—209.

Isacks, B. and Barazangi, M., 1977. Geometry of Benioff zones: lateral segmentation and downwards bending of the subducted lithosphere. In: M. Talwani and W.C. Pitman III (Editors), Island Arcs, Deep Sea Trenches, and Back Arc Basins. Am. Geophys. Union, Maurice Ewing Ser., 1: 99—114.

Knopoff, L. and Fouda, A.A., 1975. Upper-mantle structure under the Arabian Peninsula. Tectonophysics, 26: 121—134.

Kohlstedt, D.L. and Goetze, C., 1974. Low-stress high-temperature creep in olivine single crystals. J. Geophys. Res., 79: 2045—2051.

Lees, G.M., 1952. Foreland folding. Q. J. Geol. Soc. London, 108: 1—33.

Lees, G.M. and Falcon, N.L., 1952. The geographical history of the Mesopotamian Plains. Geogr. J., 118: 24—39.

McKenzie, D.P., 1972. Active tectonics of the Mediterranean region. Geophys. J. R. Astron. Soc., 30: 109—185.

Molnar, P. and Oliver, J., 1969. Lateral variations of attenuation in the upper mantle and discontinuities in the lithosphere. J. Geophys. Res., 74: 2648—2682.

Nowroozi, A.A., 1971. Seismotectonics of the Persian Plateau, eastern Turkey, Caucasus, and Hindu-Kush regions. Bull. Seismol. Soc. Am., 61: 317—341.

Nowroozi, A.A., 1972. Focal mechanism of earthquakes in Persia, Turkey, West Pakistan, and Afghanistan and plate tectonics of the Middle East. Bull. Seismol. Soc. Am., 62: 823—850.

Parrish, D.K., Kriva, A. and Carter, N.L., 1976. Finite element folds of similar geometry. Tectonophysics, 32: 183—207.

Post, R.L., 1973. The Flow Laws of Mt. Burnett Dunite. Thesis. University of California, Los Angeles.

Schmid, S.M., 1976. Rheological evidence for changes in the deformation mechanism of Solnhofen limestone toward low stresses. Tectonophysics, 31: T21—T28.

Stesky, R.M., Brace, W.F. and Robin, P.Y.F., 1974. Friction in faulted rock at high temperature and pressure. Tectonophysics, 23: 177—203.

Toksöz, M.N. and Bird, P., 1977. Modelling of temperatures in continental convergence zones. Tectonophysics, 41: 181—193.

Tsai, Y.B. and Aki, K., 1970. Precise focal depth determination from amplitude spectra of surface waves. J. Geophys. Res., 75: 5729—5743.

Tullis, J.A., 1971. Preferred Orientations in Experimentally Deformed Quartzites. Thesis. University of California, Los Angeles, Calif., 362 pp.

Weertman, J. and Weertman, J.R., 1975. High temperature creep of rocks and mantle viscosity. Ann. Rev. Earth Planet. Sci., 3: 293—315.

Wells, A.J., 1969. The crush zone of the Iranian Zagros mountains, and its implications. Geol. Mag., 106: 385—394.

Woolard, G.P., 1959. Crustal structure from gravity and seismic measurements. J. Geophys. Res., 64: 1521—1544.

Zienkiewicz, O.C., 1971. The Finite Element Method in Engineering Science. McGraw-Hill, London, 521 pp.

Geology
B 22 · 39